Living *and* Surviving *in* Harm's Way

Living *and* Surviving *in* Harm's Way

A Psychological Treatment Handbook for
Pre- and Post-Deployment of Military Personnel

Edited by
Sharon Morgillo Freeman
Bret A. Moore
Arthur Freeman

Routledge
Taylor & Francis Group
New York London

Routledge
Taylor & Francis Group
711 Third Avenue
New York, NY 10017

Routledge
Taylor & Francis Group
2 Park Square
Milton Park, Abingdon
Oxon OX14 4RN

International Standard Book Number-13: 978-0-415-98868-1 (Hardcover)

Library of Congress Cataloging-in-Publication Data

Living and surviving in harm's way : a psychological treatment handbook for pre- and post-deployment of military personnel / edited by Sharon Morgillo Freeman, Bret A. Moore, Arthur Freeman.
    p. ; cm.
Includes bibliographical references and index.
ISBN 978-0-415-98868-1 (hardback : alk. paper)
    1. Military psychiatry--Handbooks, manuals, etc. 2. War neuroses--Handbooks, manuals, etc. 3. Veterans--Mental health--Handbooks, manuals, etc. I. Freeman, Sharon Morgillo. II. Moore, Bret A. III. Freeman, Arthur, 1942-
    [DNLM: 1. Combat Disorders--therapy. 2. Cognitive Therapy--methods. 3. Military Personnel--psychology. 4. Military Psychiatry. WM 184 L785 2009]

UH629.L58 2009
616.85'2120651--dc22   2008044016

Visit the Taylor & Francis Web site at
http://www.taylorandfrancis.com

and the Routledge Web site at
http://www.routledge.com

Contents

**Part IV The Service Member's
Family and Community—Intervention**

Acknowledgments

To the brave Knights, those warriors of peace and war that serve our great nation.

In the process of writing and editing this book, I have had the great honor of meeting and working with many incredible servicemen and servicewomen who have proudly served this nation. I want to thank CSM Tom Gonzalez for the stories on the plane from Panama City to Atlanta and for the incredible gift you gave me of the Battalion Coin at the journey's end. Your gesture brings tears to my eyes to this day.

To my coeditor Bret A. Moore, Army Captain and Clinical Psychologist, and to the 85th Combat Stress unit of which he was a member—without them this book would not have been possible.

To Lt Col (Ret.) Dave Grossman, who is one of the most energetic, positive, and motivated leaders and teachers I have had the honor to meet.

To my daughter, Heather Foraker, PFC (Ret.) OIF III, and her husband, Cpl Dustin Foraker, for their service in Iraq and Afghanistan.

To Cpt Hugh Ruesser and to my uncle Jerome Kowalski, who was a soldier's soldier, an Airborne Ranger, and one of my heroes. Rest in peace, Uncle Jerry.

To my uncle Leo Adoline, an Air Force Flight Officer in World War II, who always encouraged me to follow my dreams and supported me every step of the way.

To Sgt (Ret.) Joseph Halicek, Sr., for your service in the Army in World War II.

To my first cousins and uncles, my fighting family: Michael A. Kowalski, Army Infantry, E4 Specialist; Richard D. Ammerman, Sr., Army Military Police, Cpl, Korean Conflict; Richard D. Ammerman, Jr., Army Medical Division, Sgt FC, Post Vietnam; Roger A. Ammerman, Army Infantry, PFC, post-Vietnam; Ronald J. Ammerman, Army Transportation, Cpl, post-Vietnam—gentlemen, you have made us proud.

A special thanks to my supportive friends and colleagues: Sr. Trooper Roland Purdy and his fellow heroes and paladins—the officers of the Indiana State Police Department who are deployed daily into harm's way to keep those of us on the homefront safe.

To Janet Casperson (U.S. Army, Ret.), John McBride, Lois Burry, Jan Eggiman, and Jim Dohrmann and many others too numerous to name.

To my daughters, Laura and April; their husbands, Brian and Jeremy; and our grandchildren.

Thank you to my former husband, Art, who encouraged me to make this book a reality.

And, most of all, to my father, Constantine V. Morgillo, T/Sgt, Korean Conflict, who was, and is, my hero and the person who taught me about righteous power and goodwill toward all. You taught me how to shoot a gun, how to drive a car, and how to honor my word above anything else. I never knew that you earned the Bronze Star until after you died. You were a hero to many, a friend to most everyone you met, and a loving head of the family. Sleep well, Pa.

Sharon Morgillo Freeman

Without the support and encouragement of many people in my life, my participation in this book would not have been possible. First off, I would like to thank my wife, Lori, whose love and respect provided me with the motivation and confidence to finish this work.

To my parents, David and Brynda Moore, and my brothers, David and Keith Moore, for their continuous laughter and comic relief during increased times of stress.

To my former mentors and supervisors, Jack Scobey, Richard Strait, Jacobus Donders, Edward Swanton, John Newbauer, and Thomas Vandenabell, who provided me the opportunity to grow as a professional and as an individual, and to Art and Sharon Freeman, who included me on such an important book.

And, last, but certainly not least, to the men and women of our military and their families who have sacrificed so much to preserve our rights and freedoms. Not a day goes by that I don't reflect on how fortunate we are as a nation to have so many brave men and women voluntarily strap on their armor and lace up their boots for one more mission.

In memory of Ferrell and Cecelia Abel, Charles Moore, Jack Scobey, Richard Strait, and all the men and women who have fallen on the battlefield.

Bret A. Moore

Forty-two years of clinical practice and academic life have barely prepared me for the enormity of dealing with those who continue to place themselves in harm's way. These men and women have seen and done what others can only vicariously experience in a movie or television drama. These individuals are not involved in passing debates on the political correctness of their actions. They follow a code that differs in many significant ways from the codes and beliefs of many Americans. These codes include terms such as honor, duty, and country. Their actions can be debated sometimes and looked to for safety and solace at other times. They are the men and women of the U.S. military, in all of its branches and in its myriad actions.

The goal of this book was to help the nonmilitary therapist provide the best possible informed treatment of these persons who have risked so much, have experienced what they have, and now need help in healing from their experience. My work on this book is dedicated to them, those servicemen and servicewomen who came before and those who will come after to protect us and keep our families and country free and safe.

I have been fortunate in many ways to have been at the right place at the right time. I have been able to learn from the giants in the field. Drs. Aaron T. Beck, Albert Ellis, and Emanuel F. Hammer have all, in so many ways, contributed to my work and to my success.

My family and friends have provided me with support and love. I have not always acknowledged that, so I do so here.

Arthur Freeman

Foreword

I have been graced with having a perspective of the past eight decades. In that time, I have seen the Great Depression from a child's viewpoint. I have seen many presidents and heads of state come to power and then fade away. I recall a young president, full of promise, gunned down on a sunny day in Dallas. I can recall the many breakthroughs in electronics, allowing individuals to be in contact with others around the world, not in days or in hours, but in seconds. I have seen the awful destructive power of the atom unleashed and seen the wonderful diagnostic and curative powers of the atom harnessed.

The advances in university and medical education boggle my mind. I have seen medical breakthroughs that as medical students we could not even dream about, or if we did dream of them we thought that such treatments and advances would come at some time in the far distant future. In the 1950s, the work of Drs. Jonas Salk and later Albert Sabin eradicated polio, the feared crippler of children and adults alike. Illnesses and problems that were not (or could not) be treated resulted in patients spending their lives in pain, in institutions, or at the periphery of society, functioning far below their potential. Organ transplants that lasted for days until their recipients succumbed are now a routine part of medicine, with recipients living productive lives and not merely waiting in a hospital for the transplant to be rejected. Cancers that were nearly always fatal can now be treated with combinations of treatments with good outcome. I have seen the rise of HIV and AIDS, with the medical community working toward maintaining life and bringing about both a cure and possible immunity.

As a medical student and later as a resident, I was taught that emotional problems were the result of unresolved unconscious impulses and drives. Given the lack of alternative nonpsychodynamic models, I followed the *zeitgeist* and was formally trained as a psychoanalyst at the Philadelphia Psychoanalytic Institute. From my earliest days as a medical student, however, I was interested not only in the practice of medicine but also in the science of medicine. I wanted to be able to state, with empirical support, that a particular treatment or idea could be validated. This goal steered my life and career into uncharted waters. I could not have foreseen that my early findings and formulations regarding the role of cognition as a major factor in psychopathology would have the effect that they have. What Mahoney called the "cognitive revolution" has broadened into an international movement where cognition and cognitive change are studied in the etiology of most forms of emotional distress and disorder. My early work on depression in the 1960s was only the tip of the iceberg. Cognitive therapy is shared by thousands of researchers throughout the world. The findings from various sources are similar—cognitive therapy is efficacious for a broad range of clinical populations, for a wide range of disorders, and when delivered in a variety of formats and settings.

I paid heavily for my perspective. My psychoanalytic colleagues thought me naïve for rejecting the basic psychoanalytic dicta. Cognitive therapy, they voiced, was a passing fad to be endured but not accorded much notice. Cognitive therapy, however, has not faded nor is it faddish. It has, in fact, become a required part of all psychiatric residency training programs in the United States. When I shared the stage in 2005 in a dialog with His Reverence, the Dalai Lama, at the meeting of the International Congress of Cognitive Psychotherapy, I was struck by some of the similarities between Buddhism and cognitive therapy—which gave me pause to consider how cognitive therapy has some roots in ancient traditions.

Unfortunately, there has been another thread in my life. I have been witness to Pearl Harbor; World War II; the Korean Conflict; Vietnam; the crises and resultant military interventions in Somalia, Rwanda, and Lebanon; and the first Gulf War. More recently, I have seen the tragedy and horror of 9/11, and the wars in Afghanistan and Iraq. From 1952 to 1954, I served in the military as Assistant Chief of the Department of Neuropsychiatry at Valley Forge Army Hospital, Valley Forge, PA. I saw the results of battle close up and personal. The results of battle are always the same—one side wins, one side loses—or the results of the battle are inconclusive. In any scenario, young men and women on all sides are physically or emotionally damaged, some beyond repair. As a result of the experience of the military in Vietnam, a new diagnostic entity entered our nomenclature, posttraumatic stress disorder (PTSD). It has been rewarding and satisfying to me that the developments in cognitive therapy have been able to play a powerful role in the treatment of depression, anxiety, and especially PTSD. Over the years I have worked closely with the Veteran's Health Administration to research the best treatments for a variety of disorders.

The young men and women who serve our country deserve the best possible treatment from the psychological and medical establishments. Therapists need to understand not only the nature of the military experience but also the training, mindset, culture, and battle experiences of the warrior who enters harm's way. There is a need for understanding the diversity of the military culture. At this point in time, the government is working to treat these returned warriors through VA medical centers, military hospitals, field hospitals, and combat stress units working with military just minutes from the field of battle. There is a powerful need for nonmilitary therapists to develop the understanding and the skills necessary to help pre- and postdeployment military with personal, family, relationship, substance abuse, and occupational issues.

The editors of this text have gathered the foremost mental health professionals to share their expertise. They are well suited to accomplish this goal. Sharon Morgillo Freeman has distinguished herself as a teacher, author, editor, and practitioner of cognitive therapy. As an advanced practice psychiatric clinical nurse specialist in the treatment of pain and substance use disorders, she has become an eloquent voice for cognitive therapy treatment in these and other more general areas. Bret A. Moore (CPT, U.S. Army) is part of the new generation of cognitive therapists. A former student of Art Freeman, he served two tours in Iraq as part of a Combat Stress Unit as the Officer in Charge of Clinical Operations and the Officer in Charge of Preventative Services. Having been trained in cognitive therapy, he trained his soldiers to use it directly on the battlefield. Art was a student of mine and for the past 30 years has been a colleague and collaborator. Art's creativity and enthusiasm have helped to bring cognitive therapy to new groups around the world.

The contributors to this text cover topics that are essential reading for any therapist working with members of the military, whether pre- or postdeployment. Their shared goal is to help the nonmilitary therapist better understand the military culture and to serve the needs of homecoming military and return these young men and women to more adaptive lives. I wish the young men and women who are in harm's way God speed and a safe return to their homes, families, and loved ones. I can only thank them for their dedication, work, and sacrifice.

Aaron T. Beck, MD (Captain, USAR, MC, retired)

The Contributors

Nathan D. Ainspan, PhD, is an industrial psychologist in the Washington, D.C., area. He has researched and written on improving the situation of employment for people with disabilities and returning service members and veterans with disabilities for a number of years. He worked for the Office of Disability Employment Policy at the Department of Labor in Washington, D.C., and has conducted research into and given presentations, seminars, and workshops on this topic. His research interests have focused on understanding the symptoms and treatment of PTSD, using employment as a form of positive psychology, and the impact and effect of disabilities on the friends, family members, and supporters of veterans and returning service members.

Marissa Burgoyne, MA, is a doctoral candidate in clinical psychology at Pepperdine University. She received her MA in psychology from Pepperdine University and her BA from Brown University. Marissa, with two other doctoral candidates, is currently developing a motivational interviewing group treatment protocol to promote positive health behaviors among veterans with PTSD. Marissa spent a year as a pre-intern at the Los Angeles Ambulatory Care Center (Downtown Veterans Administration) and intends to continue her work with veterans after the receipt of her doctoral degree.

Heather Campagna is a school psychologist who obtained her master's and education specialist degrees from The Citadel. She has worked in school settings from preschool to college levels, as well as the clinical setting at the Medical University of South Carolina. Heather currently serves as a Good Grief Camp Director for TAPS (Tragedy Assistance Program for Survivors), providing support groups for those children and teens nationwide who have lost a loved one in the military. She has assisted the South Carolina National Guard Family Programs with briefings for children, schools, parents, and professionals during all phases of deployment. Heather is serving full time in the South Carolina National Guard as an Apache pilot after returning from Iraq a few years ago.

Carole L. Campbell, PhD, is a clinical psychologist at the Child Abuse Program at Children's Hospital of the King's Daughters and instructor of child abuse pediatrics at Eastern Virginia Medical School in Norfolk. She is involved in multidisciplinary teams providing services to maltreated military and civilian children and training providers in evidence-based interventions.

Her interests include developing and disseminating effective interventions for traumatized youth, with particular interest in tailoring interventions to underserved groups, including Hispanic youth as well as children of military personnel.

Bonnie Carroll is Director of the Tragedy Assistance Program for Survivors (TAPS), the national veterans service organization addressing the emotional, psychological, and administrative problems that arise from the loss of a loved one in service to America. Ms. Carroll founded TAPS in 1994 after the death of her husband, Brigadier General Tom Carroll. As an Air Force Reserve Officer, Major Carroll served at HQ USAF Casualty Affairs; as a White House appointee, she served in Baghdad in the initial phase of the Iraq reconstruction. She chairs the military special interest group for the Association of Death Education and Counseling and is a trained critical incident stress debriefer. Her focus is on meeting the needs of those profoundly impacted by a death in the armed forces and developing networks of care throughout private and public sectors.

Elizabeth Casas, MA, is a doctoral candidate of Pepperdine University's Clinical Psychology Program in Los Angeles, California. Currently, she provides mental health services at the Los Angeles Ambulatory Care Center and the Pepperdine Community Counseling Center. As a member of Dr. David W. Foy's PTSD Research Lab, Elizabeth has participated in research on veterans with PTSD. Her dissertation centers on development of a motivational interviewing group treatment protocol to promote positive health behaviors among veterans with PTSD.

John R. Christian, PsyD, is deputy command psychologist at the U.S. Marine Corps Forces Special Operations Command (MARSOC). He earned his PsyD from Indiana State University and was commissioned a Lieutenant in the U.S. Navy. Following his internship at the National Naval Medical Center, Bethesda, he was posted at Naval Hospital Camp Lejeune, where he provided psychological services to Marines and sailors returning from OIF/OEF. Following that assignment, he transferred to MARSOC. As part of his current duties, he oversees the assessment and selection program for choosing Marine candidates for special operations. He also consults in the training of Marine special operations forces and is actively involved in research supporting operational issues in the military.

Jeffery A. Cigrang, Lt Col, U.S. Air Force, is a board-certified clinical health psychologist with 18 years of active-duty service in the U.S. Air Force. Lt Col Cigrang obtained his PhD in clinical psychology from the University of Memphis and completed postdoctoral training in clinical health psychology at Wilford Hall Medical Center in San Antonio, Texas. He is currently the chair of the department of psychology at Wilford Hall Medical Center. Lt Col Cigrang is an active researcher in the areas of health psychology and mental health issues in the military. He has deployed in support of Operation Enduring Freedom and Operation Iraqi Freedom, including two deployments to Iraq.

Judith A. Cohen, MD, a board-certified child and adolescent psychiatrist, is Medical Director of the Center for Traumatic Stress in Children and Adolescents at Allegheny General Hospital and Professor of Psychiatry at Drexel University College of Medicine in Pittsburgh. She is the principal author of the practice parameters for the assessment and treatment of PTSD published by the American Academy of Child and Adolescent Psychiatry (AACAP) and is a recipient of AACAP's Reiger Award for Outstanding Scientific Achievement. She has served on the boards of the International Society for Traumatic Stress Studies and the American Professional Society on the Abuse of Children and continues to train, conduct research, and write about the treatment of traumatized children as well as to treat these children in clinical practice.

Lauren M. Conoscenti, PhD, is a postdoctoral fellow at the National Center for PTSD/Boston VA Healthcare System. She received her bachelor's degree in psychology from Princeton University in 2000 and her doctorate in clinical psychology from Harvard University in 2007. She completed her clinical training at the National Crime Victims Research and Treatment Center at the Medical University of South Carolina, where she was a recipient of a National Institutes of Health training grant. Dr. Conoscenti's current research focuses on predictors, outcomes, and treatment of combat stress and interpersonal violence.

Erika Curran, LCSW, was a licensed clinical social worker with specific expertise in the family treatment of veterans with PTSD. She worked for 22 years as the primary family provider for the largest residential PTSD treatment program in the VA system. In this context, she influenced the lives of countless veterans and their families and provided training for an entire generation of PTSD clinicians. She was a gifted clinician and instructor who for many years taught regularly in the clinical training program of the National Center for PTSD, as well as at a variety of invited conferences. Erika passed away suddenly prior to the final publication of this book. She will be profoundly missed by her family, friends, and colleagues.

Kent D. Drescher, PhD, is a licensed clinical psychologist on staff at the National Center for PTSD, Dissemination and Training Division, in Menlo Park, California, part of the VA Palo Alto Health Care System. His work there involves education and research activities with two areas of special interest: trauma and spirituality, and health promotion for PTSD veterans. He is involved in a wide variety of training activities in the assessment and treatment of PTSD. For the last 8 years, he has with several colleagues been developing and piloting a treatment group called Trauma and Spirituality within a PTSD residential treatment program. Outside the VA, Dr. Drescher has served as an adjunct faculty member at the California Institute of Integral Studies in San Francisco, teaching courses in their APA-approved doctoral program. Prior to his doctoral training at Fuller Graduate School of Psychology, he received a divinity master's degree from San Francisco Theological Seminary, was ordained by the Presbyterian Church, and served as a parish minister for several years.

David W. Foy, PhD, received his doctorate in clinical psychology in 1975. He is currently professor of psychology in the Graduate School of Education and Psychology, Pepperdine University. He also holds an appointment as adjunct professor of psychology, Headington Program in International Trauma, Fuller Theological Seminary. He has many years of experience as a clinician, teacher, and researcher in the field of psychological reactions to trauma, especially combat-related PTSD. He is the editor of a widely used handbook, *Treating PTSD: Cognitive–Behavioral Strategies* (1992, Guilford Press), and served for many years as a senior research advisor to the National Center for PTSD, Menlo Park and Honolulu Divisions.

Arthur Freeman, EdD, ABPP, is a visiting professor of psychology at Governors State University, University Park, Illinois, and a clinical professor at the Philadelphia College of Osteopathic Medicine.

Sharon Morgillo Freeman, MSN, PhD, PMHCNS-BC, serves as CEO and President of the Center for Brief Therapy and the Freeman International Institute in Fort Wayne, Indiana.

Robin F. Goodman, PhD, is a licensed clinical psychologist and board-certified registered art therapist. She is director of A Caring Hand, the Billy Esposito Foundation Bereavement Center. Previously she was Director of Family Programs, Voices of September 11th; consultant to the Department of Defense Educational Opportunities Directorate and the National Child Traumatic Stress Network; and director of AboutOurKids.org. She has authored *The Day Our World Changed: Children's Art of 9/11*. Her professional service includes being past president of the American Art Therapy Association. Dr. Goodman has lectured, taught, and published on a variety of child and family mental health issues and has particular experience and interest in mental illness and bereavement related issues.

Dave Grossman, Lt Col, U.S. Army (Ret.), is director of Killology Research Group (www.killology.com). Col Grossman is a West Point psychology professor, a professor of military science, and an Army Ranger who has combined his experiences to become the founder of a new field of scientific endeavor, which has been termed *killology*. He is the author of *On Killing*, which was nominated for a Pulitzer Prize and has been translated into Japanese, Korean, and German; it is on the U.S. Marine Corps' recommended reading list and is required reading at the FBI academy and numerous other academies and colleges. Col Grossman coauthored with Gloria DeGaetano *Stop Teaching Our Kids to Kill: A Call to Action Against TV, Movie, and Video Game Violence*, which has been translated into Norwegian and German and has received international acclaim. His most recent book, with Loren Christensen, is *On Combat*, the highly acclaimed sequel to *On Killing*. Col Grossman has been called upon to write the entry on "Aggression and Violence" in the *Oxford Companion to American Military History*, three entries in the Academic Press *Encyclopedia of Violence*, and numerous entries in scholarly journals, including the *Harvard Journal of Law and Public Policy*. He has been an expert witness and consultant in state and federal courts and served on the prosecution team for *United States vs. Timothy McVeigh*. Col Grossman is an Airborne Ranger infantry officer and a prior service sergeant and paratrooper, with a total of over 23 years of experience leading U.S. soldiers worldwide. He retired from the Army in February 1998.

C. Alan Hopewell, PhD, MP, ABPP, was commissioned during the latter part of the Vietnam conflict as a Second Lieutenant in the Infantry upon his graduation from the Texas A&M Corps of Cadets. He then completed his master's degree and doctorate in clinical psychology from North Texas State University and his residency at the University of Texas Medical Branch, Galveston. Dr. Hopewell next saw Cold War service by serving as chief of psychology and neuropsychology at Moncrief Army Hospital, Ft. Jackson, South Carolina; Landstuhl Army Regional Medical Center, Germany; and Brooke Army Medical Center, Ft. Sam Houston, Texas. After earning his second master's degree in clinical psychopharmacology from the California School of Professional Psychology/Alliant University, Dr. Hopewell was commissioned for his fourth tour of duty and assigned as officer-in-chief of outpatient mental health at Carl R. Darnall Army Medical Center (CRDAMC) for the Global War on Terror. As such, he was the first medical psychologist Army officer to enter active duty with a state license as a prescribing psychologist. With a PROFIS assignment to the 785th Medical Company (CSC) at Camp Liberty, Iraq, Dr. Hopewell became the first Army prescribing psychologist to serve in a Combat Theater. As the senior Department of Defense active-duty clinical neuropsychologist, he also served as codirector of the Traumatic Brain Injury (TBI) Service at CRDAMC and as the TBI Theater Consultant for Operation Iraqi Freedom. For his service in

Operation Iraqi Freedom, he was awarded the Bronze Star. He is a fellow of the American Psychological Association and also a past president of the Texas Psychological Association.

Michael R. Hurst is a retired U.S. Army Special Forces Captain (Green Beret) and a Southwest Asia and Somalia veteran. He served in Special Forces and Airborne Infantry units as an officer and enlisted service member for approximately 18 of his 20-year military career. His counseling career includes working as a counselor for an outdoor therapeutic adolescent counseling program; as treatment coordinator for the Christian County, Kentucky, Drug Courts Program; as a probation and parole substance abuse counselor; and as an employee assistance program coordinator/employee assistance professional for the U.S. Army Substance Abuse Program at the Fort Myer Military Community, Fort Myer, Virginia.

William Isler, PhD, earned his bachelor's degree from Baylor University and his doctorage in clinical psychology from the University of North Texas. Dr. Isler entered the Air Force in 1996 and served at Wright–Patterson Air Force Base and McChord Air Force Base before completing a clinical health psychology fellowship in 2001. He then served as a primary-care behavioral health consultant at Eglin Air Force Base and at the Wilford Hall Medical Center. Dr. Isler was deployed to Balad Air Base in support of Operation Iraqi Freedom in 2006 and is currently an Air Force Deployment Behavioral Health chief. In his current position, he develops and initiates evidence-based training for both posttraumatic stress disorder and traumatic brain injury.

Megan M. Kelly, PhD, is a licensed clinical psychologist and a research psychologist at Butler Hospital in Providence, Rhode Island. She is also an assistant professor of psychiatry and human behavior at the Alpert Medical School of Brown University. Dr. Kelly received her doctorate in clinical psychology from the University at Albany, State University of New York, and completed her postdoctoral training at the Alpert Medical School of Brown University and Butler Hospital. Dr. Kelly's research interests are focused on the relationship between stress and the development and maintenance of emotional disorders, including disorders of body image, depression, and PTSD. She is currently researching neurobiological markers and psychological processes associated with stress reactivity in individuals with depression and anxiety-related disorders.

Carrie H. Kennedy, PhD, is a Lieutenant Commander in the Medical Service Corps of the U.S. Navy. She is board certified in clinical psychology and currently serves as the neuropsychologist at the Naval Aerospace Medical Institute in Pensacola, FL. She is the only dual-designated clinical psychologist/aerospace experimental psychologist in the Navy. She completed her doctorate at Drexel University and her neuropsychology postdoctoral fellowship at the University of Virginia. She serves as a member-at-large of the American Psychological Association's Division 19 (Society for Military Psychology) and as chair of the Conflict of Interest Committee for the National Academy of Neuropsychology. She is the co-editor of *Military Psychology: Clinical and Operational Applications* (2006, Guilford Press).

Barry Krakow, MD, is a board-certified internist and sleep disorders specialist who has studied and practiced for 25 years in the fields of internal medicine, emergency medicine, addiction medicine, and sleep medicine. Currently, he is medical director of two sleep facilities in Albuquerque, NM: Maimonides Sleep Arts & Sciences, Ltd., a community-based, sleep medical center, and the Sleep & Human Health Institute, a nonprofit sleep research institute. Current clinical and research interests include cognitive-imagery treatment for chronic nightmares, the role of sleep apnea in chronic insomnia, the interaction between fatigue and suicidality, and the prevalence of sleep apnea in chronic hypnotic users. He has also authored three books on sleep disorders: *Insomnia Cures*, *Turning Nightmares into Dreams*, and *Sound Sleep, Sound Mind*.

Brett T. Litz, PhD, is an associate professor in the Department of Psychiatry at Boston University School of Medicine and the Department of Psychology at Boston University. He is also Associate Director of the Behavioral Sciences Division of the National Center for Posttraumatic Stress Disorder at Boston Department of Veterans Affairs Medical Center. Dr. Litz is the principal investigator on several research studies funded by the National Institutes of Health and the Department of Defense to explore the efficacy of early intervention strategies in trauma. He is currently studying adaptation to traumatic loss as a result of 9/11. In addition to conducting research on early intervention for trauma, Dr. Litz studies the mental health adaptation of U.S. military personnel across the lifespan, the assessment and treatment of posttraumatic stress disorder, and emotional numbing in trauma.

Lauren Lovato, MA, is a doctoral candidate in clinical psychology at Pepperdine University. She received her master's degree in psychology, with an emphasis on marriage and family therapy, from Pepperdine University and her BA in psychology and BS in journalism from the University of Colorado. Lauren has worked as a pre-intern at the VA Long Beach Healthcare System and received specialized training in cognitive–behavioral and dialectical behavioral therapy at the Harbor–UCLA Medical Center. Her dissertation centers on development of a motivational interviewing group treatment protocol to promote positive health behaviors among veterans with PTSD. Lauren's main research interests include trauma and cognitive behavioral therapy.

Leslie Lundt, MD, received her degrees from The Johns Hopkins University and Rush Medical College. A board-certified psychiatrist and addiction medicine specialist, she was the chief resident at Pacific Medical Center in San Francisco. Dr. Lundt is currently a medical expert host of the *ReachMD* show on XM Radio. She has appeared on morning and evening news shows for ABC, NBC, and CBS and has appeared on Court TV. Her first book, *Think Like a Psychiatrist*, was published in 2007.

Judith A. Lyons, PhD, studied in Montreal (BA from McGill University; MA and PhD from Concordia University) and interned in Jackson, Mississippi. After serving as the founding clinical director of the National Center for PTSD at the VA Boston Healthcare System, she returned to Jackson to establish the VA Trauma Recovery Program (TRP). She serves as TRP team leader and associate professor of psychiatry and human behavior at the University of Mississippi Medical Center and conducts research with the support of the VA South Central Mental Illness Research, Education, and Clinical Center (MIRECC). Posttrauma resilience, moral conflict, and cognitive appraisal are major clinical and research interests. She has published more than 40 papers on the topic of traumatic stress. Enjoyment of animals and nature along with engagement in the Rotary Club and faith-based activities balance her trauma interests.

P. Alex Mabe, PhD, received his doctoral degree in clinical psychology from Florida State University in Tallahassee, Florida. Currently, he is Professor and Chief of Psychology in the Department of Psychiatry and Health Behavior at the Medical College of Georgia. His publications include over 40 articles in the areas of clinical child and pediatric psychology. Additionally, he has made numerous presentations at national and international professional meetings on topics related to children's mental health and family and parent management training. Dr. Mabe is licensed as a psychologist in Georgia and South Carolina and has been providing clinical psychology services to children and their families in the Central Savannah River Area for over 25 years, including extensive work with military families on assignment at Fort Gordon, Georgia.

Rosemary C. Malone, MD, is a Commander in the Medical Corps (MC) of the U.S. Navy. She is board certified in adult and forensic psychiatry and is currently the Director for the Mental Health Directorate at the National Naval Medical Center (NNMC) in Bethesda, Maryland. She deployed to Guantanamo Bay for 10 months, serving as the Behavioral Health Department Head for the Detention Hospital as well as acting Officer-in-Charge. Dr. Malone's assignments prior to NNMC included serving as a staff psychiatrist in Sigonella, Sicily, and as a fellow in the National Capital Consortium's Forensic Psychiatry

Program. Prior to becoming a MC officer, she taught chemistry, radiological fundamentals, and materials at the Naval Nuclear Power School when it was located in Orlando, Florida.

Michael D. Matthews, PhD, is a professor of engineering psychology at the U.S. Military Academy (West Point) and director of West Point's Engineering Psychology Program. He is a former Air Force officer and is past-president of the Society for Military Psychology (Division 19 of the American Psychological Association). Dr. Matthews is a Templeton Foundation Senior Positive Psychology Fellow, and his current research interests focus on the role of character strengths in soldier adaptation and performance in combat.

Donald Meichenbaum, PhD, is Distinguished Professor Emeritus, University of Waterloo in Ontario, Canada. After some 35 years, he took early retirement and he is now research director of the Melissa Institute for Violence Prevention and Treatment of Victims of Violence in Miami, Florida (see www.melissainstitute.org and www.teachsafeschools.org). He is one of the founders of cognitive–behavioral therapy; in a survey of North American clinicians reported in the *American Psychologist*, Dr. Meichenbaum was voted "one of the ten most influential psychotherapists of the 20th century." He was the honorary president of the Canadian Psychological Association and a fellow of the American Psychological Association. He has published extensively and has developed stress inoculation training programs that have been used with a wide array of groups, including soldiers. Most recently, he has been working with the National Guard to develop a website and commander's training program to bolster resilience in soldiers and their families using iPOD technology (see www.warfighterdiaries.com). He has lectured and consulted internationally and has been involved in training at various Veterans Affairs hospitals. He can be contacted at dhmeich@aol.ocm.

Bret A. Moore, PsyD, ABPP, is a board-certified clinical psychologist with the Indian Health Service, Ft. Peck, Montana. In 2008, he left active-duty service in the U.S. Army, where he served as a Captain and clinical psychologist with the 85th Combat Stress Control unit based in Fort Hood, Texas. He has extensive experience treating veterans, including two tours of duty in Iraq as an officer in charge of preventative services and officer in charge of clinical operations. Dr. Moore completed his doctoral education at the Adler School of Professional Psychology in Chicago and recently completed a master's degree in clinical psychopharmacology from Fairleigh Dickinson University in Teaneck, New Jersey. He has authored and co-authored numerous journal articles, book chapters, and books on military psychology issues, including *The Veterans and Active-Duty Military Psychotherapy Treatment Planner* (2009, Wiley). In 2007, Dr. Moore was awarded the Arthur W. Melton

Early Achievement Award by Division 19 (Society for Military Psychology) of the American Psychological Association. In addition, for his service in Iraq as part of Operation Iraqi Freedom from 2007 to 2008, he was awarded the Bronze Star.

Anthony Papa, PhD, is assistant professor of psychology at the University of Nevada, Reno. His research and publications focus on grief and reactions to bereavement and potentially traumatic events. This work has examined social and emotional processes related to risk and resilience in these contexts, research on the development of early interventions for trauma and pathological grief in both military and civilian populations, and research on Internet-delivered psychotherapy.

Walter Erich Penk, PhD, ABPP, is a professor in psychiatry and behavioral sciences at the Texas A&M College of Medicine and a consultant for the Central Texas Veterans Health Care System and the VA Center of Excellence in Stress Disorders Research. He has served as a clinical psychologist and research psychologist with the Veterans Health Administration in Houston, Dallas, and Boston, as well as in Bedford, Massachusetts, where he was chief of the psychology service. He held adjunct clinical appointments in Dallas at the University of Texas Health Sciences Center and in Boston at Tufts, Boston University, University of Massachusetts, and Harvard Medical Schools. He has served as the director of psychology for the Department of Mental Health in the Commonwealth of Massachusetts and was the associate director of the New England VA (VISN 1) Mental Illness Research, Education, and Clinical Center.

Alan L. Peterson, PhD, ABPP, is a professor in the Department of Psychiatry at the University of Texas Health Science Center at San Antonio and maintains a joint appointment at the San Antonio Military Medical Center. He served 21 years on active duty with the U.S. Air Force and completed deployments in support of Operations Enduring Freedom and Iraqi Freedom. He has authored over 100 scientific publications and presentations and has been a principal or co-investigator on research grants totaling over $50 million. He is the consortium director of the STRONG STAR Multidisciplinary PTSD Research Consortium centered in San Antonio, Texas.

Ilona Pivar, PhD, is a board-certified clinical psychologist currently working within the primary health-care setting of the New Mexico VA Health Care System in Albuquerque and Santa Fe. She completed her doctorate from the Pacific Graduate School of Psychology, Palo Alto, California, in 2000. Her dissertation at the Menlo Park division of the VA Palo Alto Health Care System addressed the measurement, prevalence, and patterns of unresolved grief experienced by veterans admitted to a PTSD residential treatment program. She was a member of the National Center for PTSD and is active in clinical work, research, and training programs addressing traumatic grief, male military sexual trauma, and disaster mental health. She has also been a consultant and trainer for the Cognitive Processing Therapy program for the Veterans Administration.

Greg M. Reger, PhD, is a licensed clinical psychologist at the National Center for Telehealth and Technology, part of the Defense Centers of Excellence for Psychological Health and Traumatic Brain Injury. Dr. Reger completed his doctorate in clinical psychology at Fuller Theological Seminary and interned at Walter Reed Army Medical Center. He subsequently served as an active-duty psychologist, deploying to Iraq with the 98th Combat Stress Control Detachment. He has worked for the Center for Deployment Psychology training military and civilian mental health providers in military psychology. Dr. Reger is currently a principal investigator for innovative research funded by the Congressionally Directed Medical Research Program exploring the potential of virtual-reality technologies to improve treatment outcomes for combat-related posttraumatic stress disorder.

David S. Riggs, PhD, is the executive director of the Center for Deployment Psychology (CDP) and a research associate professor of psychology at the Uniformed Services University of the Health Sciences. Dr. Riggs has spent much of his career treating and studying anxiety disorders and the emotional effects of trauma exposure with an emphasis on posttraumatic stress disorder (PTSD) arising from sexual assault and military combat. In his current position at the CDP, Dr. Riggs focuses on the training and education of mental health professionals to include the dissemination of evidence-based interventions for PTSD, anxiety, depression, and other stress-related problems.

M. David Rudd, PhD, ABPP, is currently professor and chair of the Department of Psychology at Texas Tech University. He graduated from Princeton University and completed his doctoral work at the University of Texas–Austin. He also completed postdoctoral training in cognitive therapy at the Beck Institute in Philadelphia under the direction of Aaron T. Beck, MD. He has published extensively in the suicide literature and his work has been recognized by multiple awards, including the American Association of Suicidology 2009 Louis I. Dublin Award for outstanding lifetime contributions in suicide-related research. During his 5 years of service as an active-duty Army psychologist, Dr. Rudd served with the 2nd Armored Division as a division psychologist during the Gulf War.

Morgan T. Sammons, PhD, ABPP, is systemwide dean of the California School of Professional Psychology at Alliant International University. He is a retired Captain in the U.S. Navy. In his 20-year naval career, he served as the Navy's Clinical Psychology Specialty Leader and as the Special Assistant to the Navy Surgeon General for Mental Health and Traumatic Brain Injury issues. He held a number of positions both in the United States and abroad, completing tours of duty in Japan, Iceland, and at the U.S. Naval Academy. In 2006, he deployed to Fallujah, Iraq. One of the first graduates of the Department of Defense's Psychopharmacology Demonstration Project, Dr. Sammons is a Fellow of the American Psychological Association, immediate past president of the National Register of Health Service Providers in Psychology, president of APA's Division 55, immediate past chair of the California Psychological Association's Division 5 (Psychopharmacology), and past president of the Maryland Psychological Association. He is an associate editor of the APA journal *Psychological Services* and a past associate editor of *Professional Psychology: Research and Practice*.

James R. Stivers is a Master Sergeant (select) in the U.S. Marine Corps. During his career with the Marines, he has served in the infantry, reconnaissance battalions, Force Reconnaissance Companies, and various special operations roles. A decorated veteran of numerous deployments, he was instrumental in the formation of the assessment and selection program for the U.S. Marine Corps Forces, Special Operations Command (MARSOC).

Edward J. Swanton, MD, is a board-certified psychiatrist and internist and is board eligible in geriatric psychiatry. He is a Lieutenant Colonel in the U.S. Army. Dr. Swanton received his medical degree from the Uniformed Services University of the Health Sciences in Bethesda, Maryland, in 1996. He completed a combined 5-year residency in internal medicine and psychiatry at Walter Reed Army Medical Center (WRAMC) in 2001. He served as the Assistant Chief of Inpatient Psychiatry and as Chief of the Outpatient Psychiatry Clinic at Landstuhl Regional Medical Center from 2001 to 2005. Following his geriatric fellowship in 2006, he served as Chief of Behavioral Health at Fort Meade, Maryland. He completed a 13-month tour in Iraq in 2008 as an Army psychiatrist; he served as the officer-in-charge of a combat stress control team. He then returned to WRAMC and today serves as assistant chief of the Psychiatry Consultation Liaison Service (PCLS).

Vera Vine is a research assistant at the University of Texas at Austin. She has researched depression and combat-related posttraumatic stress disorder, as well as the linguistic markers of personality. Her main interests are cognitive and emotional processing and emotion regulation and their influence on psychopathology.

Dawne S. Vogt, PhD, is a research psychologist in the Women's Health Sciences Division of the National Center for Posttraumatic Stress Disorder, VA Boston Healthcare System, and associate professor of Psychiatry at Boston University School of Medicine. Her primary research interests are in deployment risk and resilience factors and stigma, gender, and other barriers to VA health-care use. Dr. Vogt is co-author of the Deployment Risk and Resilience Inventory (DRRI), a suite of scales for assessing key psychosocial risk and resilience factors among military and veteran populations. She has used scales from the DRRI to examine a number of research questions regarding the impact of deployment on the health of returning veterans, and is principal investigator of a project aimed at updating DRRI scales for broader applicability across veteran cohorts.

1
Introduction*

ARTHUR FREEMAN, SHARON MORGILLO FREEMAN,
and BRET A. MOORE

Duty is the sublimest word in our language. Do your duty in all things.
You cannot do more. You should never wish to do less.

It is well that war is so terrible, or we should grow too fond of it.

General Robert E. Lee

Throughout history, men and women have, for myriad reasons, placed themselves in harm's way. Sometimes this choice was a personal one motivated by
one's beliefs, be they religious, political, or nationalistic. For some the role of
warrior is freely chosen, and for others it is a matter of being conscripted and
sent into battle. The attributed rationales and reasons may change from group
to group, time to time, and person to person:

> We lived in 31 houses, apartments, and, in one case a house trailer, had
> 20 jobs, and were always on the road—and it wasn't the road to riches,
> but when my eight-year obligation to the Army was over, I decided to
> stay. To me, there was no greater honor—no way to be nearer to the
> heart of what mattered in America—than to be serving and protecting
> the country in the United States military.

General Wesley K. Clark, Ret. (2004)
(http://www.clark04.com/about)

Each succeeding military group views their predicament as somehow new
or different. One has only to read the press or listen to media commentators
who speak of the current conflict that engages a country (whatever that might
be) as the worst, most challenging, or most evil empire or axis with which the
national interest has ever been confronted.

Through it all, for whatever the reason, war places people in harm's way.
Whether that harm is likely to be inflicted by an arrow, flaming pitch, a gun-
powder-propelled lead ball, a nuclear device, or a homemade explosive device,

---

* The views expressed herein are those of the authors and do not reflect the official position
or policy of the U.S. Army, U.S. Navy, U.S. Air Force, U.S. Department of Defense, or the
U.S. government.

the fighting corps in the field or battlefront and the civilians on the homefront are impacted in many ways. For example, the German troops at the gates of Moscow were confronted by both an implacable enemy (the Red Army) and a bitter winter. The civilians in Dresden or Berlin were confronted by both daylight and nighttime bombings that eventuated in firestorms. The workers in the World Trade Center in New York came to work on September 11, 2001, and those in the Murrah Building in Oklahoma City on April 19, 1995, with the intention of doing their job. Instead, they found themselves in harm's way.

The current conflicts in which the United States and its allies are engaged in Iraq and Afghanistan are but the latest manifestation of how harm's way impacts individuals at the front, those who train and equip them, and those in the family systems that support the service member. Harm's way is a broad blanket that stretches from the warfront to the homefront. Witness, for example, entire communities constructing supportive and patriotic billboards or electric signs, communities filled with yellow ribbons reminding all of those warriors who at every moment are in danger. This has not been the case in all conflicts where warriors were in harm's way. The bitter experience of the Vietnam era showed how those in harm's way had little support in that the war was an unpopular one in many quarters.

This is a volume on how warriors live and survive being in harm's way and the psychological impact of those actions. It addresses service members, their support systems, their interpersonal experiences, and the intrapersonal consequences of being in harm's way. It also addresses the psychological health and adjustment of warriors who have left the battlefield and are working at reintegrating back into the lives they put on hold prior to their military service. It is a book meant for the clinician who will likely be treating an individual who has been in harm's way, a member of their family, or a community member. Many returning military, of all services, may have severe and significant emotional and behavioral problems that are related to their military service. Many of these individuals will be seen by therapists who do not have military experience and are therefore not competent in military culture. Our goal in compiling this volume is to inform the nonmilitary as well as the military therapist about the broad range of problems that they might encounter with veterans or active duty military personnel.

In Chapter 2, Michael Mathews addresses the issue of the soldier's mindset. From the time soldiers complete basic training until long after the time they retire they are trained to think of their fellow soldiers as one with themselves, to act as a unit, to be mission oriented and motivated, to work tirelessly, to be respectful to the chain of command, to follow all lawful orders, to have a high moral purpose, and to act in the best interests of their country, their unit, and their service. *Loyalty, duty, respect, selfless service, honor, integrity,* and *personal courage*: These are the soldier's values that must be understood to understand the soldier.

Chapter 3, by John Christian, James Stivers, and Morgan Sammons, describes for the reader the preparation that precedes the entry into harm's way. These individuals are trained to be combatants and warriors. They are taught weapon usage, military tactics, survival techniques, and defensive strategies. They are taught to survive being in harm's way. They are also inculcated with the skills to be the saviors of a country and people that are dependent on them for survival.

Greg Reger and Bret Moore, in Chapter 4, describe the threats and challenges of deployment. Being deployed has many meanings. Some are rather simple and obvious. The individuals are being sent away from home to various sites for further training or to foreign countries where there is risk of personal losses, physical injury, and possibly death. In addition to fears related to death and injury, service members are faced with challenging environmental conditions, uncertainty about relationships back home, lack of freedom and personal space, and increased conflict between peers and superiors.

In Chapter 5, by Carrie Kennedy and Rosemary Malone, the discussion is related to the integration of women into the warrior culture, which is a central issue among today's fighting forces. In past wars, there were clear battle lines, and women generally served behind them. Now, women serve actively in combat roles. The way in which women are acculturated and treated within the military is often out of step with the traditional views and roles of women. This chapter addresses the difficulties women face as they serve in active combat and in today's military.

Megan Kelly and Dawn Vogt, in Chapter 6, discuss the effects of acute, chronic, and traumatic stress. Stress is both a motivator and an impediment to performance. This chapter addresses the various manifestations of stress experienced by military personnel who have served in both combat and noncombat situations. The military refers to the ability to deal with stress as *resilience*. The military definition of resilience and how the military identifies and utilizes resilience in its personnel are introduced in this chapter and further discussed in Chapter 7.

Each individual has a set of vulnerability factors that affect their threshold of response. The higher one's threshold, the greater the ability to cope with stressful experiences, both internally and externally generated. Vulnerability and resilience must be noted when dealing with the issues of coping and adapting to stress, whether internally or externally driven. These vulnerability factors have a summative effect so that the greater the number of vulnerability factors the lower the threshold of response. Chapter 7, by Arthur Freeman and Sharon Morgillo Freeman, provides an examination of each of these factors and how they impact the individual's response.

Chapter 8, by Lauren Conoscenti, Vera Vine, Anthony Papa, and Brett Litz, discusses how the individual who has been trained to scan for potential threats and danger adjusts to noncombat situations. In a combat environment, any shadow, movement, or sound could signal attack. In a noncombat environment, the same or similar stimuli may trigger a trained response in the combat-trained soldier. The backfire of a car or a sudden and unexpected movement may be responded to with a defensive maneuver. If the adaptive response becomes intrusive or prevents individuals from performing their usual activities they may require assistance in reducing hypervigilant intrusive responses.

In Chapter 9, by Arthur Freeman, Sharon Morgillo Freeman, and Michael Hurst, the assessment process is described and discussed relative to the overall data collection required for treatment. The assessment of emotional and behavioral disorders among military personnel can be complex. Behaviors in a civilian setting may be labeled as paranoia, while in a military setting they may be considered exceptional acumen for a sniper. A soldier's feelings of loss and hopelessness after being airlifted with an injury incurred in an incident in which that soldier was the only survivor are normal parts of a grieving process and do not necessarily represent major depression. This chapter focuses on the differential diagnosis of a broad range of disorders.

Chapter 10, by Arthur Freeman and Bret Moore, describes a clinical treatment model for intervening with military. Several key models important and useful for a comprehensive treatment program are discussed. These include cognitive behavior therapy as a key ingredient, with the additions of systemic and behavioral components.

In Chapter 11, Donald Meichenbaum describes a case conceptualization approach for treating returning soldiers. It involves several components, including a step-by-step process for developing the conceptualization, which then becomes the template for therapy and helps direct treatment.

Chapter 12, by David Riggs, discusses anxiety, which is among the most common of the emotional disorders. Anxiety is addressed as both a disorder and as part of the broader anxiety spectrum. Treatment issues for the individual who has been trained and habituated to the idea that the world is a dangerous place are discussed.

David Rudd, in Chapter 13, discusses the depressive spectrum and suicidal wishes, actions, and ideation. The nature and treatment of depression are discussed, along with reasonable responses of the soldier as opposed to unhealthy rumination or other types of responses.

In Chapter 14, by Sharon Morgillo Freeman and Michael Hurst, substance use, misuse, and abuse are targeted. Most members of the military enter the service in their early teens. This is also the age when most individuals are using or experimenting with substances, usually alcohol. It is not unusual for adolescents and young adults to relax and socialize with alcohol and other chemicals during social occasions and to use these same chemicals when upset, angry, celebrating, or grieving. The soldier is at high risk for development of a substance use disorder after deployment, and this chapter presents a variety of methods to evaluate, discuss, and if necessary treat substance use disorders in military personnel.

Chapter 15, by Bret Moore and Barry Krakow, discusses sleep disorders, including problems with sleep onset, sleep maintenance, and early waking. Additional issues for many after deployment include vivid dreaming and nightmares. Cognitive and behavioral interventions for sleep problems are discussed. In Chapter 16, Bret Moore, C. Alan Hopewell, and Dave Grossman address the issue of aggression and homicide. Aggression always has the potential for deadly consequences. The use of words and weapons, coupled with the combat skills acquired and mastered to survive in harm's way, may be deadly in the civilian milieu. The problems of impulsivity, poor executive control, and potentially dangerous behaviors that are a danger to others need to be addressed with service members.

Chapter 17, by Sharon Morgillo Freeman, Leslie Lundt, Ted Swanton, and Bret Moore, discusses the myths and realities of pharmacotherapy for military personnel. Many have misunderstandings about what pharmacotherapy can and cannot do. Members of both the military and the nonmilitary may have misperceptions about what medications the military allows soldiers to take and still be allowed to serve, to work in certain jobs, and to still be deployable. In Chapter 18, Alex Mabe describes the problems inherent in dealing with children and parental deployment. The typical issues of family and parenting can present significant problems for the military family. Many families suffer emotional trauma due to multiple deployments. Military spouses feel trapped at home, fearing they will miss a phone call. Their spouses may be depressed and grieving the loss of family gatherings, children's activities, and daily life. Every day the spouse at home fears the knock on the door heralding their worst nightmare. This chapter focuses on such problems that are common in military families.

The topic of intimate relationships and the military is addressed by Judith Lyons in Chapter 19. Spouses that have been close and intimate on a daily basis must suddenly cope with being apart for a year or more. Contact is reduced to sporadic phone calls, letters, and occasional packages. During a long deployment, there might be a brief period of leave before the soldier's return to the combat zone and being separated again. The prospect of one's spouse losing interest and finding someone else looms heavy for some. Others might find it difficult to get to know their spouses again and to have someone else regularly in their space. Moving from group identification and affiliation to one-on-one intimacy can present a number of issues that must be dealt with by the therapist.

Military children can become the sometimes orphans of war. The deployment of a parent means something different depending on the age of the child. In Chapter 20, Judith Cohen, Robin Goodman, Carole Campbell, Bonnie Carroll, and Heather Campagna review age-specific ideas about the concepts of time, separation, and the meaning of being away. In addition, this chapter discusses how frightening potential losses (be it a body part or parent) can be for children and how to discuss this loss with the child.

Chapter 21, by Walter Penk and Nathan Ainspan, notes that, unlike the widespread anger and contempt commonly encountered during the Vietnam era, the contemporary returnee may be applauded, cheered, honored, saluted, or ignored. After the parades and kind words, however, many veterans find themselves unemployed and even homeless. They may return home to find their spouses gone and their bank accounts empty. Many of our reserve and guard soldiers find that they have no benefits if they are wounded and are now unable to work. This chapter discusses the realities of coming home to a nightmare.

In Chapter 20, by Kent Drescher, Marissa Burgoyne, Elizabeth Casas, Lauren Lovato, Ilona Pivar, and Davis Foy, the issues of grief and loss, honor and remembrance, and the loss of innocence and survivor guilt are addressed within the context of the spirituality of military personnel and their families. Some young men and women do not survive in harm's way. They may be wounded, injured, disabled, or killed. Survivors have the task of returning to homes, families, and lives while leaving buddies behind. It is not unusual for the veteran to also suffer the same losses as those not in the military might, such as the death of a parent or relative who has been a major supporter. This chapter focuses on spirituality and its role in working with postdeployment veterans.

This final chapter, by Alan Peterson, Jeffrey Cigrang, and William Isler, offers a summary and describes future directions regarding trauma, resilience, and recovery research.

And I will always do my duty
No matter what the price
I've counted up the cost
I know the sacrifice
Oh and I don't wanna die for you
But if dying's asked of me
I'll bear that cross with honor
'Cause freedom don't come free
I'm an American soldier
An American
Beside my brothers and my sisters
I will proudly take a stand
When liberty's in jeopardy
I'll always do what's right
I'm out here on the front line
Sleep in peace tonight
American soldier
I'm an American soldier[*]

---

[*] "American Soldier" lyrics are by Toby Keith and Chuck Cannon and are the property and copyright of their owners (provided for educational purposes only).

I
Understanding the Service Member

2

The Soldier's Mind: Motivation, Mindset, and Attitude

MICHAEL D. MATTHEWS

Contents

To understand the motivation, mindset, and attitude of the American soldier,[*] it is critical to thoroughly understand the vital role that character and values play in shaping both individual and team adaptation and performance. The overarching importance of character and values is found in military doctrine, based on centuries of practical experience in training soldiers, preparing them for war and actual combat performance. For example, Army leadership doctrine clearly defines seven Army values (loyalty, duty, respect, selfless service, honor, integrity, and personal courage) as being of central importance to each and every soldier and Army leader, and a careful reading of *Army Field Manual 22-100* (Department of the Army, 1999) reveals at least ten additional strengths of character that underlie successful soldier adaptation and performance.

Military doctrine by itself, however, does not provide an adequate basis for the understanding of soldier adaptation and performance. There needs to be an empirically based, scientific model to organize what is known about the role of character and values to guide scientists in formulating hypotheses about soldier adaptation and performance, and to inform practitioners on

---

[*] In this chapter, the term *soldier* is used in the generic sense to refer to any member of the U.S. Armed Forces.

how to build better military teams, assist soldiers who experience adjustment difficulties, and empower soldiers to improve themselves personally. Until recently, no sufficiently broad psychological system existed for this purpose; however, the emerging field of positive psychology is especially appropriate as a theoretical framework for understanding the soldier's mind, motivation, and attitude.

Positive psychology focuses on three major areas of human adjustment: positive emotions, positive individual traits, and positive institutions (Peterson, 2006). The latter two areas of focus are particularly relevant to understanding the soldier. An explicit theory of character strengths and moral values has been developed that postulates the existence of 24 character strengths that are universal to the human species (Peterson & Seligman, 2004). A major research thrust from the author's laboratory and a growing group of military psychologists is generating sound empirical evidence that character strengths matter a great deal in a variety of military contexts, ranging from retention and military performance of West Point cadets (Duckworth, Peterson, Matthews, & Kelly, 2007) to the adjustment of Norwegian Naval Academy cadets during a 10-week mission aboard a tall-mast ship (Eid & Matthews, 2004).

Positive institutions are those that foster "better communities, such as justice, responsibility, civility, parenting, nurturance, work ethic, leadership, teamwork, purpose, and tolerance" (Positive Psychology Center, 2007). The doctrine and practice of all branches of the U.S. military support and actively promote each of these dimensions of a positive institution. Moreover, these characteristics feed directly back to understanding soldiers and what motivates them.

The military is a natural home for positive psychology. On the whole, the military population is young, physically fit, and free of severe pathologies. Individuals who are too old or in poor physical shape or who are afflicted with major physical or psychological disorders are either not recruited or admitted into the ranks of the military or are selected out of the service. Certainly, some people enter the military with personal, family, or social problems, and these may be associated with increased risk for subsequent posttraumatic stress disorder (PTSD) or other disorders (Bramson, Dirkzwager, & van der Ploeg, 2000), and following combat exposure somewhere between 10 and 20% of soldiers may show evidence of PTSD, depression, or anxiety (Litz, 2007). But, the majority (>80%) do not. Moreover, the explicit efforts of the military to be a positive institution are congruent with the positive psychology model.

It is not fair to say that a disproportionate amount of attention in contemporary military psychology has been on the negative psychology of PTSD and other behavioral disorders. With well over 150,000 military personnel assigned to combat zones in Afghanistan and Iraq, even if the pathology rate is only 10%, this translates into huge numbers of individual soldiers who need acute psychiatric care. The purpose of this chapter, however, is to provide a broader perspective on what motivates soldiers, their mental approach to the profession, and the role of attitudes in understanding their behavior. Toward this end, positive psychology will serve as the theoretical framework.

The author's work in military applications of positive psychology was stimulated by his involvement with the Medici II Conference, a working group of positive psychologists who met for several weeks during the summers of 2005 to 2007 at the University of Pennsylvania to explore and develop a variety of applications and research in the field. Through this working group, the author initiated a number of empirical studies of various military groups and worked with other military psychologists to broaden the basic research being done in this domain; key findings are summarized throughout the course of this chapter. The work in military positive psychology is new. Some of it has been published, a good deal has been presented at major scientific conferences, and even more is still being executed and analyzed. This is a dynamic and burgeoning area of inquiry. It is hoped that the work reviewed in this chapter will inspire others to look systematically at positive soldier adaptation and to explore factors that increase positive emotions, positive personal traits, and positive institutions.

To understand soldiers, it is important to understand who joins the military; thus, this chapter begins with a short overview of the demographics of military personnel. This is followed by an overview of positive psychology with a special focus on the role of character strengths and moral values in understanding soldier adaptation and performance. The chapter concludes with extensions of positive psychology to other important issues.

Soldier Demographics

The U.S. Army is representative of the military services in its demographic characteristics. The Army Family and Morale, Welfare, and Recreation Command recently published a comprehensive summary of the demographics of Army soldiers and their families (Booth et al., 2007). Major findings from this report are summarized here.

Age

Only 7% of active-duty enlisted soldiers are 40 years of age or older. A much higher percentage of active-duty officers (27%) are age 40 or older. But, this is not the whole story. The military relies more and more on the National Guard and Reserve Component to supplement the active duty force. Army National Guard enlisted personnel (20%) and officers (38%) and Army reserve enlisted personnel (19%) and officers (53%) have substantially greater proportions of personnel age 40 or over. This is an important factor in understanding the motivation and mental set of the troops. Older soldiers are more likely to have children in secondary school or college and to be career soldiers. Deployments and exposure to dangerous, life-threatening situations will thus have different implications for soldiers with children than for ones with no family obligations.

Family Status

Except for junior enlisted personnel, the majority of military members are married. In the Army, for example, the marriage rates range from 31% for junior enlisted, the majority of whom are under 25 years of age, to 97% for general officers. By comparison, only 13% of civilians under the age of 25 in the general population are married, less than half the percentage of Army soldiers in the junior enlisted pay grades. Because National Guard and Reserve soldiers are older than their active-duty counterparts, marriage rates among these soldiers is somewhat higher. Importantly, 10% of active-duty soldiers are married to someone who is also on active duty. Although only 6% of male soldiers are married to another military member, 39% of female soldiers are married to another military member. This has significant implications for child-care arrangements. As Booth et al. (2007) emphasize, dual-military couples are required to file and maintain a family care plan so that in the event of a deployment both will be able to deploy with their units. There is no requirement for military personnel married to civilian spouses to file such a plan. Previous research suggests that dual-military couples and their attendant child-care responsibilities pose a significant problem for deploying units (Berry & Matthews, 1981).

Army spouses are young, with ages similar to those of their active-duty, National Guard, or Reserve Component spouses; for example, only 9% of spouses of active-duty Army enlisted soldiers are 41 years of age or older. If one defines the Department of Defense (DoD) community as including primary family members in addition to the military member, then the majority of the military community consists of family members for all services except the Marine Corps. Booth et al. (2007) report that the ratios of family members to active-duty service members are 1.47, 1.34, 1.42, and .99 for the Army, Navy, Air Force, and Marine Corps, respectively. The importance of these statistics for understanding soldier motivation and mindset should be obvious. Family welfare is a major concern for most military personnel.

Nearly half (47%) of all Army personnel have children. This is more than the Navy (42%), Air Force (45%), or Marine Corps (31%). In the Army alone, there are over 450,000 dependent children, just in the active component, and these children are young, as over half (51%) are under the age of 7. Children and their care are, of course, a major concern for soldiers, regardless of rank. A host of issues is associated with being raised in a military family, including frequent moves, long separations from one or both parents, and the possibility of a parent being killed or wounded in combat. From the perspective of the deployed soldier, concern about their children may represent a significant factor in adjustment to long combat deployments.

Single parents represent a relatively small but important group in the military. According to Booth et al. (2007), 13.8% of females soldiers were single parents in 2005, compared to 5.7% of male soldiers. However, because males make up 85% of the overall force, in raw numbers male single parents outnumber female single parents by a wide margin.

Booth et al. (2007) conclude their review of characteristics of Army life with a brief discussion of common demands for Army families. These include deployments and long separations, risk of injury or death, frequent relocations, long and unpredictable hours, and foreign residence. Collectively, it is important for psychologists and other mental health professionals who deal with military members or their families to keep in mind the demographic characteristics of military personnel and the unique demands placed on them by their profession, because these factors play an essential role in understanding what soldiers value and find rewarding, as well as what are areas of stress and concern for them.

Positive Psychology

Toward understanding soldier adaptation and performance in combat, traditional military psychology has typically embraced the traditional negative psychology, or mental illness, model. Whereas this approach is valuable in understanding pathological responses to combat exposure, it misses the mark for the majority of soldiers who are pathology free. Current estimates suggest that perhaps 15% of soldiers returning from Operation Iraqi Freedom show clinically significant symptoms of PTSD or related disorders within 3 months after their arrival back to home units (Litz, 2007). This raises the question of what is it about the other 85% of soldiers that enables them to endure the combat rotation with varying degrees of resilience; that is, the traditional approach to understanding the soldier's mind, motivation, and attitude does little to build on strengths or to prepare soldiers for the psychological impact of deployments and combat.

In contrast, positive psychology offers an intriguing conceptual framework for understanding human adaptation and performance. This may be particularly true for understanding how humans adapt and perform in the exceptional circumstances presented to soldiers in the form of long deployments and combat. A growing body of empirical evidence suggests that human strengths contribute to the adaptation and performance of military personnel in a wide variety of challenging and sometimes dangerous training and operational contexts. In short, positive psychology provides a conceptual model for understanding how psychologically robust and healthy soldiers prepare for and adjust to the difficult circumstances encountered in military training and operations. The purpose of this chapter is to review emerging empirical and conceptual applications of positive psychology to the military.

A cornerstone of the positive psychology approach that is crucial in understanding soldier mindset and motivation is the classification of human character strengths introduced by Peterson and Seligman (2004). Utilizing a comprehensive review of the psychological literature on positive human traits and a critical analysis of current and historic worldwide religions and philosophies, Peterson and Seligman identified 24 character strengths universal to the human species. They organized these 24 strengths into 6 "core moral virtues." These core moral virtues, with their component character strengths, are *wisdom and knowledge* (creativity, open-mindedness, curiosity, love of learning, perspective), *courage* (bravery, persistence, integrity, vitality), *humanity* (love, kindness, social intelligence), *justice* (citizenship, fairness, leadership), *temperance* (prudence, self-regulation, forgiveness and mercy, humility), and *transcendence* (hope, spirituality, appreciation of beauty, gratitude, humor).

Peterson and Seligman (2004) also describe the Values in Action Inventory of Strengths (VIA-IS), which provides an empirical measure of the 24 character strengths specified in their classification scheme. The VIA-IS consists of 240 simple statements for which respondents indicate, on a five-point Likert scale, the degree to which each statement is like them. The VIA-IS can be taken online (see www.authentichappiness.org), and several hundred thousand people from around the world have done so. Peterson and Seligman report that it has sound psychometric characteristics, with alphas of greater than .70 for all 24 subscales, and good test–retest reliability of greater than .70 over a 4-month interval.

The remainder of this chapter reviews research stemming from the West Point Strengths Project (WPSP). The WPSP began in 2004 when all members of the entering West Point class of 2008 were administered a battery of tests that included some measures of character strengths. The scope of the project widened in following years to include the study of character strengths among foreign military personnel and of active-duty soldiers in the U.S. Army. Collectively, the project includes a variety of research designs, populations, and contexts. This allows for a broad and robust empirical assessment of the role of character in soldier adaptation and performance, with a special focus on challenging training and operational circumstances.

Character and Soldier Adaptation

The overarching approach of the studies reviewed in this chapter involves empirically assessing character strengths and then evaluating their relation to soldier adaptation in challenging training and operational circumstances. The emphasis is on hard, objective criterion measures wherever possible. Several major component studies that comprise this project are longitudinal, with the goal of examining the role of character across years of a soldier's career, from initial entry to well into active duty service. The project is still in its infancy, having begun in 2004, but additional cohorts/populations are being added to the project each year. The intent is to continue to expand the scope of the project with an increasing focus on active-duty soldiers performing in deployed combat assignments.

Grit

The first class of West Point cadets studied in the WPSP was the class of 2008, who were administered the grit, a test of passionate pursuit of long-term goals. Duckworth, Peterson, Matthews, and Kelly (2007) describe the origin and development of grit, along with its relationship to performance in a variety of situations. The grit was developed to allow for researchers to examine the contribution of dedicated and passionate persistence in accomplishing major goals and to compare the relative contribution of this trait vs. talent as measured by traditional aptitude measures. Duckworth et al. found that grit is a robust predictor of student performance in both public schools and at the university level. Moreover, its contribution to predicting academic performance was orthogonal to aptitude, with both aptitude and grit correlating positively with academic grades but not correlating with each other.

The purpose of the study of grit among West Point cadets, however, was to see how this trait was related to adaptation to the grueling initial training new cadets are given upon arriving at West Point. Each summer, between 1200 and 1300 new cadets begin their 47-month education and training experience at West Point that will eventually lead to commissioning as a second lieutenant in the U.S. Army and the immediate responsibility of leading a platoon, in these days, often in combat. The first 6 weeks of their training at West Point is referred to as Cadet Basic Training (CBT). It is important to understand that CBT represents, to most cadets, the most stressful and daunting experience in their lives to date. They are given military haircuts, are issued uniforms, and must immediately learn the basics of drill and military customs and courtesies. Later, they conduct field training that includes basic rifle marksmanship, hand-to-hand combatives, and long road marches with heavy packs and rifles. Sleep deprivation; dealing with heat, cold, and wet conditions while in the field; and isolation from friends and family make for a challenging situation.

The selection process at West Point is detailed and comprehensive. Applicants are assessed in three major dimensions, including aptitude, leadership potential, and physical fitness. All told, each selected cadet arrives at West Point with nearly 200 bits of information that are used in the selection process to predict successful completion of the overall 47-month program of study and instruction. The grit was administered to all members of the class of 2008 within 3 days of their arrival at West Point. The cadets were then tracked through CBT and through their first year of academic study, and the relative weight of grit vs. aptitude and other predictors was assessed. Duckworth et al. (2007) found that grit was a statistically significant and strong ($r = .41$) predictor of retention through CBT. Indeed, cadets who were a standard deviation above the mean, compared to other entering cadets, were twice as likely to complete CBT than their counterparts. Consistent with studies of other populations reported in Duckworth et al., grit was not correlated with aptitude measures. Interestingly, although grit added significantly to the prediction of academic grades during the fall and spring semesters of the freshman (or plebe) year, aptitude measures contributed toward a greater proportion in the variance in these predictions than did grit.

An analysis of new cadets from the class of 2010 showed that those with the lowest grit scores presented themselves to the West Point Cadet Counseling Unit during CBT at higher rates than grittier cadets and earned substantially lower ratings of military and physical fitness performance than higher scoring cadets. The low grit cadets who do complete CBT, however, do not differ from their higher scoring counterparts on academic grades during the plebe year. We have also found grit scores to be very consistent, with a test–retest reliability of .55 over nearly a 3-year retention interval.

We have also extended our study of grit to another military training context in a different culture. Eid and Matthews (2004) studied the adaptation and performance of Norwegian Naval Academy cadets during a 10-week sailing mission aboard a three-mast ship. The mission was characterized by demanding and sometimes dangerous tasks and social isolation from home and family. The results showed that grit combined with self- and peer ratings of performance to predict military grades received by the cadets for the mission.

The findings reported by Duckworth et al. (2007) and Eid and Matthews (2004) speak directly to the topic of this chapter—that is, understanding the soldier's mind, motivation, and attitude. An unwavering determination to succeed at a challenging task, represented by West Point's CBT or the Norwegian sailing mission, characterizes an important dimension to understanding the soldier. Note also that persistence is one of the component character strengths that together comprise the core moral virtue of courage as described by Peterson and Seligman (2004). Moreover, courage is specified by Army doctrine as one of the seven Army values that contribute to what it means to be a soldier (Department of the Army, 1999). Many military training and operational settings require this trait, and it is valued and cultivated by military culture. Soldiers who have or develop this character strength may adapt better and show better performance in a variety of challenging situations.

Character Strengths as Defined by Peterson and Seligman

Another major thrust of the WPSP involves looking at the 24 character strengths defined by Peterson and Seligman (2004) and measured by the VIA-IS or through self-ratings of the strengths. The goal of this work is to identify what character strengths are most important in facilitating soldier adaptation in challenging situations. Much of this work looks at the role of character strengths in adapting to the rigors of the West Point experience, but some work is now underway that extends the research model to active-duty soldiers.

As a starting point, it is informative to learn how military populations may differ from civilians in terms of their hierarchy of character strengths. Toward that end, Matthews, Eid, Kelly, Bailey, and Peterson (2006a) compared West Point cadets, Norwegian Naval Academy cadets, and U.S. civilians (ages 18 to 21, with some college coursework completed) on the 24 character strengths as measured by the VIA-IS. Several interesting findings emerged from this study that may give clues to understanding the warrior mind, motivation, and attitudes. First, despite obvious cultural differences between the U.S. sample and the Norwegian sample of naval cadets, the two military samples were more similar to each other in the rank ordering of their 24 character strengths (r = .82) than the West Point cadets were to U.S. civilians (r = .61). In this particular study, the VIA-IS was administered several months after cadets from both academies had begun their training. Thus, both a selection bias—that is, people with character profiles suited for the military—and indoctrination to military culture could have contributed toward the striking similarity in strength profiles between the two military samples. Further work is needed to tease out the relative contribution of these two factors.

Matthews et al. (2006a) also found that the West Point sample scored significantly higher than the U.S. civilian sample on 13 of the 24 strengths. This is not surprising, given the highly select nature of any given West Point class and the overarching importance of character in West Point's training and education doctrine. Furthermore, both the West Point cadets and the Norwegian Naval Academy cadets were particularly strong on character strengths that are congruent with military culture and, in the U.S. case, doctrine. Honesty, persistence, bravery, and teamwork were among the top strengths for both military samples. Among the U.S. civilian sample, the top strengths included kindness, humor, capacity for love, gratitude, and curiosity. These are all good character strengths to possess, to be sure, but perhaps not as important for successful adaptation to a demanding military training and education program.

We were interested in determining whether character strengths as measured by the VIA-IS changed much over the 47-month West Point experience. On the one hand, if character strengths are trait like, one would expect relatively little change over time. On the other hand, West Point emphasizes character development, and cadets are constantly reminded of the importance of character in leading others, especially in the difficult circumstances of combat and long deployments. Using a cross-sectional design, Tuite and Matthews (2006) compared VIA-IS strength scores of 132 plebes (freshmen) with a sample of 100 first-class cadets (seniors). Overall, a correlation of .87 was found between the rank order of the 24 strengths between the two samples. A comparison of mean differences for each of the 24 strengths revealed a statistically significant difference for only one strength (first-class cadets were less spiritual than plebes), a difference well within the 5 mean differences one might expect to occur by chance when making multiple comparisons. There was some shifting of ranks of character strengths, with some evidence that collectivist strengths (e.g., teamwork) were a little higher in ranking among the first-class cadets, but overall the profile of strengths was remarkably similar. A longitudinal design would be more powerful in addressing this question, but the current data suggest considerable stability across a lengthy period in the context of a very strong environment. It is not reasonable to assume, based on these findings, that character does not continue to develop as a function of military training and education. Perhaps the VIA-IS is insensitive to the subtle changes that occur over time, or perhaps already military-relevant character strengths (see Matthews et al., 2006a) are simply reinforced and become more deeply ingrained.

The next important question was to what extent might character strengths matter in predicting the performance of military personnel in challenging training and operational environments. Matthews, Peterson, and Kelly (2006b) solicited self-ratings on each of the 24 character strengths in a sample of 1208 cadets of the West Point Class of 2009. These ratings, along with three full VIA-IS subscales—optimism, persistence, and capacity to love—were collected on the day after these cadets arrived to begin CBT and their 4 years at West Point. Self-ratings of the strengths were used because testing time prohibited the use of the full 240-item VIA-IS; however, pilot work conducted at West Point with a smaller sample (132) of cadets who took the full strengths questionnaire showed fairly high (>.70) correlations between full VIA-IS scores and self-ratings on the 24 strengths. Similar to the design used by Duckworth et al. (2007), the cadets were followed through CBT and an analysis of the relationship between character strengths and retention through CBT was performed. The analysis revealed that cadets who successfully completed CBT had statistically higher scores on the following character strengths: optimism, persistence, teamwork, bravery, zest, fairness, honesty, leadership, and self-control. This same relationship held also for the full-scaled measures of optimism and persistence. A factor analysis on the 24 strengths resulted in a solution with five factors. When entered into a regression equation, one factor emerged as significantly related to the prediction of retention. This factor was comprised of the individual strengths of leadership, teamwork, and bravery.

The Matthews et al. (2006b) study provides evidence that a certain profile or constellation of character strengths may be critical in understanding how soldiers adapt to a very challenging situation. The character strengths that were related to retention through CBT are congruent with Army doctrine (Department of the Army, 1999), which emphasizes character strengths or values such as honesty/integrity, courage, and optimism. The reader may also note the similarity of these traits to the ones that Matthews et al. (2006a) found to be high among West Point (and Norwegian Naval Academy) cadets. The message for the topic of this chapter is clear: To understand the warrior, one must appreciate the importance of character strengths that are both exemplified by soldiers and are important to their successful adaptation and performance in challenging situations.

To determine whether the character strengths assessed by the VIA-IS are predictive of adaptation in another context, Matthews and associates at the University of Bergen (Eid, Johnson, Matthews, & Bartone, in preparation) administered the VIA-IS to Norwegian Naval Academy cadets prior to their setting sail on the 10-week sailing mission described previously. At the conclusion of the voyage, cadet squads conducted a series of peer evaluations on each member of their squad (each squad numbered 10 cadets). These ratings were collected on the following dimensions of adaptation and performance during the mission: robustness (how hardy/robustly the cadet handled the mission), sociability, influence (over other cadets), productivity, flexibility, creativity, responsibility, and overall performance. The six core moral virtues were entered into a stepwise multiple regression and used to predict each of the rated dimensions. The six core moral virtues did not predict two of the peer ratings of adaptation during the mission (responsibility and creativity); however, the six core moral virtues did result in significant predictions of the remaining ratings. The core moral virtue of courage was a significant predictor of robustness, productivity, flexibility, peer influence, and overall performance. Sociability was predicted by the core moral virtues of humanity and temperance, with cadets higher on the humanity virtue and lower on the temperance virtue receiving higher peer ratings of sociability. Temperance was also inversely related to peer ratings of flexibility.

In yet another context, Eid, Matthews, Johnson, Laberg, and Bartone (2008) looked at the role of character strengths in predicting the adaptation and performance of Norwegian Naval Academy cadets during a prisoner of war exercise, a component of an 8-day combat survival course that involves extreme sleep and diet restriction and continuous operations throughout each day. Cadet squads were given a mission briefing by a commanding officer and set forth on the mission. Before they could accomplish the mission, they were seized by "enemy" forces and taken into custody for approximately a 24-hour period. They were blindfolded and restrained, then confined in an underground bunker. During the confinement time, each cadet was interrogated by Norwegian intelligence personnel a minimum of three times. The objective was to see if the cadet would divulge information about the mission beyond the approved name, rank, and service number. At no time during the POW exercise were cadets allowed to talk to anyone but the interrogators, and they were forced to stand except for brief periods of rest when necessary. Two performance criteria were collected during the POW exercise: total interrogation time and overall ratings (from experienced intelligence military personnel) of performance. In this setting, the six core moral virtues were not predictive of either interrogation time or overall performance ratings; however, psychological hardiness was significantly related to both performance criteria. Hardiness correlates moderately with grit (.32) and is conceptually and empirically linked to resilience in military populations (Maddi, 2007).

During the POW exercise, cadets were sleep deprived, hungry, and under considerable stress. Eid and Morgan (2006) recently reported that cadets engaged in the POW exercise experience substantial degrees of peritraumatic dissociation, a condition associated with trauma (Bremmer & Brett, 1997) and also associated with increased risk of PTSD (Carlson & Rosser-Hogan, 1991). Symptoms include amnesia, depersonalization, and derealization. Eid and Morgan found that peritraumatic dissociation was significantly higher among cadets at the end of the POW exercise than during a training session several weeks earlier, conducted in a less threatening environment. They also found that cadets higher in psychological hardiness, specifically the challenge component, were less prone to peritraumatic dissociation than less hardy cadets.

Eid et al. (2008) used VIA strengths to predict peritraumatic dissociation. In addition to measuring peritraumatic dissociation during the POW exercise and the training session prior to the exercise, they also assessed it in a non-threatening classroom situation several weeks prior to the training session and the actual exercise. A linear increase in dissociation scores occurred under these conditions of low stress (classroom), moderate stress (training exercise), and high stress (POW exercise). Moreover, regression analyses showed that cadets who were stronger in the core moral virtue of humanity showed lower peritraumatic dissociation during the high-stress POW exercise. This is consistent with anecdotal accounts of American POWs from the Vietnam war who maintained that caring for each other and a strong social network were vital to enduring captivity.

Finally, in an attempt to uncover what character strengths are most vital to the adaptation and performance of soldiers deployed to combat operations in Afghanistan and Iraq, Matthews (unpublished data) asked 29 Army officers to complete a questionnaire following their return from a combat deployment. Respondents were asked to write a paragraph describing a situation they found most challenging during their deployment. These ranged from dramatic combat encounters to the difficulties of separation from family. They were then asked to rate, on a 5-point scale ranging from very unimportant to very important, how each of the 24 character strengths defined by Peterson and Seligman (2004) contributed to dealing effectively with that situation. The five highest rated strengths were teamwork, honesty, courage, persistence, and judgment. The five lowest were a sense of beauty, curiosity, sense of humor, prudence, and creativity.

The results of the study of the character strengths that experienced soldiers used to cope with actual combat deployments provides an exceptionally interesting glimpse into understanding what motivates a soldier under the toughest conditions. It is interesting to note that the very qualities and strengths defined by doctrine (Department of the Army, 1999) and those found to be related to successful adaptation and performance in various training situations (Duckworth et al., 2007; Matthews et al., 2006b) are the same that emerge in combat. Another interesting aspect of this is that, although these strengths are of vital importance in adapting in combat and stressful training, they may not be optimal for other situations that soldiers may encounter elsewhere in the military, in relations with their families, or in civilian settings. This may suggest that training adaptability may, in part, involve educating soldiers on how to activate various constellations of character strengths in different situations. As Duckworth et al. (2007) found, the strengths important in sticking out Cadet Basic Training at West Point are not the same as those needed to optimize academic success later on.

Clinical Applications

Although there is growing interest in applying positive psychology constructs and methods to clinical interventions, the adoption of such an approach represents a paradigm shift from the mental illness model of psychology. Clearly, soldiers who experience PTSD, depression, anxiety disorders, or substance abuse problems may benefit from traditional therapeutic approaches. As we will see, however, there is evidence that even people experiencing psychopathologic states may benefit from positive-psychology-based interventions, either supplemental to or in replacement of, traditional approaches. Moreover, positive psychology seems especially relevant to the majority of combat veterans who do not present with formally diagnosed disorders but who may nonetheless experience difficulty in adjusting to a more normal environment following a combat rotation. The purpose of this section is to provide a brief review of recent literature that describes the conceptual and empirical bases for applying positive psychology interventions to soldiers.

Peterson, Park, and Seligman (2006) report results of a retrospective, web-based study of 2087 adults who had experienced physical or psychological illness. The question of relevance for the current discussion is whether any particular patterns of character strengths, as assessed by the VIA-IS, were related to subsequent adjustment among these people. Peterson and his colleagues reported that, for physical illness, those who were relatively high in the character strengths of bravery, kindness, and humor experienced "less of a toll" on life satisfaction. For psychological illness, those relatively high in the character strengths of appreciation of beauty and love of learning fared better in terms of life satisfaction following their illness. Peterson, Ruch, Beermann, Park, and Seligman (2007) found that the character strengths most related to life satisfaction in nonclinical populations are capacity to love, hope/optimism, curiosity, and zest. Together, these studies suggest that interventions designed to target specific character strengths might be useful in helping soldiers, including those who had experienced physical or psychological trauma, to restore, maintain, or enhance a sense of life satisfaction.

Resnick and Rosenheck (2006) adopted an approach based on character strengths to assist veterans in a Department of Veteran Affairs psychiatric rehabilitation program. They viewed their approach, based on positive psychology principles, as similar to the "recovery movement," in that both approaches focus on human strengths and mental health vs. human deficiencies and mental illness. The authors administered the VIA-IS to clients enrolled in their program and reported that subjects who completed the test experienced a variety of positive outcomes, including a sense of mastery and improved mood after completing the test. Many felt that the strengths approach was very relevant to their needs. Although the reported results are qualitative, they are promising in that they show acceptance of positive psychology interventions among a military relevant psychiatric population.

Seligman, Steen, Park, and Peterson (2005) reported the results of a more systematic and methodologically sophisticated application of positive-psychology-based interventions in increasing happiness and decreasing depressive symptoms. Using web-based, random controlled trials, they recruited 577 adults to participate in a 6-month study. Subjects in the experimental groups received instructions on how to complete one of five positive-psychology-based interventions. Subjects in the control group were asked to write down early childhood memories each night for a week. Over 70% of the sample completed the exercises. Three exercises seemed especially effective: the gratitude visit, the three good things exercise, and using signature strengths in a new way. The gratitude visit involved thinking of someone who had a major and positive influence on your life, writing a letter of thanks, and reading it to that person face to face. The three good things exercise involved reflecting, at the end of each day, on three things that went well and why they went well. The using signature strengths exercise required subjects to use one of their top character strengths, as assessed by the VIA-IS, in a new way each day for a week. The results showed that the three good things exercise and the strengths exercise both increased happiness (as measured by the Steen Happiness Index) and decreased depression (as measured by the Beck Depression Inventory) for up to 6 months, compared to the placebo control condition. The gratitude visit also had large effects, but only for 1 month.

The results of Seligman et al. (2005) demonstrate that positive-psychology-based interventions are effective in increasing happiness and decreasing depression in a large convenience sample recruited through the web. Their sample was diverse in age, education, and other demographics. It would be highly informative to apply this general methodology to military relevant populations, including Department of Veterans Affairs (VA) patients, soldiers undergoing physical rehabilitation for wounds received in battle, and simply for soldiers returning from a combat rotation. If the interventions were successful in the web-administered protocol, it might be that such interventions coached and monitored in more traditional personal settings might prove more effective yet.

Adaptability and performance in the military are more than simply avoiding pathology or not making mistakes. The growing coaching movement, which focuses on ways of helping people learn to utilize positive traits and skills to excel or flourish in life, is another mental health application of positive psychology. Green, Oades, and Grant (2006) compared 28 participants assigned to a coaching protocol with 28 assigned to a waitlist control group. They found that the coaching group showed significant increases in goal striving, well-being, and hope. Some gains were sustained for up to 30 weeks. Using the character strength lexicon, goal striving is conceptually similar to grit or persistence, and hope is akin to optimism. Green et al. concluded that this sort of approach is especially promising for nonclinical populations seeking to improve adjustment and effectiveness. This is completely congruent with the needs of soldiers, who may use such interventions to improve positive psychological functioning and behavioral adaptation.

Hannah, Sweeney, and Lester (2007) introduced a model of courage designed to identify psychological states and traits that ameliorate the experience of fear in dangerous situations. Their model maintains that core values, many relevant to Army doctrine and the results of empirical studies of character previously reviewed, interact with social and situational factors to occasion courageous behavior. Their model suggests that the predisposition to behave bravely could be increased through interventions. In a very interesting paper, Fagin-Jones and Midlarsky (2007) examine the role of positive psychology variables in differentiating those who assisted Jews during the Holocaust vs. those who did not. A pattern of "social responsibility, altruistic moral reasoning, empathetic concern, and risk-taking" was associated with the courageous altruism shown by those who intervened in these dire circumstances. The work of both Hannah et al. and Fagin-Jones and Midlarsky suggests that a positive, mental-health-oriented approach aimed at increasing key character strengths could be employed to increase courage and the willingness of individual soldiers to intervene in difficult circumstances.

It would seem that positive-psychology-based therapeutic approaches hold much promise for psychologists with military clients. Traditional therapeutic and pharmacologic approaches for PTSD and related disorders may be supplemented by interventions such as those reported by Seligman et al. (2005). These approaches may have even greater application among the majority of soldiers who do not have symptoms that reach the criteria for a clinical diagnosis but nonetheless affect their quality of life. The military culture stigmatizes psychological treatment, and an approach based on more of a coaching model might well be more palatable and appropriate for many combat veterans.

Conclusions

Recently, I had the honor of addressing the commanding general of an Army combat division, all of his subordinate commanders down through battalion level, and their spouses on the topic of character in combat and posttraumatic growth. The overwhelming consensus of this group of experienced combat leaders was that character plays a critical role in how soldiers adapt to combat and how they adjust following a combat deployment. The conjunction of military doctrine and empirical research in this domain clearly suggests that an understanding of the soldier's mind, motivation, and attitude requires an appreciation of the fundamental importance of character in adaptation to combat exposure.

There are obvious training implications to this approach. As scientists delineate which combinations of character strengths are most effective in facilitating adaptation to combat, evidence-based protocols may be developed to teach soldiers how to activate critical character strengths. An important consideration in developing training approaches is that individuals may learn to uniquely apply their own inherently high character strengths to solve problems and adjust to difficult circumstances, but they may also be taught to view their hierarchy of strengths as something of a toolbox and that they can reach for the appropriate character tools to aid adaptation to given situations.

In closing, practitioners who work with military members (and their families) should consider the importance of character to this population. The military training environment, be it a service academy or basic combat training for enlisted recruits, constantly emphasizes the importance of character in adaptation and leadership. To ignore the role of character in understanding the soldier's mind, motivation, and attitudes is, to borrow a widely cited proverb from military culture, "to choose the easy wrong over the hard right."

Seminal Readings

Duckworth, A. L., Peterson, C., Matthews, M. D., & Kelly, D. R. (2007). Grit: Perseverance and passion for long term goals. *Journal of Personality and Social Psychology*, 92, 1087–1101. This paper illustrates an empirical approach to linking a positive character strength to "hard" outcome criteria in both military and civilian contexts.

Peterson, C. (2006). *A primer in positive psychology*. New York: Oxford. This book provides an excellent overview of the origins and applications of positive psychology.

Peterson, C., & Seligman, M.E.P. (2004). *Character strengths and virtues: A handbook and classification*. New York: Oxford. The development of a classification scheme that identifies 24 character strengths as universal to the species is described in depth, along with a detailed review of each of the character strengths.

References

Bartone, P. T. (1995). Development and validation of a short hardiness measure. Paper presented at the annual meeting of the American Psychological Association.

Berry, G., & Matthews, M. D. (1981). *The Air Force personnel issues study: Assessing the impact of women on mission readiness*. San Antonio, TX: Air Force Human Resources Laboratory.

Booth, B., Segal, M. W., Bell, D. B., Martin, J. A., Ender, M. G., Rohall, D. E., & Nelson, J. (2007). *What we know about Army families: 2007 update*. Washington, D.C.: ICF International.

Bramson, L., Dirkzwager, A. J. E., & van der Ploeg, H M. (2000). Predeployment personality traits and exposure to trauma as predictors of posttraumatic stress symptoms: A prospective study of former peacekeepers. *American Journal of Psychiatry*, 157, 1115–1119.

Bremmer, J. D., & Brett, E. (1997). Trauma related dissociative states and long term psychopathology in posttraumatic stress disorder. *Journal of Trauma Stress*, 10, 37–50.

Carlson, E. B., & Rosser-Hogan, R. (1991). Trauma experiences, posttraumatic stress, dissociation, and depression in Cambodian refugees. *American Journal of Psychiatry*, 148, 1548–1451.

Department of the Army. (1999). *Army leadership: Be, know, do*, FM22-100. Washington, D.C.: Author.

Duckworth, A. L., Peterson, C., Matthews, M. D., & Kelly, D. R. (2007). Grit: Perseverance and passion for long term goals. *Journal of Personality and Social Psychology*, 92, 1087–1101.

Eid, J., & Matthews, M. D. (2004). Human strengths and adaptation to a radically changed context. Paper presented at the annual meeting of the American Psychological Association.

Eid, J., & Morgan III, C. A. (2006). Dissociation, hardiness, and performance in military cadets participating in survival training. *Military Medicine*, 171, 436–442.

Eid, J., Matthews, M. D., Johnsen, B. H., Laberg, J. C., & Bartone, P. T. (2008). Character strengths and resilience during a POW exercise. Paper presented at the annual meeting of the American Psychological Association.

Eid, J., Johnsen, B. H., Matthews, M. D., & Bartone, P. T. (manuscript in preparation). Personality correlates of performance and adaptation during Norwegian naval training exercises.

Fagin-Jones, S., & Midlarsky, E. (2007). Courageous activism: Personal and situated correlates of rescue during the holocaust. *Journal of Positive Psychology*, 2, 136–137.

Green, L. S., Oades, L. G., & Grant, A. M. (2006). Cognitive–behavioral, solution-focused life coaching: Enhancing goal striving, well-being, and hope. *Journal of Positive Psychology*, 1, 142–149.

Hannah, S. T., Sweeney, P. J., & Lester, P. B. (2007). Toward a courageous mindset: The subjective act and experience of courage. *Journal of Positive Psychology*, 2, 129–135.

Litz, B. T. (2007). Research on the impact of military trauma: Current status and future directions. *Military Psychology*, 19, 217–238.

Maddi, S. R. (2007). Relevance of hardiness assessment and training to the military context. *Military Psychology*, 19, 61–70.

Matthews, M. D. (unpublished data). *Using character strengths to adapt to combat deployments*. West Point, NY: U.S. Military Academy.

Matthews, M. D., Eid, J., Kelly, D., Bailey, J. K. S., & Peterson, C. (2006a). Character strengths and virtues of developing military leaders: An international comparison. *Military Psychology*, 18(Suppl.), S57–S68.

Matthews, M. D., Peterson, C., & Kelly, D. (2006b). Character Strengths Predict Retention of West Point Cadets. Paper presented at the annual meeting of the American Psychological Society.

Peterson, C. (2006). *A Primer in Positive Psychology*. New York: Oxford University Press.

Peterson, C., & Seligman, M. E. P. (2004). *Character strengths and virtues: A handbook and classification*. New York: Oxford University Press.

Peterson, C., Park, N., & Seligman, M. E. P. (2006). Greater strengths of character and recovery from illness. *Journal of Positive Psychology*, 1, 17–26.

Peterson, C., Ruch, W., Beermann, W., Park, N., & Seligman, M. E. P. (2007). Strengths of character, orientations to happiness, and life satisfaction. *Journal of Positive Psychology*, 2, 149–156.

Positive Psychology Center. (2007). http://www.ppc.sas.upenn.edu/.

Resnick, S. G., & Rosenheck, R. A. (2006). Recovery and positive psychology: Parallel themes and potential synergies. *Psychiatric Services*, 57, 120–122.

Seligman, M. E. P., Steen, T. A., Park, N., & Peterson, C. (2005). Positive psychology: Empirical validations of interventions. *American Psychologist*, 60, 410–421.

Tuite, J., & Matthews, M. D. (2006). Developing character at the U.S. Military Academy. Paper presented at the American Psychological Association Division 19/21 Midyear Meeting, Fairfax, VA, March 2.

3
Training to the Warrior Ethos: Implications for Clinicians Treating Military Members and Their Families[*]

JOHN R. CHRISTIAN, JAMES R. STIVERS,
and MORGAN T. SAMMONS

Contents

> Greater love hath no man than this, that a man lay down his life for his friends.
>
> **—John 15:16**

Working With Military Clients: Models of Military Culture and Socialization

Forty years after the Tet Offensive (the nadir of the conflict in Vietnam), America is again in the depths of a prolonged and unpopular war that has no immediate hope of resolution. Over 4300 American service members have died and over 30,000 Americans have been wounded in combat in Iraq and

---

[*] The views expressed in this chapter are those of the authors and do not necessarily reflect the official policy position of the U.S. Department of the Navy, the U.S. Department of Defense, or the U.S. government.

Afghanistan; of those wounded, over 14,000 have been so severely wounded that they were not returned to duty (www.defenselink.mil/news/casualties.pdf). Estimates of Iraqi casualties are difficult to verify; some estimate that at least 70,000 Iraqi citizens have been killed since 2003. Americans are increasingly familiar with the effects of war on individuals and families. Over 1.5 million servicemen and -women have deployed in support of Operations Iraqi and Enduring Freedom (OIF/OEF). It is estimated that one in seven American children has had a parent deploy (Department of Defense Task Force on Mental Health, 2007). We are a culture at war and a culture that has grown used to the demands and tragedies that war brings upon us. Unlike the American experience in Vietnam, however, we have not as a society expressed our collective discomfort by vilifying service members. Whatever their opinions of the war itself, Americans' opinions toward men and women in uniform are supportive and united.

Advances in battlefield medical practices have dramatically reduced the death rates among wounded combatants. Of Americans wounded in combat in World War II, 22% succumbed to their injuries, and 16% did so in Vietnam. For OIF/OEF, the total mortality rate was 8.8% in 2006 The changes in battlefield medicine that have resulted in significantly increased survival rates are largely an outgrowth of advances in protective equipment for combatants and of trauma care in both civilian and military settings. Reductions in death rates largely result from improved battlefield aid in the form of better techniques of hemostasis and improved on-the-ground trauma care (Eastridge, Jenkins, Flaherty, Schiller, & Holcomb, 2006). As important as these advances are, they are essentially linear and not transformational, having accrued from years of accumulated experience in trauma and emergency medicine.

At the same time, however, that these linear advances assist in casualty reduction, two profoundly transformational, but as yet poorly recognized, shifts in health care have occurred. First, the demands imposed by the obligation to care for and rehabilitate warriors who survive, often with severely disabling injuries, have transformed the delivery of health care in the Department of Defense (DoD) and the Department of Veterans Affairs (VA) medical systems. In military medicine, models of acute, focused care have shifted to models emphasizing integrated care and restoration to a duty status. In the VA, models emphasizing the management of disability are shifting to models focusing on rehabilitation and reintegration.

An even more fundamental transformation has occurred in mental health care, for it can be argued with certitude that at no time in the entire history of human conflict has greater attention been paid to the mental health needs of service members and their families. The effects of this transformation are just beginning to be felt, but it is likely that this transformation, like other transformations in health care brought about by the medical and mental health community's response to war, will entirely reshape our societal attitudes and approach toward mental illness and mental health care. Within the DoD, emphasis has shifted from the diagnosis and treatment of mental health disorders to education and prevention. The role of military leaders in reducing stigma and normalizing seeking mental health care is increasingly recognized. The psychological effects of exposure to combat are being interpreted not only in terms of diagnosis-driven models of psychopathology but also as normative, if not normal, responses to the demands of conflict. We must emphatically place any attempt to educate clinicians about military culture in the context of these perceptual shifts and recognize that the lens through which we view the military experience is constantly refocusing.

Although we are a society at war, it remains true that most Americans have no direct exposure to military culture. For those entering military service, the process of acculturation to the military is unique to each individual but may be accelerated by certain events, as the following example by one of us (JRC) illustrates:

> It was during the combat casualty care rotation on internship at Bethesda National Naval Medical Center when I began to truly reflect on the different culture that I had entered as a military service member. I was assigned to provide brief interventions to wounded sailors and Marines returning from Iraq and Afghanistan while they were recovering on the surgical ward. One evening, a nurse approached me and asked that I talk to one of the newly arrived patients from Iraq. She briefly described the details of the patient: "Shawn Smith, nineteen-year-old, Fleet Marine Force Corpsman [Navy medic for the Marine Corps], eighteen months' time in service, medevac'd from Iraq after a roadside bomb blew off both of his legs below the knees."
>
> "Had my graduate training in clinical psychology prepared me to deal with the trauma this young soldier has already experienced?" I asked myself. A flood of emotions, common to many of my fellow military psychological interns, ran over me, marked by anxiety, sadness, and shock. After briefly jotting down a few notes, I entered the room, where I found a bespectacled, baby-faced man who looked much younger than his stated age, quietly reading a book. To me, he looked more suited to his high school chess team than to being a combat medic for the U.S. Marine Corps. I introduced myself and explained why I was checking up on him. After having a seat, I asked him how he was holding up:
>
> SS: "Hello, sir. To tell you the truth, sir, I've been feeling pretty anxious."
> JRC: "It's normal for someone who has just lost their legs to be feeling anxious."

SS: "It's not that, sir."

JRC: "Then what are you anxious about?"

SS: "I can't remember if I did enough," he stated with tears beginning to well up in his eyes.

JRC: "Did enough? I don't understand."

SS: "I can't remember if I did enough for my Marines after the bomb."

JRC: Still confused, I replied, "Did enough? What could you have done? You had just lost your legs!"

SS: "Sir," he stated, looking away, "I could have been calling out directions to help them save more of each other before I passed out from the bleeding. They're my brothers and I don't even know if they are safe!"

I was completely overwhelmed by this young sailor's selflessness and dedication to the members of his unit. Numerous questions filled my mind. What caused these young men to be so selfless? How did they develop such a strong commitment to the members of their unit? Where did this feeling of "brotherhood" develop? Over the ensuing months of that rotation, I heard similar stories from other service members. It was clear that the military experiences of these young individuals had a profound impact on how they viewed themselves.

The various aspects normally associated with military culture (traditions, certain values, patterns of dress, and behavioral expectations) are often in stark contrast to the popular culture of the society that the military serves. It is not unusual to encounter individuals with no military connections or experience who hold significant misconceptions about what it means to be a service member and what sort of people enter the service. One of us (JRC) specifically remembers a professor in graduate school likening military service to prison. "Why not?" she told me. "There are a lot of rules and everyone dresses the same." We have also encountered many (including colleagues) who seem to think that the military turns previously free-thinking people into aggressive, unthinking automatons with crew cuts. Such misunderstandings have potentially negative effects on the process of psychotherapy and can leave a service member feeling isolated and misunderstood.

Understanding military culture and how an individual is socialized into this culture provides clinicians with a better understanding of that individual's worldview and the difficulties he or she will encounter when returning to civilian society. This chapter presents a cultural model for viewing the military and an overview of how this culture is instilled and reinforced among its members. Before proceeding, it should be stated that, as with all cultural conceptualizations, these are generalizations and may or may not apply to the individual service member. This discussion, however, will provide a starting point for providers eager to gain a better understanding of their patients.*

A Model of Military Culture

Sue and Sue (2003, p. 106) define culture as "all those things that people have learned in their history to do, believe, value, and enjoy. It is the totality of ideals, beliefs, skills, tools, customs, and institutions into which each member of society is born." Understanding the cultural context within which our patients operate is recognized as an important aspect to ethical practice (American Psychological Association, 2003). A failure to understand a patient's worldview can not only lead to misunderstandings in therapy but can also potentially result in the breakdown of the therapeutic relationship (Sue & Sue, 2003).

One way to conceptualize the values of a culture is to characterize its orientations to their time focus, means of viewing human activity, social relations, and the relationship between people and the environment. We contend that military culture differs significantly from the majority U.S. culture in that the military expresses a collectivistic vs. individualistic ethos, has clearly defined and codified social hierarchies, explicitly regulates the expression of emotion in many circumstances, does not use material wealth as an index of social standing or power, and promotes a self-concept rooted in history.

As enculturation into the military is an active, volitional process, as opposed to the more passive adoption of a culture of origin, it is not surprising that military members often readily identify and articulate those facets of the adopted culture that distinguish them from nonmilitary members of the culture of origin, nor is it surprising that adoption of a military culture enables communication among military members. In a review of studies examining the military cultures of several countries, Soeters, Poponete, and Page (2006) concluded that many Western militaries share a separate, more hierarchical and collectivistic culture that often allows individual service members to communicate and cooperate better with one another than with civilians from their own countries. But, military culture is not monolithic. In America, each branch of the armed services instills "core values" that its members are expected to represent. Although many institutions instill values either purposefully or inadvertently (Alexander, 2005), the military attempts to do so systematically, and immense pressures are exerted to ensure compliance with these values.

---

* As noted below, each branch of the service has its own culture and tradition. Because the authors of this chapter all come from the tradition of the U.S. Naval Forces (the U.S. Navy and the U.S. Marine Corps comprise the Department of the Navy), this chapter is perforce written from that perspective. This does not imply disrespect for the culture of other service branches but simply acknowledges our cultural heritage and the limits of our expertise.

Understanding interservice differences in core values is an important part of understanding military culture; nevertheless, most facets of service culture are an accurate reflection of the larger military culture. We will now identify key cultural military values, including collectivism, rigid hierarchical orientations, a historical focus on identity, and service-oriented values.

Collectivism

It is our contention that a collectivistic value system is the cornerstone of all other aspects of military culture, and we find it difficult to underestimate the stark contrast between individualistic American cultural norms and the collectivistic military culture. As we will discuss further in this chapter, new service members are taught to subordinate the self to the group from the first day of boot camp or officer school. McGurk, Cotting, Britt, and Adler (2006) describe collectivism as consisting of defining the self as part of a group, putting group goals ahead of personal goals, and having an emotional investment in the group. The authors also associate collectivism with military culture and contrast this outlook with the emphasis on individual achievement and self-reliance by U.S. society. The communalistic value system of the military extends to putting service to the country even before one's own private life (Tziner, 1983). This difference between the greater U.S. culture and the military culture is sufficiently great that McGurk et al. (2006) question whether individuals from Asian cultures, known for being more communally oriented, would have an easier time adjusting to military life.

Diminution of the importance of the individual in favor of the organization, team, platoon, or unit is a universal military dynamic. Decisions satisfying individual needs are often seen as contradictory to the needs of the organization and are frequently sanctioned (e.g., via the imposition of explicit punishments for nonconformity in dress or attendance, with sanctions for tardiness being an early and greatly feared lesson by new enlistees). Service members are taught both explicitly and implicitly that an individual is of limited value, whereas the unit can accomplish anything. Indeed, in the U.S. Marine Corps doctrine (U.S. Marine Corps, 1997), Marines are encouraged to develop such a deep trust and familiarity with one another that they no longer have to explicitly communicate. They are encouraged to know one another so well that they can anticipate one another's actions and thoughts. Service members are taught to rely on one another from the first day of boot camp or officer school. At some point, the lesson is learned, usually through a difficult task that could not be accomplished as individuals, and through these experiences the unit forms and becomes cohesive. Extremely difficult training in harsh environments leads to very close friendships and loyalty second only to those formed in actual combat, which ideally leads to a reduction in the need for verbal communication among small task-oriented groups (e.g., fire teams on patrol).

The emphasis on a group orientation also facilitates group identification and unit cohesion. High cohesion encourages prosocial behavior among members of the group (Grojean & Thomas, 1996). This degree of cohesion is an important element of small unit combat readiness, as individuals must learn to trust group members with their very lives. As Grossman (1996) noted, the commitment and accountability to one's comrades become more powerful than the instinct of self-preservation. Concern for the members of one's unit can facilitate overcoming numerous obstacles, such as engaging in combat.

Strong unit cohesion may serve as a protective factor against stress (Stouffer, DeVinney, Star, & Williams, 1949; Mareth & Brooker, 1985; Manning & Ingraham, 1987). Various internal military documents identify unit cohesion as one of the most important factors in protecting against combat stress and maintaining unit morale in both peace and wartime (U.S. Department of the Army, 1994, 2000; Canadian Army Lessons Learned Centre, 2004; U.S. Department of the Air Force, 2008). Among the many examples of this noted in history is Marshall's (1962) account of soldiers parachuting into Normandy who became separated from their units. These separated parachutists formed impromptu units generally consisting of a few individuals who may have trained with one another but were sometimes composed entirely of strangers. According to Marshall, units made up entirely of strangers were far less effective compared to those for which even two members had trained together.

Clinical Implications

Any clinician working with members of the military should pay particular attention to how this influences their patients; for example, service members leaving the military may have difficulty adjusting to working with civilians, whom the service members may perceive as "selfish." This can cause numerous adjustment issues as they transition into the civilian workplace and can leave them feeling disillusioned, depressed, and angry. Moreover, service members from the Reserves or National Guard may find that they miss the camaraderie and trust in fellow service members once they return from the war zone. These individuals may benefit from reconnecting with the members of their units to ease their transition back into the civilian world.

Hierarchical Orientation

The objective of Marine Corps leadership is to develop the leadership qualities of Marines to enable them to assume progressively greater responsibilities to the Marine Corps and society.

—*Marine Corps Common Skills (MCCS) Handbook*
(**U.S. Marine Corps, 2001**)

As Soeters et al. (2006) noted, a defining feature of military culture is the emphasis on hierarchical relationships with clear power differences. Social status is well defined within the military and determines everything, from peer group identity to obedience to orders. Social relationships are often limited or determined by the rank structure, and a promotion in rank is equivalent to an elevation in social status (Tziner, 1983). Most civilians are aware through movies, television shows, books, and other elements of popular culture that the military has a rigid rank structure with certain courtesies (e.g., saluting superior officers, coming to attention when a commanding officer enters the room) that accompany this structure. The relationships between ranks and the subtle implications of these relationships go far beyond giving and receiving orders, however, as the true focus of this rigid rank structure is the perpetuation of effective leadership.

The concept of leadership is firmly interwoven with the hierarchical relationships within the military. Junior military members entering the service are quickly encouraged to take on leadership roles to prepare them for ever greater levels of responsibility. Assigned to a work detail, four privates will always defer to whoever is senior to provide guidance. This may be whoever has been in the military longest if they are all the same rank; however, if they all have the same amount of time in service and are the same rank, then the eldest of the group may be expected to take charge of the situation. It is difficult for those outside of the military to understand how little power or authority the most junior rankings (basic airman, private, or seaman recruit) hold within the organization. While many civilians may find this stifling or be puzzled and even appalled at these displays of authority, they serve an important purpose. Young service members gain valuable experience in both leading and following. Each military member is expected to set an example for those junior to them. Senior military members often fondly remember the frustration they experienced in submitting to such junior authority or their awkward attempts at leading.

For a fuller understanding of the relationships between members of the different ranks, a brief explanation of the different roles of officers and enlisted service members must be presented. In general, officers are required to have a college degree and enter a commissioning program, which places them on a career path somewhat equivalent to management within a large company. Enlisted service members of higher rank are generally viewed as the technical experts in their fields. As such, officers hold authority over enlisted members but trust them to carry out the majority of the duties required to accomplish the mission.

One of the more complex relationships to understand is that between officers and enlisted service members. Young Marines, for example, are taught that officers are to be respected and looked up to, almost reverently. An almost blind loyalty is taught and must be adhered to due to the nature of the job and potentially the requirement to successfully accomplish the mission at hand (e.g., "Private, attack that machine gun bunker!"). As a private, that person is not allowed to talk to an officer sitting down or, if walking, must stop and stand at attention and speak. The higher the rank of the officer, the more formal the relationship becomes. The distance is considerable between a private in the Marine Corps and a corporal (the fourth enlisted rank in the Marines and the one at which a Marine earns his or her blood stripe* and is formally identified as a leader). Most enlisted Marines will never attain the rank of corporal before they leave the service. Throughout their military career, which on average is less than 4 years, enlisted Marines will very rarely come into daily contact with any commissioned officer of greater rank than captain (the third commissioned officer rank). Both junior noncommissioned officers (NCOs) and officers are learning and require mentoring, generally acquired at the hands of senior NCOs who are career military personnel, often with many years of service. Although of lower rank, senior NCOs are informally relied upon as trainers of junior commissioned officers due to their many years of leadership experience and familiarity with the culture of their service. In this role, senior NCOs (SNCOs)[†] must balance their informal expert authority against the defined positional authority of commissioned officers, a balancing act that maintains formal military protocol while imparting wisdom and management skill to those of higher rank. Additionally the SNCO has the responsibility of maintaining good order and discipline within the unit, often relying on junior NCOs to do so at their level as well. Thus, although SNCOs salute the junior officer, call him/her "Sir" or "Ma'am," and defer to the officer's authority, SNCOs can often be viewed as having more influence with both subordinates and senior officers.

Because the majority of military training is conducted by junior officers and noncommissioned officers, effective leadership rests on an understanding of the subtleties of the positional/expert authority relationship between officers and enlisted personnel. A key element of effective leadership in this complex situation is the development of trust. The U.S. Marine Corps document *MCDP 1: Warfighting* (1997, p. 58) states: "Consequently, trust is an essential trait among leaders—trust by seniors in the abilities of their subordinates and by juniors in the competence and support of their seniors. Trust must be earned, and actions which undermine trust must be met with strict censure."

---

* The blood stripe is a crimson stripe worn on the trousers of Marines of the rank of corporal and above. Historically, the blood stripe has its origin in the battle of Chapultepec in the Mexican–American war of 1848, when, in an attack on that city, junior Marines showed superior leadership and bravery (Alexander, 1999).

† "Senior" and "staff" NCOs are somewhat interchangeable terms. Formally, enlisted personnel of the rank of E-6 through E-9 (staff sergeant through sergeant major) are referred to as "staff" NCOs.

Thus, trust is a vital element of positional authority, as it is essential to the development of respect, without which those with positional authority have diminished effectiveness. Within small elite units such as a Marine Corps Force Reconnaissance Team, an individual's reputation and past performance are key elements in instilling trust and are determining factors in how effective the individual is as a leader/authority figure. In such teams, even though an individual outranks others in the team, that individual can be vetoed if he or she makes unsound decisions or displays behavior that jeopardizes mission success. Respect for authority can be as steadfast as the loyalty between individuals once a team has formed. This dynamic is usually formed after difficult training or combat, where decisions have a high price for failure.

It is easy to misunderstand how these relationships function in the military. Many hold the misconception that subordinates are to act as unthinking robots, automatically carrying out the orders of their superiors. But, effective military leadership depends on the ability of superior officers to foster initiative, responsibility, and independent judgment in their subordinates so they can carry out the intent of orders. The *Marine Corps Common Skills (MCCS) Handbook* (U.S. Marine Corps, 2001) explicitly states the expectation that leaders are to contribute to the success of their subordinates.

Familiarity between officers and enlisted personnel is a complex aspect of a leader's relationships with subordinates, and numerous rules, both formal and informal, govern these relationships. Without some degree of closeness and understanding of the personal situations of subordinates, effective small group leadership is difficult. At the same time, undue familiarity is destructive to leadership. Fraternization between ranks is strictly prohibited and may be enforced under penalty of the Uniformed Code of Military Justice (the legal code governing the armed services). In other words, it is illegal to have too familiar of a relationship with an individual of a lower rank. The existence of these sanctions is often surprising to civilians who do not understand that it is literally a criminal offense to become too familiar with those of subordinate rank.

The emphasis on maintenance of the rigid boundaries between ranks serves several purposes. First, leaders, both officers and NCOs, may one day be called upon to send people to their potential deaths. The purpose of placing individuals in increasingly greater positions of responsibility is to build their confidence to perform in life-threatening situations. Similarly, if leaders were overly familiar with their subordinates, their subordinates may fail to obey them or the leaders might feel reluctant to send individuals into harm's way. As Grossman (1996) observed, leaders must maintain the will to sacrifice their subordinates in order to continue being effective in combat. If they lose this will, then their force will be defeated. These lessons are considered fundamental to all in the military; therefore, the importance of leadership and discipline is instilled in every service member (Cutler, 2002).

Time Orientation

Although the military can be said to be future oriented in its emphasis on goals and understanding future conflicts, there is also a strong sense of history. Service members are taught to understand the history of their service as a means of instilling pride and identification with the group. Especially in the U.S. Marine Corps, knowledge of the history of the Corps is a key component of early socialization and training for all Marines, and retention of this knowledge becomes a matter of pride for all Marines. This emphasis on history and tradition plays an important part in shaping an individual's identification with the service, thus further cementing their pride in and identification with their particular service. There are numerous examples of how an individual comes to strongly identify with the history of their unit. A particularly striking example of this was experienced by one of us (JRC) who was attending training at an Army installation and noticed that several of the soldiers in the course who had deployed with the 10th Mountain Division of the Army spoke with particular pride at having been part of this unit due to its long history of successes in battle and numerous deployments.

Service Values

> I am an American, fighting in the forces which guard my country and our way of life. I am prepared to give my life in their defense.

> **—Article I of the Code of Conduct for Members of the Armed Forces of the United States (Cutler, 2002)**

Each of the services has clearly articulated core values that are expected to guide both the institution and the actions of individual service members; for example, the Navy and Marine Corps share the core values of "honor, courage, and commitment." *The Bluejacket's Manual* (Cutler, 2002), a manual of basic seamanship and shipboard life for sailors, states the purpose of core values in this way: "Before you make a decision or do something, consider whether your action will reflect a loss of honor, a failure of courage, or a lack of commitment" (pp. 9–10).

The power of these values lies perhaps in their ability for young service members to identify with an ideology. As Grojean and Thomas (2006) note, the military seeks to make these values an integrated and stable part of their service members' identities. In this way, service members' actions will reflect the ethos of the organization. Underlying the core values is an overall commitment to the service of one's country and a recognition that this service could ultimately lead to the giving up of one's own life in that service. The recognition that military leaders may be called upon to sacrifice the lives of young men and women implies adherence to a higher moral standard (Ficarrota,

1997) than that used to measure the population at large. The emphasis on duty to one's country plays an important role in motivating service members to perform their duties despite gaining very little reward or recognition for their service. This willingness to sacrifice for the greater good of others also works toward fostering the collectivistic value system (Tziner, 1983). A belief in the ideals of one's service and country not only serves to motivate individuals further but may also be a protective factor in maintaining unit morale and thus may be protective against combat stress (Gal, 1986).

From Civilian to Soldier

All officers and enlisted Marines undergo similar entry-level training which is, in effect, a socialization process. This training provides all Marines a common experience, a proud heritage, a set of values, and a common bond of comradeship. It is the essential first step in the making of a Marine.

—*MCDP 1: Warfighting* (U.S. Marine Corps, 1997, p. 59)

How does one make the change from civilian to an effective member of the Armed Forces, sharing in the values and culture of the service that are likely in contrast to the larger U.S. popular culture? To understand this process, it is important to first begin with the initial training that a new service member experiences. Referred to as boot camp or basic training for enlisted service members or as officer school for the officers, this training is where the military attempts to indoctrinate their members with the values that represent the service and equip them for being successful. As is noted in the above quote from the U.S. Marine Corps document *MCDP 1: Warfighting* (U.S. Marine Corps, 1997), the emphasis in this training is on socializing the new service members rather than equipping them with a large set of technical skills for their job. For this reason, it is useful to view the goal of entry-level military training as being to instill the values and norms of each service.

Although we recognize that individuals, to a certain extent, choose a career or job that is consistent with their prior self-concept, we agree with Tziner's (1983) contention that changes occur in an individual's identity as a result of his or her career choice and experience within his or her career. Moreover, we believe that this is especially true within the military, which makes a conscious effort to systematically change an individual's values and behavior to conform to the organization's culture (Stevens, Rosa, & Gardner, 1994). The process by which individuals are changed from civilians to soldiers, sailors, airmen, or Marines has been characterized in the academic literature as both a socialization process (Grojean & Thomas, 2006) and an indoctrination process (McGurk et al., 2006). For the purposes of this chapter, we find both models to be useful for creating a better understanding of how an individual is changed and the environmental forces that work to create that change. We will provide a brief overview of each model and proceed to illustrate the models with personal experiences, primarily relying on JRS's experience in Marine Corps boot camp. Finally, we will provide some comments on how the values that are imparted in boot camp continue to be reinforced throughout a service member's career. As Kennedy (2006) noted, however, the experiences of individuals in one military service may not translate perfectly into the experiences of another service. Despite this, we believe that our description highlights the themes common throughout the military branches' trainings and provides a basic understanding of the process.

Military Training as a Socialization Process

Anticipatory Stage   Grojean and Thomas (2006) argue that the military socializes individuals into a value system with the goal of creating a new identity within the individual, whereby that individual incorporates the organization's values and thereby performs more successfully within the organization. In their review of the socialization literature, they apply a three-stage model of socialization to explain this process. The first stage of the socialization process is the anticipatory stage. During the anticipatory stage, potential service members learn all that they can about the military organization, and established military members are actively engaged in attempting to recruit. If both the recruit and the established member feel that there is a good fit, then the recruit enters the organization; however, some potential recruits are also screened out or self-select out.

Encounter Stage   During the encounter stage, the new service member begins to actually experience the organization. This results in a shift in their values, skills, and attitudes. Grojean and Thomas (2006) suggest that it is here that an individual begins to develop an identity associated with their role as a service member and that this identity is associated with his or her role in the larger military organization. During boot camp or officer school, students struggle as they discover many of their behaviors or attitudes may not fit with the organization. They may have to make a conscious effort to begin to conform to the group.

Acquisition Stage   The final stage of socialization is known as the change and acquisition stage. This is when individuals make large adjustments in their values to fit the organizations and an individual internalizes the role identity of being a service member. Individuals must struggle with several factors at this stage. First, service members must find a way to deal with conflicts between work and family life, because, as Tziner (1983) noted, private life comes secondary to an individual's military duties. Second, individuals must master the actual tasks that they will be required to perform for their military jobs. Finally, socialization is influenced by the extent to which an individual internalizes the values and norms of the organizations into a coherent and consistent role identity. Values consistent with the military that individuals already held prior to entering service may also be clarified and solidified (Stevens et al., 1994). This change may require adjustments in their previously held behaviors and values. Individuals will strive for congruity between their personal values and the organization's values (Tziner, 1983). The development of a role identity that combines an individual's values with the internalization of the organization's values is the successful end-state of socialization into the military culture (Grojean & Thomas, 2006).

Military Training as Indoctrination

McGurk et al. (2006) view military basic training within the framework of indoctrination similar to that used by cults; however, the authors make clear that several important distinctions exist between the indoctrination of the military and that of cults. The primary difference between the two is that, unlike cults, the military develops traditionally socially accepted values and behaviors within individuals, such as honor, selfless service, and duty. Also, although the military does train individuals for combat and to sacrifice one's own life, they also teach individuals to avoid harming noncombatants and to consider the orders of superiors within the context of a value system that supersedes those orders.

McGurk et al. (2006) posit that it is useful to view military training as an indoctrination process in addition to a socialization process because indoctrination implies a more intense form of persuading individuals to adopt behaviors that are far outside of their previously held worldview. Specifically, the authors cite being called upon to kill another person in the service of one's country and being willing to put group goals ahead of one's own survival as being radical departures from the mindset an individual held as a civilian. The authors draw on the work of several authors to create an integrated model of military indoctrination. Based mostly on the work of Baron (2000), they describe the process consisting of four stages, including the softening-up stage, the compliance stage, the internalization stage, and the consolidation stage.

Softening-Up Stage   The first stage, known as the softening-up stage, prepares the individual to begin incorporating the values of the group. Key processes within this stage include separating the individual from family and friends, exposing the individual to physical and psychological stress, and the use of fear. This stage also consists of attempts to remove aspects of the individual's previous identity that are not conducive to the group. Specifically, attempts are made to remove the new service member's individualistic mindset and move them to thinking in terms of the group. Johnson and Wilson (1993) described the stress during Navy Officer Indoctrination School due to separation from one's friends and family and much of one's identity as an individual. Other stressors may include drastically reduced sleep, long physical training sessions, and the near-constant yelling and criticism of drill instructors. Needless to say, the experience can be very unsettling.

In my (JRS) experience at Marine Corps boot camp, the stress of the initial arrival and training closely matches Baron's (2000) description. Beginning with initial training, I was immersed in the culture of collectivism; it is actually the first lesson taught during the first minutes of in-processing. Recruits are taught that the unit and, more importantly, the mission at hand are more important than any one person and that one individual's actions can affect the entire unit. These lessons are reinforced through physical punishment (in the form of extra exercise), peer pressure, or, if need be, nonjudicial punishment (a form of military administrative punishment) if the behavior is severe enough or is a recurring issue. An example of this is getting all of your issued gear into the large issued duffle bag (known as a "sea bag") in a timely manner. It is planned so there is never enough time given to complete the task and the last person done receives the wrath of the drill instructors.

The phrase "pain is memory" is often used in the U.S. Marine Corps and rings true for most who enter the military. Lessons taught in boot camp are always reinforced by some kind of physical cost if they are not learned quickly. Once we were put into our training platoon in boot camp we were taught everything from how to lace our boots and get dressed to how to shower. Once again, most of these events have a time limit that never seems to be long enough to get the task completely accomplished. When observing a platoon moving from their barracks to the chow hall for breakfast, it is not unusual to see at least one or two individuals still trying to finish the process of dressing. During the initial stages of training, this method of instruction is done to keep recruits on their heels as well as to enforce instant obedience to orders.

Methods of stripping individual values differ among the branches, but the end goal is the same: to strip every individual of the desire to fulfill self-interest and to focus on the collective success of the group. First names are never used in boot camp or officer school; rather, individuals are referred to by title (e.g., Private or Candidate). In Marine Corps boot camp, this principle is taken to the extreme. The use of the word "I" is forbidden, and all recruits are forced to talk in the third person (e.g., "this recruit"), and ownership of anything except one's own actions is very limited.

Compliance Stage The second stage of indoctrination is known as the compliance stage. It is here that the new service member begins to experiment with the new behaviors of the group. Although individuals may be complying with the group, they have not yet internalized the group's values and instead are performing to avoid punishment. The models of the new behaviors (i.e., drill instructors) further encourage individuals to think as part of the group by rewarding or punishing as a group. The transition between civilian and military lifestyles can be as smooth or as awkward as the individual makes it, but in my (JRS) experience it becomes more about conformity than anything else. However, I have experienced incentive training (IT) because of the mistakes of individuals that caused the platoon to fail an objective. In most cases, these are simple mistakes, but any flaw is taken advantage of by the instructors.

My most memorable experience occurred when the platoon was learning to march. We were attempting a maneuver to turn as a unit and march in a different direction. I was turning on the wrong foot and it caused me to fall out of step with the rest of the platoon. The senior drill instructor and his assistant were unhappy about the mistake and took us to one of the many sand pits on base at Marine Corps Recruit Depot San Diego to give us some incentive training, which consisted of plenty of push-ups, flutter kicks, and other exercises. I was unpopular for about a day after that until someone else got us in trouble and took the heat off me.

Identity with the group and the larger military organization may be further encouraged through being required to memorize and recite word for word various trivia including knowledge of the history of one's service, the core values, service songs, or other information related to the military. If a recruit or officer candidate cannot recite this information exactly, the group may be punished. Although the primary motivation is to avoid punishment by instructors and the wrath of your peers, the method is effective in training an individual to pay more attention to how one's actions affect the group. The memorization of history, songs, and other information of the military further serves to instill the values and beliefs of the service.

Internalization Stage   During the third phase, recruits begin to more actively integrate the values of the group into their own individual worldviews. Conforming to the group is encouraged through social pressure and the continued emphasis on group norms. Because the individual now has incorporated the groups values, norms, standards, and behaviors into his or her own self-concept, inclusion into the group becomes important. Individuals are now more internally motivated to behave in a way that is acceptable to the group (Baron, 2000; McGurk et al., 2006). Symbols of inclusion into the group take on special significance. Individuals in U.S. Marine Corps boot camp or officer candidacy school may begin to yearn for the moment when they graduate and can pin on the Eagle, Globe, and Anchor emblem that is symbolic of being a U.S. Marine.

Consolidation Stage   During the consolidation stage, individuals achieve the last step of indoctrination. The consolidation stage consists of the individual becoming totally committed to the group. Identification with the group is now central with their self-concept. A result is the development of an in-group vs.

out-group mindset. McGurk et al. (2006) observed that within indoctrination service members are also taught to dehumanize and deindividuate the enemy (i.e., stop thinking of the enemy as individuals). As Grossman (1996) noted, the emotional distance created by these processes facilitates the ability to kill in the context of combat. The processes of dehumanization and deindividuation may appear strange or even revolting to mental health professionals, who are taught to empathize with others and to engender a concern for the well-being of people. The moral ambiguities associated with a combat role, however, are central to understanding the issues of combatants who enter psychotherapy. Guilt associated with killing, even when this occurs in a sanctioned context such as combat, may be a psychotherapy issue, but the absence of remorse does not imply characterological deficits associated with such acts in civilians. It is more likely that the unintended death of noncombatants is likely to be a focus of psychotherapy. To effectively accomplish their jobs, service members must create an emotional distance from their enemies. If not, then they may freeze up in combat and thereby put themselves and their entire unit in danger. It has also been our experience that, when individuals return from combat and begin to consider the people they killed as being similar to friends or neighbors, these service members experience immense guilt and suffering, even if it is clear that they were placed in a position where that person was attempting to kill them.

Aggression beyond what would normally be acceptable within society is a fact of life that is encouraged in the military, particularly among infantry and Special Operations Forces units, where close combat is more likely to occur. This acceptance of controlled aggression begins with drill instructors and becomes part of the culture within the military. During combat training, it may seem as if every action or word is laced with aggression. Individuals who cannot tolerate or adapt to this environment will not succeed in most military organizations. Furthermore, individuals who do not succeed in training are often deemed to be personally defective by the group (Schroeder, 1984).

Learning controlled aggression progresses through training until service members reach a basic skills phase where most field-type training occurs. Close combat, rifle marksmanship, offensive and defensive tactics, and other skills are focused on destroying the enemy. The training continues as service members become more senior and are trusted with more responsibilities. The culture of combat arms (e.g., infantry, artillery, special operations) is that killing and violence are part of the job and not something to be avoided. Noncombat arms military jobs support combat arms professions. Individuals within the military must learn to feel comfortable with knowing that directly or indirectly their actions support an inherently violent enterprise.

The emphasis on dehumanizing the enemy and killing should not be mistaken with an overall attempt to demonize humanity in general. As McGurk et al. (2006) noted, democratic countries such as the United States go to great lengths to train their service members to distinguish enemy combatants from civilians and noncombatants. Service members entering combat are expected to have knowledge of rules of engagement that describe when and whom they are allowed to engage in combat. Moreover, it is common to find service members of all ranks gaining an interest in foreign cultures to learn how to better interact with the local populations in a country where they may be operating.

Baron (2000) and McGurk et al. (2006) suggested that throughout these stages several psychological processes facilitate an individual moving through the stages of indoctrination. First, psychological processes of persuasion play a role in convincing service members to adopt new behaviors and attitudes that they previously would have rejected outright. Values such as collectivism, extreme self-sacrifice, and killing other people are thought to be incompatible with most recruits' worldviews prior to entering service. Due to the physical and psychological demands of basic training, recruits are thought to have little mental energy left for processing the messages that they are sent. Instead, this leaves recruits more open to the messages being sent from authority figures (i.e., drill instructors) (Baron, 2000).

The pressure for group conformity also facilitates indoctrination (Baron, 2000; McGurk et al., 2006). Specifically, recruits are pressured to conform to the values and behaviors of the group to avoid exclusion by the group and punishment by superiors. Conformity is further facilitated by the physical and mental stressors that accompany training. As described above, intensive exercise, verbal tirades by instructors, and the anger of the group are all powerful stressors during basic training.

Identification with the group further facilitates indoctrination (McGurk et al., 2006). Throughout indoctrination, an individual's group identification becomes a more important part of their self-concept. This facet of social identification theory may explain an important part of the process of military indoctrination (McGurk et al., 2006). Social identification theory discusses the part of an individual's identity that results from being part of a group (Tajfel & Turner, 1986). Individuals with strong group identification ascribe positive attributes to their group and may make negative attributions to other groups. McGurk and colleagues tie this into the individual's need to improve himself or herself through identification with the presumed elite or distinguished nature of the military services.

Although we would agree that these are important psychological processes that undoubtedly facilitate military indoctrination, we agree with Baron (2006), who posited that resolution of a more existential conflict is also an important element of enculturation—specifically, the individual's ability to confront and integrate aspects of his or her identity as a military member that are inconsistent with previously held self-concepts. We would add that there is also an existential element to the military indoctrination process. Baron (2006) discusses the importance of existential pressure during indoctrination when individuals confront aspects about themselves that are inconsistent with their previously held self-concepts. As Baron noted, groups that offer a transcendent cause offer a path for individuals to improve both self-esteem and feelings of insignificance. The individualism and materialism present in popular culture can leave many feeling that they are not contributing to a purpose higher than themselves. The military, with its emphasis on service, collectivism, values, and self-sacrifice for others, gives many young people an opportunity to connect with a higher cause. Although individuals often may enter the military for the educational opportunities, steady income, or travel, we have found that many later find fulfillment in identifying with the values represented by the institution. Thus, service members are able to cope with low pay, frequent separation from families, anonymity, the possibility of death, and subordinating oneself to follow the orders of superiors because they are content to know that they are contributing to a cause higher than themselves.

Clinical Implications

- Therapists who treat military clients ought to focus on their own assumptions, experiences, and biases toward military culture; for example, many civilians find the elaborate hierarchies of the military to be strange or, in the words of one colleague, "classist." In military culture, however, such beliefs play an important role in accomplishing the mission and in protecting individuals against stress. Thus, the development of trust and confidence between superiors and subordinates serves more than the purpose of accomplishing the mission. Having confidence in one's leadership has been found to be a protective factor for combat and operational stress (Canadian Army Lessons Learned Centre, 2004; Campise, Geller, & Campise, 2006; U.S. Department of the Air Force, 2008; U.S. Marine Corps, 2008).
- Therapists should consider conducting a military history with clients who have a service record and should explore the cultural nuances of the client's military experience. This can be done both informally and formally.
- Therapists need to give special consideration to the therapeutic alliance when working with military clients. Given the emphasis on a communalistic value system and the importance of group cohesion as a protective factor against stress, the relationship between a patient and his unit must be considered in treatment. Considering the bonds that service members develop with one another and the ethic of being responsible to the group, service members can often have very intense grief reactions and feelings of guilt when they lose someone in their unit. Individuals within a unit often prefer to rely on one another instead of turning to a mental health provider when

dealing with these issues. Furthermore, individuals who are sent back home due to injuries sustained in combat will often feel guilty that they left their fellow soldiers and may be very anxious about the safety of their unit.

- Therapists consider common combat-related stressors when working with military clients. This can begin by listening for issues related to experiences of rejection and isolation when working with military clients. One must consider the stress that individuals can experience when rejected by members of their units. A parallel may be drawn to the significant distress that individuals from Asian–American cultures whose identity is closely tied to the family are disowned by their family group. While not a perfect parallel, we have often observed significant distress stemming from a rejection by members of one's unit.

- Therapists providing services to military clients are aware of common transitional problems that stem from returning to civilian life.

- Therapists who treat military clients are sensitive to the emotions associated with injuries sustained while serving in the military. Individuals who leave the military, especially due to circumstances beyond their control such as an injury that prevents further service, often experience distress, guilt, and anxiety about returning to the civilian world. It is interesting to note that when service members leave the military they do not "quit their job." Instead, it is referred to as being "separated" from the military. Recognizing these transition issues and normalizing them for the patient have often been very therapeutic in our experience.

- Therapists look for ways to empower clients by utilizing cultural aspects of military clients' experiences.

- Therapists help military clients become aware of and utilize military resources.

- Due to the unavoidable association between the role of the combatant and killing, civilian therapists may mistakenly assume that training to kill is the most salient feature of enculturation into the military. This, however, is a misplaced assumption. Teamwork, mission accomplishment, conflict avoidance, and placement of the needs of others over self are, rather, the fundamental elements of assimilation into this environment. Thus, the essential presupposition is acceptance of the principle of selflessness. Service members must understand that their willingness to give their own lives in the service of their country or in defense of their comrades is a necessary first step before accepting the moral responsibilities associated with taking the life of another and that the preservation of life is indeed a higher value than its removal.

Conclusion

As with any culture or organization, the degree to which individuals represent the ideals, values, or generalities varies greatly. Not all service members are highly motivated, honorable, service-oriented people who are willing to lay down their life for their comrades in arms. Similarly, not all fully adopt the values of the service or become fully indoctrinated into the military life; therefore, just as it is important to understand a minority patient's level of acculturation to effectively conduct psychotherapy (Sue & Sue, 2003), it is also useful to consider the degree to which a service member or former service member has become socialized into the military before treating him or her. In our experience, we have encountered service members who felt they no longer had much in common with civilians and service members who could not wait to leave the military system and regain the freedom they felt they had lost. But, regardless of where a patient falls along the spectrum of adoption of the military culture, it is important to understand the system of which they were part and how that system attempts to incorporate its members.

Some may view the processes of changing young people from civilians into members of the armed services that we have described to be disturbing and possibly wrong. Admittedly, as members of the armed services, we are biased to be promilitary in the sense that we uphold many of the values to be good and necessary; however, we recognize that the military is not perfect, and it can be a difficult system within which to work and live. Regardless of one's feelings about the process, it is an important component of developing an understanding of our patients and their experiences. It is hoped that it will provide a starting point for better treatment.

References

Alexander, H. A. (2005). Education in ideology. *Journal of Moral Education*, 34, 1–18.

Alexander, J. H. (1999). *The battle history of the U.S. Marines*. New York: HarperCollins.

American Psychological Association. (2003). Guidelines on multicultural education, training, research, practice, and organizational change for psychologists. *American Psychologist*, 58(5), 377–402.

Baron, R. S. (2000). Arousal, capacity, and intense indoctrination. *Personality and Social Psychology Review*, 4(3), 238–254.

Campise, R. L., Geller, S. K., & Campise, M. E. (2006). Combat stress. In C. H. Kennedy & E. A. Zillmer (Eds.), *Military psychology: Clinical and operational applications* (pp. 215–240). New York: Guilford Press.

Canadian Army Lessons Learned Centre. (2004). Stress injury and operational deployments. *Dispatches: Lessons Learned for Soldiers*, 10, 1–39.

Cutler, T. J. (2002). *The Bluejackets manual*. Annapolis, MD: Naval Institute Press.

Department of Defense Task Force on Mental Health. (2007). *An achievable vision*. Washington, D.C.: Author.

Eastridge, B. J., Jenkins, D., Flaherty, S., Schiller, H., & Holcomb, J. H. (2006). Trauma system development in a theater of war: Experiences from Operation Iraqi Freedom and Operation Enduring Freedom. *Journal of Trauma*, 61, 1366–1373.

Ficarrotta, J. C. (1997). Are military professionals bound by a higher moral standard? *Armed Forces and Society*, 24(1), 59–75.

Gal, R. (1986). Unit morale: From theoretical puzzle to an empirical illustration—an Israeli example. *Journal of Applied Social Psychology*, 16, 549–564.

Grojean, M. W., & Thomas, J. L. (2006). From values to performance: It's the journey that changes the traveler. In T. W. Britt, A. B. Adler, & C. A. Castro (Eds.), *Military life: The psychology of service in peace and combat*. Vol. 4. *Military culture* (pp. 34–59). Westport, CT: Praeger Publishers.

Grossman, D. (1996). *On killing: The psychological cost of learning to kill in war and society*. New York: Little, Brown.

Johnson, W. B., & Wilson, K. (1993). The military internship: A retrospective analysis. *Professional Psychology: Research and Practice*, 24(3), 312–318.

Kennedy, C. H. (2006). Understanding one of our nation's greatest assets: Military personnel. *PsycCRITIQUES*, 51(38).

Manning, F. J., & Ingraham, L. H. (1987). An investigation into the value of unit cohesion in peacetime. In G. Belenky (Ed.), *Contemporary studies in combat psychiatry* (pp. 47–68). Westport, CT: Greenwood Press.

Mareth, T. R., & Brooker, A. E. (1985). Combat stress reaction: A concept in evolution. *Military Medicine*, 150(4), 186–190.

Marshall, S. L. A. (1962). *Night drop: The American airborne invasion of Normandy*. Boston, MA: Little, Brown.

McGurk, D., Cotting, D., Britt, T., & Adler, A. (2006). Joining the ranks: The role of indoctrination in transforming civilians to service members. In T. W. Britt, A. B. Adler, & C. A. Castro (Eds.), *Military life: The psychology of service in peace and combat*. Vol. 2. *Operational stress* (pp. 13–31). Westport, CT: Praeger Publishers.

Schroeder, J. E. (1984). Attribution and attrition in the U.S. Army basic training program. *The Journal of Psychology*, 117, 149–157.

Soeters, J. L., Poponete, C., & Page, J. T. (2006). Cultures consequences in the military. In T. W. Britt, A. B. Adler, & C. A. Castro (Eds.), *Military life: The psychology of service in peace and combat*. Vol. 4. *Military culture* (pp. 13–34). Westport, CT: Praeger Publishers.

Stevens, G., Rosa, F. M., & Gardner, S. (1994). Military academies as instruments of value change. *Armed Forces and Society*, 20, 473–484.

Stouffer, S. A., DeVinney L. C., Star, S. A., & Williams, R. M. (1949). *Studies in social psychology in World War II*. Vol. 2. *The American soldier: Combat and its aftermath*. Princeton, NJ: Princeton University Press.

Sue, D. W., & Sue, D. (2003). *Counseling the culturally diverse: Theory and practice* (4th ed.). New York: John Wiley & Sons.

Tajfel, H., & Turner, J. C. (1986). The social identity theory of inter-group behavior. In S. Worchel & L. W. Austin (Eds.), *Psychology of intergroup relations* (pp. 7–24). Chicago, IL: Nelson-Hall.

Tziner, A. (1983). Choice and commitment to a military career. *Social Behavior and Personality*, 11, 119–128.

U.S. Department of the Air Force. (2008). *Leader's guide for managing personnel in distress*, http://www.airforcemedicine.afms.mil/idc/groups/public/documents.

U.S. Department of the Army. (1994). *Field manual no. 22-51: Leader's manual for combat stress control*. Washington, D.C.: Author.

U.S. Department of the Army. (2000). *Field manual no. 6-22.5: Combat stress*. Washington, D.C.: Author.

U.S. Marine Corps. (1997). *MCDP 1: Warfighting*. Washington, D.C.: Author.

U.S. Marine Corps. (2001). *Marine Corps common skills (MCCS) handbook*. Book 1A. *All Marines individual training standards*. Albany, GA: Author.

U.S. Marine Corps. (2008). *The leader's guide for managing Marines in distress*, http://www.usmc-mccs.org/leadersguide/index.htm.

4
Challenges and Threats
of Deployment

GREG M. REGER and BRET A. MOORE

Contents

Military personnel who deploy in support of combat operations serve in a variety of unusual contexts and face a wide range of personal and professional challenges. Some of these challenges are common to most, if not all, service members. Other challenges are unique to the circumstances, roles, and responsibilities of a particular soldier, sailor, airman, or Marine. The most obvious and lethal challenge deployed service members face is combat itself. Data from four U.S. combat infantry units who deployed to Afghanistan and Iraq provide evidence for high levels of combat experience (Hoge et al., 2004). Of Army personnel who deployed to Iraq, 93% reported being shot at or receiving small arms fire and 95% saw dead or seriously injured Americans. Out of this group, 50% reported handling or uncovering human remains, and 48% were responsible for the death of an enemy combatant. In a follow-up, population-based study of soldiers returning from Iraq (Milliken, Auchterlonie, & Hoge, 2007), researchers again documented high rates of difficult combat experiences. Specifically, 70% of reserve and 66% of active-duty soldiers reported potentially traumatic combat experiences during their deployment.

Previous research suggests that the vast majority of individuals exposed to trauma experience posttraumatic stress symptoms immediately following the events, yet relatively few meet diagnostic criteria for posttraumatic stress disorder (PTSD) 3 months later (Rothbaum, Foa, Riggs, Murdock, & Walsh, 1992; Riggs, Rothbaum, & Foa, 1995). Accordingly, it is not surprising that most service members adapt well following combat deployments, despite high prevalence rates of exposure to difficult experiences; however, for a significant minority, these exposures can exact a devastating mental health toll, impacting work, marriage, and family life. Indeed, 3 to 4 months after returning from Iraq, more than 17% of soldiers met strict screening criteria for depression, anxiety, or PTSD (Hoge, Auchterlonie, & Milliken, 2004). Over 19% of service members who deployed to Iraq reported a mental health problem within 1 to 2 weeks of deployment (Hoge et al., 2006), and problems persist much longer for some. Three to 6 months after returning from a deployment to Iraq, nearly 17% of soldiers screened positive for PTSD, and more than 10% screened positive for depression (Milliken et al., 2007).

An analysis of 4 years of healthcare records at the Department of Veterans Affairs found that more than 32,000 Operation Iraqi Freedom (OIF) and Operation Enduring Freedom (OEF) veterans had a mental health or psychosocial problem, and 25% of the study's population had a mental health visit (Seal, Bertenthal, Miner, Sen, & Marmar, 2007). In spite of significant barriers to care (Hoge et al., 2004), many of these wounded warriors are seeking help. Preventing these difficulties would be even better.

Although there are no quick fixes to the challenges service members and their families face, military personnel encounter common challenges, and planning for those challenges may help mitigate the negative impact. Some of these challenges present long before the unit leaves for deployment. Other challenges occur during the deployment, and still others follow the return home. The purpose of this chapter is to review some of these challenges and to offer practical suggestions that can be utilized to assist those going into harm's way. Other discussions of combat stress may also be relevant to the reader (U.S. Department of the Army, 1994; U.S. Marine Corps, 2000; Nash, 2007). This chapter is not intended to be a comprehensive review of all deployment challenges but rather to highlight the most common challenges, based on the experience of the authors.

Predeployment Challenges and Personal Preparation

Military personnel who are expecting to deploy to a combat environment are well acquainted with the professional preparation that is required. Field training exercises, live fire simulations, and weeks or months spent at military training centers are all expected parts of preparing to deploy. Accordingly, a large amount of a military unit's training calendar is dedicated to knowledge acquisition and skill rehearsal to be tactically and technically prepared for the requirements of the mission. Much less time is spent on personal preparation. Yet, as noted above, significant numbers of military personnel are negatively impacted by their experiences, and personal predeployment preparation can play a significant role in supporting success in the combat zone long before boots ever hit the ground.

Staying Connected

For service members to persist through lengthy deployments with high rates of exposure to potentially traumatic events (Hoge et al., 2004), it is common for many military personnel to experience a certain degree of emotional detachment. As it relates to interpersonal relationships, this may take the form of decreased frequency of communication, reduced disclosure during communication, or feeling an absence of intimacy or love for those with whom one was previously close. To continue doing difficult work over a prolonged period of time, this detachment may be adaptive for the service member.

Yet, many service members experience difficulties feeling close to loved ones even before leaving for a deployment. This can be particularly frustrating to those who want to take advantage of precious moments together in the days, weeks, and months leading up to the departure. Although the anticipation of a lengthy separation can drive a wedge between loved ones, there are things that can be done prior to deploying to prepare for that separation and to use the time that is available to productively plan to stay connected. In fact, the very act of planning for the separation often enhances intimacy and helps reduce emotional detachment.

Proactively developing a strategy for staying connected with loved ones requires intentionality and planning. The content of these plans should be developed in age-appropriate ways when considering children and can involve a great deal of creativity. First, deploying service members and their loved ones might plan the frequency of communication and the form it will take. This planning involves thinking through how children and loved ones will remain in touch across the many miles that separate them. This might begin with a realistic assessment of communication expectations and an evaluation of how congruent these expectations are with the realities of the service member's job, anticipated location, and operational tempo; for example, one might assess how realistic it is for the service member to be able to call home everyday. Furthermore, even if this is possible, is it a good idea? Too much and too little contact can be disruptive to some families, so finding a balance is critical. Talking with children and other loved ones about their expectations and various anticipated barriers helps to undercut assumptions and reduces the risk of resentment and anger.

The frequency and form of communication is likely to be impacted by modern computer advances, which have dramatically improved the communication opportunities of today's military personnel relative to previous generations. E-mail, text messaging, web cameras, cell phones, Internet communities, and other applications all make communication with loved ones easier than during any previous military conflict.

These technologies can help deployed service members stay in touch, despite potentially large differences in time zones. The requirements of a working parent's schedule and children's school and extracurricular activities can make e-mail and Internet communities an attractive option for staying connected without disrupting day-to-day schedules. In addition, teenagers (and many school-age children) are more technologically savvy than their parents, making technology-based communication plans attractive to many young people. Today's service members in Iraq and Afghanistan have relatively good access to phones and Internet service at most forward operating bases (FOBs), although many personnel serve in remote locations or are frequently on the move, making access unpredictable. Some personnel are taking their own laptops, and certain locations even provide the opportunity to access the web from personal living areas.

A second issue related to staying connected with children involves the impact of a lengthy deployment on a child's basic familiarity with his or her parent. Concerns about simply being recognized by young children after the return home are common. Finding ways to maintain familiarity with the deployed parent is particularly important for younger kids but certainly is relevant to children of all ages. As a result, many families place pictures of the deploying parent around the home at the eye level of the child. Others have made videorecordings of bedtime stories or other personal messages prior to departing. Some have obtained "talking photo albums," which record brief audio messages of the deploying parent's voice and include corresponding photographs of the parent with the child.

Third, predeployment planning to express one's care to loved ones reduces the inconvenience and stress of attempting to put these plans together while deployed; for example, planning in advance for birthdays, anniversaries, Mother's or Father's Day, graduations, and other celebrations might involve purchasing flowers, gifts, or writing cards months ahead of the events. Some have coordinated in advance for monthly or even weekly expressions of care. Others have planned for scheduled surprises by hiding gifts or cards prior to leaving and later providing intermittent hints or clues for where to look for them. Another technique involves placing a dollar each day into a jar for the purpose of spending a romantic evening together upon the service member's return home. Creative, thoughtful planning makes the best out of a difficult situation and assists loved ones in feeling cared for in spite of the geographical separation.

Finally, providing key information about the deployment to loved ones helps some cope with the separation; for example, a family might hang up a map of the country to which the service member is deploying and discuss facts about that location. A globe might assist school-age children in understanding, at least in part, where a loved one is deployed relative to home. Some children have difficulty grasping when a parent will be coming home, and tools to assist in the count down may be of some help. A jar filled with 365 chocolate Kisses has been used by some deploying parents to assist their child in understanding how many days remain in a year-long deployment. This method also allows the primary caregiver to give the child a "kiss" from daddy or mommy each day of the deployment. Many adults and children cope with difficult circumstances by seeking additional information. Regardless of the methods used, creatively finding ways to communicate about the deployment may help all concerned.

Facing Mortality

Some of the early formulations of combat operational stress reactions (referred to earlier as *war neurosis*) conceptualized the problem as a conflict between the innate will to live and one's conscious personal dedication to duty. An early pioneer in military psychiatry, Colonel Thomas Salmon described the etiology as follows:

> … the moving demands of the instinct of self-preservation stirring deep and strong affective currents vs. the conscious expectations, desires, and requirements of "soldierly ideals" embedded in an emotional matrix of discipline, patriotism, and the like … .

> **—as cited in Jones (1995, p. 40)**

Although many contemporary military mental health providers might not find this analytic conceptualization theoretically instructive or useful, the psychological tension between self-preservation and dedication to duty continues to be observed frequently among combat operational stress casualties. As it relates to predeployment preparation, one must first fully accept the looming deployment and all the large and small inconveniences that come with it. For most service members, this is a foundational assumption of the job. For some, however, this is no small feat. Faced with the potential threat to their own lives or the inconveniences of a combat deployment, a small minority of service members seek ways to avoid deploying. Some seek physical or psychological restrictions to duty that will keep them at home. Others attempt to time pregnancies in a manner that will interfere with their deployment. Still others engage in misconduct (e.g., drug use, disrespect to superiors) in an attempt to obtain an administrative separation from the service prior to deploying. Although the powerful conflict between self-preservation/self-interest and military service is understandable, both personal growth and honorable service are facilitated by squarely facing the potential risks associated with supporting combat operations and continuing to serve. Indeed, this is the very definition of courage.

Consciously coming to terms with this risk provides an opportunity for service members to face their own mortality at an unusual developmental stage. Many military personnel are in their 20s when they deploy and most likely would not be preparing for their own death if they were working in the civilian sector. Indeed, some might argue that it is uncommon for people, regardless of their age, to reflect upon their own mortality. Regardless, it is our belief that facing one's mortality and preparing for the possibility of one's injury or death can promote personal preparation for deployment as well as psychological resiliency. This is most aptly described in the highly acclaimed book, *On Combat* (Grossman & Christensen, 2004).

So they can focus their attention on the mission, military personnel preparing to deploy should strive to achieve confidence that their loved ones will be taken care of in case they are seriously injured or killed. The psychological freedom that comes from knowing loved ones will be secure regardless of what happens to oneself arguably decreases the likelihood that this will come to pass. Financial stability is one important factor. Addressing what would be required to provide financial stability is a personal question, and financial counseling is clearly beyond the scope of this chapter (and competence of the authors!). Nonetheless, it is appropriate for deploying service members to consider what financial resources would be required to provide for their loved ones. For some, the required resources may be covered by Servicemember's Group Life Insurance (SGLI); for others, financial security may require that their dependents are able to continue their current lifestyle for many years without placing an undue burden on the surviving spouse. It may be necessary to purchase additional life insurance in addition to SGLI, but service members will want to explore options that do not include policy disqualification for deployment-related deaths (the combat clause).

Predeployment preparation should also include personal relationship planning. Obviously, different kinds of relationships require different kinds of planning, but the impact of deployment may prompt arrangements that otherwise would not be considered. Married personnel with children might discuss various potential courses of action following a death. Where might the survivors live and what challenges might be encountered? How might these challenges be mitigated in advance? Would the surviving spouse remarry; if so, what weight do the deploying spouse's opinions carry, if any? What arrangements need to be made on behalf of children to prepare for the remote possibility of both parents dying during a deployment? Are godparents nominal in nature, or are they realistically prepared for the possibility of taking custody of children?

Deploying service members may also want to discuss last wishes or review end-of-life planning with loved ones either directly or indirectly. Any personal preferences for elements of a funeral or ceremony should be reviewed. Preferences for burial or cremation might be discussed, and any religious or spiritual commitments impacting these plans might be shared. Deploying military personnel could consider a "last letter" to loved ones to be read in the case the need arises. Alternatively, deploying service members might consider whether they have things left unsaid or undone that should be personally taken care of prior to departing.

Challenges to Single Service Members

Single or dating service members have their own set of questions to consider, including the impact of a lengthy combat deployment on any dating relationship. The length of the relationship prior to deployment and the level of emotional commitment between the couple are important variables when considering future relationship plans (i.e., splitting up, getting engaged, or getting married before deploying). Considering that many service members have only recently reached young adulthood, emotional and physical ties may be strong, which in turn may create an intensity and a sense of urgency in the relationship. This may be less of an issue for the more mature service member. Hasty relationship decisions that seem rational and logical prior to leaving may turn out to be disastrous within months or even weeks after deploying. In our experience, these individuals make up a substantial portion of a mental health provider's caseload. If possible, single service members who are involved in significant relationships and are about to deploy should be encouraged to discuss these issues openly and honestly with their partners. Those considering an expedited marriage prior to deploying might consider premarital counseling or marriage mentoring to reduce the likelihood of a bad outcome.

Deployment Challenges

When service members are in the deployed environment, a wide range of stressors is expected and routinely encountered. Many of these challenges are predictable, and military leaders train their units to sustain high levels of performance in spite of them. Other challenges are less predictable, or more devastating, and even the best prepared service member can be affected. The stressors of combat operations are commonly categorized as being comprised of environmental, physiological, cognitive, and emotional challenges (U.S. Department of the Army, 1994, 2006). The discussion in the following section of this chapter follows this categorization, although it is acknowledged that they may not represent mutually exclusive categories. Also, as the above-referenced *Field Manual* (U.S. Department of the Army, 2006) states, a stressor in one category may also increase the risk of, or directly cause, stressors in other categories.

Environmental Stressors

Environmental stressors can cause immediate discomfort to deployed personnel. Although not lethal in most instances, environmental stressors can negatively impact job performance, resulting in decreased readiness and negatively impacting mission accomplishment. In more extreme cases, mere environmental inconvenience can intensify and legitimately threaten the lives of service members.

The most frequently encountered environmental stressors are weather-related challenges. Heat in Iraq can top 130°F during the summer months. By itself, this can potentially challenge the physical integrity of the service member; however, when the service member is required to don an additional 40 to 50 pounds (in some cases, much more) of body armor and equipment in addition to the heat, dehydration and heat stroke become serious threats. Other environments challenge military personnel with mountainous terrain that is rugged, cold, and taxing to the body and disruptive to military operations. Wind storms in desert environments blind personnel and blast exposed skin with painful sand and debris. Some locations in Iraq endure an uncomfortable rainy season that turns a burned desert landscape into gelatinous mud that can interfere with movement or frustrate dismounted personnel.

Other common environmental challenges include unusual or dangerous insects and reptiles; exposure to noxious agents resulting from burning trash, diesel vehicles, and oil fires; and harsh and sustained noise. Whatever the source, long-term exposure to challenging environmental factors can take its toll on deployed personnel.

Physiological Stressors

Physiological stressors directly challenge one's body. One of the most common physiological stressors is decreased sleep. Anecdotal observation suggests that sleep deprivation and sleep problems are typical during combat deployments. The majority of deployed personnel seem to have difficulties getting good sleep. Others sleep hard when given the opportunity but sustain an operational tempo that negatively impacts regular or predictable sleep. Either way, the result can be a significant acute sleep deprivation or an ongoing pattern of restricted sleep that results in a sleep debt over time. As it relates to deployment, the effects can be devastating, as previously published vignettes suggest (Reger & Moore, 2006; Wesensten, Belenky, & Balkin, 2006).

The research literature on the impact of sleep deprivation on cognitive performance suggests significant consequences (Pilcher & Huffcut, 1996; Van Dongen, Maislin, Mullington, & Dinges, 2003). Obviously, the importance of maximizing cognitive performance during military operations cannot be overstated. Many military personnel work in positions that require sustained attention over long periods of time. Some must persist through routine sets of repeated steps under low stress, whereas others make complex decisions involving the integration of multiple sources of information under high levels of stress. In either case, sleep deprivation can take its toll. Service members quickly learn to sleep when they can; they take advantage of naps and try to increase their total sleep time, even if it is broken up into numerous shorter sleep periods. Obviously, this is not ideal considering that sleep quality is as important, if not more so, than total sleep time (see Chapter 15); however, conditions in a combat environment are rarely ideal.

Another physiological stressor is the physically taxing nature of combat deployments. Many service members must carry significant weight on their person in the form of weapons, ammunition, personal protective equipment, and other gear. Movement can be challenging under these conditions, particularly during sustained operations. Over time, physical weariness or exhaustion can result. Personnel who have deployed often obtain a greater appreciation for the emphasis that the military services place on physical fitness.

Cognitive Stressors

Cognitive stress does not directly threaten the physical integrity of service members but does impact the body indirectly (U.S. Department of the Army, 2006). This category encompasses challenges involving difficult decision making and one's cognitive capabilities; for example, many service members are faced with difficult decisions related to the rules of engagement and how and when to escalate the use of force. In Iraq and Afghanistan, many of these decisions involve enemies who do not operate according to the Law of War. Accordingly, combatant identification can be extremely difficult, placing some service members in the unenviable position of attempting to determine potential friend from foe. These life and death decisions are typically made by service members under time constraints and may have to be made while the service member is in mortal danger.

Other information may impinge on these difficult decisions resulting in additional complexities. Hindsight is 20/20. Although military personnel make the best decisions possible with the information available at the time, some reflect on past events in light of information obtained after incident investigations. Service members may subsequently regret that their use of force escalated too quickly or too slowly. Needless to say, this perception can present a cognitive stress when the service member is again exposed to an ambiguous situation of great consequence. Similarly, command investigations following some incidents are intended to ensure that service members act in accordance with the rules of engagement; however, the perceived threat of such an investigation can also stress some service members, thus presenting an additional cognitive challenge.

Another cognitive challenge involves the inevitable downtime. Service members often experience intermittent periods of high-intensity action separated by sometimes lengthy periods of little activity. This common experience is often frustrating to personnel and is sometimes characterized as "hurry up and wait." The resulting boredom can be stressful to deployed personnel who have limited access to many of the pastimes they engaged in at home. Lengthy deployments can, as a result of this cognitive stressor, begin to feel even longer.

The resourceful service member will plan to pursue personal goals during any available downtime to avoid boredom. This might involve pursuing additional military or civilian education through correspondence courses or pursuing personal goals such as increased physical fitness.

Emotional Stressors

Emotional stress during deployment is likely the most difficult for service members to endure. During predeployment training, many service members have received some degree of preparation for the environmental, physiological, and cognitive stressors of a deployment; however, little training is typically provided with regard to warding off the emotional stress of combat and everyday deployed life. Obviously, one of the most difficult emotional challenges is the grief that follows the serious injury or death of a fellow service member. At the time of this writing, Operations Iraqi and Enduring Freedom have claimed more than 4400 lives, and more than 30,800 service members have been injured (U.S. Department of Defense, 2008), making this emotional challenge more prevalent than one would like. As noxious as the death of a buddy is, the most common source of emotional stress during deployments to Iraq and Afghanistan is homefront issues.

Homefront issues represent a broad category of emotional stressors that include problems that may be caused or amplified by separation. A common homefront issue that service members face is marital problems. Psychologist Dan Wile wrote: "When choosing a long-term partner … you will inevitably be choosing a particular set of unsolvable problems that you'll be grappling with for the next ten, twenty, or fifty years" (quoted in Gottman & Silver, 1999, p. 131). The good news is that many couples can sustain satisfying relationships in spite of these inevitable problems. The bad news is that many couples who fail to solve these problems when living together at home continue to attempt to solve them while under the stress and separation of a combat deployment. Needless to say, this is not the most optimal environment for mending long-standing, preexisting relationship problems.

A frequent challenge to relationships during deployment is feared and actual infidelity. Thoughts of infidelity are a common and distressing concern with which many service members and their nondeployed partners struggle. Moreover, it doesn't matter if the thoughts are real or imagined. Both can be emotionally distressing and can occupy significant amounts of time in rumination about potential scenarios. Although time, future positive experiences, and the development of trust are the best cures for this problem, service members and their partners can attempt to manage their anxiety by objectively considering the evidence that does and does not support their fears.

Financial problems are another homefront issue that service members face when deployed. Although active-duty service members tend to make more money while deployed, poor financial planning, financial irresponsibility, and unforeseen expenses can wreak havoc on the emotional state of the service member. Couples often develop a power of attorney that gives the nondeployed spouse significant control with regard to financial affairs. Soured relationships have left some unfortunate service members in difficult situations. Some have seen entire savings wiped out, houses and cars sold, and credit ruined without the service member even knowing. In many cases, there is very little recourse for the service member. Other homefront issues that can create emotional distress for the service member include problems with children (e.g., behavioral problems at home and school) and health and psychological problems of the nondeployed partner or their children. Being so far away from home and unable to assist with child-rearing and caretaking responsibilities can create a sense of helplessness and may even challenge their sense of worth.

Postdeployment Challenges

The process of reintegrating back into one's life after being deployed is oftentimes a difficult transition. This may seem counterintuitive, as the idea of being reunited with loved ones, reclaiming previous life roles, and carrying on with future plans appears overwhelmingly positive at first glance; however, many changes occur for the family during deployment that can lead to very stressful situations upon the service member's return home. Expectations of all parties can be exceedingly high prior to the service member's return; consequently, the potential for disappointment and hurt feelings can also be high. Undue stress can be placed on service members and their loved ones by residual psychological problems resulting from combat experiences. Being faced with difficult decisions, such as whether or not to stay in the military, start or finish college, or enter the workforce, can also be emotionally draining. In the following section, we discuss some of the challenges that service members and their families face after deployment.

Role Changes

Military or not, most families have specific roles that each member adopts as a means to create a sense of order, fairness, and efficiency within the family. In some families, the male takes on the financial responsibilities (e.g., pays bills, files taxes), and the female handles the domestic duties (e.g., cleaning, child rearing). Other families share these roles, or the roles are reversed. Regardless of who performs what duties, family members are often required to shift roles and take on new and unfamiliar responsibilities when the service member deploys. Although this can be stressful, most families are able to adapt and function with minimal disruption. When the service member returns, however, conflicting expectations regarding family members' role responsibilities can create unexpected emotional stress.

As mentioned above, in many families the nondeployed spouse is forced to take on additional responsibilities when the service member deploys. Tasks previously managed by a team of two become the burdens of just one. This can result in personal growth, and in many cases the nondeployed spouse develops a new-found sense of independence, confidence, and self-reliance. Problems may arise if the service member returns and expects things to go back to the way things were before deployment. The spouse may feel unappreciated, used, or angry, particularly if he or she wants to maintain some of these new responsibilities. If these issues are not dealt with, bitterness and resentment can develop and place an undue strain on the entire family.

Resuming parenting roles can also be difficult. Establishing one's role in the day-to-day care of children can be challenging, particularly when children have developed strong bonds with the nondeployed spouse. Similarly, anger and resentment can develop on the part of school-age children if the service member returns with the intent to immediately resume the disciplinarian role. Parental styles that differ significantly may lead to confusion and rebellion on the child's part, creating even more stress on the family.

Another change that can be difficult to handle is going from the role of combatant to peacetime service member. For the combat soldier, sailor, airmen, or Marine, turning off many months of intense physiological and emotional arousal can be difficult. Many service members literally go from the battlefield to the unit's stateside motor pool in a matter of days. Increased hypervigilence and hyperarousal or diminished levels of impulse control can cause problems for the service member at work, at home, and in his or her social life.

Future Plans

Uncertainty about the appropriate course for one's future is a common conflict for many young adults. For the service member, decisions about career, education, and family often coincide with the return home and can present additional stress during the adjustment to the homefront. Army personnel typically spend anywhere from 12 to 15 months in a deployed setting. During this time, it is not uncommon for an individual's active-duty service obligation to expire; consequently, the soldier is faced with the decision of reenlisting for an additional period of obligated service or leaving the Army soon after returning home. This may not be an easy or straightforward decision. Deciding to stay in or leave the military can be complicated by many variables. Questions arise related to civilian occupational opportunities and additional education. If the service member is married, the spouse's views about the prospect of making the military a career must be considered, including the potential impact of continued service on the spouse's career and the family. Alternatively, military service provides job security that may be uncomfortable to relinquish. Decisions about reenlistments can also be complicated by enticements such as signing bonuses, which can reach tens of thousands of dollars, and offers of appealing duty assignments (e.g., Germany, Japan, Italy, or back near family).

Choosing to remain in the military can raise its own set of questions. Opportunities to move into new military specialties might be considered, and some evaluate options for military-funded civilian education. With additional education, some enlisted military personnel are given the opportunity to be commissioned as an officer. Others might change duty stations shortly after the return home, only to join a unit preparing to deploy again in the near future. Depending on how long one has been in the military, it can be difficult adjusting to the freedom, social norms, and looser structure of the civilian world. Such tasks as finding and interviewing for jobs and buying a home may be unfamiliar to the service member.

Psychological Adjustment

Some would argue that few service members return from a combat deployment unaffected. As the preceding chapter explained, change does not necessarily equate with a negative outcome within this context. One example of positive change is the potential for growth after a traumatic experience (Solomon & Dekel, 2007; Westphal & Bonanno, 2007). For some, however, returning from a combat deployment does unfortunately involve negative change. The myriad psychological problems service members encounter after returning home are discussed in greater detail in subsequent chapters, but some discussion of postdeployment psychological adjustment is warranted at this point.

Adjusting psychologically after returning from a deployment can be a difficult process. As previously mentioned, expectations can be high, which may lead to disappointment and hurt feelings on all sides; therefore, it is important for the friends and family members of the returning service member to allow the psychological reintegration process to happen naturally. It may take time for the service member to feel connected again with his or her loved ones. Well-intentioned attempts to force emotionally laden discussions about issues that arose while the service member was gone may backfire and lead to alienation and further communication and emotional gaps. Although probing questions about combat experiences by loved ones may be an attempt to reconnect and empathize with the service member, it may inadvertently arouse negative feelings and propagate withdrawal behavior.

Friends, family, and coworkers may notice an increase in agitation and possibly aggressive behavior. In a later chapter, Moore, Hopewell, and Grossman outline various possible reasons for this phenomenon, but, within this context, it should be mentioned that it is important for loved ones to be aware of this tendency in some service members so they can intervene when necessary. This may include calling a family member, fellow service member, commanding officer, or, in extreme circumstances, military or civilian police.

To some degree, change in the service member is to be expected, and it is important for loved ones to be prepared. It is also equally important for loved ones to understand that this change is usually temporary. Psychological adjustment can be difficult, but with time, understanding, and patience most families will stabilize and, if necessary, develop a new normal.

Conclusions

Predeployment, deployment, and postdeployment difficulties tax most service members who go to war. In spite of the many challenges reviewed above, most veterans come home proud of their accomplishments and with a sense of honor from having served one's country. Self-confidence is often strengthened from having sustained the rigors of a combat deployment, and some service members develop the courage to address postdeployment pursuits and challenges. Those needing help with the transition home can draw on the strength that got them through their deployment to take advantage of available treatment and resources.

Whether the challenges are great or small, they can be overcome through thoughtful planning and caring support. Just as service members protect and sustain one another through the deployment, the love of family, friends, community, and the nation can support veterans after their many challenges and sacrifices. This support may be as important as any coping strategy or practical advice. Psychiatrist Larry Dewey, in his book *War and Redemption* (2004), states that, "Love makes it possible for good men to wage war and survive." Perhaps it is also love that can propel these veterans into the future, motivating them to use their experience to contribute to society and to the next group of brave men and women who will be asked to go into harm's way.

References

Dewey, L. (2004). *War and redemption: Treatment and recovery in combat-related post-traumatic stress disorder.* Burlington, VT: Ashgate.

Gottman, J. M., & Silver, N. (1999). *The seven principles for making marriage work.* New York: Three Rivers Press.

Grossman, D., & Christensen, L. W. (2004). *On combat: The psychology and physiology of deadly conflict in war and peace.* Millstadt, IL: PPCT Research Publications.

Hoge, C. W., Castro, C. A., Messer, S. C., McGurk, D., Cotting, D. I., & Koffman, R. L. (2004). Combat duty in Iraq and Afghanistan: Mental health problems, and barriers to care. *New England Journal of Medicine*, 351, 13–22.

Hoge, C. W., Auchterlonie, J. L., & Milliken, C. S. (2006). Mental health problems, use of mental health services, and attrition from military service after returning from deployment to Iraq or Afghanistan. *Journal of the American Medical Association*, 295, 1023–1032.

Jones, F. D. (1995). Traditional warfare combat stress causalities. In F. D. Jones, L. R. Sparacino, V. L. Wilcox, J. M. Rothberg, and J. W. Stokes (Eds.), *War psychiatry* (pp. 133–148). Washington, D.C.: Borden Institute.

Milliken, C. S., Auchterlonie, J. L., & Hoge, C. W. (2007). Longitudinal assessment of mental health problems among active and reserve component soldiers returning from the Iraq war. *Journal of the American Medical Association, 298*, 2141–2148.

Nash, W. P. (2007). The stressors of war. In C. R. Figley and W. P. Nash (Eds.), *Combat stress injury: Theory, research, and management* (pp. 11–32). New York: Routledge.

Pilcher, J. J., & Huffcut, A. I. (1996). Effects of sleep deprivation on performance: A meta-analysis. *Sleep, 19*, 318–326.

Reger, G. M., & Moore, B. A. (2006). Combat operational stress control in Iraq: Lessons learned during operation Iraqi freedom. *Military Psychology, 4*, 297–307.

Riggs, D. S., Rothbaum, B. O., & Foa, E. B. (1995). A prospective examination of symptoms of posttraumatic stress disorder in victims of nonsexual assault. *Journal of Interpersonal Violence, 2*, 201–214.

Rothbaum, B. O., Foa, E. B., Riggs, D. S., Murdock, T., & Walsh, W. (1992). A prospective examination of post-traumatic stress disorder in rape victims. *Journal of Traumatic Stress, 5*, 455–475.

Seal, K. H., Bertenthal, D., Miner, C. R., Sen, S., & Marmar, C. (2007). Bringing the war back home: Mental health disorders among 103,788 U.S. veterans returning from Iraq and Afghanistan seen at Department of Veterans Affairs facilities. *Archives of Internal Medicine, 167*(5), 476–482.

Solomon, Z., & Dekel, R. (2007). Posttraumatic stress disorder and posttraumatic growth among Israel ex-POWs. *Journal of Traumatic Stress, 20*, 303–312.

U.S. Department of Defense. (2008). *Operation Iraqi Freedom, Operation Enduring Freedom casualty status: Fatalities as of January 25, 2008*, http://www.defenselink.mil/news/casualty.pdf.

U.S. Department of the Army. (1994). *Field manual no. 22-51: Leader's manual for combat stress control*. Washington, D.C.: Author.

U.S. Department of the Army. (2006). *Field manual no. 4-02.51: Combat and operational stress control*. Washington, D.C.: Author.

U.S. Marine Corps. (2000). *MCDP 1: Combat stress*. Washington, D.C.: Author.

Van Dongen, H. P., Maislin, G., Mullington, J. M., & Dinges, D. F. (2003). The cumulative cost of additional wakefulness: Dose–response effects on neurobehavioral functions and sleep physiology from chronic sleep restriction and total sleep deprivation. *Sleep, 26*, 117–126.

Wesensten, N. J., Belenky, G., & Balkin, T. J. (2006). Sleep loss: Implications for operational effectiveness and current solutions. In T. W. Britt, C. A. Castro, and A. B. Adler (Eds.), *Military life: The psychology of serving in peace and combat*. Vol. 1. *Military performance* (pp. 128–154). Westport, CT: Praeger.

Westphal, M., & Bonanno, G. (2007). Posttraumatic growth and resilience to trauma: Different sides of the same coin or different coins? *Applied Psychology: An International Review, 56*, 417–427.

5

Integration of Women Into the Modern Military[*]

CARRIE H. KENNEDY and ROSEMARY C. MALONE

Contents

> Women. You can't live with them; you can't live without them.
>
> **Unknown**

Historically, the military has had a wavering relationship with women, building up their numbers in times of need and then reducing or eliminating them when peace resumed (Thomas, 1994); however, the civil rights movement, the end of the draft, the proven meritorious service of military women, and the lack of clear battle lines in modern warfare have changed this. Women are no longer serving behind the front line. They are becoming increasingly integrated into combat roles, and the proportion of the military that is female is rising steadily. The notion of living and surviving in harm's way is a dynamic

---

[*] The views expressed in this chapter are those of the authors and do not reflect the official policy or position of the Department of the Navy, Department of Defense, or the United States Government.

one, and one that is fundamentally different for female service members as compared to their male counterparts. This chapter addresses the evolution of women's ongoing integration into the military, the arguments for and against women's service in the military, and the unique difficulties women face as they serve their country.

Historical Overview of Women's Service in the U.S. Military

Let the generations know that women in uniform also guaranteed their freedom.

Dr. Mary Edwards Walker
(U.S. Army, Medal of Honor Recipient, 1866)

Women have a long history of U.S. military service dating back to the Revolutionary War, when one woman was assigned to every 15 men to manage rations. The majority of these women were the wives and mothers of soldiers (Solaro, 2006), and some served on the front lines in artillery units swabbing out the cannons. George Washington called upon women to serve as nurses and paid them the same as male soldiers (25¢ per day) (Hoiberg, 1991).

For both the Confederacy and the Union during the Civil War, women served as spies, saboteurs, cooks, gunrunners, blacksmiths, scouts, drivers, guides, laundresses, and nurses, and at least 400 women cross-dressed and served as regular soldiers (Perlin, Mather, & Turner, 2005; Solaro, 2006; Sherrow, 2007). Military nurses during the Civil War pioneered battlefield treatment methods and placed themselves directly in the line of fire on the front line (Murdoch et al., 2006). One of these nurses, Clara Barton, founded the American Red Cross after the war (Sherrow, 2007). The Civil War saw the first and only woman to be awarded the Medal of Honor (Hoiberg, 1991), physician Mary Walker, who treated soldiers while under fire at Bull Run (Sherrow, 2007). The Spanish–American War clinched the need for military nurses and facilitated the creation of the U.S. Army Nurse Corps in 1901 and the U.S. Navy Nurse Corps in 1908 (Hoiberg, 1991; Pierce, 2006).

During World War I (WWI), 36,000 women (Committee on Women in the NATO Forces, 2001) served in the military. Of these, 6000 women were physicians (Gillmore, 1918), 20,000 served as nurses, and the remainder filled such roles as yeomen, telephone operators, and translators (Pierce, 2006). Female fatalities in WWI numbered 200, and 80 women were taken prisoner of war (Committee on Women in the NATO Forces, 2001). No women held military rank during the war, and they were neither provided military pay or retirement benefits nor considered to be veterans (Spector, 1985).

World War II

I moved my WACs forward early … because they were needed and they were soldiers in the same manner that my men were soldiers. Furthermore, if I had not moved my WACs when I did, I would have had mutiny … as they were eager to carry on where needed.

General Douglas MacArthur

Prior to the start of World War II (WWII), the only official capacity for women was as nurses in the Army and Navy nurse corps, although these women were not provided military rank or benefits. Once WWII was underway, the Women's Army Auxiliary Corps was created to replace men in noncombatant roles with women so the men could be deployed to the front line (Anon., 1942). These women did not attain true military status until 1943, when the qualifier "auxiliary" was dropped and the Women's Army Corps (WACs) was formed (Spector, 1985). These women attended gender-specific training and took a gender-specific selection examination—the Women's Classification Test (Goodenough, 1942; Zeidner & Drucker, 1988), and they were held to higher standards than men (Murdoch et al., 2006). The Navy, Coast Guard, and Marines soon followed with the implementation of WAVES (Women Accepted for Volunteer Emergency Service), SPARs (an acronym taken from the Coast Guard motto *Semper Paratus* and the translation "always ready"), and Marines in order to replace shore-based men. These women were provided military rank, and female officers in the ranks of Ensign, Lieutenant Junior Grade, and Lieutenant Senior Grade were provided the same pay and allowances as male officers (Anon., 1942).

While the Army deployed some women overseas, the Navy initially restricted women's service to within the continental United States. As need grew, this policy was relaxed somewhat to allow women to serve in Alaska, Hawaii, and the Caribbean (Spector, 1985). As measured by comments in the mail from soldiers in the Pacific, favorable remarks regarding military women "rose from less than 15% in August 1944, to over 50% by March of 1945; by the end of the war over 70 percent spoke highly of them" (Spector, 1985, p. 396). Approximately 400,000 women served in World War II (Committee on Women in the NATO Forces, 2001); 54 nurses were taken prisoner of war, and 400 women were killed by hostile fire (Pierce, 2006). At the end of the war, the Women's Armed Services Act of 1948 was signed by President Truman, establishing permanency for women in the military (Hoiberg, 1991), with the restrictions that only 2% of the U.S. military could be female, no women could serve in combat roles, and women could not advance beyond the rank of O-5 (i.e., Army, Air Force, or Marine Lieutenant Colonel or Navy or Coast Guard Commander).

As compared to WWII, few women saw service in Korea, and vocal military leaders and enlisted men engaged in an active campaign to keep women out of military service (Murdoch et al., 2006). Efforts to recruit sufficient numbers of women in this environment were unsuccessful, and women who did serve were largely restricted from the training and opportunities available during WWII. Only 7500 women served in Vietnam, most of them as nurses (Hoiberg, 1991), although toward the end of this war the increasing trend of women in U.S. military service began.

The Integration of Women Into the Modern Military

My decision to register women confirms what is already obvious throughout our society—that women are now providing all types of skills in every profession. The military should be no exception.

President Jimmy Carter

Following Vietnam and the end of the draft, the ceiling placed on military women (2%) was lifted, and the military, with the exception of combat occupations, was more fully opened to interested women (Yoder & Naidoo, 2006). In 1975, women were given the option of remaining on active duty during and following pregnancy (Hoiberg, 1991). The service academies were opened to women in 1976, and separate female units were eliminated in 1978 (Committee on Women in the NATO Forces, 2001). Women began receiving shipboard assignments in 1978 following the trial gender integration of the hospital ship *USS Sanctuary* (Holm, 1982; Hoiberg, 1991). In 1987, women began to be stationed on ammunition and supply ships that traveled to and from combat zones, as well as on P-3 Orion reconnaissance planes (McGuire, 1990). In 1994, all surface ships were opened to women (Committee on Women in the NATO Forces, 2001).

By the end of the 1980s, 10.8% of the military was female. At the end of 2004, women constituted 14.7% of the active-duty Army, 14.6% of the Navy, 6.1% of the Marine Corps, and 19.6% of the Air Force (Maxfield, 2004). Women comprise a higher percentage in the reserves than in the active-duty force; for example, the Army Reserve is composed of 23.3% women (Maxfield, 2005).

Beginning in the 1980s, women began to be more fully integrated into traditional military roles. In 1983, 200 women were involved in Grenada as pilots, flight engineers, and loadmasters. In 1986, women served on tankers that fueled bombers flying to Libya, in addition to performing aircraft carrier landings. In the Persian Gulf in 1987, 248 women were assigned to the destroyer tender that responded to the *Stark*. In 1989, 800 women participated in the invasion of Grenada (Hoiberg, 1991). During the Gulf War, over 33,000 women served as pilots and worked in key combat support functions; 4% of battlefield casualties were women, 4 female Marines received Combat Action

Figure 5.1 Search and Rescue (SAR) swimmer AWAN Kathryn Apostolina glides through the water as she tows fellow SAR swimmer William Quadrina during a SAR Fitness Test. Both swimmers are assigned to Helicopter Anti-Submarine Combat Squadron 85. SAR swimmers are required to perform two semiannual evaluations designed to test their endurance, strength, and stamina. Men and women are held to the same physical standards. (U.S. Navy photograph by Mass Communication Specialist 1st Class Mark A. Rankin.)

Ribbons, and 2 women were captured by the enemy and held as prisoners of war (Murdoch et al., 2006).

Women in Aviation and Space

> This is not a time when women should be patient. We are in a war and we need to fight it with all our ability and every weapon possible. Women pilots, in this particular case, are a weapon waiting to be used.
>
> **Eleanor Roosevelt (1942)**

The skills of women were first used in military aviation during WWII, when 1074 WASPs (Women's Airforce Service Pilots) ferried aircraft, towed gunnery targets, and instructed male Air Force pilots (Holm, 1982). Of these women, 38 gave up their lives in the line of duty (Hoiberg, 1991). It was not until 1973, however, that a branch of service designated any female military pilots. At that time, the Navy winged six female Naval Aviators. The Army followed a year later, graduating its first female helicopter pilot. The Air Force followed in 1976 (Holm, 1982). It was not until 1993, however, that aviation truly opened to women when the aviation combat role restriction was lifted (Committee on Women in the NATO Forces, 2001; Sherrow, 2007).

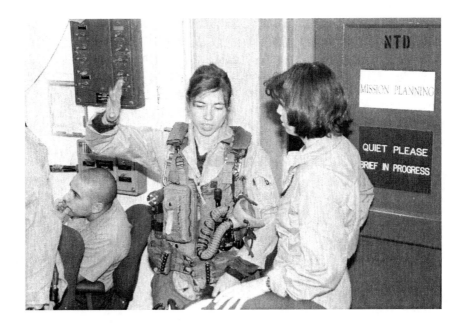

Figure 5.2 A Naval fighter pilot from Strike-Fighter Squadron Thirty-Seven (VFA-37), the "Ragin' Bulls," discusses her mission into Iraq with another member of her squadron. Her mission was launched from the deck of the *USS Enterprise* in the opening stages of Operation Desert Fox. (U.S. Navy photograph by Photographer's Mate 3rd Class Tedrick E. Fryman III.)

Women have historically been interested not only in aviation careers but also in space travel; however, in the 1950s, when the first astronauts were being chosen, one of the basic requirements was that each candidate must have at least 1500 hours of flight time as a test pilot (John, 2004). This presented a problem for women, as the military did not permit women to fly as test pilots at that time. One of the doctors, Randolph Lovelace, tasked with screening the male applicants became aware that the U.S.S.R. was training a female cosmonaut. This led to him to screen one female pilot, Jerri Cobb, with the same physical tests as the *Mercury* astronauts. She passed, and Lovelace screened 25 additional female pilots, 13 of whom passed the tests. Lovelace proposed that not only could women be astronauts but they might also be superior to men, given their significantly lower usage of food and oxygen in addition to lower body mass. These women went on to train under the same conditions as the men. Unfortunately, NASA cut off access of these women to their facilities in 1961. Despite congressional hearings and the successful performance of the Russian cosmonaut Valentina Tereshkova in 1963, women continued to be restricted from the astronaut corps (John, 2004). Finally, in 1983, Sally

Ride served as a mission specialist, and Eileen Collins piloted the space shuttle in 1995. Collins went on to become the first female shuttle commander in 1999 (Ackmann, 2004), and in 2007 NASA assigned Peggy Whitson as its first female commander of the international space station (Yembrick & Cloutier-Lemasters, 2007).

Tackling the Issue of Women and Ground Combat Roles: The International Stance

The issue of deploying women, particularly for the purpose of ground combat, is contrary to traditional views of women and roles held by women. There are arguments for and against further integration of women in the U.S. Armed Forces; however, the United States is not alone in making this decision, and the actions of other militaries warrant exploration.

Some countries have integrated their armed forces, to include placing women in combat roles. In Denmark, in 1978, combat unit restrictions were lifted, effectively integrating their armed forces (Committee on Women in the NATO Forces, 2001). In 1981, Belgium opened all jobs to women. Norway opened combat positions to women in 1985. Women were integrated into the Luxemburg Army in 1987 with no combat restrictions. In 1989, a Human Rights Review Tribunal in Canada found that excluding women from combat duty was discriminatory, and the tribunal mandated the full integration of women into combat assignments (Hoiberg, 1991) with the exception of submarine duty (Committee on Women in the NATO Forces, 2001). Canada integrated its submarines in 2001, assigned its first female executive officer to a Halifax-class frigate in 2006 and suffered its first female combat casualty in Afghanistan in 2006 (Committee on Women in the NATO Forces, 2007). Hungary opened combat assignments to women in 1996, although as of 2006 none had yet to be selected for the Special Forces (Committee on Women in the NATO Forces, 2007).

Despite the fact that Spain did not open military service for women until 1992, that country ended their combat restriction on women in 1999. Germany opened all positions, including combat roles, to women in 2000. Women continue to be restricted from combat roles in France, Greece, the Netherlands, Poland, Portugal, Turkey, and the United Kingdom (Committee on Women in the NATO Forces, 2001; Woodward & Winter, 2006). Italy only recently opened military service to women (1999) and has not integrated women into operational roles (Committee on Women in the NATO Forces, 2001). Israel is often cited as being the most progressive nation with regard to equal treatment of men and women, as both are drafted; however, even though the combat exclusion ended for women in 1995, no women have been assigned to combat units (Sherrow, 2007).

Figure 5.3 A Marine searches an Iraqi woman for weapons and explosives at a checkpoint in Iraq. (Photograph courtesy of Michael Franks.)

The Arguments Regarding Further Integration of Women

Cost Prohibitive

One argument for not integrating women into some roles within the U.S. military is the cost of refitting living environments to provide sufficient privacy and gender separation; for example, submarine duty continues to be closed to women "due to the prohibitive cost to modify these ships for appropriate berthing and privacy arrangements" (Department of the Navy, 2007, p. 3). Other countries that have addressed this specific issue include Norway, Australia, Sweden, and Canada. Using Canada as an example, this Navy officially integrated women in 2001, and the first woman qualified for submarine duty in 2003. This was done at the same time that a new class of submarines was acquired that allowed for segregated changing areas and shower facilities. Men and women share berthing (Dick, 2001), a policy contrary to the current U.S. military stance. Solutions to the privacy problem have ranged from refitting submarines to having all female crews, with none of these proposals gaining momentum for a wide variety of reasons. The U.S. military, however, is an organization that maintains rigorous selection procedures for a multitude of specialized jobs (e.g., aviation, drill instructor, diving duty, sniper, special operations), and it is likely that a select group of men and women could be chosen to participate in a trial integration that allows for showering and changing privacy but cohabitation in the berthing areas.

Women Are Unable to Handle the Physical and Psychological Rigors

One can argue that men, on average, are bigger and stronger than women, and this is often the primary argument against ending the combat exclusion. Although this argument is not insensible, the reality is that some women can pass the more stringent physical requirements of the more specialized ground combat forces. Also true is that women have been placed in combat situations without the feared disastrous results, a phenomenon that appears to be increasing at this time given the continuing shifting of the front line and repeated attacks on military members in vaguely drawn battle zones. For example, Army Sgt. Leigh Ann Hester and Army Spc. Monica Lin Brown received Silver Stars in Iraq and Afghanistan, respectively, for gallant actions while under attack. History is replete with additional examples of successful female warriors: In the 13th century B.C., a northern African female cavalry was documented; in 100 B.C., Celtic women warriors were regaled for their physical battle skills; in the 2nd century, female Vietnamese warriors fought against the Chinese; and in the 1500s Latin American female soldiers fought against Pizarro (for more details and examples, see Sherrow, 2007). Although men may be stronger on average, it has not yet been shown that this is a singularly valid argument to exclude physically qualified and interested women.

With regard to managing the psychological rigors of war, comparisons of coping effectiveness of groups of military men and women suggest that women are just as effective in managing stress in the military as men and in some cases may have an edge. With regard to substance use, military men are more likely to abuse alcohol than women as a result of stress, with both genders being equally likely to smoke cigarettes or utilize illegal drugs (Bray, Fairbank, & Marsden, 1999). In addition to alcohol abuse, men are also more likely to violate military laws and regulations than women (Hoiberg, 1991). Levels of psychological distress following service in a war zone during Operation Desert Storm were the same for men and women (Sutker, Davis, Uddo, & Ditta, 1995), although these rates rose in the years following, with 16% of women experiencing posttraumatic stress disorder (PTSD) compared to 8% of men (Wolfe, Erickson, Sharkansky, King, & King, 1999). Research investigating the impact of deployment length and deployment experience on male and female soldiers deployed to Bosnia–Herzegovina found that deployment experience (i.e., prior deployments) was related to lower rates of depression and PTSD for both men and women; however, only men experienced increased rates of those disorders due to longer deployment lengths (Adler, Huffman, Bliese, & Castro, 2005). A recent Veterans Affairs study found no gender-related differences in long-term physical and mental health status between male and female veterans (Frayne et al., 2006), although Katz, Bloor, Cojucar, and Draper (2007) found that military sexual trauma places women at significant risk of problems following wartime deployments.

Women Cannot Effectively Lead

Research with actual military leaders is lacking in this area; subsequently, the majority of information available regarding this phenomenon comes from the military academies. A study conducted with cadets at West Point found that general intellectual abilities as well as the personality traits of conscientiousness and agreeableness predict leader performance. Of interest in this study is that women performed higher in these realms than men. This was hypothesized to be due to the fact that women who seek to attend the U.S. Military Academy at West Point are a highly select group that surpasses their male peers in academics, athletics, and extracurricular activities prior to entering the Academy (Bartone, Snook, & Tremble, 2002). Another West Point study found that, although chain-of-command rating performances favored female cadets in the areas of duty motivation and supervising, peers rated male cadets higher on the attribute of influencing others, and subordinates rated males higher in influencing others, professional ethics, and developing subordinates (Morgan, 2004). Women may not be objectively different from men in their actual abilities, but gender stereotypes continue to negatively impact women.

One study evaluated gender stereotypes and their impact on evaluations of men and women in military training. Boldry et al. (2001) studied a group of 12 female and 13 male cadets at Texas A&M to examine performance and the impact of gender stereotypes. No differences were found between the men and women in performance measures (i.e., grade point average, physical training scores, science grades, and rank positions). When controlling for objective variables, however, gender stereotypes emerged, with men being perceived as higher in masculinity, motivation, and leadership than women. An Air Force Academy study found that "experience with leadership and seniority at the military academy maintained or increased masculine stereotyping," as stereotyped beliefs about effective military leaders increased as cadets progressed through their military training (Boyce & Herd, 2003). To combat some of these problems, the U.S. Naval Academy teaches midshipmen to build unifying visions and give public support to minority groups while reinforcing words with action, acknowledging problems, apologizing and seeking reconciliation when necessary, and creating opportunities for exposure to diverse groups (Johnson & Harper, 2005). As the number of women in the military continues to grow, more women will begin to see leadership positions. For example, the Army recently appointed the first female four-star general, General Ann E. Dunwoody. As this trend increases, it is anticipated that gender stereotypes in military leadership will decline.

Negative Impact on Military Effectiveness

Concerns abound about reductions in effectiveness if the military were to be fully integrated. Some of these revolve around possible distractions women bring to a combat unit and the detriment women have on a unit's ability to bond in survival situations. Other concerns include a potential reduction in motivation for men to serve in the military if their primary motivation previously was to provide defense for women and children (Sherrow, 2007). Another problem is that women have higher attrition rates than men, costing the military more money in training dollars. This is hypothesized to be due to issues related to job satisfaction, deployments resulting in PTSD, and family and childcare issues (Pierce, 2006).

Another concern focuses on the fact that women become pregnant and then are lost to their units during the pregnancy and maternity leave, which impacts manpower. Pregnancy is often cited as a reason by women to consider leaving the military, but childrearing may not be the most salient issue in this decision. Studies have cited leadership's responses to and support of pregnant women to be a major determinant in a woman's decision to leave any job (Pierce, 2006).

Despite the multiple opinions regarding women's potential negative impact on military effectiveness, studies looking at the impact of women on the operational effectiveness of a variety of military missions (ships, support units, ground-support units, combat exercises, and extended deployments) have not shown any detriment from integrating women into the work force (Hoiberg, 1991). Some argue that military effectiveness is decreased by assigning jobs based on the gender of individuals as opposed to professional qualifications (Sherrow, 2007).

Combat Roles Place Women at an Increased Risk

Another argument against integrating women into combat roles is that women are placed at increased risk of not only being killed by the enemy, partly due to their decreased physical abilities, but also being sexually assaulted by their male counterparts (Jeffreys, 2007). Evidence for this is presented in examples of the military's use of prostitutes and subsequent fueling of the human trafficking trade (Hughes, Chon, & Ellerman, 2007), the sexual violence carried out by militaries historically (for an example, see Chang, 1997), high rates of the sexual harassment of military women (see below), and incidents such as the Tailhook scandal in 1991 and rapes of women in military academic institutions (Jeffreys, 2007).

The U.S. military is actively addressing the concerns noted above. In 2004, the use of prostitutes was made a violation of the Uniform Code of Military Justice (UCMJ), and all service members are required to receive training on the trafficking of human beings (Department of Defense, 2007). Areas and establishments that are known for these activities have been made off-limits to overseas service members, and the frequenting of a prostitute in any locale, even in places where this activity is legal, may be punishable by 5 years in prison, a dishonorable discharge, and forfeiture of all pay and benefits. Punishments for sexual crimes against women are significant, and efforts have been made recently to expand and further define these types of crimes. Stalking was added as a UCMJ violation in 2006, for example, and a rape committed by a military member can result in the death penalty. In 2005, the military changed its policies to allow for confidential reporting of rapes so active-duty sexual assault victims can receive medical and mental health treatment from military providers without mandatory initiation of a legal investigation (Department of Defense, 2005). Regulations to maintain good order within military ranks include rules regarding fraternization. Although fraternization rules were not necessarily put in place to protect women from abuses, these rules do act to prevent inappropriate contact. There is quite a way to go with regard to the safe and equitable integration of women into the male-dominated military, but concerted efforts are in place to significantly improve upon the history noted above.

Sexual Harassment

The above-mentioned issues are obvious and extreme, but the issue of sexual harassment remains a more subtle and pervasive problem. The integration of women into the predominantly male societal structure of the military has not been without its challenges in this regard. Sexual harassment is an unfortunate byproduct of the integration of women into the U.S. workforce as a whole and is known to increase in environments that have a large proportion of male workers and supervisors and encompass traditional male occupations (Vogt, Bruce, Street, & Stratford, 2007). In addition to the safety of women in the workforce, sexual harassment negatively impacts productivity of the organization, as well as the mental health, physical health, morale, and family relationships of the affected women (Pryor, 1995). Women who have experienced sexual harassment are much more likely to leave the military than those who have not (Sims, Drasgow, and Fitzgerald, 2005).

Problems with sexual harassment have led researchers to search for risk factors and preventive measures. Research has shown that, within the military, positive attitudes toward women are higher among ethnic and racial minorities and those who have undergone sexual harassment training, as well as toward those women participating in any branch of service other than the Marine Corps (Vogt et al., 2007). In addition, these authors found that gender (specifically women), higher educational attainment, ethnic/racial minority status, and more training on sexual harassment were associated with decreased tolerance of sexual harassment in the military workplace.

The military has conducted routine surveys of personnel since 1988 with regard to sexual harassment experiences (Lipari & Lancaster, 2003). In the 1988 survey, 64% of women reported some type of sexual harassment. This appears to be declining steadily, as 24% of women in 2002 reported sexual harassment. The greatest declines were in the Marine Corps, with rates of 57% in 1995 and 27% in 2002; however, sexual harassment of women is lowest in the Air Force (18% reported in 2002). Reports of rape/attempted rape declined from 6% in 1995 to 3% in 2002. The military requires sexual harassment training as well as routine command climate surveys and integrated command-managed equal opportunity programs and resources. Increasing numbers of women within the military are also thought to be contributing to the declining trend of sexual harassment.

Conclusion

The integration of women into the military has not been without its hurdles, some of which remain in place today. Some maintain that combat roles require a warrior mentality that is irrevocably contrary to implementing a gender-equal combat force. Others believe that the arguments against further integration of women is now a moot point and that the characteristics of the current war have drawn women unavoidably into ground combat roles where women are proving "that gender distinctions are irrelevant" (Wise & Baron, 2006). Of interest is the fact that some believe that the military is one of the last bastions of male-dominated occupations, but the opposite may turn out to be true. The military has historically brought women into nontraditional occupations at rates greater than male-dominated occupations in the civilian sector (Yoder & Naidoo, 2006). With 95% of all military jobs now open to women, the steadily increasing numbers of women comprising the military will likely determine the status of that remaining 5%.

The modern soldiers, sailors, airmen, and Marines who also happen to be women are writing history during this current war. They are serving in positions never before held by women and with clearly positive results. At the same time, these same service members continue to face inequality regarding careers due to some jobs being closed to them, continued sexual harassment, and negative impacts of gender stereotypes. At this time, living and surviving in harm's way is a dramatically different experience for women than for men. The further evolution of the current war, the continued direct exposure of women to war, and the expansion of women's military roles will have a significant impact on the next chapter for American women.

References

Ackmann, M. (2004). *The Mercury 13: The true story of thirteen women and the dream of space flight*. New York: Random House.

Adler, A. B., Huffman, A. H., Bliese, P. D., & Castro, C. A. (2005). The impact of deployment length and experience on the well-being of male and female soldiers. *Journal of Occupational Health Psychology*, 10, 121–137.

Anon. (1942). Psychology and the war: Notes. *Psychological Bulletin*, 39, 300–304.

Bartone, P. T., Snook, S. A., & Tremble, T. R. (2002). Cognitive and personality predictors of leader performance in West Point cadets. *Military Psychology*, 14, 321–338.

Boldry, J., Wood, W., & Kashy, D. A. (2001). Gender stereotypes and the evaluation of men and women in military training. *Journal of Social Issues*, 57, 689–705.

Boyce, L. A., & Herd. A. M. (2003). The relationship between gender role stereotypes and requisite military leadership characteristics. *Sex Roles*, 49, 365–378.

Bray, R. M., Fairbank, J. A., & Marsden, M. E. (1999). Stress and substance use among military women and men. *American Journal of Drug and Alcohol Abuse, 25,* 239–256.

Chang, I. (1997). *The Rape of Nanking.* London: Penguin Books.

Committee on Women in the NATO Forces. (2001). *Year in review 2001.* Brussels, Belgium: NATO.

Committee on Women in the NATO Forces. (2007). *Meeting records 2007.* Brussels, Belgium: NATO (http://www.nato.int/ims/2007/win/year-in-review.htm).

Department of Defense. (2005). *Directive 6495.01: Sexual assault prevention and response program.* Washington, D.C.: U.S. Department of Defense.

Department of Defense. (2007). *Instruction 2200.01: Combating trafficking in persons (CTIP).* Washington, D.C.: U.S. Department of Defense.

Department of the Navy. (2007). *OPNAV instruction 1300.17A: Assignment of women in the Navy.* Washington, D.C.: Office of the Chief of Naval Operations.

Dick, R. (2001). Women in submarines: The last bastion falls. *Legion Magazine,* May/June (http://www.legionmagazine.com/frontline/journal/01-05.asp).

Frayne, S. M., Parker, V. A., Christiansen, C. L., Loveland, S., Seaver, M. R., Kazis, L. W. et al. (2006). Health status among 28,000 women veterans: The VA women's health program evaluation project. *Journal of General Internal Medicine, 21,* S40–46.

Gillmore, E. W. (1918). War bulletin: An unprecedented opportunity for women. *Journal of Abnormal Psychology, 13,* 197–198.

Goodenough, F. L. (1942). The selection of candidates for the officer candidate school at the Women's Army Auxiliary Corps training center. *Psychological Bulletin, 39,* 634–637.

Hoiberg, A. (1991). Military psychology and women's role in the military. In R. Gal and A. D. Mangelsdorff (Eds.), *Handbook of military psychology* (pp. 725–739). Chichester: John Wiley & Sons.

Holm, J. (1982). *Women in the military: An unfinished revolution.* Novato, CA: Presidio Press.

Hughes, D. M., Chon, K. Y., & Ellerman, D. P. (2007). Modern-day comfort women: The U.S. military, transnational crime, and the trafficking of women. *Violence Against Women, 13,* 901–922.

Jeffreys, S. (2007). Double jeopardy: Women, the U.S. military and the war in Iraq. *Women's Studies International Forum, 30,* 16–25.

John, M. (2004). NASA: Making space for women. *MIT Undergraduate Research Journal, 11,* 19–25.

Johnson, W. B., & Harper, G. P. (2005). *Becoming a leader the Annapolis way: 12 combat lessons from the Navy's leadership laboratory.* New York: McGraw-Hill.

Katz, L. S., Bloor, L. E., Cojucar, G., & Draper, T. (2007). Women who served in Iraq seeking mental health services: Relationships between military sexual trauma, symptoms, and readjustment. *Psychological Services, 4,* 239–249.

Lipari, R. N., & Lancaster, A. R. (2003). *Armed Forces 2002 sexual harassment survey.* Arlington, VA: Defense Manpower Data Center.

Maxfield, B. D. (2004). *FY04 profile of Air Force.* Washington, D.C.: Office of Army Demographics.

Maxfield, B. D. (2005). *Army profile FY05.* Washington, D.C.: Office of Army Demographics.

McGuire, F. L. (1990). *Psychology aweigh: A history of clinical psychology in the United States Navy, 1900–1988.* Washington, D.C.: American Psychological Association.

Morgan, M. J. (2004). Women in a man's world: Gender differences in leadership at the military academy. *Journal of Applied Social Psychology, 34,* 2482–2505.

Murdoch, M., Bradley, A., Mather, S. H., Klein, R. E., Turner, C. L., & Yano, E. M. (2006). Women and war: What physicians should know. *Journal of General Internal Medicine*, 21, S5–S10.

Perlin, J. B., Mather, S. H., & Turner, C. L. (2005). Women in the military: New perspectives, new science. *Journal of Women's Health*, 14, 861–862.

Pierce, P. F. (2006). The role of women in the military. In T. W. Britt, A. B. Adler, and C. A. Castro (Eds.), *Military life: The psychology of serving in peace and combat* (pp. 97–118). Westport, CT: Praeger Security International.

Pryor, J. B. (1995). The psychosocial impact of sexual harassment on women in the U.S. military. *Basic and Applied Social Psychology*, 17, 581–603.

Sherrow, V. (2007). *Women in the military*. New York: Chelsea House.

Sims, C. S., Drasgow, F., & Fitzgerald, L. F. (2005). The effects of sexual harassment on turnover in the military: Time-dependent modeling. *Journal of Applied Psychology*, 90, 1141–1152.

Solaro, R. (2006). *Women in the line of fire: What you should know about women in the military*. Emeryville, CA: Seal Press.

Spector, R. H. (1985). *Eagle against the sun: The American war with Japan*. New York: Vintage Books.

Sutker, P. B., Davis, J. M., Uddo, M., & Ditta, S. R. (1995). Assessment of psychological distress in Persian Gulf troops: Ethnicity and gender comparisons. *Journal of Personality Assessment*, 64, 415–427.

Thomas, P. J. (1994). Preface to the special issue. *Military Psychology*, 6, 65–67.

Vogt, D., Bruce, T. A., Street, A. E., & Stafford, J. (2007). Attitudes toward women and tolerance for sexual harassment among reservists. *Violence Against Women*, 13, 879–900.

Wise, J. E., & Baron, S. (2006). *Women at war: Iraq, Afghanistan, and other conflicts*. Annapolis, MD: Naval Institute Press.

Wolfe, J., Erickson, D. J., Sharkansky, E. J., King, D. W., & King, L. A. (1999). Course and predictors of posttraumatic stress disorder among Gulf War veterans: A prospective analysis. *Journal of Consulting and Clinical Psychology*, 67, 520–528.

Woodward, R., & Winter, P. (2006). Gender and the limits to diversity in the contemporary British army. *Gender, Work and Organization*, 13, 45–67.

Yembrick, J., & Cloutier-Lemasters, N. (2007). *First woman station commander arrives for historic spaceflight*, Release 07-226. Washington, D.C.: National Aeronautics and Space Administration (www.nasa.gov/home/hqnews/2007/oct/HQ_07226_ISS_Docking_prt.htm).

Yoder, J. D., & Naidoo, L. (2006). Psychological research with military women. In A. D. Mangelsdorff (Ed.), *Psychology in the service of national security* (chapter 15). Washington, D.C.: American Psychological Association.

Zeidner, J., & Drucker, A. J. (1988). *Behavioral science in the Army: A corporate history of the Army Research Institute*. Washington, D.C.: U.S. Army Research Institute for the Behavioral and Social Sciences.

II
On Being a Service Member

6

Military Stress: Effects of Acute, Chronic, and Traumatic Stress on Mental and Physical Health

MEGAN M. KELLY and DAWNE S. VOGT

Contents

Stress is a common part of military life. Although combat is often the first stressor that comes to mind in terms of military stressors, service members may experience many other conditions of stress, including high work demands and family separations due to deployment (Bartone, 2006). While some military personnel suffer from physical and mental health problems related to the stress of military life, many other military members show a high resiliency to stress (Bonanno, 2004). In this chapter, we provide a description of the different types of stressors that military personnel are exposed to in their profession. In the first section, we begin with a description of different classes of stressors (i.e., acute, chronic, traumatic) and focus on how individuals, and specifically military personnel, respond to these stressors. The next section provides an overview of the mental and physical health consequences of stress. In the third section, we describe the various types of stressors that are common to military personnel and provide an overview of research findings on the consequences of these stressors in terms of life functioning and mental health for military members. Finally, we discuss other important factors that may influence how military members respond to stress.

Stress and Adaptation

Although stress is common, it is often difficult to define. Stressors can be classified in several different ways. One helpful way of distinguishing between different types of stressors is to label them based on their duration and course (Elliot & Eisdorfer, 1982); for example, acute or brief stressors involve a person confronting a time-limited challenge, such as completing a particularly difficult training task during boot camp. On the other hand, chronic stressors involve situations that do not have a clear end in sight and cause substantial changes in lifestyles and identity. Adapting to and finding one's place in the hierarchical organizational structure of the military or being deployed without a clear timeframe for returning home are examples of chronic stressors. Traumatic stressors include events that involve the perceived threat of harm to oneself or others, which often results in the experience of intense fear, helplessness, or horror. Examples of this type of stressor include combat exposure and sexual assault exposure. Traumatic experiences that continue to have effects on individuals' physical and emotional functioning long after the event are considered distant stressors (Baum, Cohen, & Hall, 1993); for example, combat experiences can have wide-ranging effects on mental and physical health, sometimes lasting many years beyond the actual experience of war. Military life, especially during periods of deployment, involves a variety of acute, chronic, and traumatic stressors.

Stress: Helpful or Harmful?

Stress can be helpful and adaptive at certain levels but harmful if chronic, persistent, and severe. The Yerkes–Dodson law is a model that explains the relationship between arousal and performance in the face of a stressor (Yerkes & Dodson, 1908). According to this principle, during simple tasks, such as rote filing or cleaning military equipment, the relationship between arousal and performance is linear. Performance increases as arousal increases; that is, the more geared up one is for the task at hand, the better the performance on that task. For more difficult and stressful tasks, however, such as planning how to engage an enemy, the relationship between arousal and performance does not function in the same manner. For these challenging tasks, arousal is related to performance in a bell-curve-shaped relationship; thus, at the lowest and highest ends of the arousal continuum, performance on the task will be poorer. In this case, a moderate amount of arousal is necessary for optimal performance. For challenging tasks, experiencing some stress is normal, healthy, and related to positive outcomes. Too much arousal, however, can be problematic and lead to difficulties in adapting to stressful conditions.

Physiological Adaptation to Stress

Several theories have been put forth to account for how one reacts or adapts to stress. One important theory was postulated by Selye (1956). In fact, the term *stress* was first popularized by Selye and was defined as a nonspecific bodily response to demands placed on an organism (Selye, 1976). He proposed that there are three stages of adaptation to a stressor, collectively referred to as the *general adaptation syndrome*. First, individuals who face a stressor meet the demands of that stressor with an alarm reaction, characterized by the body's fight-or-flight response. If the stressor persists beyond this alarm phase and becomes chronic, such as might be the case for a combat deployment, then an individual enters the resistance stage, in which the body attempts to adapt and cope with the stressor. If the stressor still continues, the body eventually runs out of resources to cope, and an individual enters into the third stage, exhaustion. In this stage, illnesses and mental health problems result as the body is no longer able to maintain normal functioning.

Stress, Appraisals, and Coping

Other theories have been put forth to account for how individuals adapt to psychological stressors based on their perception or appraisal of the stressor. According to Lazarus and Folkman (1984), when an individual encounters a potential stressor the event is appraised via a two-step process. A primary appraisal is made of the extent to which the stressor is relevant and threatening to the individual. A secondary appraisal is made of the extent to which the individual possesses the coping resources to deal with the stressor. This suggests that stress is in the eye of the beholder, such that events that are stressful for one person may be routine or even exciting for another person. As will be discussed later in this chapter, many factors may influence how stressors are appraised and moderate the impact of potential stressors on the health and well-being of military personnel.

Lazarus (1993) also focused on how two different styles of coping affect how individuals manage these stressors. These coping styles can be labeled *problem-focused coping* and *emotion-focused coping*. In problem-focused coping, the individual attempts to actively manage the stressor, resulting in a change in the individual's relationship to the stressor. An example would be when a service member handles a conflict with a coworker by directly addressing the problem until the stressor is no longer a threat. In contrast, emotion-focused coping is characterized by efforts to manage the emotions associated with stressful events and includes coping strategies such as reappraisal, acceptance, seeking social support, and venting. If a military member views a conflict with a coworker as a result of the coworker's heavy work load and makes a conscious decision not to take it personally, this would be an emotion-focused coping strategy. The goal of both of these approaches, though different, would be to downgrade the service member's negative emotional reaction, whether fear or anger.

Although both problem-focused and emotion-focused strategies may be helpful in managing the negative impact of stress exposure, each can be problematic at times (Suvak, Vogt, Savarese, King, & King, 2002). For example, problem-focused coping can be problematic when one attempts to actively problem-solve situations that are beyond personal control, such as coping with high levels of combat exposure. In this case, efforts to actively change the situation are likely to be fruitless and frustrating. Emotion-focused coping can be problematic when applied to lower level stressors that are amenable to active problem-solving strategies; for example, the application of emotion-focused coping strategies, such as acceptance, may serve to maintain a stressor that could be easily eliminated with a more active problem-solving approach. Overall, problem-focused coping has been found to be most adaptive for events that are perceived as controllable, whereas emotion-focused coping is most suited to stressors that are not easily changed and viewed as beyond personal control (Suvak et al., 2002).

Another way of classifying coping strategies is to view them as either approach based or avoidance based. Approach-based coping strategies are those that involve actively confronting the problems related to the stressful event, and avoidance-based strategies involve denying, minimizing, or ignoring the stressor (Holahan & Moos, 1987). Although approach-based coping overlaps in many ways with problem-focused coping, and avoidant-based coping shares some characteristics with emotion-focused coping, these two coping classifications do not have a one-to-one correspondence. Approach-based and avoidant-based coping may also be adaptive or problematic, depending on the situation; however, approach-based coping is generally associated with greater resilience and better psychological adjustment, whereas avoidant-based coping is linked to greater psychological distress (Holahan & Moos, 1987).

Consequences of Stress

When stressors are particularly intense or prolonged and effective methods of regulating stress are not in place, individuals have more difficulties adapting, and physical and mental health problems may develop or worsen (Schneiderman, Ironson, & Siegel, 2005). Stressful and traumatic events are associated with the onset of depressive episodes and anxiety-related problems, as well as other mental health problems and trauma-related disorders. High levels of stress have also been associated with significant physical health conditions, including cardiovascular problems, immunosuppression, and risk for medical illnesses. We will briefly describe each of these consequences below.

Mental Health Consequences of Stress

A substantial body of research has demonstrated that high levels of stress and exposure to negative life events predict and precede the onset of mental health problems and worsening of existing psychopathology. Stress has figured prominently in many psychiatric models of mental health problems, including depressive disorders (Hammen, 2005; Kessler, 1997), anxiety disorders (Barlow, 2002; Chorpita & Barlow, 1998), and addictive disorders (Piazza & Le Moal, 1998; Sinha, 2001). In addition, extensive attention has been directed toward how exposure to traumatic events may place individuals at risk for the development of trauma-related disorders, particularly acute stress disorder and posttraumatic stress disorder (Keane & Barlow, 2002).

Acute Stress Disorder   Acute stress disorder (ASD) is a clinical diagnosis for individuals who show a characteristic pattern of maladaptive responding and functional impairment within one month after exposure to a traumatic stressor (APA, 2000). The *Diagnostic and Statistical Manual of Mental Disorders*, 4th edition (DSM-IV) criteria state that individuals with ASD have been exposed to a traumatic event in which actual or threatened harm to self or others was experienced along with high levels of fear, helplessness, or horror. The symptoms of ASD must be experienced within 2 days to 4 weeks after exposure to the traumatic stressor and are briefly summarized below. A series of at least three dissociative symptoms must be experienced during or after the traumatic event, including emotional numbing, reduced awareness of surroundings, derealization, depersonalization, and reduced ability to recall important details of the traumatic event. The traumatic event is also reexperienced in one of the following ways: recurrent images, thoughts, dreams, illusions, flashbacks, feeling as if one is reliving the event, or severe distress upon being exposed to a reminder of the trauma. Individuals with ASD display avoidance of cues that remind them of the traumatic event, and they experience increased arousal and anxiety. If clinically significant symptoms continue, an individual may meet criteria for posttraumatic stress disorder.

Posttraumatic Stress Disorder   Posttraumatic stress disorder (PTSD) is characterized by a similar maladaptive symptomatic profile following exposure to an extreme traumatic stressor; however, PTSD can only be diagnosed when these symptoms are present for more than one month (APA, 2000). At least one of the previously described dissociative symptoms must be experienced. Further significant avoidance of stimuli associated with the traumatic event or diminished responsiveness must be present, as indicated by at least three of the following: avoidance of thoughts, feelings, and conversations about the traumatic event; avoidance of activities and other cues associated with the trauma; inability to remember aspects of the event; reduced interest or participation in activities; feelings of detachment from others; emotional numbing; and a sense of a foreshortened future. Individuals with PTSD must also demonstrate at least two symptoms of increased arousal: trouble falling or staying asleep, increased irritability and anger outbursts, difficulty concentrating, hypervigilance, or an exaggerated startle response. Acute PTSD is defined by symptoms of less than 3 months' duration, whereas chronic PTSD is characterized by symptoms lasting for 3 months or longer. In addition, individuals may display delayed-onset PTSD, which is the onset of PTSD symptoms at least 6 months following the traumatic event (APA, 2000). PTSD is associated with a variety of other mental health problems, including substance abuse, depression, anxiety, relationship difficulties, and other forms of functional impairment (Keane & Barlow, 2002), and is a significant problem for many who have been exposed to traumatic life situations.

Physical Consequences of Stress

The majority of literature on the health effects of stress exposure, and especially exposure to extreme or traumatic stressors, has focused on mental health consequences. A much smaller body of literature has addressed how stress and trauma exposure may translate into physical health problems (Schnurr & Green, 2004). Research on exposure to lower level stressors of daily living has focused primarily on documenting the impact of stress exposure on heart disease (Rozanski, Blumenthal, & Kaplan, 1999) and immune functioning (Segerstrom & Miller, 2004). Within the traumatic stress literature, findings indicate that exposure to traumatic events is associated with a number of adverse physical health outcomes, including poor self-reported health status, a greater number of self-reported medical problems, and increased morbidity and mortality (Friedman & Schnurr, 1995).

Types of Military-Related Stress

Military personnel may be exposed to a number of stressors in their occupation. These stressors can be broadly grouped into three categories: job demands, work–family conflict, and sexual harassment and assault. In addition, military personnel may experience a number of stressors during deployment. Below we provide an overview of these different categories of military stressors.

Job Demands

Just as for other occupations, job demands are a significant concern for military members. Frequently cited stressors associated with military service include long work hours, heavy work load, low autonomy, and unpredictable schedules (Pfanlz & Ogle, 2006; Tucker, Sinclair, & Thomas, 2005), as well as conflict with supervisors and other military members. Indeed, conflict with superiors in a military context may be particularly stressful, as conflicts may tend to be resolved in favor of the superior, and military members generally do not have the freedom to leave their jobs if conflict exists (Pfanlz & Ogle, 2006). Studies have shown that military personnel report greater work stress than civilian workers (Pfanlz & Sonnek, 2002). In turn, job stress in military members is associated with emotional distress and mental health problems (Pfanlz, 2001; Pfanlz & Sonnek, 2002; Williams, Hagerty, Yousha, Hoyle, & Oe, 2002), as well as greater absenteeism and poorer work performance (Pfanlz & Ogle, 2006).

Work–Family Conflict

Another source of stress for military personnel is work–family conflict. Although work–family conflict is an issue for many occupations, conflict between one's job commitments and the demands of raising a family may be especially salient for service members given that the military role has historically been considered to supersede other roles in a military member's life (Segal, 1989). Sources of work–family conflict may include frequent separations from family, transfers, the threat of deployment and war, significant family stressors (e.g., illness, divorce), and long and unpredictable schedules (Westman & Etzion, 2005). These stressors may be particularly difficult for women, who often take on a greater burden of the responsibilities of family life and therefore may have more difficulty balancing family responsibilities and the demands of military life (Vinokur, Pierce, & Buck, 1999). In turn, research has shown that work–family conflict is related to lower job and life satisfaction (Adams, King, & King, 1998).

Sexual Harassment and Assault

Sexual Harassment   Another concern that may be especially relevant for women in the military is sexual harassment. Sexual harassment is defined as unwanted attention of a sexual nature that contributes to an intimidating or hostile work environment, sexual coercion or *quid pro quo*, and criminal sexual misconduct. Evidence suggests that sexual harassment may be more common in the military than in civilian work settings (Ilies, Hauserman, Schwochau, & Stibal, 2003; Niebuhr, 1997; USMSPB, 1995), and this may be associated with the high value placed on "masculine" qualities such as aggressiveness and toughness in the military (Vogt, Bruce, Street, & Stafford, 2007). Between 55 and 78% of female military personnel report having experienced sexual harassment within the previous year (Bastian, Lancaster, & Reyst, 1996; Firestone & Harris, 1999), and between 51 and 93% of female veterans have reported experiencing sexual harassment at some point during their military service (Murdoch, Polusny, Hodges, & Cowper, 2006; Murdoch, Polusny, Hodges, & O'Brien, 2004; Sadler, Booth, Mengeling, & Doebbeling, 2004). In addition, approximately 32% of female military personnel have reported that they were sexually harassed by their boss or supervisor (Bostock & Daley, 2007).

Sexual harassment is also a significant issue for male service members—38% of men have reported military sexual harassment in the past year (Bastion et al., 1996), and 42% of male military personnel report having experienced sexual harassment at some point during their military service (Street, Gradus, Stafford, & Kelly, 2007). In addition, there is some evidence that men exposed to sexual harassment in the military may experience more negative health consequences relative to women (Magley, Waldo, Drasgow, & Fitzgerald, 1999; Street et al., 2007; Vogt, Pless, King, & King, 2005), perhaps because sexual harassment may be more unexpected and have a more stigmatizing effect for male relative to female service members. Many negative outcomes are associated with the experience of sexual harassment, including greater job dissatisfaction, turnover, absenteeism, and interpersonal conflict (Fitzgerald, Drasgow, & Magley, 1999; Schneider, Swan, & Fitzgerald, 1997; Williams, Fitzgerald, & Drasgow, 1999). In addition, the stress associated with sexual harassment has been implicated in a number of physical and mental health problems (Frayne et al., 1999; Murdoch & Nichol, 1995; Murdoch, Polusny, Hodges, & Cowper, 2006; Rosen & Martin, 1998; Street et al., 2007; Wolfe et al., 1998).

Sexual Assault   Sexual assault can also be a source of significant stress for some military personnel, especially for women. Military sexual assault is typically defined as completed oral, anal, or vaginal penetration as a result of the use of threat or physical force during the course of military service (Zinzow, Grubaugh, Monnier, Suffoletta-Maierie, & Frueh, 2007). The literature indicates that rates of sexual assault are significantly higher for women veterans compared to female civilians (Frayne et al., 1999), with most studies reporting prevalence rates ranging from 30 to 45% (Benda, 2006; Murdoch et al., 2006; Sadler, Booth, Cook, Torner, & Doebbeling, 2001; Suris, Lind, Kashner, Borman, & Petty, 2004; Zinzow et al., 2007). For male military personnel, rates of sexual assault range from 1 to 4% (Martin, Rosen, Durand, Knudson, & Stretch, 2000; Murdoch et al., 2004). In turn, military personnel with a history of military sexual assault report more mental health problems, including PTSD, anxiety, and depression (Kang, Dalager, Mahan, & Ishii, 2005; Martin et al., 2000; Wolfe et al., 1998; Zinzow et al., 2007), as well as physical health complaints (Golding, 1994; Koss, Koss, & Woodruff, 1991; Martin et al., 2000; Sadler et al., 2004; Skinner et al., 1999; Stein et al., 2004; Waigandt, Wallace, & Phelps, 1990). In addition, individuals who have experienced sexual assault are more frequent users of health care (Koss et al., 1991; Golding, Stein, Siegel, Burnam, & Sorenson, 1988).

Deployment Stress

Most military personnel experience one or more deployments during their military career, and in recent years military personnel have experienced more frequent deployments, longer deployments, and less time between deployments. Deployments, and war-zone deployments in particular, may place considerable stress on military personnel. Deployments typically engender a number of potential stressors, but perhaps most notable among these stressors are combat-related experiences.

Combat-Related Experiences   Combat-related experiences may include being physically attacked or ambushed, being fired on by enemies, being wounded, and witnessing injury and death of other military personnel, enemies, and civilians (Hoge et al., 2004; King et al., 2006). In addition to combat-specific stressors, experiences associated with witnessing the destruction that often occurs following battle, such as observing and handling human remains, is also a significant source of trauma (King et al., 2006). In terms of prevalence, the National Vietnam Veterans Readjustment Study (NVVRS) found that 64% of veterans who served in the Vietnam theater experienced one or more traumatic events associated with combat (Kulka et al., 1990). Rates of exposure for combat veterans deployed in support of Operation Iraqi Freedom (OIF) in Iraq and Operation Enduring Freedom (OEF) in Afghanistan are estimated at approximately 67 to 70% (Milliken, Auchterlonie, & Hoge, 2007).

Combat exposure has been found to be associated with depression, anxiety, and other negative mental health problems (Bond, 2004; Hoge et al., 2004; Hoge, Auchterlonie, & Milliken, 2006), particularly high rates of PTSD. Findings from the NVVRS indicate that 15% of male Vietnam theater veterans experienced PTSD, and 30% met criteria for lifetime PTSD (Kulka et al., 1990). Among female Vietnam theater veterans, 9% experienced PTSD and 27% had lifetime PTSD (Kulka et al., 1990). PTSD prevalence rates for veterans of the Gulf War range from 4 to 23% (Kang, Natelson, Mahan, Lee, & Murphy, 2003; Keane, Cadell, & Taylor, 1988; Peconte et al., 1993; Wolfe, Brown, & Kelley, 1993), with rates of PTSD increasing with greater time between the war and the assessment (Southwick et al., 1995). Among the most current cohort of Iraq War veterans, rates of PTSD have been found to range from 12 to 13% (Hoge et al., 2004). In addition to PTSD and other mental health problems, combat exposure is associated with physical health problems (Bond, 2004; Iowa Persian Gulf Study Group, 1997), and higher rates of health care utilization (Hoge et al., 2004; Rosenheck & Massari, 1993). As these findings indicate, exposure to combat can have widespread and long-term deleterious effects on the lives of military personnel.

A number of factors increase the risk that an individual who experiences combat will meet criteria for PTSD. Individuals who experience injury as a consequence of combat exposure are more likely to develop PTSD than those who do not (Grieger et al., 2006; Koren, Norman, Cohen, Berman, & Klein, 2005). In addition, research has demonstrated a positive dose–response relationship between combat exposure and PTSD (Brewin et al., 2000), and findings indicate that witnessing or participating in atrocities contributes to the severity of PTSD symptoms above and beyond other experiences of combat exposure (Beckham, Feldman, & Kirby, 1998).

Perceived Threat   In addition to these more objective deployment stressors is the subjective experience of perceived threat that often, but not always, accompanies deployment to a war zone. Perceived threat in the war zone has been widely studied as a contributing factor to negative mental health outcomes. Perceived threat is the fear of harm that a service member may experience in the war zone, including threat or fear of attack, entrapment, injury, and death (King et al., 2006; Vogt and Tanner, 2006). It also encompasses fear of harm that might befall other unit members, including friends, leaders, and other military personnel (King et al., 2006). Many military personnel report these experiences as the most prevalent and salient in the war zone (Gifford, Ursano, Stuart, & Engel, 2006; Ikin et al., 2005). In turn, perceived threat is related to a host of mental and physical health problems following deployment (King et al., 2006; Vogt & Tanner, 2006). In both Vietnam veterans and Gulf War veterans, the majority of the impact of direct combat exposure on PTSD has been found to be mediated by levels of perceived threat (King, King, Foy, & Gudanowski, 1996; King, King, Foy, Keane, & Fairbank, 1999; Vogt & Tanner, 2007). Indeed, according to the current criteria for PTSD in DSM-IV, perceived threat or fear of serious harm to oneself or another is a critical element of the PTSD diagnosis (APA, 2000). The DSM-IV criteria for PTSD require that an individual "experienced, witnessed, or was confronted with an event or events that involved actual or threatened death or serious injury, or a threat to the physical integrity of self or others" (p. 467). Thus, the perception of threat is an integral part of the definition of PTSD and, as expected, is highly associated with PTSD symptomatology in combat veterans.

Difficult Living and Working Conditions   Although combat exposure has received the majority of attention in the deployment stress literature, findings indicate that difficult living and working conditions during deployment (i.e., "malevolent environment"), though low in magnitude when compared to combat stressors, can cause significant stress and discomfort (King, King, Gudanowski, & Vrevren, 1995; Litz, King, King, Orsillo, & Friedman, 1997). Specific difficult living and working conditions include lack of privacy, uncomfortable climate, lack of adequate food, undesirable living situations, working long hours, heavy workload, lack of sleep, unpredictable schedules, and other pressures related to deployments overseas and in the war zone (Gifford et al., 2006; King et al., 2006). In turn, these lower level stressors have been linked to negative mental health outcomes, including depression, anxiety, and PTSD (King et al., 1995, 1996; King, King, Vogt, Knight, & Samper, 2006). Findings from the NVVRS revealed that a war-zone environment that is harsh and difficult is an important factor in the development of PTSD along with other stressors that are more directly related to combat, including exposure to combat, violence, and the atrocities of war and perceived threat in the war zone (Kulka et al., 1990).

Interpersonal Stressors   In addition to stressors specific to the war zone are interpersonal stressors introduced by deployment, especially those associated with being separated from family members and other loved ones (King et al., 2006). Concerns about life and family disruptions can be especially potent for women, who as primary caregivers may experience significant concerns surrounding separation from children (McNulty, 2005; Ryan-Wenger, 1992; Vogt et al., 2005). Not surprisingly, separation from family members is associated with problems adjusting to war-zone deployments. A survey of U.K. military personnel deployed to Iraq in 2003 found that 85% of soldiers evacuated from the war zone experienced low mood due to being separated from family and friends (Turner et al., 2005). In addition, concern about life and family disruptions is more strongly associated with symptoms of PTSD in National Guard/Reservist personnel compared to active-duty personnel, most likely because National Guard/Reservist personnel are less accustomed to family separations due to work-related duties than are active-duty personnel (Vogt, Samper, King, King, & Martin, 2008). Vogt and colleagues (2005) also found that concerns about life and family disruptions during the Gulf War were related to significantly greater anxiety for female veterans compared to male veterans.

Other interpersonal stressors include sexual harassment and assault during deployment. These stressors can be particularly challenging during deployment because service members exposed to sexual harassment and assault often have to continue to interact with the perpetrator (Zinzow et al., 2007). In addition, these experiences may compound the negative effects of other stressors during deployment, including combat exposure. Research on Gulf War veterans has found that sexual trauma is predictive of PTSD for both male and female military personnel exposed to high levels of combat exposure (Kang et al., 2005).

Other Factors to Consider in Understanding the Impact of Military Stressors on the Health and Well-Being of Military Personnel

Exposure to a highly stressful or traumatic situation, such as combat exposure, does not necessarily lead to the development of PTSD or other health problems. In fact, the majority of individuals exposed to traumatic stressors recover after a short period (Bonanno, 2004) and do not develop PTSD (Green, 1994); therefore, other factors, in some combination with military stressors, must play a significant role in influencing the health outcomes of service members. A review of the literature reveals that a number of factors may be important, such as preexisting characteristics of the individuals exposed to trauma, including a history of prior exposure to traumatic events; internal variables, such as coping, appraisals of trauma, and hardiness; and external variables, including posttrauma social support. Below we provide a brief overview of the literature on each, focusing on PTSD as a primary consequence of traumatic stress exposure.

Preexisting Characteristics

Within military samples, several characteristics of individuals exposed to trauma have been consistently linked to higher rates of PTSD, including younger age, minority racial/ethnic status, female gender, less education, lower socioeconomic status, and lower intelligence (Brewin, Andrews, & Valentine, 2000).

Findings generally indicate that both younger age and minority racial/ethnic status confer an increased risk for PTSD in military veterans exposed to combat trauma (King, Vogt, & King, 2004). In addition, some evidence suggests that women exposed to combat may be at greater risk for PTSD than men (Tolin & Foa, 2006). Findings from the literature also indicate that personal and family histories of psychiatric problems are risk factors for PTSD (Brewin et al., 2000; Bromet, Sonnega, & Kessler, 1998; Keane, Marshall, & Taft, 2006). In particular, studies have shown that a history of psychiatric disorders, including depression, early conduct problems, and antisocial personality disorder, may constitute vulnerabilities for PTSD in military veterans (King et al., 1996; Schnurr, Friedman, & Rosenberg, 1993; O'Toole, Marshall, Schureck, & Dobson, 1998).

Exposure to Multiple Traumatic Experiences

Another factor that is strongly implicated in the development of PTSD among combat veterans is the incidence of adverse childhood experiences (Keane et al., 2006). Several studies have shown that traumatic and abusive experiences in childhood are associated with PTSD in veteran samples (Bremner, Southwick, & Charney, 1995; Donovan, Padin-Rivera, Dowd, & Blake, 1996; King et al., 1996; Schnurr, Lunney, & Sengupta, 2004; Stein et al., 2005). Exposure to conditions of instability, lack of adequate support, and mental health problems often associated with childhood adversity may increase maladaptive stress responding to subsequent trauma exposure (Bremner et al., 1995). Gahm and colleagues (2007) found that the number of adverse childhood experiences, including emotional abuse, physical abuse, sexual abuse, and interpersonal violence between parents, was predictive of PTSD in military veterans. Results have demonstrated that childhood adversity contributes to the prediction of PTSD above and beyond combat exposure (Cabrera, Hoge, Bliese, Castro, & Messer, 2007).

Although findings indicate that other types of repeated trauma exposure increase the risk of subsequent trauma exposure (Breslau, Davis, & Andreski, 1995; King et al., 1998), few studies have addressed whether previous traumatic and negative life events, independent of traumas associated with childhood adversity, increase the risk for PTSD among military personnel (Clancy et al., 2006). However, there is some evidence to suggest that the combination of other traumas (e.g., sexual victimization) and combat experiences is associated with a greater likelihood of developing PTSD (Rosenman, 2002).

Appraisals and Coping

Appraisals of military stressors are important factors in cognitive models of PTSD. According to these models, negative appraisals of trauma may become overgeneralized, resulting in perceptions of threat in response to nontraumatic as well as traumatic stimuli (Ehlers & Clark, 2000). Indeed, appraisals appear to moderate the effects of combat exposure on the development of PTSD symptoms. More negative appraisals of combat increase the likelihood of PTSD symptoms, whereas more positive appraisals of combat experiences decrease the relationship between combat and PTSD symptoms (Aldwin, Levenson, & Spiro, 1994). Positive appraisals of war, characterized by finding positive meaning in deployment experiences, are associated with better psychological adjustment, including reduced risk for PTSD in combat veterans (Fontana & Rosenheck, 1998; Spiro, Schnurr, & Aldwin, 1999).

Findings in the literature have shown that veterans who use approach-based coping strategies to deal with combat stress, including problem-focused strategies for resolving problems, social support, and seeking positive reappraisals, have better postdeployment psychological adjustment than those who do not use these strategies (Sharkansky et al., 2000). In one study, approach-based coping was more strongly associated with lower levels of PTSD symptoms in veterans who reported higher levels of combat stress, demonstrating that approach-based coping has a greater impact for individuals experiencing higher levels of traumatic stress (Sharkansky et al., 2000). Similar findings were observed for Vietnam War veterans, for whom problem-focused coping strategies in the war zone were related to greater lifetime adaptation and achievement (Suvak et al., 2002). In contrast, more frequent use of emotion-focused coping strategies (in response to war-zone stressors) such as blunting and venting was associated with lower levels of post-war adjustment among Vietnam War veterans (Suvak et al., 2002).

Research has also demonstrated that avoidant coping is more strongly associated with PTSD symptomatology for military veterans with higher levels of combat exposure (Stein et al., 2005). It has been suggested that avoidant coping may be relatively adaptive for lower levels of stress but problematic in response to traumatic and chronic stressors (Stein et al., 2005), underscoring the argument that the usefulness of different coping strategies may depend on the goodness of fit between the strategy and the situation (Folkman, Schaefer, & Lazarus, 1979; Park, Folkman, & Bostrom, 2001; Suvak et al., 2002).

Hardiness

Another internal factor that may be implicated in how military personnel respond to stress and trauma exposure is hardiness (Sutker, Davis, Uddo, & Ditta, 1995; Taft, Stern, King, & King, 1999). Hardiness is defined as consisting of three components: commitment, or a sense of purpose and meaning; control, defined by the belief that one can influence the events; and challenge, the perception that change is a challenge and a normal part of life (Kobasa, Maddi, & Courington, 1981). Research findings indicate that hardy individuals experience less stress-related sequelae following exposure to stressful events (Florian, Mikulincer, & Taubman, 1995; Maddi, 1999; Zakin, Solomon, & Neria, 2003), especially under highly stressful or traumatic circumstances (Bartone, 1999; Klag & Bradley, 2004; Pengilly & Dowd, 2000; Waysman, Schwarzwald, & Solomon, 2001). Other findings indicate that hardiness may mediate the relationship between combat exposure and PTSD (King et al., 1998).

Social Support

A consistent factor linked to positive outcomes in the face of military stressors, and especially combat exposure, is perceived social support following the stressful or traumatic event (Brewin et al., 2000). Though some studies have shown that posttrauma social support is not significantly related to the severity of PTSD (Gold et al., 2000), many more studies have demonstrated that posttrauma social support is associated with fewer symptoms of PTSD in combat veterans (Brewin et al., 2000; Engdahl, Dikel, Eberly, & Blank, 1997; King et al., 1998; Taft et al., 1999; Vogt & Tanner, 2007). King and colleagues (1998) demonstrated that the presence of both structural support (i.e., the size and complexity of the social support base) and functional social support (i.e., the amount of emotional support provided by social supports) were directly related to fewer PTSD symptoms in Vietnam veterans. In addition, they found that social support mediated the link between hardiness and PTSD symptomatology, such that more hardy individuals are able to create a larger social support base, which, in turn, may be critical in reducing risk of PTSD (King et al., 1998). Thus, multiple interrelationships between factors such as hardiness, social support, and many other risk and resilience factors are likely responsible for the emergence of PTSD symptoms and other negative health consequences of stress and trauma exposure in military personnel.

Summary

Stress is ubiquitous to military life. Stress can have beneficial effects on performance, but chronic and especially traumatic stressors, including combat exposure and sexual assault, can be difficult to handle and increase the risk of serious mental and physical health consequences. Most military personnel effectively manage these stressors, but some conditions, including the severity of the stressful or traumatic life event, preexisting personal characteristics, a prior history of highly stressful or traumatic experiences, little social support, and poor internal resources, may increase the likelihood of deleterious consequences such as PTSD or other mental health problems. Though our knowledge of the effects of stress on military personnel is greatly expanding, further attention to conditions that put service members at risk for adverse outcomes can inform the development of prevention and intervention programs targeted at the key factors involved in the development of maladaptive stress reactions in members of the military.

Acknowledgments

Preparation of this manuscript was supported by funding from the Department of Veterans Affairs (Project #DHI 05-130-3, "Further Development and Validation of the DRRI: Phase I," Dawne Vogt, Principal Investigator). Additional support was provided by the National Center for PTSD and the

Massachusetts Veterans Epidemiology Research and Information Center (MAVERIC), VA Boston Healthcare System. Correspondence concerning this book chapter should be addressed to Dawne Vogt, PhD, VA Boston Healthcare System (116B-3), 150 S. Huntington Avenue, Boston, MA 02130. Electronic mail may be sent via Internet to Dawne.Vogt@va.gov.

References

Adams, G. A., King, L. A., & King, D. W. (1998). Relationships of job and family involvement, family social support, and work–family conflict with job and life satisfaction. *Journal of Applied Psychology, 81*, 411–420.

Aldwin, C. M., Levenson, M. R., & Spiro, A. (1994). Vulnerability and resilience to combat exposure: Can stress have lifelong effects? *Psychology and Aging, 1*, 34–44.

APA. (2000). *Diagnostic and statistical manual of mental disorders* (4th ed., text revision). Washington, D.C.: American Psychiatric Association.

Barlow, D. H. (2002). *Anxiety and its disorders* (2nd ed.). New York: Guilford Press.

Bartone, P. T. (1999). Hardiness protects against war-related stress in Army Reserve forces. *Consulting Psychology Journal: Practice and Research, 51*, 72–82.

Bartone, P. T. (2006). Resilience under military operational stress: Can leaders influence hardiness? *Military Psychology, 18*, S131–S148.

Bastian, L. D., Lancaster, A. R., & Reyst, H. E. (1996). *Department of Defense 1995 sexual harassment survey*, Rep. No. 96-014. Arlington, VA: Defense Manpower Data Center.

Baum, A., Cohen, L., & Hall, M. (1993). Control and intrusive memories as possible determinant of chronic stress. *Psychosomatic Medicine, 55*, 274–286.

Beckham, J. C., Feldman, M. E., & Kirby, A. C. (1998). Atrocities exposure in Vietnam combat veterans with chronic posttraumatic stress disorder: Relationship to combat exposure, symptom severity, guilt, and interpersonal violence. *Journal of Traumatic Stress, 11*, 777–785.

Benda, B. B. (2006). Survival analyses and social support and trauma among homeless male and female veterans who abuse substances. *American Journal of Orthopsychiatry, 76*, 70–79.

Bonanno, G. A. (2004). Loss, trauma, and human resilience: Have we underestimated the human capacity to thrive after extremely aversive events? *American Psychologist, 59*, 20–28.

Bond, E. F. (2004). Women's physical and mental health sequellae of wartime service. *Nursing Clinics of North America, 39*, 53–68.

Bostock, D. J. & Daley, J. G. (2007). Lifetime and current sexual assault and harassment victimization rates of active duty United States Air Force women. *Violence Against Women, 13*, 927–944.

Bremner, J. D., Southwick, S. M., & Charney, D. S. (1995). Etiological factors in the development of posttraumatic stress disorder. In C. M. Mazure (Ed.), *Does stress cause psychiatric illness? Progress in psychiatry* (Vol. 46, pp. 149–185). Washington, D.C.: American Psychiatric Association.

Breslau, N., Davis, G. C., & Andreski, P. (1995). Risk factors for PTSD-related traumatic events: A prospective analysis. *American Journal of Psychiatry, 152*, 529–535.

Brewin, C. R., Andrews, B., & Valentine, J. D. (2000). Meta-analysis of risk factors for posttraumatic stress disorder in trauma-exposed adults. *Journal of Consulting and Clinical Psychology, 68*, 748–766.

Bromet, E., Sonnega, A., & Kessler, R. C. (1998). Risk factors for DSM-III-R posttraumatic stress disorder: Findings from the National Comorbidity Survey. *American Journal of Epidemiology*, 147, 353–361.

Cabrera, O., Hoge, C. W., Bliese, P. D., Castro, C. A., & Messer, S. C. (2007). Childhood adversity and combat as predictors of depression and posttraumatic stress in deployed troops. *American Journal of Preventive Medicine*, 33, 77–82.

Chorpita, B. F., & Barlow, D. H. (1998). The development of anxiety: The role of control in the early environment. *Psychological Bulletin*, 124, 3–21.

Clancy, C. P., Graybeal, A., Tompson, W., Badgett, K. S., Feldman, M. E., Calhoun, P. S. et al. (2006). Lifetime trauma exposure in veterans with military-related posttraumatic stress disorder: Association with current symptomatology. *Journal of Clinical Psychiatry*, 67, 1346–1353.

Donovan, B. S., Padin-Rivera, E., Dowd, T., & Blake, D. D. (1996). Childhood factors and war zone stress in chronic PTSD. *Journal of Traumatic Stress*, 9, 361–368.

Ehlers, A., & Clark, D. M. (2000). A cognitive model of post-traumatic stress disorder. *Behaviour Research and Therapy*, 38, 319–345.

Elliott, G. R., & Eisdorfer, C. (1982). *Stress and human health*. New York: Springer.

Engdahl, B., Dikel, T. N., Eberly, R., & Blank, A. (1997). Posttraumatic stress disorder in a community group of former prisoners of war: A normative response to severe trauma. *American Journal of Psychiatry*, 154, 1576–1581.

Firestone, J. M., & Harris, R. (1999). Changes in patterns of sexual harassment in the U.S. military: A comparison of the 1988 and 1995 DoD surveys. *Armed Forces and Society*, 25, 613–632.

Fitzgerald, L. F., Drasgow, F., & Magley, V. J. (1999). Sexual harassment in the armed forces: A test of an integrated model. *Military Psychology*, 11, 329–343.

Florian, V., Mikulincer, M., & Taubman, O. (1995). Does hardiness contribute to mental health during a stressful real-life situation? The roles of appraisal and coping. *Journal of Personality and Social Psychology*, 68, 687–695.

Folkman, S., Schaefer, C., & Lazarus, R. S. (1979). Cognitive processes as mediators of stress and coping. In V. Hamilton and D. M. Warburton (Eds.), *Human stress and cognition* (pp. 265–298). Chichester: John Wiley & Sons.

Fontana, A., & Rosenheck, R. (1998). Psychological benefits and liabilities of traumatic exposure in the war zone. *Journal of Traumatic Stress*, 3, 485–503.

Frayne, S. M., Skinner, K. M., Sullivan, L. M., Tripp, T. J., Hankin, C. S., Kressin, N. R. et al. (1999). Medical profile of women Veterans Administration outpatients who report a history of sexual assault occurring while in the military. *Journal of Women's Health and Gender-Based Medicine*, 8, 835–845.

Friedman, M. J., & Schnurr, P. P. (1995). The relationship between trauma, post-traumatic stress disorder, and physical health. In M. J. Friedman, D. S. Charney, and A. Y. Deutch (Eds.), *Neurobiological and clinical consequences of stress: From normal adaptation to post-traumatic stress* (pp. 507–524). Philadelphia, PA: Lippincott Williams & Wilkins.

Gahm, G. A., Lucenko, B. A., Retzlaff, P., & Fukuda, S. (2007). Relative impact of adverse events and screened symptoms of posttraumatic stress disorder and depression among active duty soldiers seeking mental health care. *Journal of Clinical Psychology*, 63, 199–211.

Gifford, R. K., Ursano, R. J., Stuart, J. A., & Engel, C. C. (2006). Stress and stressors of the early phases of the Persian Gulf War. *Philosophical Transactions of the Royal Society B*, 361, 585–591.

Gold, P. B., Engdahl, B. E., Eberly, R. E., Blake, R. J., Page, W. F., & Frueh, B. C. (2000). Trauma exposure, resilience, social support, and PTSD construct validity among former prisoners of war. *Social Psychiatry and Psychiatric Epidemiology*, 35, 36–42.

Golding, J. M. (1994). Sexual assault history and physical health in randomly selected Los Angeles women. *Health Psychology*, 13, 130–138.

Golding, J. M, Stein J. A., Siegel, J. M., Burnam, M. A, & Sorenson, S. B. (1988). Sexual assault history and use of health and mental health services. *American Journal of Community Psychology*, 16, 625–644.

Green, B. L. (1994). Psychosocial research in traumatic stress: An update. *Journal of Traumatic Stress*, 7, 341–362.

Grieger, T. A., Cozza, S. J., Ursano, R. J., Hoge, C., Martinez, P. E., Engel., C. C. et al. (2006). Posttraumatic stress disorder and depression in battle-injured soldiers. *American Journal of Psychiatry*, 163, 1777–1783.

Hammen, C. (2005). Stress and depression. *Annual Review of Clinical Psychology*, 1, 293–319.

Hoge, C. W., Castro, C. A., Messer, S. C., McGurk, D., Cotting, D. I., & Koffman, R. L. (2004). Combat duty in Iraq and Afghanistan, mental health problems, and barriers to care. *New England Journal of Medicine*, 351, 13–22.

Hoge, C. W., Auchterlonie, J. L, & Milliken, C. S. (2006). Mental health problems, use of mental health services, and attrition from military service after returning from deployment to Iraq or Afghanistan. *JAMA*, 295, 1023–1032.

Holahan, C. J., & Moos, R. H. (1987). Personal and contextual determinants of coping strategies. *Journal of Personality and Social Psychology*, 52, 946–955.

Ikin, J. F., McKenzie, D. P., Creamer, M. C., McFarlane, A. C., Kelsall, H. L., Glass, D. C. et al. (2005). War zone stress without direct combat: The Australian naval experience of the Gulf War. *Journal of Traumatic Stress*, 18, 193–204.

Ilies, R., Hauserman, N., Schwochau, S., & Stibal, J. (2003). Reported incidence rates of work-related sexual harassment in the United States: Using meta-analysis to explain reported rate disparities. *Personnel Psychology*, 56, 607–631.

Iowa Persian Gulf Study Group. (1997). Self-reported illness and health status among Gulf War veterans: A population-based study. *JAMA*, 277, 259–261.

Kang, H. K., Natelson, B. H., Mahan, C. M., Lee, K. Y., & Murphy, F. M. (2003). Posttraumatic stress disorder and chronic fatigue syndrome-like illnesses among Gulf War veterans: A population-based survey of 30,000 veterans. *American Journal of Epidemiology*, 157, 141–148.

Kang, H. K., Dalager, N., Mahan, C., & Ishii, E. (2005). The role of sexual assault on the risk of PTSD among Gulf War veterans. *Annals of Epidemiology*, 15, 191–195.

Keane, T. M., & Barlow, D. H. (2002). Posttraumatic stress disorder. In D. H. Barlow (Ed.), *Anxiety and its disorders* (2nd ed., pp. 418–453). New York: Guilford Press.

Keane, T. M., Cadell, J. M., & Taylor, K. M. (1988). Mississippi scale for combat-related post-traumatic stress disorder: Three studies in reliability and validity. *Journal of Consulting and Clinical Psychology*, 56, 85–90.

Keane, T. M., Marshall, R. P., & Taft, C. (2006). Posttraumatic stress disorder: Etiology, epidemiology, and treatment outcome. *Annual Review of Clinical Psychology*, 2, 161–197.

Kessler, R. C. (1997). The effects of stressful life events on depression. *Annual Review of Psychology*, 48, 191–214.

King, D. W., King, L. A., Gudanowski, D. M., & Vrevren, D. L. (1995). Alternative representations of war zone stressors: Relationships to posttraumatic stress disorder in male and female Vietnam veterans. *Journal of Abnormal Psychology*, 104, 184–196.

King, D. W., King, L. A., Foy, D. W., & Gudanowski, D. M. (1996). Prewar factors in combat-related posttraumatic stress disorder: Structural equation modeling with a national sample of female and male Vietnam veterans. *Journal of Consulting and Clinical Psychology*, 64, 520–531.

King, L. A., King, D. W., Fairbank, J. A., Keane, T. M., & Adams, G. A. (1998). Resilience-recovery factors in post-traumatic stress disorder among female and male Vietnam veterans: Hardiness, postwar social support, and additional stressful life events. *Journal of Personality and Social Psychology*, 74, 420–434.

King, D. W., King, L. A., Foy, D. W., Keane, T. M., & Fairbank, J. A. (1999). Posttraumatic stress disorder in a national sample of female and male Vietnam veterans: Risk factors, warzone stressors, and resilience-recovery variables. *Journal of Abnormal Psychology*, 108, 164–170.

King, D. W., Vogt, D. S., & King, L. A. (2004). Risk and resilience factors in the etiology of chronic PTSD. In B. Litz (Ed.), *Early intervention for trauma and traumatic loss* (pp. 24–64). New York: Guilford Press.

King, L. A., King, D. W., Vogt, D. S., Knight, J., & Samper, R. E. (2006). Deployment risk and resilience inventory: A collection of measures for studying deployment-related experiences of military personnel and veterans. *Military Psychology*, 18, 89–120.

Klag, S., & Bradley, G. (2004). The role of hardiness in stress and illness: An exploration of the effects of negative affectivity and gender. *British Journal of Health Psychology*, 9, 137–161.

Kobasa, S. C., Maddi, S. R., & Courington, S. (1981). Personality and constitution as mediators in the stress-illness relationship. *Journal of Health and Social Behavior*, 22, 368–378.

Koren, D., Norman, D., Cohen, A., Berman, J., & Klein, E. M. (2005). Increased PTSD risk with combat-related injury: A matched comparison study of injured and uninjured soldiers experiencing the same combat events. *American Journal of Psychiatry*, 162, 276–282.

Koss, M. P., Koss, P. G., & Woodruff, W. J. (1991). Deleterious effects of criminal victimization on women's health and medical utilization. *Archives of Internal Medicine*, 151, 342–347.

Kulka, R. A., Schlenger, W. E., Fairbank, J. A., Hough, R. L., Jordan, B. K., Marmar, C. R. et al. (1990). *Trauma and the Vietnam War generation: Report of findings from the National Vietnam Veterans Readjustment Study.* New York: Brunner/Mazel.

Lazarus, R. S. (1993). From psychological stress to the emotions: A history of changing outlooks. *Annual Review of Psychology*, 44, 1–21.

Lazarus , R. S., & Folkman, S. (1984). *Stress, appraisal and coping.* New York: Springer.

Litz, B. T., King, L. A., King, D. W., Orsillo, S. M., & Friedman, M. J. (1997). Warriors as peacekeepers: Features of the Somalia experience and PTSD. *Journal of Consulting and Clinical Psychology*, 65, 1001–1010.

Maddi, S. R. (1999). The personality construct of hardiness. I. Effects on experiencing, coping, and strain. *Consulting Psychology Journal: Practice and Research*, 51, 83–94.

Magley, V. J., Waldo, C. R., Drasgow, F., & Fitzgerald, L. F. (1999). The impact of sexual harassment on military personnel: Is it the same for men and women? *Military Psychology*, 11, 283–302.

Martin, L., Rosen, L. N., Durand, D. B., Knudson, K. H., & Stretch, R. H. (2000). Psychological and physical health effects of sexual assaults and nonsexual traumas among male and female United States Army soldiers. *Behavioral Medicine*, 26, 23–33.

McNulty, P. A. (2005). Reported stressors and health care needs of active duty Navy personnel during three phases of deployment in support of the war in Iraq. *Military Medicine*, 170, 530–535.

Milliken, C. S., Auchterlonie, J. L., & Hoge, C. W. (2007). Longitudinal assessment of mental health problems among active and reserve component soldiers returning from the Iraq War. *JAMA*, 18, 2141–2148.

Murdoch, M., & Nichol, K. L. (1995). Women veterans' experiences with domestic violence and with sexual harassment while in the military. *Archives of Family Medicine*, 4, 411–418.

Murdoch, M., Polusny, M. A., Hodges, J., & O'Brien, N. (2004). Prevalence of in-service and postservice sexual assault among combat and noncombat veterans applying for Department of Veterans Affairs posttraumatic stress disorder disability benefits. *Military Medicine*, 169, 392–395.

Murdoch, M., Polusny, M. A., Hodges, J., & Cowper, D. (2006). The association between in-service sexual harassment and posttraumatic stress disorder among Department of Veterans Affairs disability applicants. *Military Medicine*, 171, 166–173.

Niebuhr, R. E. (1997). Sexual harassment in the military. In W. O'Donahue (Ed.), *Sexual harassment: Theory, research, and treatment* (pp. 129–151). Needham Heights, MA: Allyn & Bacon.

O'Toole, B. I., Marshall, R. P., Schureck, R. J., & Dobson, M. (1998). Risk factors for posttraumatic stress disorder in Australian Vietnam veterans. *Australian and New Zealand Journal of Psychiatry*, 32, 21–31.

Park, C. L., Folkman, S., & Bostrom, A. (2001). Appraisals of controllability and coping in caregivers and HIV+ men: Testing the goodness-of-fit hypothesis. *Journal of Consulting and Clinical Psychology*, 69, 481–488.

Peconte, S. T., Wilson, A. T., Pontius, E. B., Dietrick, A. L., Lirsch, C., & Sparacino, C. (1993). Psychological and war stress symptoms among deployed and non-deployed reservists following the Persian Gulf War. *Military Medicine*, 158, 516–521.

Pengilly, J. W., & Dowd, E. T. (2000). Hardiness and social support as moderators of stress. *Journal of Clinical Psychology*, 56, 813–820.

Pfanlz, S. (2001). Occupational stress and psychiatric illness in the military: Investigation of the relationship between occupational stress and mental illness among military mental health patients. *Military Medicine*, 166, 457–462.

Pfanlz, S., & Ogle, A. D. (2006). Job stress, depression, work performance, and perceptions of supervisors in military personnel. *Military Medicine*, 171, 861–865.

Pfanlz, S., & Sonnek, S. (2002). Work stress in the military: Prevalence, causes, and relationship to emotional health. *Military Medicine*, 167, 877–882.

Piazza, P. V., & Le Moal, M. (1998). The role of stress in drug self-administration. *Trends in Pharmacological Sciences*, 19, 67–74.

Rosen, L. N., & Martin, L. (1998). Psychological effects of sexual harassment, appraisal of harassment, and organizational climate among U.S. Army soldiers. *Military Medicine*, 16, 63–67.

Rosenheck, R., & Massari, L. (1993). Wartime military service and utilization of VA health care services. *Military Medicine*, 158, 223–228.

Rosenman, S. (2002). Trauma and posttraumatic stress disorder in Australia: Findings in the population sample of the Australian National Survey of Mental Health and Wellbeing. *Australian and New Zealand Journal of Psychiatry*, 36, 515–520.

Rozanski, A. Blumenthal, J. A., & Kaplan, J. (1999). Impact of psychological factors on the pathogenesis of cardiovascular disease and implications for therapy. *Circulation*, 99, 2192–2217.

Ryan-Wenger, N. M. (1992). Physical and psychosocial impact of activation and deactivation on Army Reserve nurses. *Military Medicine*, 157, 447–452.

Sadler, A. G., Booth, B. M., Cook, B. L., Torner, J. C., & Doebbeling, B. N. (2001). The military environment: Risk factors for women's non-fatal assaults. *Journal of Occupational and Environmental Medicine*, 43, 262–273.

Sadler, A. G., Booth, B. M., Mengeling, M. A., & Doebbeling, B. N. (2004). Life span and repeated violence against women during military service: Effects on health status and outpatient utilization. *Journal of Women's Health*, 13, 799–811.

Schneider, K. T., Swan, S., & Fitzgerald, L. F. (1997). Job-related and psychological effects of sexual harassment in the workplace: Empirical evidence from two organizations. *Journal of Applied Psychology*, 82, 401–415.

Schneiderman, N., Ironson, G., & Siegel, S. D. (2005). Stress and health: Psychological, behavioral, and biological determinants. *Annual Review of Clinical Psychology*, 1, 607–628.

Schnurr, P. P., & Green, B. L. (2004). Understanding relationships among trauma, posttraumatic stress disorder, and health outcomes. In P. P. Schnurr and B. L. Green (Eds.), *Trauma and health: Physical health consequences of exposure to extreme stress* (pp. 247–275). Washington, D.C.: American Psychological Association.

Schnurr, P. P., Friedman, M. J., & Rosenberg, S. D. (1993). Premilitary MMPI scores as predictors of combat-related PTSD symptoms. *American Journal of Psychiatry*, 150, 479–483.

Schnurr, P. P., Lunney, C. A., & Sengupta, A. (2004). Risk factors for development versus maintenance of posttraumatic stress disorder. *Journal of Traumatic Stress*, 17, 85–95.

Segal, M. W. (1989). The nature of work and family linkage: A theoretical perspective. In G. K. Bowen and D. K. Orthner (Eds.), *The organization family: Work and linkages in the U.S. military* (pp. 3–36). New York: Praeger.

Segerstrom, S. C., & Miller, G. E. (2004). Psychological stress and the human immune system: A meta-analytic study of 30 years of inquiry. *Psychological Bulletin*, 130, 601–630.

Selye, H. (1956). *The stress of life*. New York: McGraw-Hill.

Selye, H. (1976). Forty years of stress research: Principal remaining problems and misconceptions. *Canadian Medical Association Journal*, 115, 53–56.

Sharkansky, E. J., King, D. W., King, L. A., Wolfe, J., Erickson, D. J., & Stokes, L. R. (2000). Coping with Gulf War combat stress: Mediating and moderating effects. *Journal of Abnormal Psychology*, 109, 188–197.

Sinha, R. (2001). How does stress increase risk of drug abuse? *Psychopharmacology*, 158, 343–359.

Skinner, K., Sullivan, L. M., Tripp, T. J., Kressin, N. R., Miller, D. R., Kazis, L. et al. (1999). Comparing the health status of male and female veterans who use VA health care: Results from the VA Women's Health Project. *Women and Health*, 29, 17–33.

Southwick, S. M., Morgan, C. A., Darnell, A., Bremner, D., Nicolaou, A. L., Nagy, L. M. et al. (1995). Trauma-related symptoms in veterans of Operation Desert Storm: A 2-year follow-up. *American Journal of Psychiatry*, 152, 1150–1155.

Spiro, A., Schnurr, P. P., & Aldwin, C. M. (1999). A life-span perspective on the effects of military service. *Journal of Geriatric Psychiatry*, 30, 92–128.

Stein, A. L., Tran, G. Q., Lund, L. M., Haji, U., Dashevsky, B. A., & Baker, D. G. (2005). Correlates for posttraumatic stress disorder in Gulf War veterans: A retrospective study of main and moderating effects. *Journal of Anxiety Disorders*, 19, 861–876.

Stein, M. B., Lang, A. J., Laffaye, C., Satz, L. E., Lenox, R. J., & Dresselhaus, T. R.. (2004). Relationship of sexual assault history to somatic symptoms and health anxiety in women. *General Hospital Psychiatry, 26,* 178–183.

Street, A. E., Gradus, J. L., Stafford, J., & Kelly, K. (2007). Gender differences in experiences of sexual harassment: Data from a male-dominated environment. *Journal of Consulting and Clinical Psychology, 75,* 464–474.

Suris, A., Lind, L., Kashner, T. M., Borman, P. D., & Petty, F. (2004). Sexual assault in women veterans: An examination of PTSD risk, health care utilization, and cost of care. *Psychosomatic Medicine, 66,* 749–756.

Sutker, P. B., Davis, J. M., Uddo, M., & Ditta, S. R. (1995). War zone stress, personal resources, and PTSD in Persian Gulf War returnees. *Journal of Abnormal Psychology, 104,* 444–452.

Suvak, M.K., Vogt, D.S., Savarese, V.W., King, L.A., & King, D.W. (2002). Relationship of warzone coping strategies to long-term general life adjustment among Vietnam veterans: Combat exposure as a moderator variable. *Personality and Social Psychology Bulletin, 28,* 974–985.

Taft, C. T., Stern, A. L., King, L. A., & King, D. W. (1999). Modeling physical health and functional status: The role of combat exposure, posttraumatic stress disorder, and personal resource attributes. *Journal of Traumatic Stress, 12,* 3–25.

Tolin, D. F., & Foa, E. B. (2006). Sex differences in trauma and posttraumatic stress disorder: A quantitative review of 25 years of research. *Psychological Bulletin, 132,* 959–992.

Tucker, J. S., Sinclair, R. R., & Thomas, J. L. (2005). The multilevel effects of occupational stressors on soldiers' well-being, organizational attachment, and readiness. *Journal of Occupational Health Psychology, 10,* 276–299.

Turner, M. A., Kiernan, M. D., McKechanie, A. G., Finch, P. J., McManus, F. B., & Neal, L. A. (2005). Acute military psychiatric casualties from the war in Iraq. *British Journal of Psychiatry, 186,* 476–479.

U.S. Merit Systems Protection Board (USMSPB). (1995). *Sexual harassment in the federal workforce: Trends, progress, continuing challenges.* Washington, D.C.: U.S. Government Printing Office.

Vinokur, A. D., Pierce, P. F., & Buck, C. L. (1999). Work-family conflicts of women in the Air Force: Their influence on mental health and functioning. *Journal of Organizational Behaviour, 20,* 865–878.

Vogt, D. S., & Tanner, L. R. (2007). Risk and resilience factors for posttraumatic stress symptomatology in Gulf War I veterans. *Journal of Traumatic Stress, 20,* 27–38.

Vogt, D. S., Pless, A. P., King, L. A., & King, D. W. (2005). Deployment stressors, gender, and mental health outcomes among Gulf War I veterans. *Journal of Traumatic Stress, 18,* 272–284.

Vogt, D. S., Bruce, T. A., Street, A. E., & Stafford, J. (2007). Attitudes toward women and tolerance for sexual harassment among reservists. *Violence Against Women, 13,* 879–900.

Vogt, D.S., Samper, R. E., King, D. W., King, L. A., & Martin, J. (2008). Deployment stressors and posttraumatic stress symptomatology: Comparing active duty, National Guard, and reserve personnel from Gulf War I. *Journal of Traumatic Stress, 21(1),* 66–74.

Waigandt, A., Wallace, D., & Phelps, L. (1990). The impact of sexual assault on physical health status. *Journal of Traumatic Stress, 3,* 93–102.

Waysman, M., Schwarzwald, J., & Solomon, Z. (2001). Hardiness: An examination of its relationship with positive and negative long term changes following trauma. *Journal of Traumatic Stress, 14,* 531–548.

Westman, M. & Etzion, D. (2005). The crossover of work-family conflict from one spouse to the other. *Journal of Applied Social Psychology, 35*, 1936–1957.

Williams, J. H., Fitzgerald, L. F., & Drasgow, F. (1999). The effects of organizational practices on sexual harassment and individual outcomes in the military. *Military Psychology, 11*, 303–328.

Williams, R. A., Hagerty, B. M., Yousha, S. M., Hoyle, K. S., & Oe, H. (2002). Factors associated with depression in navy recruits. *Journal of Clinical Psychology, 58*, 323–337.

Wolfe, J., Brown, P. J., & Kelley, J. M. (1993). Reassessing war stress: Exposure and the Persian Gulf War. *Journal of Social Issues, 49*, 15–31.

Wolfe, J., Sharkansky, E., Read, J., Dawson, R., Martin, J., & Ouimette, P. (1998). Sexual harassment and assault as predictors of PTSD symptomatology among U.S. female Persian Gulf War military personnel. *Journal of Interpersonal Violence, 13*, 40–57.

Yerkes, R. M., & Dodson, J. D. (1908). The relation of strength of stimulus to rapidity of habit formation. *Journal of Comparative Neurology and Psychology, 18*, 459–482.

Zakin, G., Solomon, Z., & Neria, Y. (2003). Hardiness, attachment style, and long term psychological distress among Israeli POWs and combat veterans. *Personality and Individual Differences, 34*, 819–829.

Zinzow, H. M., Grubaugh, A. L., Monnier, J. Suffoletta-Maierie, S., & Frueh, B. C. (2007). Trauma among female veterans: A critical review. *Trauma, Violence, and Abuse, 8*, 384–400.

7

Vulnerability Factors: Raising and Lowering the Threshold for Response

ARTHUR FREEMAN and SHARON MORGILLO FREEMAN

Contents

Introduction

In Chinese philosophy, the concept of *yin/yang* describes the mutual relationships from a human perspective of a broad range of phenomena in the natural world. These phenomena combine to create a unity of opposites, even when this unity occurs in constructs that may appear antithetical one to the other. The concept of *yin* and *yang* (or earth and heaven) describes two opposing and, at the same time, complementary (completing) aspects of any one phenomenon (object or process) or can be used to compare any two phenomena. There are four laws of *yin/yang*:

- *Yin/yang* are opposing. *Yin/yang* describe the polar effects or impacts of phenomena. When viewing any one phenomenon (or comparing two phenomena), *yin/yang* describe the opposing qualities inherent in it; for example, winter and summer would be the *yin/yang*, respectively, of the year.
- *Yin/yang* are mutually rooted. *Yin/yang* are not adversarial, but are complementary qualities in a gestalt. That is to say, the *yin* aspect and the *yang* aspect of any one phenomenon will, when put together,

form the entire phenomenon, with the combination being more than each component part. *Yin/yang* is a philosophy of duality. This is the reason why the Chinese term has no "and" between *yin/yang*—the term always expresses the two parts making up the one. In the example above, winter plus summer makes up the whole year.

- *Yin/yang* mutually transform each other—The maximum effect of one quality will be followed by the transition toward the opposing quality. In other words, once the maximum *yang* aspect has manifest, such as the long days of summer, this will be followed by the transition toward the yin aspect, with the shortening of the days as winter approaches.
- *Yin/yang* mutually wax and wane—The *yin/yang* aspects are in dynamic equilibrium. As one aspect declines, the other increases to an equal degree. For example, in the cycle of the year, the long days of summer gradually shorten and the nights gradually lengthen as winter approaches. Throughout the process, however, the length of each day is a constant 24 hours (the equilibrium), and it is only the relative length of light and darkness that changes (the dynamic).

Yin/yang are neither substances nor forces. They are the terms used in a system of dualistic qualification that can be applied universally. Further dividing *yin/yang* into their respective *yin* and *yang* aspects yields four combinations: the *yin* of the *yang*, the *yang* of the *yang*, the *yin* of the *yin*, and the *yang* of the *yin*. This allows an endless scale of universally defined qualities that is foundational to classical Chinese thought. It is represented by this symbol:

Two analogous factors to *yin/yang* are vulnerability and resilience. They are complementary when an increase in one would signal a decrease in the other; for example, as an individual becomes more vulnerable, that person's ability to be resilient would decrease. Masten and Powell (2003) define resilience as "positive adaptation in the context of significant risk or adversity" (p. 4). Masten and Coatsworth (1998) suggested that: "Resilience is an inference about a person's life that requires two fundamental judgments: (1) that a person is 'doing okay' and (2) that there is now or has been significant risk or adversity to overcome." In his discussion of risk research, Garmezy (1996, p. 9) stated that risk factors "may eventuate in disease or disorder (which identifies *vulnerability*), but others, in many instances, may be overcome and lead to positive adaptive behavior (which identifies *resilience*)."

Although the terms *adversity* and *risk* are generally understood, what makes an experience a risk for one individual does not make that same experience or situation adverse or risky for another person. One person's ability to successfully and adaptively navigate through a situation may be thought of as resulting from their personal construction of the experience. The experience is filtered through the lens of the individual's past experience, previous training, and perception of the type and degree of danger to oneself or to one's domain. For example, an individual who perceives being outdoors as potentially dangerous might be compelled to cope by staying at home, which in its extreme state might result in agoraphobia. Being outdoors is, in fact, dangerous according to that individual's construction of the myriad factors that might serve to inflict injury. If, however, one perceives being outdoors as liberating, joyful, and wonderful, then being outdoors becomes, for that individual, a very positive experience. The sequelae of being in a situation of prolonged and significant danger may create for some individuals an acute stress reaction, which, if continued over time, might become a posttraumatic stress reaction. One might infer that the development of the posttraumatic stress reaction would be a result of vulnerability. It would also signal that the person's resilience, coping ability, or adaptability has been limited or curtailed.

Factors that have been identified as placing an individual at risk for posttraumatic stress disorder (PTSD) are both numerous and complex following a traumatic event or events. These factors might be referred to as "psychological vulnerability factors for PTSD," or "PTSD vulnerability." General vulnerability factors that have been hypothesized to be operative in PTSD include limited social support or coping skills, a family history of psychological problems, previous exposure to other traumatic events, and preexisting psychological problems (Agaibi & Wilson, 2005). Individuals who have PTSD vulnerability, however, may never encounter a perceived traumatic event and therefore may never develop PTSD.

Several physiological factors must be considered. Chronic stress, for example, has been shown to cause dendritic atrophy, cell death, and inhibition of neurogenesis in the hippocampus (McEwen, 1999) and prefrontal cortex (Radley et al., 2004). The importance of these findings in PTSD (Liberzon & Phan, 2003; Rauch et al., 2003) have led some researchers to propose that traumatic events or the resulting chronic stress cause physiologic damage to the underlying brain structures. Ongoing combat actions in areas such as Iraq and Afghanistan have focused the attention of researchers and the general public on an impending crisis of significant mental health problems among military members returning from deployments, sometimes multiple times. Recent figures estimate that 17% or more of returning service members develop PTSD, depression, generalized anxiety, and other mental health problems after deployment which is more than double the estimates for predeployment service members (Hoge et al., 2004). It does not account for problems in families, between relationship partners, and among the children of those postdeployment individuals.

Brief History of PTSD in Combat Situations

The first known documentation of PTSD dates back to the early 1800s when military physicians identified soldiers with "exhaustion" following the stress of battle. They described this "exhaustion" as mental shutdown following an individual or group trauma (Agaibi & Wilson, 2005). Given that soldiers were expected not to show any fear in the heat of battle, the only treatment for this "exhaustion" was to move the afflicted soldiers to the rear for awhile until they were deemed cured and then send them back into battle. As a result of extreme and often repeated stress, these soldiers became fatigued as a part of their body's natural adaptation to repeated shock and stress (Agaibi & Wilson, 2005), thus the term "battle fatigue."

A similar syndrome was documented in 1876 when Mendez DaCosta published a paper diagnosing Civil War combat veterans with what he coined as "soldier's heart." The symptoms included startle responses, hypervigilance, and heart arrhythmias. Another descriptor, "shell shock," emerged during World War I and was followed in World War II by the term "combat fatigue." The official designation of "posttraumatic stress disorder" did not come about until 1980, when the third edition of the *Diagnostic and Statistical Manual of Mental Disorders* (DSM) was published. The first definition of PTSD included a description of a psychological condition experienced by a person who had faced a traumatic event that caused a catastrophic stressor outside the range of usual human experience (an event such as war, torture, rape, or natural disaster). This definition separated PTSD stressors from "ordinary" stressors that were characterized in DSM-III as "adjustment disorders," such as divorce, failure, rejection, and financial problems.

Threshold for a Stress Response

Individuals have threshold points above which they can control their responses to external or internal stressors. The higher the threshold, the less likely it is that an individual will respond to a particular stimulus. One may think of the threshold as a trigger point. A gun is, in and of itself, an innocuous mechanism, except when a sufficient amount of pressure is put on the trigger. When sufficient pressure is exerted on the trigger to fire the weapon, it now becomes a potentially lethal mechanism. The metaphor follows that when the stress on the organism is sufficient to trigger a response, the organism will react. When considering the trigger of a gun, we can think of a continuum from 1 to 100. The trigger point for an individual's particular stressor may sit anywhere on that scale; for example, the trigger point for an emergency-room medical worker should be quite high on the continuum or that worker would burn out after only a few days or weeks on the job. In effect, an individual with a low threshold would frequently be triggered. Ideally, the trigger point for individuals in battle would be relatively high, helped by the training they have received. A typical part of basic military training involves soldiers being exposed to a gas attack that requires that they quickly and effectively don gas masks. Without such training, the impact of a gas attack could create panic. With training, even though the exercise creates discomfort, soldiers learn adaptive responses to the threat.

Suppose an individual's trigger point is 75 (on a scale of 1 to 100). If the usual stress of military experience is 60, they are in good control inasmuch as they have a cushion of 15 units. If a confluence of factors serves to either raise the stressor (increased danger) or lower the threshold trigger point, that individual will experience a stress reaction. If the increased stress increases to 70, the individual is still in control, but with a smaller buffer. If the individual's trigger point or threshold falls to 50, even for a brief moment, the buffer is gone and the individual will again experience a stress reaction. A vulnerability factor may lower the threshold only slightly and thereby not affect the trigger point, but if several vulnerability factors are in place, they will have a cumulative effect. A series or combination of vulnerability factors can make the individual more reactive and open to stress, which may have the effect of giving that person only tenuous control. If another vulnerability factor comes into play that lowers that individual's threshold even further, the individual could be triggered and respond.

When several vulnerability factors are operative, the resulting combination will lower the individual's ability to cope and any further lowering could become the proverbial straw that breaks the camel's back. These specific *vulnerability factors* are environmental circumstances, situations, or deficits that have the effect of decreasing the individual's ability to effectively cope with any of a number of life stressors, to lose options for more effective coping, or fail to see available options. A vulnerability factor has the effect of decreasing the patient's threshold so events that were previously ignored or never noticed now shout loudly for attention. Vulnerability factors also serve to make the individual more sensitive to internal and external stimuli that may affect their adaptive actions. Vulnerability factors include

1. Acute illness or injury
2. Chronic illness
3. Deterioration of health
4. Hunger
5. Anger
6. Fatigue
7. Loneliness
8. Major life loss
9. Poor problem solving ability
10. Substance abuse
11. Chronic pain
12. Poor impulse control
13. New life circumstance

The threshold may be variable, depending on circumstances. The same stimulus may at one point elicit a response and at another time be subthreshold. In fact, different activities may have very different thresholds so it is essential to be aware of one's vulnerability factors.

Resilience

Two terms are relevant to a discussion of resilience. The first is *resilience* and the second is *recovery*. Resilience is the ability not to respond to internal or external stimuli. Recovery is when the individual has become vulnerable but later returns to his or her previous baseline adaptive performance. We can think of the spinal column as an organ of resilience. In addition to its many functions, the spinal column offers the individual support by giving structure the body. Made up of hard bone, the vertebrae, the spinal column has soft flexible disks placed between the hard bone to provide a cushioning effect. The soft disks allow the individual to absorb shock and to bend forward, backward, and to each side, and they prevent the bones from rubbing together. When the disks are injured or missing, the individual experiences pain and a loss of flexibility or resilience.

Expected Stressors During Wartime

According to the *Iraq War Clinician's Guide* (National Center for Posttraumatic Stress Disorder, 2004), the serviceman is expected to be prepared for, psychologically deemed fit for, and qualified to serve in an environment in which

> The destructive force of war creates an atmosphere of chaos and compels service members to face the terror of unexpected injury, loss, and death. The combat environment (austere living conditions, heavy physical demands, sleep deprivation, periods of intense violence followed by unpredictable periods of relative inactivity, separation from loved ones, etc.) is itself a psychological stressor that may precipitate a wide range of emotional distress and/or psychiatric disorders. Psychological injury may occur as a consequence of physical injury, disruption of the environment, fear, rage, or helplessness produced by combat, or a combination of these factors. (p. 11)

The range of emotional responses in the context of the multiphasic traumatic stress response include (Benedek, Holloway, & Becker, 2002; Ursano, McCaughey, & Fullerton, 1994)

- *Immediate phase* characterized by strong emotions, disbelief, numbness, fear, and confusion accompanied by symptoms of autonomic arousal and anxiety
- *Delayed phase* characterized by persistence of autonomic arousal, intrusive recollections, somatic symptoms, and combinations of anger, mourning, apathy, and social withdrawal

- *Chronic phase* characterized by continued intrusive symptoms and arousal for some, disappointment or resentment or sadness for others, and for the majority a refocus on new challenges and the rebuilding of lives

Preexisting Mental Health Issues as Vulnerability Factors

The Accession Medical Standards Analysis & Research Activity (AMSARA) was established in 1996 with a mission to support the development of evidence-based accession standards by guiding the improvement of medical and administrative databases, conducting epidemiologic analyses, and integrating relevant operational, clinical, and economic considerations into policy recommendations. According to the 2002 AMSARA report, more than 6% of active-duty service members sought treatment for mental health disorders in 1998 and 1999. Furthermore, psychiatric disorders were cited as the leading cause of hospitalizations among enlisted personnel within 1 year of accession from 1995 to 1998 (AMSARA, 2000, p. 23). It was estimated that approximately 4% of all new accessions were hospitalized within their first 6 months on active duty, and of these 26% were for mental health disorders. In a follow-up study of 303,433 accessions from 1999 through 2000, of the 6527 (2.2%) who were hospitalized, approximately 40.1% of the hospitalizations were for psychiatric conditions (AMSARA, 2002).

The 2000 AMSARA annual report describes 5-year, retrospective cohort studies of the relationship between waivers for mental health and attrition during the first 2 years of service. One study compared 502 first-time enlistees across the services who were granted waivers (permission to enter military service despite having a history of a condition usually preventing one from service) for depression with a matched group of 1501 recruits who were qualified on all physical, medical, and mental criteria. The results of that study revealed that the recruits with depression waivers had a significantly lower likelihood of remaining in military service compared to the nonwaivered group; the probability of retention was .62 vs. .72 at the 2-year mark (AMSARA, 2000). It should be noted that it is up to the individual recruit to reveal the presence or absence of a history of mental health issues. During processing at a military entrance processing station, recruits complete a medical history questionnaire; however, although the recruits undergo thorough medical evaluations, they are not subject to a formal psychiatric assessment.

The experience of prior stressful life events among servicemen yet to be deployed must also be taken into consideration. Given that previous exposure to stressful life events and even a history of PTSD can increase the risk for psychological problems after deployment, it is important to evaluate the likelihood that servicemen may have already been exposed before deployment. The incidence of PTSD in the general population is approximately 1 to 4%; therefore, it should be expected that there would be a similar percentage among those recruits entering into military service. Researchers have learned that prior assault contributes to increased vulnerability and decreased resilience against PSTD symptoms by almost twofold among military professionals deployed to recent combat operations (Smith et al., 2008). This information is critical given the psychological symptoms inherent in the disorder and the greater likelihood of these symptoms worsening when such soldiers find themselves in the combat theater.

Predeployment Psychological Screening

In 2002, British military service personnel were asked to complete psychological screening questionnaires in a study named Operation Telic for the purposes of assessing premorbid mental health prior to deployment to Iraq and Afghanistan. Over 2800 personnel completed the assessments as well as a second questionnaire between June 2004 and March 2006 after redeployment (Rona et al., 2006). The full psychological screening battery included at baseline (in 2002) the civilian version of the posttraumatic stress disorder checklist (PCL-C) (Blanchard, 1996), the General Health Questionnaire 12 (GHQ-12) (Goldberg & Williams, 1988), assessment of 15 physical symptoms, a self-assessment of health status from the SF-36® quality of life survey (Ware, Snow, Kosinski, & Gandek, 1993), and three questions from the World Health Organization's Alcohol Use Disorders Identification Test Questionnaire (AUDIT) (Barbor, Higgins, Biddle, Saunders, & Monteiro, 2001).

The researchers concluded that there is little empirical support for the use of mental health screening before deployment to identify vulnerability to mental disorders after deployment (Rona et al., 2006). In other words, for every psychological assessment conducted, either the positive predictive value or the negative predictive value was too low to be useful as an indicator of postdeployment mental health outcomes. The positive likelihood ratio for posttraumatic stress disorder was higher than for any other psychological assessment; however, the disorder was uncommon (<3.2%). Even with a relatively high positive likelihood ratio, the positive predictive value was low (Rona et al., 2006). The researchers concluded that: (1) psychological screening before deployment has a low predictability for most common mental health conditions, (2) the predictability of screening for posttraumatic stress disorder is higher than for any other mental health problem, and (3) because the prevalence of posttraumatic stress disorder is low before deployment, screening for the condition would be inappropriate, despite a moderately high predictability, due to the cost and limited sensitivity and specificity of current available measures (Rona et al., 2006). A similar neuropsychological study conducted in the United States indicated that deployment effects on sustained attention, learning, and memory suggest that negative neuropsychological outcomes following Iraq deployment could not be attributed to preexisting dysfunction (Vasterling et al., 2006). The findings further indicated that it was unlikely that intervening variables influenced the service members' performances significantly due to the relatively abbreviated interval between redeployment and assessment (Vasterling et al., 2006).

From an evolutionary perspective, the findings in this study seem to contraindicate the memory and attentional outcome information one would expect; however, these results are consistent with neurobiological responses directed toward survival (Habib, Gold, & Chrousos, 2001). In other words, when an individual is confronted with a life-threatening situation, basic physiological responses take over in preparation for life-preserving action (Vasterling et al., 2006). This "flight, fight, or freeze" response is activated through the autonomic neurotransmitter systems associated with increased arousal. Specifically, the noradrenergic system becomes activated, while neuroendocrine responses, via the hypothalamic–pituitary–adrenal axis, produce heightened behavioral reactivity. During this heightened reactive time, the body is in survival mode, not in learning mode, so attention, learning, and memory functions are dampened down as they are not relevant to dealing with the current event (Habib et al., 2001). The vast majority of service members are exposed to life-threatening traumatic events when in the combat theater along with prolonged exposure to a hypervigilant, high-arousal environment that would be categorized as imminently life threatening. It would stand to reason, therefore, that these individuals would exhibit signs or symptoms consistent with prolonged hypervigilant survival mode states (Vasterling et al., 2006).

In-Theater Vulnerability Factors

In keeping with the concept of social support as a factor for the development of PTSD, some researchers have evaluated unit cohesiveness, including trust within a unit, as a potential contributor to the development of PTSD. The level of connection, trust, and support that service members had with other members of their unit (i.e., unit cohesiveness) affected the likelihood of developing PTSD. Individuals who reported stronger unit cohesion had less severe PTSD symptoms (Brailey et al., 2007), and these findings suggest that unit cohesion may decrease vulnerability for psychological problems among those at high risk.

Postdeployment Vulnerability Factors Related to Development of PTSD

The mental health effects of exposure to combat duty in Iraq or Afghanistan for U.S. combat infantry units (three Army units and one Marine Corps unit) were investigated by Hoge et al. (2004). Among these groups, exposure to combat was significantly greater among those deployed to Iraq than those deployed to Afghanistan. After duty in Iraq, 15.6 to 17% of the military personnel met screening criteria for major depression, generalized anxiety, or posttraumatic stress disorder compared with 9.3% before deployment to Iraq. After duty in Afghanistan, 11.2% met criteria for these disorders. Posttraumatic stress disorder accounted for the largest difference in the pre- and postdeployment rates. Importantly, only 23 to 40% of these military personnel sought mental health care. Concern about stigmatization as well as other barriers (e.g., difficulty scheduling an appointment, difficulty getting time off from work, transportation problems) were cited by military personal as reasons for not seeking mental health care.

In addition, the barrier related to the stigma of mental illness, including the label of PTSD, created a barrier itself with regard to the assessment of vulnerability to development of PTSD. Researchers and clinicians have noted a dose–response relationship between the severity of the trauma and the subsequent development of PTSD. As individuals are exposed to greater levels of horror and catastrophe along the trauma spectrum, this dose–response factor may become particularly important in the evaluation of vulnerability factors prior to exposure to a trauma-prone environment such as the combat theater. The criteria for PTSD include threat to life; however, because of the qualitative and temporal differences between a motor vehicle accident and purposeful torture and threat of death, physical and psychological injury vulnerability factors may be more significant for the development of PTSD in response to torture (Yehuda, 1998).

Positive and Negative Trajectories

An individual's reaction to stressors must be viewed as part of a process, not simply as a diagnosed outcome: "A process view investigates trajectories of functioning over time from pre-trauma levels to the present" (Moreland, Butler, & Leskin, 2008, p. 40). A positive trajectory would be indicated by the ways in which individuals maintain, return to, recover from, or even exceed their previous levels of functioning. Lepore and Revenson (2006) identified three reactions, *recovery, resistance,* and *reconfiguration,* that are components of the positive trajectory. In recovery, the flexibility and plasticity of the emotional, cognitive, and behavioral "spinal column" allow individuals to return to their baseline behavior. In the throes of an attack or other traumatic experience, these individuals may experience a diminution of their pretraumatic function, and there may be a temporary loss of function that then returns. Resistance indicates that the traumatic experience was below an individual's threshold and therefore did not elicit a response. The experience was quickly integrated and did not become a focus of attention for an extended period of time. Finally, reconfiguration conveys the notion of a permanent change and adaptation following the traumatic event. As a result of reconfiguration of the traumatic event, the individual experiences what has been termed *posttraumatic growth* (Linley & Joseph, 2004). Positive trajectory individuals who recover from the traumatic stress are less likely to be seen by clinicians inasmuch as they appear to have overcome and mastered the traumatic experience.

A negative trajectory would be marked by disturbance, impairment, decline, and disability. If individuals remain symptomatic and maladaptive in their behavior, then the behavior may be described as *survival with impairment* (Moreland, Butler, & Leskin, 2008). Such individuals are generally referred for therapeutic intervention. They can continue to function, albeit at a markedly diminished level that shows significant impairment. When the negative trajectory is more severe, traumatized individuals may not survive their traumatic experiences and may self-harm through the use of drugs or impulsive and reckless behavior. They may be psychologically impaired by severe and unremitting psychological disorders (e.g., depression, anxiety, phobia), or they may attempt to escape the emotional pain by committing suicide. This is labeled as *succumbing* (Moreland et al., 2008).

Treatment of Vulnerability

After assessing the patient and identifying the problem, the clinician can then develop a treatment protocol. Clinicians must attempt to reveal the vulnerability factors that have reduced their patients' thresholds and negatively impacted their resilience and coping:

1. *Limited social support*—The use of support groups, chat rooms, and couples and family treatment modalities can accomplish several tasks: (1) Service members can experience the comradeship that they had while in service; (2) they can be helped to see that they do not stand alone but are part of a community and family system or team; (3) such an approach mitigates loneliness; and (4) social support can serve as the crutch or brace that is necessary when a limb has been damaged and is repairing itself.
2. *Limited coping skills*—A planned and structured problem-solving plan can be instituted to help individuals to generalize what they have learned and to enhance basic social and relational skills.
3. *Family history of psychological problems*—Individual therapy can help individuals to overcome or distance themselves from any family pathology. This is an exceedingly difficult task inasmuch as the basic rules and schema that were acquired in childhood now may stand at the forefront of the individual's functioning. Individual or group therapy would be recommended.
4. *Previous exposure to other traumatic events*—Rather than inoculating individuals against any future reoccurrence of the stress or stressors, their earlier exposure to traumatic events has reduced their threshold and made them more vulnerable. In this case, threshold-raising strategies are required.
5. *Preexisting psychological problems*—Here, again, both individual and group therapies can work to reduce the impact of the preexisting

problems. The nature, breadth, impact, pervasiveness, and chronicity of the preexisting conditions will all play a role in the severity of the individual's postdeployment functioning.

6. *Acute illness or injury*—In cases of acute medical illness or injury, whether service-related or not, immediate medical intervention is recommended. When necessary, rehabilitative procedures might be indicated. Wounds, exposure to toxins, or injury incurred as part of the military experience would fall into this category. When the acute illness is reduced, the individual's threshold will most likely rebound to its baseline.

7. *Chronic illness*—In situations where the health problem is chronic, individuals may experience acute exacerbation of their reactivity on an ongoing (and chronic) basis. Long-term health problems such as diabetes, fibromyalgia, lupus, or asthma may cause a reduction in the threshold and lower the trigger point for reacting to stress or other stimuli.

8. *Deterioration of health*—Chronic or acute illness, injury, or the normal aging process can result in a loss of health and adaptability, and individuals may find that they are unable to perform up to previous expectations. They may not be able to run as fast or lift as much weight, or they may experience decreased stamina.

9. *Hunger*—During times of food deprivation, individuals are more vulnerable to a variety of stimuli, particularly food-related stimuli; for example, it is not a good idea for individuals who are hungry to shop for food because they are likely to buy too much. The tendency to respond to stimuli in the grocery may overwhelm the individual's good intentions. Also, food-deprived individuals may be more responsive to verbal, physical, physiological, behavioral, or environmental stimuli. The easiest treatment is regular meals or carrying nutritive products to avoid the hunger-related decrease in threshold.

10. *Anger*—When individuals are angry, they appear to lose appropriate problem-solving abilities. They may lose their impulse control or respond to stimuli that are more usually ignored.

11. *Fatigue*—In a similar fashion, fatigue decreases both problem-solving strategies and impulse control. Often, individuals' sleep cycles are affected, or they try to eliminate sleep altogether to avoid their frightening dreams. Substance abuse disrupts the sleep cycles of some; for others, a lack of structure and focus in their lives leads to broken sleep, early rising, or sleep-onset problems. Helping to structure the sleep patterns of these individuals can allow them to become more rested and increase their resilience.

12. *Loneliness*—When individuals see themselves as isolated and apart from others, the lack of a support system will serve to lower their

thresholds. Being isolated becomes especially important for postdeployment individuals who have spent months or years living, working, fighting, and surviving shoulder to shoulder with their battle buddies. Individuals can feel isolated even when in the midst of a crowd or their family. Connections made through veterans groups would be important to establish and sustain.

13. *Major life loss*—The loss of a significant other through death, divorce, or separation can lower an individual's threshold. These significant others would include the individual's relatives as well as the soldier's military family. When this occurs, individuals may see themselves as having reduced options or may give up caring for what happens to them.

14. *Poor problem-solving ability*—Not all individuals have the same level of problem-solving ability. Some individuals have an impaired problem-solving ability, a deficit that may not become obvious until the individual is placed in situations of great stress. Having to deal only with minor problems in the past may have never truly tested an individual's problem-solving ability. Problem-solving training would be a recommended and reasonable intervention.

15. *Substance abuse*—Substance abuse can increase vulnerability. In acute cases the patient's judgment is compromised primarily during periods of intoxication, whereas in chronic cases the patient's judgment is more generally impaired. Substance abuse treatment programs, group interventions, and twelve-step programs are recommended for these individuals (see Chapter 14).

16. *Chronic pain*—Chronic pain may have a similar effect as chronic illness when individuals devote all of their time and energy to trying to eliminate the pain. Pain management strategies, pharmacotherapy, exercise, physical therapy, or occupational therapy will all contribute to the individual's recovery.

17. *Poor impulse control*—Many individuals have poor impulse control because of organic or functional problems. Patients with bipolar illness or borderline, antisocial, or histrionic personality disorders may all have impulse control deficits. Control problems can also be due to injury to the executive portions of the brain. Self-control strategies, pharmacotherapy, skill-building, and anger management will all be of value.

18. *New life circumstance*—Changing jobs, marital status, homes, assignment, battle experience, or family status are all stressors that are vulnerability factors. Coming back from deployment (whether or not the deployment involved combat) creates a new circumstance. Reintegrating into a family, cohort of friends, a job, a church, or any social or family group will all be stressful.

The threshold may be variable, depending on circumstances. The same stimulus may at one point elicit a response and at another time be subthreshold. In fact, different activities may have very different thresholds so it is essential to be aware of an individual's vulnerability factors.

Summary

Civilian practitioners treating military service members who have served in a highly traumatic environment must remember that these men and women are highly trained and exceptionally courageous individuals who have a great deal of pride in their service to their country. The development of PTSD or any other mental health disorder is often met with feelings of frustration, shame, and guilt that somehow they have not measured up to the same standards that they have been trained to meet. The development of PTSD as a result of exposure to a life-threatening situation is not a guarantee but is a likelihood; therefore, a thorough assessment must be undertaken. In addition, the development of PTSD as described in this chapter does not necessarily point to an underlying preexisting psychological weakness, flaw, or inability to cope. Often these disorders are multifactorial and combine a dose–response effect with neurological factors, psychological factors, and such social factors as unit trust and cohesiveness and a multitude of others that remain undiscovered. Service members must be evaluated and treated as individuals to better understand these factors. Individuals who have a positive posttraumatic experience may experience recovery (a return to baseline behavior), resistance (a limited response to the traumatic event), or reconfiguration (growth and adaptation that take place as a consequence of the traumatic event). A negative trajectory would include survival with impairment (the individual continues though life but with a significant loss of functioning) or succumbing (the individual seeks self-harming or self-destructive alternatives).

The vulnerability factors of service members should be identified and addressed so they will not be as likely to react or respond to particular stimuli. Further, by assessing and understanding the vulnerability factors, the clinician can develop a treatment focus to lower stress (both internal and external) and raise the threshold, as needed.

References

Agaibi, C. E., & Wilson, J. P. (2005). Trauma, PTSD, and resilience: A review of the literature. *Trauma, Violence, and Abuse*, 6, 195–216.

AMSARA. (2000). *Accession medical standards analysis & research activity annual report.* Washington, D.C.: U.S. Department of Defense (http://www.amsara.amedd.army.mil/reports/2000/Waivers.htm).

AMSARA. (2002). *Accession medical standards analysis & research activity annual report.* Washington, D.C.: U.S. Department of Defense (http://www.amsara.amedd.army.mil/reports/2002/2002 report.asp?strReportChoice=7).

Barbor, T. F., Higgins-Biddle, J. C., Saunders, J. B., & Monteiro, M. G. (2001). *The alcohol use disorders test: Guidelines for use in primary care* (2nd ed.). Geneva, Switzerland: World Health Organization.

Benedek, D. M., Holloway, H. C., & Becker, S. M. (2002). Emergency mental health management in bioterrorism events. *Emergency Medicine Clinics of North America*, 20(2), 393–407.

Blanchard, E. B., Jones-Alexander, J., Buckley, T. C., & Forneris, C. A. (1996). Psychometric properties of the PTSD checklist (PCL). *Behavioral Research and Therapy*, 34, 669–673.

Brailey, K., Vasterling, J. J., Proctor, S. P. et al. (2007). PTSD symptoms, life events, and unit cohesion in U.S. soldiers: Baseline findings from the Neurocognition Deployment Health Study. *Journal of Traumatic Stress*, 20, 495–503.

Bremner, J. D. (1999). Does stress damage the brain? *Biological Psychiatry*, 45(7), 797–805.

Garmezy, N. (1996). Reflections and commentary on risk, resilience, and development. In R. Haggerty, L. R. Sherrod, N. Garmezy, and M. Rutter (Eds.), *Stress, risk, and resilience in children and adolescents: Processes, mechanisms, and interventions* (pp. 1–18). Cambridge, U.K.: Cambridge University Press.

Goldberg, D., & Williams, P. (1988). *A user's guide to the general health questionnaire*. Windsor, U.K.: NFER-Nelson.

Habib, K. E., Gold, P. W., & Chrousos, G. P. (2001). Neuroendocrinology of stress. *Endocrinology Metabolism Clinics of North America*, 30, vi–viii, 695–728.

Hoge, C. W., Castro, C. A., Messer, S. C., McGurk, D., Cotting, D. I., & Koffman, R. L. (2004). Combat duty in Iraq and Afghanistan, mental health problems, and barriers to care. *New England Journal of Medicine*, 351, 13–22.

Lepore, S., & Revenson, T. (2006). Relationships between posttraumatic growth and resilience: recovery, resilience, and reconfiguration. In L. G. Calhoun and R. G. Tedeschi (Eds.), *Handbook of posttraumatic growth* (pp. 24–46). Mahwah, NJ: Erlbaum.

Liberzon, I., & Phan, K. L. (2003). Brain-imaging studies of posttraumatic stress disorder. *CNS Spectrums*, 8, 641–650.

Linley, P. A., & Joseph, S. (2004). Positive change following trauma and adversity: A review. *Journal of Traumatic Stress*, 17, 11–21.

Makris, N. (2003). Selectively reduced regional cortical volumes in post-traumatic stress disorder. *NeuroReport*, 14, 913–916.

Masten, A. S., & Coatsworth, J. D. (1998). The development of competence in favorable and unfavorable environments: Lessons from successful children. *American Psychologist*, 53, 205–220.

Masten, A. S., & Powell, J. L. (2003). A resilience framework for research, policy, and practice. In S. S. Luthar (Ed.), *Resilience and vulnerability: Adaptation in the context of childhood adversities* (pp. 1–25). New York: Cambridge University Press.

McEwen, B. S. (1999). Stress and hippocampal plasticity. *Annual Review of Neuroscience*, 22, 105–122.

Moreland, L. A., Butler, L. D., & Leskin, G. A. (2008). Resilience and thriving in a time of terrorism. In S. Joseph and P. A. Linley (Eds.), *Trauma, recovery, and growth* (pp. 39–61). Hoboken, NJ: John Wiley & Sons.

National Center for Posttraumatic Stress Disorder. (2004). *Iraq War clinician's guide* (2nd ed.). Washington, D.C.: U.S. Department of Veterans Affairs.

Radley J. J., Sisti, H. M., Hao, J., Rocher, A. B., McCal, T., Hof, P. R., McEwe, B. S., & Morrison, J. H. (2004). Chronic behavioral stress induces apical dendritic reorganization in pyramidal neurons of the medial prefrontal cortex. *Neuroscience*, 125, 1–6.

Rauch, S. L., Shin, L. M., Segal, E., Pitman, R. K., Carson, M. A., McMullin, K., Whalen, P. J., Resnick, H.S., Kilpatrick, D. G., Best, C. L., & Kramer, T. L. (1992). Vulnerability stress factors in the development of posttraumatic stress disorder. *Journal of Nervous and Mental Disorders*, 23, 424–430.

Rauch, S. L., Shin, L. M., Segal, E., Pitman, R. K., Carson, M. A., McMullin, K., Whalen, P. J., & Makris, N. (2003). Selectively reduced regional corticol volumes in post-traumatic stress disorder. *NeuroReport*, 14, 914–916.

Rona, R. J., Hooper, R., Jones, M., Hull, L., Browne, T., Horn, O., Murphy, D., Hotopf, M., & Wessely, S. (2006). Mental health screening in armed forces before the Iraq war and prevention of subsequent psychological morbidity: follow-up study. *British Medical Journal*, 333, 991–994A.

Sapolsky, R. M. (1996) Why stress is bad for your brain. *Science*. 273, 749–750.

Smith, T. C., Wingard, D. L., Ryan, M. A., Kritz-Silverstein, D., Slymen, D. J., & Sallis, J. F. (2008). Prior assault and posttraumatic stress disorder after combat deployment. *Epidemiology*, 19(3), 505–512.

Ursano, R. J., McCaughey, B. G., & Fullerton, C. S. (1994). Trauma and disaster in Ursando. In B. G. McCaughey and C. S. Fullerton (Eds.), *Individual and community responses to trauma and disaster: The structure of human chaos* (pp. 3–27). Cambridge, U.K.: Cambridge University Press.

Vasterling, J. J., Proctor, S. P., Amoroso, P., Kane, R., Heeren, T., & White, R. F. (2006). Neuropsychological outcomes of Army personnel following deployment to the Iraq war. *Journal of the American Medical Association*, 296, 519–529.

Ware, J., Snow, K., Kosinski, M., & Gandek, B. (1993). *SF-36 health survey manual and interpretation guide*. Boston, MA: Health Institute, New England Medical Center.

Yehuda, R. (1998). Resilience and vulnerability factors in the course of adaptation to trauma. *National Center for Posttraumatic Stress Disorders Clinical Quarterly*, 8(1).

8

Scanning for Danger: Readjustment to the Noncombat Environment

LAUREN M. CONOSCENTI, VERA VINE,
ANTHONY PAPA, and BRETT T. LITZ

Contents

In war zones, not only are service members at risk for being exposed to specific life-threatening traumatic experiences, but they also need to maintain high levels of alert, alarm, and arousal because of potential threats. The latter is particularly true in the context of guerilla wars such as those in Iraq and Afghanistan, where civilians and combatants are often indistinguishable and covert aggression, such as sabotage and terrorism, is ubiquitous. Lengthy deployment to such dangerous and uncertain contexts can lead not only to acute symptoms of combat-related posttraumatic stress disorder (PTSD) and acute stress disorder (ASD) but also to enduring changes in habit, lifestyle, style of communication, ways of relating to others, and wellness behaviors (e.g., leisure, sleep, eating, and hygiene behaviors). In the harsh and demanding world of combat, symptoms of PTSD and changes in demeanor and manner are occupational hazards. Although the majority of combat veterans do not go on to develop PTSD or other behavioral problems associated with poor adjustment to garrison life and beyond, no one is completely immune; all must adapt and reintegrate and slowly reclaim their predeployment repertoires to return to homeostasis.

Loved ones, care providers, peers, and friends can testify to the changes in demeanor associated with deployment to combat. Imagine a soldier—let's call him RJ—who is a 22-year-old man driving in dangerous convoys and patrolling crowded city centers in the Red Zone. Over the course of a 12-month deployment, he is in harm's way on a nearly daily basis. On multiple occasions, roadside bombs have destroyed other vehicles in his convoy, and although he has never been injured himself he is constantly afraid that the next bomb will tear into his own vehicle. He learns to be alert to warning signs of improvised explosive devices by the side of the road, and he learns to stay away from people in the city who might be suicide bombers. These survival reflexes are reinforced by the military culture, in which superiors drill repeatedly, "If you let your guard down, you die."

Now bring that soldier home, to his safe, suburban community in the United States, where he must resume the roles, responsibilities, and relationships that were once part of his civilian lifestyle. It is understandable that after many months of combat and life threats in an uncertain environment, many service members find it hard to shut off high levels of alertness. For some time, even the most resilient of military returnees may be unable to shake the defensive reflexes that were so appropriate in the war zone. When he sees suspicious-looking refuse on the side of the road in his suburban neighborhood, RJ's heart begins to race, and he involuntarily ducks and swerves his car. Even though it is an innocuous sofa discarded on garbage pick-up day to most people, RJ responds to it just as he did in Iraq.

This hypervigilance is a natural reaction to combat, and similar problems have been observed across many generations of warriors. Combat-related adjustment problems have been documented since as early as the 19th century, although they have historically gone by other names. During the Boer War, soldiers were diagnosed with "nervous shock," which included functional somatic symptoms such as head, neck, back, and limb pain; general feebleness; tremors; and anxiety. During Word War I, the term "shell shock" was used to describe similar nervous system and cardiovascular symptoms among large numbers of psychiatric casualties. High rates of psychiatric casualties in World War II spurred the development of a dose–response conception of combat-related psychiatric illness based on observed correlations between battle intensity and combat exposure, on the one hand, and rates of psychiatric distress on the other (Campise, Geller, & Campise, 2006; Jones & Wessely, 2001; Mareth & Brooker, 1985). Psychiatrists and doctors began to look for measurable causes of this "battle fatigue" or "combat exhaustion," as it was then called.

Following the Vietnam War, combat-related distress, conceptualized as "posttraumatic stress disorder," entered into the nosology in 1980, undergoing further revisions in 1990 and 1994 (APA, 1980, 1990, 1994). Currently, PTSD is thought to be composed of four major symptom clusters (King, Leskin, King, & Weathers, 1998a): *reexperiencing, hyperarousal, avoidance*, and *emotional numbing*. PTSD is also associated with significant levels of psychosocial impairment (Schnurr, Hayes, Lunney, McFall, & Uddo, 2006), sleep problems (Krystal & Davidson, 2007), and anger (Novaco & Chemtob, 2002). In 1994, acute stress disorder (ASD) was introduced in the DSM-IV-R (APA, 1994) to capture maladaptive posttraumatic responses occurring 2 to 30 days after a traumatic event. The diagnostic criteria for ASD overlap highly with PTSD and focus on reexperiencing, hyperarousal, and avoidance, they but also highlight the role of early dissociative symptoms with subsequent maladaptation (Harvey & Bryant, 2002).

The persistence of combat-related problems, despite changes in the modalities of warfare, has urged the development of more nuanced models of post-combat adjustment. The emergence of severe distress among service members deployed to unconventional combat environments, such as peacekeeping missions (Litz, King, King, Orsillo, & Friedman, 1997a), Vietnam (Kulka et al., 1990), the Gulf War (Stein, Tran, Lund, Haji, Dashevsky, & Baker, 2005), and the current conflicts in Iraq and Afghanistan (Hoge et al., 2004), has challenged researchers to look for new deployment-related factors that may account for risk and resilience. For example, perceived threats of bodily harm during deployment (King, King, Foy, Keane, & Fairbank, 1999) or a "malevolent environment" (Litz et al., 1997a) can be powerful determinants of combat-related psychopathology, even in the absence of discrete traumatic experiences.

Today, we understand that the behavioral, cognitive, social, and biological transformation that takes place as a result of sustained exposure to combat is not reducible to a single construct, syndrome, or predictable course for all service members. Only 6.2 to 12.9% of service members exposed to combat in Iraq or Afghanistan have developed PTSD (Erbes, Westermeyer, Engdahl, & Johnsen, 2007; Hoge et al., 2004), and trajectories of postcombat adjustment have been found to vary dramatically (Litz, 2007). The older dose–response view of posttraumatic adjustment has been replaced by a diathesis–stress model that places trauma exposure as a necessary but in no way sufficient condition. Modern trauma research has identified numerous individual difference characteristics that influence postcombat outcome. For example, genetic predisposition, personality, coping capacity, and social support affect adaptation to combat and life after combat. The field has also recognized that combat trauma is unique because it typically entails not only life threat but also traumatic loss and challenges to moral beliefs resulting from various acts of omission and commission or bearing witness to carnage and human maliciousness (Kubany, 1994). In other words, combat is not just about the terror of the threat of annihilation; grief, sorrow, anguish, shame, or guilt may dominate or greatly color deployment experiences and postdeployment adaptation. Nevertheless, the priority of hardwired structural responses to life threats makes it safe to assume that, for many, danger-related behavioral repertoires and routines predominate functionally, both in the war zone and beyond.

In this chapter, we focus on combat-related hypervigilance problems, which involve undue alertness (e.g., scanning the environment for threats, sleeping with a handgun under the pillow) and the tendency to respond behaviorally as if threats are imminent in any setting (e.g., reacting defensively when startled). These problems are typically associated with specific behavioral and cognitive responses (e.g., irritability, restlessness, impaired concentration). Hypervigilance has salient physiological concomitants as well, including elevated levels of arousal (hyperarousal) and stress hormones, as well as a strong startle reflex. Hypervigilance can be nonconscious; because vigilance in the face of danger is evolutionarily hardwired and can become overlearned or habitual during the course of a deployment, service members may act in excessively vigilant ways once redeployed without knowing it. Hypervigilance can also be deliberate; service members may strategize about ways to reduce their sense of possible harm to themselves and loved ones. Thus, chronic hypervigilance problems can also inspire maladaptive coping strategies, such as avoidance of situations reminiscent of the war zone (e.g., driving, crowded places).

Because its elements resemble the PTSD symptoms of intrusive thoughts, hyperarousal, and avoidance, hypervigilance may indicate the presence of PTSD; however, even in the absence of PTSD, hypervigilance can serve as a gateway to other forms of posttraumatic distress. Severe hyperarousal is associated with reexperiencing symptoms (Hopwood & Bryant, 2006; Nixon, Nehmy, & Seymour, 2007; Schnell, Marshall, & Jaycox, 2004) and emotional numbing symptoms (Litz et al., 1997b; Schnell et al., 2004). Other problems associated with hyperarousal include depression (Erickson, Wolfe, King, King, & Sharkansky, 2001), dissociation (Kastelan et al., 2007; Nixon & Bryant, 2006), and sleep difficulties (Germain & Neilsen, 2003). Further, military personnel with severe hyperarousal are predisposed to aggressive behavior (Taft et al., 2007) and problematic substance use, possibly as a form of self-medication (Price, Risk, Haden, Lewis, & Spitznaagel, 2004).

Thus, treating arousal problems and hypervigilance may prevent many forms of combat-related posttraumatic adjustment difficulties. Thankfully, arousal and hypervigilance are readily targeted. In the following sections, we summarize theoretical frameworks that illustrate how hypervigilance in the wake of combat can translate into more pervasive difficulties. Next, we review topics pertaining to the prevention of combat-related PTSD, barriers to care among service members, and the efficacy of existing treatments for hypervigilance problems. Finally, we offer suggestions for clinicians working with redeployed service members and consider some future directions for research and clinical work devoted to returning service members.

Theoretical Perspectives on Hypervigilance

Behavioral models, based on classical and operant conditioning principles derived from animal research (Pavlov, 1927; Skinner, 1938), are the most parsimonious account of hypervigilance-related phenomena. Mowrer's (1956) two-factor learning theory illustrates how stimuli associated with a traumatic event can acquire fear-eliciting properties. Fear can generalize via higher order conditioning, such that a wider variety of stimuli acquire the same fear-eliciting properties, resulting in persisting hypervigilance across a variety of contexts following trauma exposure (Keane, Zimering, & Caddell, 1985). The second factor in Mowrer's model demonstrates how fear reactions and hypervigilance (as well as avoidance behaviors) can become habitual; each time avoidance of fear-eliciting stimuli provides relief, the avoidance behavior becomes reinforced, and the person fails to learn that conditioned cues are not harmful, leading to the maintenance of conditioned fear behaviors.

The conditioning model is particularly germane to postcombat hypervigilance. Unlike other types of trauma involving an isolable index event, combat exposure occurs over prolonged deployments to malevolent environments that provide ample opportunity for conditioning to take place. Repeatedly pairing a shock with the sound of a bell would teach a dog to expect the shock any time it hears the bell. Similarly, repeated suicide bomb explosions in crowded commercial centers in Iraq would teach our soldier RJ to associate such crowded places with suicide bomb threats. As a result, he will learn to be on high alert and scan for potential insurgents when entering a crowded area. Upon returning home, the fear may generalize to other crowded environments where he might be approached from behind, such as malls, movie theaters, or social gatherings, and he may reflexively become uneasy around people in baggy clothing that could conceal a bomb. Higher order conditioning may take place when RJ learns to fear not only bomb explosions but also the experience of hypervigilance itself. He may begin to avoid any situations that might elicit hyperalertness. Each time his avoidance successfully provides relief, the avoidance strategy will become more ingrained, and the failure to challenge the behavioral association between the stimulus and the fear emotion will serve to reinforce the hypervigilant response.

RJ may be conditioned with fear from his deployment even if he never experienced a direct threat to his life. The environment itself is so volatile and so pervaded by instability and imminent threat that service members are constantly keyed up and on edge. Vicarious conditioning, or the acquisition of fear associated with specific stimuli through watching the fearful reactions of others (Cook, Mineka, Wolkenstein, & Laitsch, 1985), may be pervasive in deployment settings. Word of mouth within a squad, training by superiors, and witnessing others' misfortune all serve to vicariously condition service members, and for good reason. With lives at stake in the war zone, service members are well served by learning to identify danger signals from others' experiences before encountering them themselves.

The conditioning model remains illuminating for problems associated with fear and avoidance learning, but it has been criticized for being overly simplistic in explaining individual differences in the etiology and course of fear-related pathology. Contemporary learning theorists (Mineka & Zinbarg, 2006) address this shortcoming in their review of newer research indicating that fear learning is mediated by genetic variables (e.g., differences in the serotonin transporter gene) (Lee et al., 2005) and personality variables (e.g., anxiety sensitivity) (Taylor, 1999), as well as by pre-, peri-, and posttrauma appraisals. Previously made appraisals of a particular stimulus as being safe and not dangerous can attenuate the fear conditioning that takes place during a future potentially traumatic experience with that stimulus (Kent, 1997). Peritraumatic appraisals, such as the feeling of helplessness, have been linked to more severe PTSD years later (Basoglu & Mineka, 1992). Posttrauma reappraisals regarding the danger of the stimulus and frequent mental rehearsal of the stimulus–trauma pairing, such as might occur during reexperiencing, can exaggerate the strength of fear conditioning (Davey, 1997; Davey & Matchett, 1994). These new modifications to the traditional model provide promising leads for the development of innovative interventions after stressful events (Wald & Taylor, 2007).

Biological research adds to the overall picture of how hypervigilance can create permanent impairment by illustrating trauma's consequences for the body and brain. During traumatic experiences, rapid releases of adrenaline and norapenephrine mobilize the fight-or-flight reaction, resulting in greater cardiac output and muscle tension. For service members with hypervigilance, the body remains in this activated, fear-primed state, even when danger is no longer present. When confronted with reminders of their traumatic experience or other fear-eliciting stimuli, arousal spikes. People with PTSD, for example, have elevated heart rate and blood pressure, even at resting baseline, as well as heightened muscle tension and startle reactions, in reaction to trauma-related cues (Pole, 2007).

Over time, elevated physiological arousal can cause lasting damage to the brain and nervous system which can translate into pervasive behavioral, emotional, and cognitive problems. Fear-eliciting stimuli activate neural networks in the amygdala (Fanselow & Poulos, 2005). With chronic hypervigilance, these regions of the brain can become hyperactive, impairing affect modulation. Psychological assessment of a situation is bypassed and fight-or-flight reactions can be activated immediately (Emilien et al., 2000; Pissiota et al., 2002). When affect modulation by these networks is compromised, significant behavioral problems may result. For example, hyperactivity in these neural networks has been implicated in aggressive behavior among war veterans (Pavic, 2003). Prolonged hyperarousal also releases a cascade of hormones in the brain's hypothalamus, pituitary, and adrenal glands (known as the HPA axis), culminating in elevated levels of glucocorticoids, a type of stress hormone. Elevated levels of this stress hormone may cause atrophy in the hippocampus of the brain and damage the prefrontal cortex over the long term (Emilien et al., 2000; Lupein et al., 2005), which can have severe consequences. Reduced hippocampal volume has been observed in combat veterans with chronic PTSD (Emilien et al., 2000; Kitayama, Vaccarino, Kutner, Weiss, & Bremner, 2005; Shin et al., 2004) and has been associated with memory problems (Glisky, Rubin, & Davidson, 2001; Kensinger & Corkin, 2003). Damage to the prefrontal cortex is linked with deficits in emotion regulation, inhibition of unhealthy thoughts and behaviors, and impairment of other executive functions.

Cognitive models of trauma have been developed to explain hyperarousal and hypervigilance problems. Some cognitive frameworks are more concerned with automatic memory processing and some with memory processed at higher levels; others unite both levels of processing (for a review, see Brewin & Holmes,

2003). According to information-processing models of posttraumatic pathology, service members may essentially become stuck seeing the world through a combat lens. Information-processing models suggest that traumatic experiences activate cognitive networks associating the contextual elements of the scene with physiological arousal, fear-related affective states, defensive/avoidance behaviors, and emotions, thoughts, or interpretations related to the event. Because these cognitive nodes activate simultaneously during the traumatic event, subsequent memory for the event is encoded within these networks of associations. For service members with hypervigilance, trauma-related stimuli automatically trigger this network of trauma-related responses, overriding other networks that would produce healthy responses, such as positive emotion or approach behaviors (Foa, Steketee, & Rothbaum, 1989; Litz & Keane, 1989). These fear networks may remain weakly activated at all times and may guide interpretations of even mundane events (Chemtob, Roitblat, Hamada, & Carlson, 1988).

In other words, traumatic experiences have the potential to freeze chronically activate cognitive networks that, while protective during true danger, tend to be maladaptive in normal conditions; for example, RJ experiencing a traumatic explosion in a crowded Baghdad marketplace might establish a strong mental association between crowded spaces, the notion of being attacked, fear, and the physiological responses to fear. After retuning from his deployment, going to a crowded place such as a shopping mall might activate this network, eliciting the experience of fear, the associated physiological reactions, and the idea that there might be someone dangerous inside. Even activation of the *idea* of a crowded place might activate the fear network. The activation of these networks would be so unpleasant that it might lead to the habituation of maladaptive coping strategies; in this instance, RJ may begin staying home to avoid crowded places, which may interfere with his ability to go to work, socialize, and contend with everyday life demands. Alternatively, he may start drinking excessively or abusing drugs to dull or distract from the emotional and physiological distress activated by the fear network.

Other cognitive models of trauma take into account knowledge and memory that take place at higher levels of cognitive processing and within a longer time frame. Whereas the information-processing view addresses automatic, lower levels of processing concerned with stimulus input into the brain, these theories are concerned with conceptual-level meaning making. Foa and Rothbaum's (1998) emotional processing model proposes that memories of traumatic events remain pathological as long as they are at odds with an individual's views of the self and the world (e.g., the self as competent and the world as secure). Appraisals that take place after trauma—about the event itself, about the individual's own reactions and symptoms, and about the responses of others—may interact with prior self and world schemas to produce negative outcomes. Ehlers and Clark (2000) extended this model by suggesting that the appraisal of the self as being helpless to influence one's own fate and being vulnerable to danger in the present (a frame of mind they term "mental defeat") may be particularly toxic for posttraumatic outcomes. Consistent with this prediction, feelings of mental defeat as early as 4 months following traumatic political imprisonment have been found to predict later PTSD severity (Ehlers, Maercker, & Boos, 2000). Ehlers and Clark (2000) further suggest that traumatic memories involve a failure of integration into autobiographical knowledge about the self, which would contextualize and bind them within a temporal time frame. For RJ, the human suffering and carnage witnessed during his deployment may violate his previous convictions about the world. No longer is it a safe and moral place but rather a dangerous place full of people who cannot be trusted. Or, perhaps RJ tried to help save a dying buddy amidst the rubble, but his inability to save his buddy's life has left him feeling persistently helpless and defeated. Both experiences increase his sense of vulnerability in everyday situations, leading to increased hypervigilance and hyperarousal, and subsequently, avoidance.

Dual representation theory can be seen as a marriage between the information-processing and cognitive models discussed above (Brewin, Dalgleish, & Joseph, 1996). This theory posits two memory systems that can have both direct and indirect effects on conscious experience. One system is concerned with memories that have been consciously processed, can be represented within a personal temporal context, and are available for deliberate retrieval and written or oral recitation. These verbally accessible memories (VAMs) include primary emotions (those that happened at the time of the traumatic event) and secondary emotions (those that have occurred since the event). The second memory system is concerned with memories processed automatically at lower levels of consciousness that can be triggered involuntarily, such as flashbacks, and have no verbal code by which they can be communicated. These situationally accessible memories (SAMs) are stored in associative networks such as those proposed by information-processing theories, which include fear-eliciting stimuli, primary emotions, and bodily responses. According to this theory, RJ may have strong SAMs of a particular bombing incident. These SAMs manifest themselves as strong, bodily experiences of the fear that he felt at the time; however, when he tries to recount the experience in words, the story may sound disorganized and lack a narrative structure that would weave it into the fabric of the rest of his life.

Treating Hypervigilance and Preventing PTSD

There is evidence to suggest that when posttraumatic routines such as hypervigilance and hyperarousal become habitual and chronic individuals face risk for lifelong problems and resistance to various treatments (Kessler, Sonnega, Bromet, Hughes, & Nelson, 1995). Consequently, secondary prevention of chronic PTSD is critical (Litz & Gray, 2003), and providing service members the help they need early on is a critical challenge for the Departments of Defense and Veterans Affairs. To that end, researchers and healthcare professionals alike have increased their focus on the development and application of early interventions for PTSD; however, for many veterans and service members, the intensity and debilitating nature of combat stress symptoms may not become apparent until a good deal of time passes. RJ may not realize the degree to which he is on high alert until he finds himself overwhelmed with fear while driving on the freeway for the first time. Or perhaps it will take him 6 months to realize that he is not yet back to normal; it may take his friends and family pointing out to him that he seems to avoid leaving the house or continues to startle at sudden noises, such as the neighbor's car backfiring.

When a service member recognizes the need for care, various strategies are available, but which ones are most appropriate? In general, cognitive–behavioral therapies (CBTs) have been the most widely studied treatments and have shown the greatest effects (Bradley, Greene, Russ, Dutra, & Westen, 2005; van Etten & Taylor, 1998). The evidence supporting CBT as an effective treatment has led such organizations as the International Society for Traumatic Stress Studies (ISTSS) and the American Psychiatric Association to recommend CBT as the treatment of choice for PTSD (Litz & Bryant, 2008; Ursano et al., 2004); however, CBT is not regularly used by clinicians to treat PTSD in most clinical practices (Becker, Zayfert, & Anderson, 2004; Foy et al., 1996; Rosen et al., 2004). Even many clinicians formally trained in CBT techniques do not routinely use those techniques and instead rely on alternative treatments for which little evidence supports their effectiveness (Becker et al., 2004). This discrepancy, however, does not appear to be due to a lack of acceptance by the patients; in fact, several studies have shown that patients may be more receptive to CBT strategies for PTSD than clinical practice would suggest (Becker, Darius, & Schaumberg, 2007; Feeny & Zoellner, 2004; Tarrier, Liversidge, & Gregg, 2006). Lack of professional training and experience in CBT techniques have been cited as barriers to clinician delivery of these treatments (Becker et al., 2004). Although a variety of clinician and patient factors undoubtedly influences the course of care, it is vital that clinicians with limited CBT experience seek out training so they may offer a full range of effective strategies to veterans and service members. It is also important that this effort be supported and promoted by the institutions with which providers are affiliated.

As described earlier, untreated hypervigilance and hyperarousal symptoms are often harbingers of chronic problems even in the absence of a full PTSD symptom profile (Schnell et al., 2004; Taft et al., 2007); therefore, it is speculated that targeting hyperarousal and hypervigilance may be most effective for reducing overall distress. Some individuals may present exclusively with these reactions, but others may experience a constellation of other problems, such as anger, substance use, or depression. For both groups of patients, data suggest that treatment should address hypervigilance and hyperarousal early in the therapeutic process to reduce the likelihood of ongoing, chronic distress; to reduce the intensity of other symptoms; and to improve overall coping.

The most basic and widely applicable tools for addressing hyperarousal are the collection of cognitive and behavioral techniques designed to help individuals manage stress. These techniques can specifically and effectively target hyperarousal symptoms related to PTSD, such as sleeplessness, jitteriness, and irritability. Two of the more common stress management techniques are deep breathing and progressive muscle relaxation. Both are designed to quell the fight-or-flight response to help individuals manage and cope with reactions to trauma-related (conditioned) triggers. Deep breathing and progressive muscle relaxation retrain the body to respond calmly to previously fear-eliciting stimuli by decoupling the fight-or-flight response from the nonthreatening situation. Deep breathing techniques use slow, deep, diaphragmatic breaths to help reduce arousal. Progressive muscle relaxation works to reduce tension in muscle groups. Individuals are instructed to isolate a muscle group, increase the tension in that group, and then release the muscles. By systematically tensing and relaxing muscle groups or engaging in deep breathing, individuals can increase their physiological awareness of the differences between states of tension and relaxation and are better able to notice these changes within the body and counter them. Learning these strategies may help RJ remain calm after driving past the discarded sofa mentioned earlier. As he notices his breathing change, he can use his skills to slow down his respiration, which ultimately leads to increased relaxation and greater capability to manage reactions to these memory triggers.

Stress management techniques may be especially palatable for military personnel as they transition to civilian life. Many military personnel have strong reservations about seeking mental health care. Stress management is often more acceptable because it focuses on concrete skills that promote self-mastery while avoiding the trappings of popular ideas of psychotherapy. In order to be of maximal value in moments of distress, these techniques require training and practice. In other words, they are only useful when rehearsed. From their military training experience, service members understand the importance of routine practice, preparation, and skill development, so the emphasis within the stress management model on disciplined skill development and mastery will seem both familiar and acceptable. Stress management also promotes self-efficacy or confidence in coping and self-management abilities. Although military culture values camaraderie and teamwork, the demands of combat teach service members the importance of self-sufficiency. Learning stress management skills may initially require guidance from an outside source such as a counselor, but practicing and application of these skills can take place anywhere and are entirely dependent on the veteran's initiative. Additionally, because veterans can learn a variety of techniques, having a number of interchangeable stress management strategies for unexpected situations is consistent with the need for self-efficacy.

Stress management techniques may be helpful on their own, but they are also often presented in conjunction with other treatment strategies. Exposure therapy is based on the principles of basic fear conditioning and extinction. Exposure therapy instructs individuals to confront an avoided stimulus to extinguish the conditioned response. By repeatedly exposing themselves to various conditioned stimuli, service members will attenuate (extinguish) their intense emotional reactions and arousal (the conditioned response). In stress inoculation training (SIT), therapists work with individuals to create a hierarchy of exposure tasks that are increasingly more challenging. Breathing techniques or other stress management strategies are typically taught early in the treatment to give individuals tools for coping with the distress inherent in each intermediate confrontation. Exposure therapy has been shown to be effective in reducing symptoms of PTSD, particularly hyperarousal and reexperiencing symptoms (Keane, Fairbank, Caddell, & Zimering, 1989).

Working within the exposure model, RJ may wish to conquer his fear of crowded places so he can shop at the mall during peak holiday hours. First, he might attempt going to the mall parking lot during off-peak hours. Although RJ may experience distress while sitting in the mall parking lot, the stress management strategies learned early in the process now help him endure the experience. Once he has mastered the first step, he may challenge himself by walking into the atrium of the mall on an off-peak day. Gradually, RJ will be able to complete behaviors that bring him closer to his ultimate goal.

Although stress management techniques can be useful, as described above, they may not be necessary for the effective treatment of posttraumatic hyperarousal and reexperiencing. Prolonged exposure (PE) is a particular form of exposure therapy designed specifically for PTSD (Foa, Rothbaum, Riggs, & Murdock, 1991). PE is based on the premise that it is the memory of the traumatic event itself that triggers symptoms. When a stimulus is reminiscent of the original traumatic event, the individual is motivated to avoid the stimulus and, in turn, the memory of the event. In PE, patients are guided by a therapist to mentally relive the experience as vividly as possible by repeating a narrative describing the event, without the use of stress management skills. Over time, the patient's emotional response to the narrative decreases, and the patient is better able to confront safe but otherwise fear-evoking trauma reminders (Foa & Rothbaum, 1998). Similar to the exposure therapy described above, PE instructs individuals to confront an avoided stimulus—in this case, the memory of the event. In both types of exposure, the repeated process of confronting the avoided stimulus in a safe environment, such as a mall parking lot or the therapist's office, gradually produces new learning—the association between the previously fear-eliciting stimulus and the idea of safety. Although most studies of PE focus on victims of sexual assault, PE can also be applied to service members returning from war who struggle with hyperreactivity related to combat trauma. A recent study of female veterans that compared PE against present-focused treatments that deal with current life problems as manifestations of PTSD found PE to be a more effective method of treating PTSD symptoms (Schnurr et al., 2007).

With the prolonged exposure approach, RJ would be instructed to confront the memory of patrolling the marketplace. Over several sessions, he would relive his experience of being on guard on the day the bomb went off. Although, at first RJ would experience high fear and distress while reciting the story, these feelings would lessen over time. When the memory of the traumatic experience no longer elicits strong emotional reactions in the therapeutic setting, he should be able to confront the crowded shopping mall. By that time, the mall would have less salience as a trigger of traumatic memories of patrolling the marketplace, and RJ would be less likely to experience hypervigilance and distress there.

A third version of CBT is cognitive processing therapy (CPT) (Resick & Schnicke, 1992). CPT combines both cognitive and exposure components and can be conducted in individual or group formats. Similar to PE, CPT instructs individuals to create a vivid narrative of the traumatic experience and repeat the narrative until it is no longer distressing; however, it adds another component drawn from the information processing theory outlined earlier. CPT proposes that changing the meaning of the traumatic event will change the response to the memory of the event, making the memory less distressing (Resick & Schnicke, 1992). In CPT, therapists guide patients in identifying schema conflicts, maladaptive beliefs, or "stuck points" and reinterpreting memories and appraisals to reduce these forms of conflict. Through this process, the individual's beliefs accommodate the new experience in a healthy and adaptive way, and distress is reduced. A recent trial of CPT for veterans with PTSD showed that, compared to veterans in the waitlist condition, those receiving CPT experienced significant improvements in PTSD symptoms (Monson et al., 2006).

With the cognitive processing therapy approach, RJ would be instructed to write a detailed narrative of the suicide bombing that killed his buddy, including thoughts and feelings he experienced at the time of the event. With the help of the therapist, he would identify cognitive distortions, often called "stuck points," in the narrative. One such stuck point might be that he is a worthless person because he was unable to save his dying buddy. This feeling of worthlessness may contribute to his hypervigilance in crowded places, as he does not want to again find himself unable to help others if an emergency arose. In CPT, the therapist would challenge RJ on this belief by asking him to consider other people's courses of action in the same situation or by identifying times when he went above and beyond the call of duty to help others. He may also read statistics about the relative safety of the local shopping mall. After processing this stuck point, RJ would feel better equipped to venture to the crowded mall.

Treatments for PTSD symptoms, including hypervigilance, often include a range of FDA-approved pharmacological options. Selective serotonin reuptake inhibitors (SSRIs), such as sertraline and paroxetine, are the most extensively studied medications for PTSD, and research shows that they are more effective than placebo (Martenyi, Brown, Zhang, Prakash, & Koke, 2002; Tucker et al., 2001). Other medications, such as risperidone, have shown promise for alleviating symptoms (David, De Faria, & Mellman, 2006), but only SSRIs have enough evidence to support their recommendation as a treatment specific for PTSD. Even though medication generally has demonstrated effectiveness in treating PTSD symptoms, two caveats remain. First, discontinuing medication is often associated with decompensation; thus, it is important to carefully monitor medication with the help of a doctor to maintain clinical gains (Ursano et al., 2004). Second, it is recommended that pharmacological treatments be paired with CBT; however, medications are often prescribed without adjunct therapies (Rosen et al., 2004).

Another important consideration in treating hyperarousal is possible comorbidities that may reinforce and maintain hyperarousal symptoms. Chronic pain and panic attacks, two common comorbid problems in returning service members and veterans, may have profound effects on arousal levels. Chronic pain, especially from combat-related injuries, may trigger persistent low levels of anxiety and heighten attentional biases to physiological status and environmental threat, two features at the core of hypervigilance symptoms, by reminding the individual of possible traumatic events and by chronically triggering threat-related schemas, as described above (Sharp & Harvey, 2001). Both panic attacks and hyperarousal are characterized by high levels of dysregulated physical reactivity (Blechert, Michael, Grossman, Lajtman, & Wilhelm, 2007), and panic can play an important role in the development and maintenance of hyperarousal symptoms (Hinton, Pich, Safren, Pollack, & McNally, 2005; Fikretoglu et al., 2007). In both cases, treatment of pain or panic symptoms may be a key factor in treating hyperarousal symptoms and preventing later PTSD symptoms.

Although we have described a sample of potentially helpful and empirically validated treatments, many veterans are unable to get the help they need (Kessler et al., 1995; Kulka et al., 1990; Hoge et al., 2004; Litz & Gray, 2004). Often, those most in need of help are least likely to receive it, due to a variety of logistical, financial, systemic, cultural, and personal barriers (Maguen & Litz, 2006). Fear of stigmatization is perhaps the most significant barrier to obtaining care. One recent survey showed that 70% of military respondents feared being labeled "crazy" if they sought treatment, and 55% worried about how they would be perceived at work if people knew they were in treatment (Stecker, Fortney, Hamilton, & Ajzen, 2007). Officers or others in authority positions within the military may be particularly concerned about stigma, as they may worry that others will question their leadership abilities, that they will be denied promotions, or that they may be asked to leave the service altogether.

Other service members may feel that they ought to handle struggles with mental health on their own or that seeking help would be an admission of weakness (Stecker et al., 2007). Furthermore, lack of confidence in the ability of the healthcare system to address concerns sensitively and adequately is also cited as a barrier to care (Grant, 1997). Healthcare costs are also common concern of individuals seeking care (Maguen & Litz, 2006; Newman, 2000). Veterans who do not qualify for service-connected healthcare benefits often must either pay out-of-pocket for mental health services or rely on insurance programs that pay limited benefits. Other barriers include living in rural areas that have few mental-health resources, lacking transportation, having limited social support, or an inability to make time for treatment due to work or school schedules.

New approaches for expanding access to treatments are emerging to address this growing problem (Lange et al., 2003; Litz, Williams, Wang, Bryant, & Engel, 2004). These approaches include Internet-based treatment programs, self-help protocols, and minimal-contact therapies designed to help alleviate symptoms of traumatic stress. Internet-based treatments refer to treatments delivered via the Internet, such as CBT presented to the patient on a specially designed website instead of face to face. Typically, these programs are monitored by professionals. Self-help interventions include materials that are not monitored by professionals, such as workbooks, DVDs, or computer programs that do not interface with a professional. Minimal-contact treatments include programs where the therapist guides the patient at various intervals on the phone, in person, or via email, but between meetings patients are largely responsible for following through with the instructions given by the therapist on their own.

Although these treatment methods have not yet been rigorously tested, the increasing need for services, especially among veterans, has piqued national interest. These nascent therapeutic modalities carry several advantages over traditional techniques. They may be more cost effective than traditional face-to-face therapy for two reasons. First, they generally require fewer therapist resources. Therapists are typically available on an as-needed basis for emergencies or questions but are otherwise monitoring progress from afar and checking in with the patient at specified intervals. Second, although the startup costs may be significant (i.e., programming a website), once the material has been designed it is easy to disseminate. These programs may be completed in the privacy of the home at the service member's convenience, allowing access to individuals who live in remote locations, who have mobility or transportation difficulties, or who are unable to engage in treatment due to other responsibilities. For veterans who suffer from hypervigilance outside of the home, treatments available in a comfortable, safe location are likely to be welcomed. The privacy afforded by these forms of treatment also reduces stigma. Finally, these programs are typically designed in a format that respects service members' ability to help themselves with proper structure, guidance, and support. Frequently framed as "trainings" or "self-management," these formats can empower the patient to take a proactive role in his or her care.

One such program recently has shown promise in a randomized controlled trial. Self-management CBT (SM-CBT) is an 8-week, self-managed, cognitive–behavioral therapy program delivered online to participants. SM-CBT is designed to teach patients strategies to help them cope and manage reactions to situations that trigger recall of traumatic experiences (Litz, Engel, Bryant, & Papa, 2007). In particular, it teaches stress management strategies, requires participants to self-monitor situations that trigger trauma-related distress, and assists participants in creating a hierarchy of triggers for graduated *in vivo* exposure. Additionally, it features a writing component, where participants write about a particularly salient and troubling traumatic experience. Service members with PTSD participated in a trial of SM-CBT; compared to an Internet-delivered, nondirective, supportive counseling program, participants in SM-CBT had greater reductions in PTSD, depression, and anxiety symptoms at 6 months' follow-up (Litz et al., 2007).

Caution is warranted, however. Because these treatments have not yet been tested stringently, it is unclear if a specific program or treatment is effective in reducing symptoms. Ethical concerns about the rapid proliferation and commercialization of unproven programs have been raised (Rosen, 1987). Finally, it is important to emphasize that not all treatments will work for all patients. Self-help and similar programs may be helpful for some people, but others will benefit most from face-to-face contact with a mental health professional. Individuals who do not find the programs helpful may be even less motivated to seek out traditional in-person treatment as a next course of action.

Treatment Matching and Other Considerations

Empirically based techniques can be effective tools for addressing current problems and preventing development of chronic PTSD; however, they must be employed within the context of the individual's current life challenges and abilities. For example, access to positive social support resources in the aftermath of combat is an important predictor of the development of PTSD (Brewin, Andrews, & Valentine, 2000; Ozer, Best, Lipsey, & Weiss, 2003; Fontana, Rosenheck, & Horvath, 1997). A client with few social supports, therefore, may be at greater risk for developing chronic PTSD; thus, treatment may focus on avoidance related to social situations in an attempt to improve the support network.

In addition to treatment matching issues, more general concerns arise when beginning to work with military service members. Certain approaches consistent with good therapy practices in general may be particularly salient for treatment of military service members. The first is the importance of taking a collegial, contractual approach. This approach is important in most, if not all, therapeutic endeavors but may be especially useful in treating service members for whom honor and commitment are serious ideals. It will be important to convey accurate expectations about the treatment process, such that it takes hard work and time and can be quite challenging.

Additionally, service members should enter into treatment with realistic expectations. Therapists should emphasize that successful treatment of hypervigilance problems may take the form of constant self-management rather than erasure of fear. In fact, extinction training is not erasure of fear but rather *new learning*. In this context, it may be helpful to frame treatment as training to improve mental fitness and learning how to manage reactions in any situation so these reactions do not have long-term negative consequences. This is particularly important because of the significant cognitive and institutional barriers to care, leading to high rates of treatment drop out. Unrealistic or overly optimistic expectations may lead to greater disillusionment with treatment results, which may increase susceptibility to drop out.

Service members tend to be goal oriented; therefore, treatment plans should involve hierarchies of goals that will be achieved and coping skills that will be learned, and clinicians should try to frame things positively and constructively. Failure experiences, or patient perceptions of failure in treatment, should be avoided. It is often helpful to stress the goal of *returning to functioning*, or, achieving family, work, and life satisfaction. The emotional and experiential aspects of hypervigilance may stick around for some time, but it is possible to teach tools that enable the patient to get back to work, improve family relationships, and carry on with life.

For service members who remain committed to the military, this might mean being a better, stronger soldier, sailor, airman, or Marine. Civilian clinicians should be mindful of their own ideologies and thoughts about the military and specific military conflicts and should bear in mind that for many service members struggling with posttraumatic adjustment and hypervigilance problems the goal is to get back to high occupational functioning as a member of the military.

The above recommendations can be illustrated further using the case of RJ. Following his return home, RJ's relationship with his wife has taken a turn for the worse. In addition to not being able or willing to relax his guard since returning home, RJ is constantly worried about his family's safety, something he feels his wife does not understand or take seriously enough. The couple argues often, and RJ typically retreats to his bedroom to avoid the arguments; however, because he stays home most of the time, his wife is his main source of social support, and when they fight he feels particularly isolated.

RJ presents for treatment after having a panic attack at the mall. He went to the mall to try to make amends with his wife by keeping her company during a shopping trip, but when he got there he found it to be too crowded and exposed. He noticed he was getting wound up, but pushed himself to stay to please his wife. As he became more anxious, feeling trapped and helpless, he began to experience what he thought were heart palpitations, and he began sweating and trembling. This experience frightened him greatly. Since then, he has withdrawn further and no longer invites friends and family to the home.

A therapist evaluating RJ's current difficulties finds that his main reaction to the combat stressors he experienced in Iraq include high levels of hyper-arousal symptoms, including extreme irritability, anger, inability to relax or sleep well (unrelated to nightmares), exaggerated startle response, and hyper-vigilance (constant checking the door locks at night, scanning the environment outside the house, constant worry about all the "what-ifs"). Although he does report some reexperiencing and avoidance symptoms, they all seem secondary to his hyperarousal symptoms. Further evaluation also indicates that he has depression symptoms, feels shame and guilt over his angry outbursts at home due to thinking about the death of his friend in Iraq, thinks that his friends and family are avoiding him because of his anger, is experiencing panic symptoms (as described above), and above all is feeling "weak."

Earlier in this chapter, we mentioned how stress management, exposure techniques, and CPT might be employed with RJ in a straightforward fashion; however, combat-related hyperarousal symptoms have had a rippling effect on RJ's life. They have impacted many areas of functioning and eroded many coping resources, including his own beliefs about his ability to cope (e.g., his feeling "weak") and his access to social support resources.

In the case of RJ, it would be important to assess his level of panic symptoms, as they may interfere with his ability to benefit from exposure techniques. It may be necessary to begin intervention with interoceptive exposure for panic symptoms prior to beginning stress inoculation training or prolonged exposure. Additionally, it may be very important to address coping beliefs and any depression-related cognition that may interfere with active engagement with intervention using cognitive reframing techniques.

Finally, it is critically important to identify the main social resources available within RJ's life and to develop strategies for how to utilize them, whether via adjunct cognitive restructuring regarding RJ's ability to reach out to others or directly within an exposure framework as an exposure hierarchy item. But, whereas hyperarousal symptoms may represent normative responses to the experience of acute or chronic stressors, they may also have significant inter- and intrapersonal consequences for service members. Ironically, the symptoms may drive away the very social support that is so critical to the service member's recovery.

Conclusions

In sum, returning home from combat may present considerable challenges for service members returning from combat theaters. Many will easily make the adjustment back to living and working in noncombat environments; others will have more difficulty. Almost all will display some level of hypervigilance upon return. For those who display acute or chronic functional difficulties, and especially those whose hypervigilance symptoms begin to interact with their environment to cause downward spirals of functional adjustment, CBT-based interventions can be quite effective. At the same time, much research remains to be done on the development of refined secondary prevention treatment packages. Vital tasks for further research include identifying who needs prevention interventions, when prevention or treatment might be most effective to prevent chronic mental illness, how to deliver this treatment broadly and effectively, and how to overcome significant barriers to mental health care, especially in military contexts.

References

APA. (1980). *Diagnostic and statistical manual of mental disorders* (3rd ed.). Washington, D.C.: American Psychiatric Association.

APA. (1990). *Diagnostic and statistical manual of mental disorders* (3rd ed., revised). Washington, D.C.: American Psychiatric Association.

APA. (1994). *Diagnostic and statistical manual of mental disorders* (4th ed.). Washington, D.C.: American Psychiatric Association.

Basoglu, M., & Mineka, S. (1992). The role of uncontrollable and unpredictable stress in post-traumatic stress responses in torture survivors. In M. Basoglu (Ed.), *Torture and its consequences: Current treatment approaches* (pp. 182–225). New York: Cambridge University Press.

Becker, C. B., Zayfert, C., & Anderson, E. (2004). A survey of psychologists' attitudes towards and utilization of exposure therapy for PTSD. *Behaviour Research and Therapy*, 42, 277–292.

Becker, C. B., Darius, E., & Schaumberg, K. (2007). An analog study of patient preferences for exposure versus alternative treatments for posttraumatic stress disorder. *Behaviour Research and Therapy*, 45, 2861–2873.

Blechert, J., Michael, T., Grossman, P., Lajtman, M., & Wilhelm, F. H. (2007). Autonomic and respiratory characteristics of posttraumatic stress disorder and panic disorder. *Psychosomatic Medicine*, 69, 935–943.

Bradley, R., Greene, J., Russ, E., Dutra, L., & Westen, D. (2005). A multidimensional meta-analysis of psychotherapy for PTSD. *American Journal of Psychiatry*, 16, 214–227.

Brewin, C., & Holmes, E. (2003). Psychological theories of posttraumatic stress disorder. *Clinical Psychology Review*, 23(3), 339–376.

Brewin, C., Dalgleish, T., & Joseph, S. (1996). A dual representation theory of posttraumatic stress disorder. *Psychological Review*, 103(4), 670–686.

Brewin, C. R., Andrews, B., & Valentine, J. D. (2000). A meta-analysis of risk factors for posttraumatic stress disorder in adults exposed to trauma. *Journal of Consulting and Clinical Psychology*, 68, 748–766.

Bryant, R., & Litz, B.T. (2009). Early mental health interventions. In Y. Neria, S. Galea, & F. Norris (Eds.), *Mental health and disasters*. New York: Cambridge University Press.

Campise, R., Geller, S., & Campise, M. (2006). Combat stress. In C. H. Kennedy & E. A. Zillmer (Eds.), *Military psychology: Clinical and operational applications* (pp. 215–240). New York: Guilford Press.

Chemtob, C., Roitblat, H., Hamada, R., & Carlson, J. (1988). A cognitive action theory of post-traumatic stress disorder. *Journal of Anxiety Disorders*, 2(3), 253–275.

Cook, M., Mineka, S., Wolkenstein, B., & Laitsch, K. (1985). Observational conditioning of snake fear in unrelated rhesus monkeys. *Journal of Abnormal Psychology*, 94(4), 591–610.

Davey, G. C. L. (1997). A conditioning model of phobias. In G. C. L. Davey (Ed.), *Phobias: A handbook of theory, research, and treatment* (pp. 107–127). Chichester: John Wiley & Sons.

Davey, G. C. L., & Matchett, G. (1994). Unconditioned stimulus rehearsal and the retention and enhancement of differential "fear" conditioning: effects of trait and state anxiety. *Journal of Abnormal Psychology*, 103, 708–718.

David, D., De Faria, L., & Mellman, T. A. (2006). Adjunctive risperidone treatment and sleep symptoms in combat veterans with chronic PTSD. *Depression and Anxiety*, 23, 489–491.

Ehlers, A., & Clark, D. (2000). A cognitive model of posttraumatic stress disorder. *Behaviour Research and Therapy*, 38(4), 319–345.

Ehlers, A., Maercker, A., & Boos, A. (2000). Posttraumatic stress disorder following political imprisonment: The role of mental defeat, alienation, and perceived permanent change. *Journal of Abnormal Psychology*, 109, 45–55.

Emilien, G., Penasse, C., Charles, G., Martin, D., Lasseaux, L., & Waltregny, A. (2000). Post-traumatic stress disorder: Hypotheses from clinical neuropsychology and psychopharmacology research. *International Journal of Psychiatry in Clinical Practice*, 4, 3–18.

Erbes, C., Westermeyer, J., Engdahl, B., & Johnsen, E. (2007). Post-traumatic stress disorder and service utilization in a sample of service members from Iraq and Afghanistan. *Military Medicine*, 172(4), 359–363.

Erickson, D., Wolfe, J., King, D., King, L., & Sharkansky, E. (2001). Posttraumatic stress disorder and depression symptomatology in a sample of Gulf War veterans: A prospective analysis. *Journal of Consulting and Clinical Psychology*, 69(1), 41–49.

Fanselow, M. S., & Poulos, A. M. (2005). The neuroscience of mammalian associative learning. *Annual Review of Psychology*, 56, 207–234.

Feeny, N. C., & Zoellner, L. A. (2004). Prolonged exposure versus sertaline: Primary outcomes for the treatment of chronic PTSD, paper presented at the International Society for Traumatic Stress Studies, New Orleans, LA.

Fikretoglu, D., Brunet, A., Best, S. R., Metzler, T. J., Delucchi, K., Weiss, D. S., Fagan, J., Liberman, A., & Marmar, C. R. (2007). Peritraumatic fear, helplessness and horror and peritraumatic dissociation: do physical and cognitive symptoms of panic mediate the relationship between the two? *Behaviour Research and Therapy*, 45, 39–47.

Foa, E. B., & Rothbaum, B. O. (1998). *Treating the trauma of rape: Cognitive–behavioral therapy for PTSD*. New York: Guilford Press.

Foa, E. B., Steketee, G., & Rothbaum, B. (1989). Behavioral/cognitive conceptualizations of post-traumatic stress disorder. *Behavior Therapy*, 20(2), 155–176.

Foa, E. B., Rothbaum, B. O., Riggs, D. S., & Murdock, T. B. (1991). Treatment of post-traumatic stress disorder in rape victims: A comparison between cognitive–behavioral procedures and counseling. *Journal of Consulting and Clinical Psychology*, 59, 715–723.

Fontana, A., Rosenheck, R., & Horvath, T. (1997). Social support and psychopathology in the war zone. *Journal of Nervous and Mental Disease*, 185(11), 675–681.

Foy, D. W., Kagan, B., McDermott, C., Leskin, G., Sipprelle, R. C., & Paz, G. (1996). Practical parameters in the use of flooding for treating chronic PTSD. *Clinical Psychology and Psychotherapy*, 3, 169–175.

Germain, A., & Nielsen, T. (2003). Sleep pathophysiology in posttraumatic stress disorder and idiopathic nightmare sufferers. *Biological Psychiatry*, 54(10), 1092–1098.

Glisky, E. L., Rubin, S. R., & Davidson, P. S. R. (2001). Source memory in older adults: an encoding or retrieval problem? *Journal of Experimental Psychology: Learning, Memory, and Cognition, 27*, 1131–1146.

Grant, B. F. (1997). Prevalence and correlates of alcohol use and DSM-IV alcohol dependence in the United States: results of the National Longitudinal Alcohol Epidemiologic Survey. *Journal of Studies on Alcohol, 58*, 464–473.

Harvey, A. G., & Bryant, R. A. (2002). Acute stress disorder: a synthesis and critique. *Psychological Bulletin, 128*, 886–902.

Hinton, D. E., Pich, V., Safren, S. A., Pollack, M. H., & McNally, R. J. (2005). Anxiety sensitivity in traumatized Cambodian refugees: A discriminant function and factor analytic investigation. *Behaviour Research and Therapy, 43*, 1631–1643.

Hoge, C., Castro, C., Messer, S., McGurk, D., Cotting, D., & Koffman, R. (2004). Combat duty in Iraq and Afghanistan: Mental health problems, and barriers to care. *New England Journal of Medicine, 351*(1), 13–22.

Hopwood, S., & Bryant, R. (2006). Intrusive experiences and hyperarousal in acute stress disorder. *British Journal of Clinical Psychology, 45*(1), 137–142.

Jones, E., & Wessley, S. (2001). Psychiatric battle casualties: An intra- and interwar comparison. *British Journal of Psychiatry, 178*, 242–247.

Kastelan, A., Franciskovic, T., Moro, L., Roncevic-Grzeta, I., Grkovic, J., Jurcan, V. et al. (2007). Psychotic symptoms in combat-related post-traumatic stress disorder. *Military Medicine, 172*(3), 273–277.

Keane, T. M., Zimering, R. T., & Caddell, J. M. (1985). A behavioral formulation of post-traumatic stress disorder in Vietnam veterans. *The Behavior Therapist, 8*(1), 9–12.

Keane, T. M., Fairbank, J. A., Caddell, J. M., & Zimering, R. T. (1989). Implosive (flooding) exposure reduces symptoms of PTSD in Vietnam combat veterans. *Behavior Therapy, 20*, 245–260.

Kensinger, E. A., & Corkin, S. (2003). Neural changes in ageing. In L. Nadel (Ed.), *Encyclopedia of cognitive science* (pp. 697–711). London: Macmillian.

Kent, G. (1997). Dental phobias. In G. C. L. Davey (Ed.), *Phobias: A handbook of theory, research, and treatment* (pp. 107–127). Chichester: John Wiley & Sons.

Kessler, R. C., Sonnega, A., Bromet, E., Hughes, M., & Nelson, C. B. (1995). Post-traumatic stress disorder in the National Comorbidity Survey. *Archives of General Psychiatry, 52*, 1048–1060.

King, D. W., Leskin, G., King, L. A., & Weathers, F. (1998a). Confirmatory factor analysis of the clinician-administered PTSD scale: Evidence for the dimensionality of posttraumatic stress disorder. *Psychological Assessment, 10*(2), 90–96.

King, D. W., King, L. A., Foy, D. W., Keane, T. M., & Fairbank, J. A. (1999). Posttraumatic stress disorder in a national sample of female and male Vietnam veterans: Risk factors, war-zone stressors, and resilience–recovery variables. *Journal of Abnormal Psychology, 108*(1), 164–170.

King, L. A., King, D. W., Fairbank, J. A., Keane, T. M., & Adams, G. A. (1998b). Resilience–recovery factors in post-traumatic stress disorder among female and male Vietnam veterans: Hardiness, postwar social support, and additional stressful life events. *Journal of Personality and Social Psychology, 74*, 420–434.

Kitayama, N., Vaccarino, V., Kutner, M., Weiss, P., & Bremner, J. D. (2005). Magnetic resonance imaging (MRI) measurement of hippocampal volume in posttraumatic stress disorder: A meta-analysis. *Journal of Affective Disorders, 88*, 79–86.

Krystal, A., & Davidson, J. (2007). The use of prazosin for the treatment of trauma nightmares and sleep disturbance in combat veterans with post-traumatic stress disorder. *Biological Psychiatry, 61*(8), 925–927.

Kubany, E. S. (1994). A cognitive model of guilt typology in combat-related PTSD. *Journal of Traumatic Stress, 7*, 3–19.

Kulka, R. A., Schlenger, W. E., Fairbank, J .A., Hough, R. L., Jordan, B. K., Marmar, C. R., & Weiss, D. S. (1990). *Trauma and the Vietnam War generation: Report of the findings of the National Vietnam Veterans Readjustment Survey.* New York: Brunner/ Mazel.

Lange, A., Rietdijk, D., Hudcovicova, M., van de Ven, J., Schrieken, B., & Emmelkamp, P. M. G. (2003). Interapy: A controlled randomized trial of the standardized treatment of posttraumatic stress through the Internet. *Journal of Consulting and Clinical Psychology, 71*, 901–909.

Lee, H., Lee, M., Kang, R., Kim, H., Kim, S., Kee, B., Kim, Y. H., Kim, Y., Kim, J. B., Yeon, B. K., Oh, K. S., Oh, B., Yoon, J., Lee, C., Jung, H. Y., Chee, I., & Paik, I. H. (2005). Influence of the serotonin transporter promoter gene polymorphism on susceptibility to posttraumatic stress disorder. *Depression and Anxiety, 21*, 135–139.

Litz, B. T. (2007). Research on the impact of military trauma: current status and future directions. *Military Psychology, 19*(3), 217–238.

Litz, B. T., & Bryant, R. (2008). Early intervention for trauma in adults: cognitive–behavioral therapy. In E. Foa, T. Keane, M. Friedman, & J. Cohen (Eds.), *Effective treatments for PTSD: Practice guidelines from the International Society for Traumatic Stress Studies* (2nd ed., pp. 117–136). New York: Guilford Press.

Litz, B. T., & Gray, M. J. (2004). Early intervention for trauma in adults: A framework for first aid and secondary prevention. In B. T. Litz (Ed.), *Early Intervention for Trauma and Traumatic Loss* (pp. 87–111). New York: Guilford Press.

Litz, B. T., & Keane, T. (1989). Information processing in anxiety disorders: Application to the understanding of post-traumatic stress disorder. *Clinical Psychology Review, 9*(2), 243–257.

Litz, B. T., King, L. A., King, D. W., Orsillo, S., & Friedman, M. (1997a). Warriors as peacekeepers: Features of the Somalia experience and PTSD. *Journal of Consulting and Clinical Psychology, 65*(6), 1001–1010.

Litz, B. T., Schlenger, W., Weathers, F., Caddell, J., Fairbank, J., & LaVange, L. (1997b). Predictors of emotional numbing in posttraumatic stress disorder. *Journal of Traumatic Stress, 10*(4), 607–618.

Litz, B. T., Williams, L., Wang, J., Bryant, R. A., & Engel, C. C. (2004). A therapist-assisted internet self-help program for traumatic stress. *Professional Psychology: Research and Practice, 35*, 628–634.

Litz, B. T., Engel, C. C., Bryant, R. A., & Papa, A. (2007). A randomized, controlled proof-of-concept trial of an Internet-based, therapist-assisted self-management treatment for posttraumatic stress disorder. *American Journal of Psychiatry, 164*, 1676–1683.

Lupein, S. J., Fiocco, A., Wan, N., Maheu, F., Lord, C., Schramek, T., & Tu, M. T. (2005). Stress hormones and human memory function across the lifespan. *Psychoneuroendocrinology, 30*, 225–242.

Maguen, S., & Litz, B. T. (2006). Predictors of barriers to mental health treatment in Kosovo and Bosnia peacekeepers: A preliminary report. *Military Medicine, 171*, 454–458.

Mareth, T., & Brooker, A. (1985). Combat stress reaction: a concept in evolution. *Military Medicine, 150*(4), 186–190.

Martenyi, F., Brown, E. B., Zhang, H., Prakash, A., & Koke, S. C. (2002). Fluoxetine vs. placebo in prevention of relapse in posttraumatic stress disorder. *British Journal of Psychiatry, 181*, 315–320.

Mineka, S., & Zinbarg, R. (2006). A contemporary learning theory perspective on the etiology of anxiety disorders: It's not what you thought it was. *American Psychologist*, 61(1), 10–26.

Monson, C. M., Schnurr, P. P., Resick, P. A., Friedman, M. J., Young-Xu, Y., & Stevens, S. P. (2006). Cognitive processing therapy for veterans with military-related posttraumatic stress disorder. *Journal of Consulting and Clinical Psychology*, 74, 898–907.

Mowrer, O. H. (1956). Two-factor learning theory reconsidered, with special reference to secondary reinforcement and the concept of habit. *Psychological Review*, 63, 114–128.

Newman, M. G. (2000). Recommendations for a cost offset model of psychotherapy allocation using generalized anxiety disorder as an example. *Journal of Consulting and Clinical Psychology*, 68, 549–555.

Nixon, R., & Bryant, R. (2006). Dissociation in acute stress disorder after a hyperventilation provocation test. *Behavioural and Cognitive Psychotherapy*, 34(3), 343–349.

Nixon, R., Nehmy, T., & Seymour, M. (2007). The effect of cognitive load and hyperarousal on negative intrusive memories. *Behaviour Research and Therapy*, 45(11), 2652–2663.

Novaco, R., & Chemtob, C. (2002). Anger and combat-related posttraumatic stress disorder. *Journal of Traumatic Stress*, 15(2), 123–132.

Ozer, E. J., Best, S. R., Lipsey, T. L., & Weiss, D. S. (2003). Predictors of posttraumatic stress disorder and symptoms in adults: a meta-analysis. *Psychological Bulletin*, 129(1), 52–73.

Pavic, L. (2003). Alterations in brain activation in posttraumatic stress disorder patients with severe hyperarousal symptoms and impulsive aggressiveness. *European Archives of Psychiatry and Clinical Neuroscience*, 253(2), 80–83.

Pavlov, I. P. (1927). *Conditioned reflexes*, G. V. Anrep (trans.). Oxford: Oxford University Press.

Pissiota, A., Frans, O., Fernandez, M., Knorring, L. Fischer, H., & Fredrikson, M. (2002). Neurofunctional correlates of posttraumatic stress disorder: PET symptom provocation study. *European Archive of Clinical Neuroscience*, 252, 68–75.

Pole, N. (2007). The psychophysiology of posttraumatic stress disorder: a meta-analysis. *Psychological Bulletin*, 133(5), 725–746.

Price, R., Risk, N., Haden, A., Lewis, C., & Spitznagel, E. (2004). Post-traumatic stress disorder, drug dependence, and suicidality among male Vietnam veterans with a history of heavy drug use. *Drug and Alcohol Dependence*, 76, S31–S43.

Resick, P. A., & Schnicke, M. K. (1992). Cognitive processing therapy for sexual assault victims. *Journal of Consulting and Clinical Psychology*, 60, 748–756.

Riggs, D., Byrne, C., Weathers, F., & Litz, B. (1998). The quality of the intimate relationships of male Vietnam veterans: Problems associated with posttraumatic stress disorder. *Journal of Traumatic Stress*, 11(1), 87–101.

Rosen, C. S., Chow, H. C., Finney, J. F., Greenbaum, M. A., Moos, R. H., Sheikh, J. I., & Yesavage, J. A. (2004). VA practice patterns and practice guidelines for treating posttraumatic stress disorder. *Journal of Traumatic Stress*, 17, 213–222.

Rosen, G. M. (1987). Self-help treatment books and the commercialization of psychotherapy. *American Psychologist*, 42, 46–51.

Ruscio, A., Weathers, F., King, L., & King, D. (2002). Male war-zone veterans' perceived relationships with their children: The importance of emotional numbing. *Journal of Traumatic Stress*, 15(5), 351–357.

Schnell, T., Marshall, G., & Jaycox, L. (2004). All symptoms are not created equal: the prominent role of hyperarousal in the natural course of posttraumatic psychological distress. *Journal of Abnormal Psychology*, 113(2), 189–197.

Schnurr, P. P., Hayes, A., Lunney, C., McFall, M., & Uddo, M. (2006). Longitudinal analysis of the relationship between symptoms and quality of life in veterans treated for posttraumatic stress disorder. *Journal of Consulting and Clinical Psychology*, 74(4), 707–713.

Schnurr, P. P., Friedman, M. J., Engel, C. C., Foa, E. B., Shea, M. T., Chow, B. K., Resick, P. A., Thurston, V., Orsillo, S. M., Haug, R., Turner, C., & Bernardy, N. (2007). Cognitive–behavioral therapy for posttraumatic stress disorder in women: a randomized controlled trial. *Journal of the American Medical Association*, 297, 820–830.

Sharp, T. J., & Harvey, A. G. (2001). Chronic pain and posttraumatic stress disorder: mutual maintenance? *Clinical Psychology Review*, 21, 857–877.

Shin, L. M., Shin, P. S., Heckers, S., Krangel, T. S., Macklin, M. L., Orr, S. P. et al. (2004). Hippocampal function in posttraumatic stress disorder. *Hippocampus*, 14(3), 292–300.

Skinner, B. F. (1938). *The Behavior of Organisms*. New York: Appleton-Century.

Stecker, T., Fortney, J. C., Hamilton, F., & Ajzen, I. (2007). An assessment of beliefs about mental health care among veterans who served in Iraq. *Psychiatric Services*, 58, 1358–1361.

Stein, A., Tran, G., Lund, L., Haji, U., Dashevsky, B., & Baker, D. (2005). Correlates for posttraumatic stress disorder in Gulf War veterans: A retrospective study of main and moderating effects. *Journal of Anxiety Disorders*, 19(8), 861–876.

Taft, C., Kaloupek, D., Schumm, J., Marshall, A., Panuzio, J., King, D. et al. (2007). Posttraumatic stress disorder symptoms, physiological reactivity, alcohol problems, and aggression among military veterans. *Journal of Abnormal Psychology*, 116(3), 498–507.

Tarrier, N., Liversidge, T., & Gregg, L. (2006). The acceptability and preference for the psychological treatment of PTSD. *Behaviour Research and Therapy*, 44, 1643–1656.

Taylor, S. (Ed). (1999). *Anxiety sensitivity: Theory, research, and treatment of the fear of anxiety*. Mahwah, NJ: Lawrence Erlbaum Associates.

Tucker, P., Zaninelli, R., Yehuda, R., Ruggiero, L., Dillingham, K., & Pitts, C. (2001). Paroxetine in the treatment of chronic posttraumatic stress disorder: Results of a placebo-controlled, flexible-dosage trial. *Journal of Clinical Psychiatry*, 62, 860–868.

Ursano R. J., Bell, C., Eth, S., Friedman, M., Norwood, A., Pfefferbaum, B., Pynoos, R., Zatzick, D. F., & Benedek, D. M. (2004). *Practice guidelines for the treatment of patients with acute stress disorder and posttraumatic stress disorder*. Washington, D.C.: American Psychiatric Association.

van Etten, M. & Taylor, S. (1998). Comparative efficacy of treatments for posttraumatic stress disorder: A meta-analysis. *Clinical Psychology and Psychotherapy*, 5, 126–145.

Wald, J., & Taylor, S. (2007). Efficacy of interoceptive exposure therapy combined with trauma-related exposure therapy for posttraumatic stress disorder: a pilot study. *Journal of Anxiety Disorders*, 21, 1050–1060.

9
Assessment and Evaluation: Collecting the Requisite Building Blocks for Treatment Planning

ARTHUR FREEMAN and SHARON MORGILLO FREEMAN

Contents

Introduction

Returning military personnel are a growing segment of the U.S. population that will require services from mental health professionals. A key service will be the assessment of skills, strengths, weaknesses, problems, supports, and losses for these individuals and their families. Without such data, counseling or therapy services will be vague and amorphous. Vague treatment goals lead to vague treatment which results in vague treatment outcomes.

For psychotherapy, institutional placement, vocational training, pharmacotherapy, or assessment of benefits to be most effective, they must all derive from, and be based on, a collection of data. The data collection process is labeled *assessment* or *evaluation*. It is the essential ingredient in any treatment plan inasmuch as it provides both the foundation and also the structural elements for treatment conceptualization. Without data, the formulations and strategies for helping an individual, couple, or family are based on conjecture, hope, theory-based hypotheses, or magical thinking. Ideally, mental health professionals receive information regarding the patient's personal history, military experience, physical evaluations, medical records, occupational history and experience, and psychological assessments and evaluations. Sometimes mental health professionals will be gathering this information in their role as case managers; other times they will do so in their role as a treating professional being asked to become therapeutically involved with an individual or family. The data contained in the assessment report are an essential element in the overall treatment planning process. To not use the data, to not understand the data, or to not understand the evaluation process by which the data were collected robs the clinician of important diagnostic material. Assessment is a multifaceted activity that serves as the substrate for the clinician's perspective of the patient, the nature of the problem, the frequency of the problem's occurrence, the amplitude or severity of the problem, and the associated antecedent and consequential social, physical, affective, and environmental factors as they relate to the development and maintenance of the disorders or problem constellation.

The role of the assessment specialist is to perform the assessment and give other treating professionals the data they need to perform the work of intervention and treatment necessary for these soldiers and their families and support systems. Successful treatment requires an organized, meaningful, and systematic data collection process. Our goals in this chapter are fourfold: (1) to place the assessment process within an historical context; (2) to describe the rationale for the assessment, both generally and specifically; (3) to describe the assessment tools seen as most valuable; and (4) to outline the steps for reporting the data garnered from an assessment. The following offers a view of the military perspective on the need and purposes of assessment (Bailey, 1929):

Circular No. 22. War Department, Office of the Surgeon General

Washington, August 1, 1917

Examinations in Nervous and Mental Disease

1. For the safety, efficiency, and economy of the military service it is highly essential that nervous and mental disease be recognized at the earliest possible moment. Nervous and mental diseases may, and frequently do, exist in persons who are strong, active, and apparently healthy and who make no complaints of disability. Such persons are, however, more than useless as soldiers, for they cannot be relied on by their commanders, break down under strain, and become an encumbrance to the Army and an expense to the Government. Disorders of this character are often demonstrable only as the result of a painstaking and special examination directed toward the mind and nervous system. This circular is published for the special purpose of calling the attention of medical officers to the particular diseases most frequently overlooked on general examination, and the symptoms most important to their diagnosis, and to certain characteristics in personality and in the behavior which might raise the question of the existence of mental disease.

2. The duties of the examiner are to be familiar with the symptoms and significance of nervous and mental disease and the means of eliciting them, and to recommend for rejection from service all those in whom any of the evidences mentioned in paragraph 4 are demonstrated. He should determine the importance of slight variations from the ordinary normal standard and recommend acceptance or rejection on the basis thereof. He should search for symptoms or tendencies which may be concealed for the purpose of obtaining service, and he should recognize symptoms which are feigned for the purpose of avoiding service. Organic nervous disease cannot be feigned in a way to deceive a skillful and careful examiner. To demonstrate feigned insanity a period of several weeks' observation may be necessary.

3. It is assumed that the examiner is familiar with the current methods of examination in neurology and psychiatry, and that he will make careful employment of them in all cases referred

to him for consultation. But, in addition to acting as a consultant to whom cases are referred, he must also himself select cases for special examination. To this end, he is directed to be present as often as possible when the recruits are gathered together at times of instruction and training and for such general medical purposes as vaccinations, inoculations, group examinations of the heart, lungs, etc. At such times he should discriminatingly observe the appearance and behavior of the recruits, pass in and out among them, converse with them when possible, and report to the camp surgeon the names of any whom his observations have led him to consider as requiring a special neurological and psychiatric examination. By thus learning, in a way, to know the recruits personally his special training should enable him now and then to pick out one who might pass the general medical examination and yet whom special examination would clearly prove to be a hazard to the Army.

Queerness, peculiarities, and idiosyncrasies, while not inconsistent with sanity, may be the beginnings or surface markings of mental disease. A soldier is too important a unit for such variations from a standard of absolute normality not to be looked into before the recruit who presents them is accepted for service. To aid the neurologist and psychiatrist in these ways the camp surgeon shall direct all medical officers, dental surgeons, instructors, hospital sergeants, and others who come in close contact with recruits to refer to him (the camp surgeon) all recruits who persistently show any of the following characteristics: irritability, seclusiveness, sulkiness, depression, shyness, timidity, overboisterousness, suspicion, sleeplessness, dullness, stupidity, personal uncleanliness, resentfulness to discipline, inability to be disciplined, sleepwalking, nocturnal incontinence of urine, and any of the various characteristics which gain for him who displays them the name of "boob," "crank," "goat," "queer stick," and the like.

The reaction of the pupils to light should be part of every medical examination, and if this is not systematically provided for the neurologist and psychiatrist should be directed to determine it. This could be done at the time of group inoculations and with the help of a hospital sergeant could be made rapidly. Electric light should be used. It is especially important in the examination of officers and recruits above 25 years of age.

It is further recommended to camp surgeons to provide neurological examinations for all cases of syphilis.

4. The following are causes of rejection for military service:
   a. Organic nervous diseases
   b. Mental defect
   c. Mental disease and pathological mental states
   d. Confirmed inebriety (alcohol or drugs)

As can be seen by the above 90-year-old War Department memorandum, by World War I army surgeons or consultants were actively working toward identifying those soldiers who might not fit well in military service.

History of Military-Related Assessment

Alfred Binet developed the first tests of intelligence in France, where he devised a series of verbal and nonverbal measures to identify children in need of special education. Binet believed that children who were in need of extra help could be identified by these tests, but importantly he argued that intelligence is not a fixed quantity and that it can be improved by further help. Subsequently, Binet's ideas have led to development of the commonly used concept of the intelligence quotient (IQ). Binet's notion was that a "normal" child had a mental age (MA) that was equal to their chronological age (CA); that is, a child 6 years of age would have a mental age of 6 years. This number was multiplied by 100 to determine the IQ. For our example of the 6 year old, MA/CA × 100 = IQ, and 6/6 = 1 × 100 = an "average" IQ of 100. In 1905, Simon and Binet made public their famous Binet–Simon Intelligence Scale, the first intelligence measuring device ever devised. Throughout his life after this point, Simon always remained critical of immoderate and improper use of the scale. He believed that its overuse and inappropriate use prevented other psychologists from achieving Binet's ultimate goal of understanding human beings, their nature, and their development. The scale was revised in 1908 and again in 1911.

Robert Yerkes believed that intelligence was a fixed quantity, and he set out to conduct one of the largest tests of intelligence in history. In 1915, mental testing did not enjoy much credibility, but Yerkes tried to change this. Yerkes was concerned with establishing psychology as a "hard" science and believed that using such a scientific approach to mental testing might be a promising route to achieve this. He believed that intelligence testing should be as rigorous as any other science, and he equated science with numbers and quantification. During World War I (1914–1918), Yerkes found an opportunity to promote the use and status of mental testing and therefore the status of psychology as a serious science. The American military gave Yerkes permission to carry out mental tests on over 1.75 million army recruits. By 1917, he had devised three types of mental tests.

The *Army Alpha* was a written test designed for literate recruits. It required writing and the use of a pencil, a skill that uneducated men might have lacked. The Alpha test had eight parts, such as analogies, filling in the missing number, and unscrambling a sentence. These types of test tasks have become common in modern intelligence testing and are a major part of the various Wechsler scales. The *Army Beta* test was a pictorial test for men who were deemed illiterate or who failed the Alpha. The Beta test had seven parts, including running a maze, numbers work, and a picture completion task that asked the examinee to identify parts missing from common objects (e.g., a rabbit with one ear). This more performance-based test and the data derived from it were considered adequate for the Army's purposes of recruit classification. Finally, Yerkes developed an *individual examination*, which was a spoken test for men who failed the Beta. The Alpha and Beta tests could be administered to large groups and took less than an hour to complete. The tests were scored A+ to E–, and it was assumed that men scoring lower than "C" would be inappropriate for officer training.

A major force in the development of intellectual assessment process and procedures was Lewis Terman. In 1910, he joined the faculty of Stanford University as a professor of cognitive psychology and remained associated with the university until his death. During World War I, Terman served in the United States military while conducting psychological tests. In 1916, he published the *Stanford Revision of the Binet–Simon Scale* (1916), and with Maud A. Merrill he published a second revision (1937) and a third revision (1960) based on previous work by Binet and Simon. Terman promoted his test, known colloquially as the *Stanford–Binet test*, as an aid for evaluating developmentally disabled children. It is used today as a general intelligence test for children and adults, although not without some controversy. The test underwent its fifth revision in 1999.

In the early 1920s, Hermann Rorschach, a Swiss Freudian psychoanalyst published what has become perhaps the most famous of all assessment tools, the *Rorschach Psychodiagnostik*. It was based on the premise that one could classify and codify a subject's interpretations of ambiguous inkblots. These were then seen to offer information into the subject's needs, wants, conflicts, personality style, attachment style, social style, and psychopathology. The scoring system was improved after his death by Bruno Klopfer (1946) and others. John E. Exner (1997, 2001, 2003) summarized some of these later developments in the comprehensive Exner system, at the same time trying to make the scoring more statistically rigorous.

In 1927, at the age of 33, Henry Murray became assistant director of the Harvard Psychological Clinic. Murray developed the concepts of *latent needs* (not openly displayed), *manifest needs* (observed in people's actions), *press* (external influences on motivation), and *thema* (comprised of the subject's perception of press and the personal needs that emerge from particular experiences). Murray used the term *apperception* to refer to the process of projecting fantasy imagery onto an objective, although sometimes vague, stimulus. The concept of apperception and the assumption that everyone's thinking is shaped by subjective processes provide the rationale behind the *thematic apperception test* (TAT). During World War II, Murray left Harvard and worked as a lieutenant colonel for the Office of Strategic Services (OSS), the forerunner of the CIA. James Miller, in charge of the selection of secret agents at the OSS during World War II, reported that Murray was the originator of the term *situation test*. This type of assessment, based on practical tasks and activities, was pioneered by the British Military and used for the selection of spies.

Lauretta Bender worked at Bellevue Hospital in New York City from 1930 to 1956. She was a contemporary of David Wechsler, who was Chief Psychologist at Bellevue Hospital in New York City. Bender is best known for her *visual–motor Gestalt test* (Bender, 1938). The test consists of the subject reproducing nine figures that are printed on cards. The *Bender–Gestalt* test, as it is now often called, is typically among the top five tests used as an assessment tool by clinical psychologists. It measures perceptual motor skills and perceptual motor development and gives an indication of neurological intactness. It has been used a personality test and a test of emotional problems.

In the late 1940s, Wechsler developed an IQ test for adults. Inasmuch as he worked at Bellevue Hospital, he called his measure the *Wechsler–Bellevue*. Wechsler's test became the basis for several subsequent measures that were more specialized, including the current Wechsler Adult Intelligence Scale, third revision (WAIS-III); the Wechsler Scale for Children, fourth revision (WISC-IV); and the Wechsler Preschool and Primary Scale of Intelligence, second revision (WPPSI-II). The Wechsler intelligence scales describe two areas of concern: verbal and the performance scales. The verbal scales subtests include Information, Comprehension, Arithmetic, Similarities, Digit Span, and Vocabulary. The performance subtests include Digit Symbol, Picture Completion, Block Design, Picture Arrangement, and Object Assembly.

The *Minnesota Multiphasic Personality Inventory* (MMPI) is one of the most frequently used personality tests in mental health. The test is used by trained professionals to assist in identifying personality structure and psycho-pathology. The original MMPI was developed in the late 1930s and early 1940s and was published in its final form in 1943.

The 1940s and 1950s witnessed the development and introduction of a large number of projective measures—that is, tests asking subjects to project their ideas about self and the world onto ambiguous stimuli. These included drawing tests such as the *House–Tree–Person test* (Hammer, 1958) or the *Draw a Person in the Rain test*, wherein the subject draws a person reacting to the rain, an external stressor. The *Rosenzweig Picture Frustration test* is a projective test administered to assess personality characteristics. The subject is shown scenes depicting moderately frustrating situations and asked what the frustrated person depicted would probably do or how the subject would react in such situations. Literally hundreds of tests, scales, measures, and inventories are available that can be used to assess virtually any aspect of human performance. Some are more well known than others, some are more easily used and scored than others, and some are limited in use by virtue of laws, ethics, and training. Psychological testing is most often seen as the province of psychologists and to a lesser degree counselors. Most test publishing companies request that purchasers of such tests submit their credentials for review by the company. Agencies and institutions may generate their own measures, but it is essential to recognize that these measures, although interesting, may lack validity (ability to measure what they purport to measure) and reliability (ability to consistently evaluate the same trait). Further, instruments may be designed for use with a particular population and cannot simply be used with all populations.

The General and Specific Rationale for Assessment

When Lewis Carroll's Alice came to a fork in the road, she wonders which way to go. The Cheshire Cat questions Alice: "Where do you wish to go?" "I don't know," she answers. "In that case," the Cat says, "any road will get you there."

Without a goal or target, any successful intervention will be a product of good luck. The conceptualization and resultant treatment planning require a set of discrete data to offer guidance and direction. Data must be used to build a conceptual model of the patient. This would include their problems, their environment, their ongoing and situational reinforcers, and the individual's goals and readiness to change. Part of the assessment model must include what the patient has done to unlearn his or her maladaptive behavior, what has worked, and what has not worked in favor of change.

The assessment process provides data about what the person does, the circumstances under which the behavior commonly and reliably occurs, and how often the behavior occurs. It will clarify what has been used by the individual or the system to decrease the behavior, what circumstances or timing are commonly and reliably involved in the increase in the behavior, how long the behavior lasts, and what is obtained, escaped, or avoided as a result of the symptoms (Bellack & Hersen, 1998).

The referral for assessment or treatment must be clear, unambiguous, targeted, discrete, and realistic. Referrals based on vague patient comments such as "I want to feel better," "We are having marital difficulty," "I just feel lousy about everything," or "Everything I touch turns to dirt" do not offer the clinician any sort of direction. Professional referrals based on mental-health jargon are equally problematic; for example, "low self-esteem," "poor self-image," "posttraumatic stress disorder," and "chronic depression" are terms that are vague and amorphous. Empathetic clinicians who respond with a heartfelt "uh huh" are still far from where they need to be. The assessment, then, is a focused, useful, parsimonious, systematic approach designed to facilitate the understanding of behavior and to provide a sound basis for clinical decision making and the development of effective behavior change strategies (Haynes et al., 1997; Haynes & Williams, 2003). Prior to the beginning of any assessment, it is necessary to clarify for the patient the purposes of the evaluation, the particular types of tools to be used, and the final disposition of the data. Allowing the patient to offer informed consent to the evaluation is an ethical responsibility. The exception to this is when the patient is mandated to be evaluated.

The Clinical Interview

It is in the clinical interview that the clinician's personal style, experience, training, and expertise will be valuable. By using the interview as a snapshot or microcosm of the patient's world, the patient's response speed, latency, response style, and mode of response to the interviewer all offer opportunities for data gathering. We find that information forms given to the patient prior to the interview and reviewed by the clinician offer a structure to the interview that is very important.

The clinician must differentiate between patient *complaints* and patient *problems*. Generally, when individuals seek therapy or are referred or remanded for therapy they often do so because of broadly painted symptom pictures that take the form of complaints. In fact, we may go so far as to state that complaints are far less treatable than are problems. The complaint of "anxiety," for example, may be viewed as far less treatable than a targeted list of discrete behaviors that negatively impact individuals, their world, and their future and form the syndrome that we have come to call "anxiety." The same can be said of the presenting complaint or referral for "depression." Depression is not a single phenomenon, but a constellation of feelings, thoughts, behaviors, and physical reactions. What becomes the focus of the assessment, and later the treatment that derives from the assessment, are the problems the patient has in effectively coping with an anxiety attack or a depressive episode. For one patient, the predominant complaint of "depression" might involve an unplanned weight change, sleep changes, and sadness. For another individual, the complaint of "depression" may involve suicidal thoughts or actions. For yet a third individual, "depression" might impact their work because they are having problems making decisions or acting on their decisions.

In some cases, the specific measures to be used are mandated by the referring agent or agency. Many of these measures are limited use only by psychologists who are trained and experienced in the selection, administration, data compilation, interpretation, and reporting of the assessment data. Many assessment tools are sold only to those who meet accepted minimal standards of training. The use of many of these instruments by unqualified individuals may be seen as a breech of ethical conduct. It is not enough to administer a test, even one scored by a computer, and then to use the computer interpretation to assess the patient's problem, nor is it enough to simply report data. The essential skill of the clinician performing the assessment is to integrate all of the accumulated data into a meaningful and linear "story." The clinician doing the assessment must account for data variations and differences, conflicts in data, or anomalous data. Because the assessment process has been initiated to clarify issues, the clinician must make sense of any conflicting data and may need to administer additional tests to clarify issues. As an example, a veteran who had completed 3 years of college prior to his military service and had maintained a grade point average of 3.3 was evaluated to have an IQ of 82. The veteran's success in college and his apparent low-average intellectual functioning seemed contradictory. Given the axiom in testing that it is very difficult to fall *uphill*, the evaluator in this case had to make sense of the disparate data. Evaluators are expected to find answers, not use terms such as "diagnosis deferred" or "can rule out [other related disorders]." It is our position that it is the job of the evaluator to make sense of the data and not pass off a final decision regarding the diagnosis to another professional.

Interpreting assessment results must take into account the purpose of the assessment, the patient's test-taking abilities (visual deficits, auditory loss, lack of facility in English, brain damage), or even malingering. The assessment report must address any limitations to the testing or the psychologist's ability to use the data. The data and test results are then reported in language that is nonjudgmental and avoids the use of abbreviations, jargon, and verbal shortcuts. The reports are to be brief but complete, understandable, and useable for all clinicians involved in the care of that patient.

The evaluator must be sensitive to issues regarding cultural diversity and differences. The choice of assessment tools may have to be dictated by the language proficiency of the patient. Using a verbal tool with individuals whose grasp of English is minimal would be inappropriate. The clinician must have access to a broad range of tests that will allow as complete an evaluation as possible, given any cultural factors that may prevent patients from doing their best. For example, use of the Test of Nonverbal Intelligence (TONI) would be preferred to the Wechsler Adult Intelligence Scale (WAIS) for individuals whose first or most used language is not English.

Many aspects of a patient's responses are judged subjectively, with the potential for the evaluator's own cultural exposure and background to color these assessments; for example, there is a major distinction between "different" and "abnormal." Correct or expected responses that are a function of upbringing or background may in fact have nothing to do with an individual's intellectual function. A question typically asked of children as part of an intellectual evaluation is: "What should you do if a child much smaller than you starts a fight with you?" Ideally, the child would answer, "Ignore him and just walk away." In some subcultures, though, allowing a smaller individual to get away with starting a fight identifies the bigger child as an easy mark and someone who may then be targeted by other children. An incorrect, although reasonable, response might be: "I would punch him in the face so he would never try that again." Similarly, tests of memory that require the subject to know U.S. history may not be an appropriate measuring tool depending on a person's country of origin, language skills, educational level, etc. These situations are unavoidable in the extremely diverse community in which we live. Although it is reasonable to expect people to be aware of certain basic facts (e.g., name, the year, the reason for visiting the hospital), it is also important to recognize that observation and interpretation of patient behavior and responses are colored by the evaluator's own life experiences.

The Mental Status Examination

The mental status interview has three main purposes: (1) organized collection of new data, (2) clarification of already collected data, and (3) behavioral observation of the patient. Achieving all three is integrated into the flow of the interview.

Appearance

This category covers the physical aspects of the person, including age, height, and weight and how the person is dressed and groomed. Is the individual dressed appropriately for the season (e.g., is the person wearing three coats and a sweater in the summer?).

Attitude and Response

The patient's dominant attitude as presented in the interview is important. Was the patient cooperative, poised, and comfortable or combatative, angry, sarcastic, and demeaning of the evaluator, agency, facility, or military?

Physical Behavior

How does the individual move and the position his or her body? Does the person have any unusual or abnormal movements such as tics or chorea? Also note the freedom of movement.

Speech and Language

Although speech reflects the patient's thoughts, both speech and language must be assessed with regard to the volume of speech, clarity of speech, enunciation, rate and flow of speech itself, any identifiable accent, speech handicap (lisp, stuttering, stammering), stress or lack of it, and hesitations. Evaluating the patient's use of language should include noting the range and use of vocabulary, correct usage of words, use of intimidating language, and pronunciation. Vocabulary is especially important in that the Vocabulary subtest on the WAIS is the test most highly correlated with overall verbal intelligence.

Affect and Mood

Affect is the outward show of emotions, and mood is the general pervasive emotional state as reported by the patient. A person's affect may range from depression to elation to anger to normality, but if the overall sense from the examination is one of depression then that is used to describe the mood. The range of the affect describes whether the person shows a full or even expanded range or if that person's affect is blunted or restricted. Cultural considerations are important in this evaluation. Appropriateness of the affect is also important. Is the emotion shown consistent with the topic being discussed? A patient with an inappropriate affect may cry talking about a parking ticket and show little or no emotion when discussing the recent death of a loved one.

Thought

This factor can be divided into the way a person thinks (thought process) and the thought content (what that person is thinking). The thought evaluation notes such features as the rate of thoughts and the way in which the patient's thoughts are connected and integrated. A formal thought disorder includes such processes as pressure of thought (thoughts are excessively rapid and appear to be forced), flight of ideas (one thought quickly and unexpectedly shifts to a related or unrelated idea), thought block (a stream of thoughts simply stops), disconnected thoughts (thoughts exhibit a loosening of association and derailment), tangentiality, and circumstantial thoughts (thinking may be overly inclusive and cover all areas even distantly related to the main point or direction of the thinking, or the thought process is slow to get to the point). Thought content includes those things discussed in the interview and the beliefs a person has. The person may be preoccupied with certain thoughts (e.g., ideas of reference, obsessions, ruminations, phobias) or may have overvalued ideas, first-rank symptoms (e.g., delusions of control, thought alienation comprised of thought insertion, thought withdrawal, broadcasting of thoughts, delusional perception, somatic passivity), or delusions (e.g., paranoid, persecutory, religious, erotomanic grandiose, reference, somatic). Depressed persons may have ideas that may seem to border on delusional (e.g., delusions of hopelessness, helplessness, and worthlessness).

Perceptions

Perceptions involve the way in which people relate to the world through their senses. Findings here might include such distortions as illusions, delusions, or hallucinations. Auditory hallucinations are common in schizophrenia, and visual disturbances are more common in organic problems. In addition, a person can have gustatory, olfactory, tactile, somatic, or kinesthetic hallucinations. Also included in the realm of perceptions are depersonalization, when the person feels unreal, and derealization, when the person feels that his or her surroundings are unreal.

Cognition

An evaluation of cognition looks at a number of areas, such as the level of abstract thought, which declines or is absent in a number of conditions such as dementia and schizophrenia; the level of general education and intelligence; and the degree of concentration, which is often tested by asking the patient to count down by sevens from 100 (i.e., 100, 93, 86, 79, ...). There is an expectation that adults would be operating at the level of what Piaget (1955 and Rosen (1985) termed *formal operations* which involves the ability to generalize and use abstract thinking. In point of fact, Nigro (2006) found that many adults have not achieved formal operational thinking and operate at a level that Piaget termed *concrete operations*.

Consciousness

The level of the patient's conscious state is assessed in terms of whether the individual's consciousness is steady or fluctuating. A second evaluation would be whether the consciousness is clouded or clear.

Orientation

An evaluation of orientation looks at whether the person knows the time (including the date), place (where he or she is), person (who he or she is), and situation (that he or she is in); for example, the author was asked to interview a patient who had been admitted the previous evening to the psychiatry unit of a general hospital. The patient had been interviewed and was judged to be selectively mute, likely as a result of psychosis. When interviewed, the patient's response to questions regarding orientation to person, time, and place were lacking. When asked these orientation questions, the patient would simply stare at the floor. The author then tried a different approach. Rather than asking for the data, the interviewer asked whether the patient knew who he was. This was to be indicated by the patient shaking his head in the affirmative or negative. The question "Do you know your name?" elicited a head nod. Similarly, questions regarding date and place were affirmed by a nod. On a hunch, the author asked, "Are you trying to figure out the best or most complete answer?" This also was affirmed. Similar questioning revealed this patient's internal process when he was asked for his name: The patient's name was Alan James Jones. He generally signed his name as A. James Jones, but he preferred to be called Jim. As he was trying to decide the best answer to give regarding his name, he was asked another question: "Do you know where you are?" Should he say that he was in the emergency room, the emergency room of the county hospital, or the psychiatry office in the emergency room? The answers were not quick in coming because Jim was trying to decide which answers were best.

Memory

Memory is tested in several ways. The evaluator first evaluates the person's immediate recall, or short-term memory (the ability to remember several things after 5 minutes) and then the person's long-term memory (the ability to remember distant events such as the childhood experiences). Memory can also be tested for verbal stimuli and for visual stimuli.

Judgment and Decision Making

The evaluator must look at both the process and the content of the patient's judgment and decision making. Does the patient make judgments according to a problem-solving strategy or through a process of trial and error? Is the judgment reasoned and logical or is it idiosyncratic and unusual? Does the patient look at a number of possibilities before making a final judgment or use only a limited number of coping strategies?

Insight

Insight describes how much understanding or awareness the person has of his or her own psychological functioning or disturbance and the reactions of others.

Impulse Control

How well the individual is able to control both specific and broader types of impulses must be noted. If there are any situations or circumstances that are more likely to trigger specific impulsive behavior, the evaluator must get as complete a list of these stimuli as possible.

Vulnerability and Threshold of Response

Environmental circumstances, situations, or deficits can have the effect of decreasing a patient's ability to effectively cope with life stressors or to see available options. These vulnerability factors work to lower the patient's threshold or tolerance for life stress situations and alone or in summative combination may serve to lower the patient's threshold for stress or increase the patient's vulnerability to depressogenic thoughts and situations (Freeman & Simon, 1989). A vulnerability factor has the effect of decreasing the threshold so events that were previously ignored or never noticed now shout loudly for attention. The vulnerability factors also serve to make the individual more sensitive to internal and external stimuli that may prompt suicidal thoughts and actions.

Individuals have threshold points above which they control their responses to external or internal stressors. There are factors that serve to lower the threshold and thereby increase the individual's vulnerability and acting-out potential. When several factors are operative, they have a summation effect that lowers the individual's ability to cope until any further lowering becomes the proverbial straw that breaks the camel's back. It is at this point that suicidal action can occur. A tenet of Alcoholics Anonymous is to make members aware of the acronym HALT (hungry, angry, lonely, and tired). AA emphasizes that these are conditions under which an individual is most likely to lose control and resume drinking.

Suppose an individual's trigger point to act on a suicidal thought is 65 (on a scale of 1 to 100). If this person's usual threshold for response is 75 and the usual stress of life is 60, they are in good control, but the presence of a vulnerability factor could lower the individual's threshold to 67, which leaves this individual in tenuous control. The introduction of another vulnerability factor could lower the threshold to 63, thus triggering a response.

The vulnerability factors include

1. *Acute illness or injury*—Acute illnesses may run the range from being severe and debilitating to more transient, such as headaches or a viral infection.
2. *Chronic illness*—Chronic health problems can exacerbate suicidal thinking.
3. *Deterioration of health*—Aging can result in a loss of activity and an inability of the body to perform up to the person's expectations.
4. *Hunger*—During times of food deprivation, individuals are more vulnerable to a variety of stimuli, particularly food-related stimuli. In literature, Jean Valjean was willing to break the conventions of society. Studies have indicated that individuals should not shop for food when they are hungry because they are likely to purchase more than they intended.
5. *Anger*—Angry individuals appear to lose appropriate problem-solving abilities. They may lose impulse control or may respond to stimuli that might normally be ignored.
6. *Fatigue*—In a similar fashion, fatigue affects both problem-solving strategies and impulse control.
7. *Loneliness*—When individuals see themselves as isolated and apart from others, leaving this unhappy world may seem to be a reasonable option.
8. *Major life loss*—After losing a significant other through death, divorce, or separation, individuals often see themselves as having reduced options or they no longer care about what happens to them.
9. *Poor problem-solving ability*—Some individuals may simply have an impaired problem-solving ability. This deficit may not be obvious until the individual is placed in situations of great stress, as dealing with everyday, minor problems may never test an individual's ability.

10. *Substance abuse*—Substance abuse can increase suicidality. In acute cases the patient's judgment is compromised primarily during periods of intoxication, whereas in chronic cases the patient's judgment is more generally impaired.
11. *Chronic pain*—Chronic pain may have the effect of causing the individual to view suicide as a means of eliminating the pain.
12. *Poor impulse control*—Many individuals have poor impulse control because of organic or functional problems. Patients with bipolar illness or borderline, antisocial, or histrionic personality disorders may all have impulse control deficits.
13. *New life circumstances*—Changing jobs, marital status, homes, or family status are all stressors that are vulnerability factors.

General Information Regarding the Patient

General information about the patient includes the reason for referral, chief complaint, history of problems, medical history, substance abuse history, any abuse of drugs, psychiatric history, social history, family history, legal history, family history regarding any and all of the above issues, and military history. When possible, corroboration of the patient report can be done through clinical interviews with family members or significant others. Previous assessment data, school records, military records, legal records and complaints, and hospital or institutional records can all offer clarifying data. The DSM-IV-TR (APA, 2000) requires that for a diagnosis of antisocial personality disorder (ASPD) the patient must have a childhood or adolescent history of conduct disorder. In this case, the patient report may not be accurate, because ASPD patients are likely to lie about their past to appear to have a less problematic history or to appear to be tougher or more threatening than they really are. Among the most important outcomes for the assessment is a tentative diagnosis based on a DSM-IV-TR diagnosis. It must be recognized by the clinician that *all* diagnoses are provisional and can change with the collection of additional data. Further, a diagnosis may be changed as a result of the passage of time, circumstances different than those for the previous evaluations (e.g., an earlier evaluation was done in an emergency room at a point of crisis for the soldier).

Drug Use and Abuse

It is essential to review the patient's previous records. Such records would include medical, arrest, and criminal records; substance abuse treatment; court reports; police reports; and previous evaluations. The evaluator must be familiar with various substances of abuse and their regional variations, availability, mode of administration, effects on functioning, and costs. Further knowledge of central nervous system (CNS) stimulants (amphetamines, cocaine, caffeine, nicotine) and CNS depressants (barbiturates, nonbarbiturates, tranquilizers, alcohol, narcotics, hallucinogens, inhalants) is also necessary. Refer to Chapter 14 for more information about substance use disorders.

Defining the Target Behavior

A useful tool for assessment of baseline and future change is the Freeman Diagnostic Profile (FDPS) (Freeman & Freeman, 2005). This profile is based on DSM-IV-TR criteria (APA, 2000) and requires that clinicians review each of the criteria with their patients. Ranking each criterion on a scale from 0 to 10 will result in a profile of the contribution of a particular criterion (problem) to the overall depression (complaint). The FDPS is DSM-IV-TR based and is easily used. It focuses on problems rather than complaints and separates problems into component parts. This aids the clinician's ability for treatment planning and can be used to show patient areas of greatest concern. It can also be useful for evaluating change and progress and to help supervisors identify patients' problems.

Patient Report

The patient's report of his or her life experience is important in that it is often the first statement of the complaint and is told from the patient's view. Patients' experiences are the direct result of their construction of their world, their understanding of themselves, and their view of the future. This cognitive triad is a useful tool for treating the depression (Beck et al., 1979; Freeman, Pretzer, Fleming, & Simon, 2004).

Report of Significant Others

The information gleaned from the patient's significant others is not meant to replace or challenge the patient's view but rather to corroborate the patient's view, add information that the patient may not have or remember, and offer an outsider's view. A patient may report experiencing depression since his teenage years, but a parent or older sibling may recall depressive episodes at an even earlier age. These differing perspectives can offer multiple views that are, in effect, contradictory. To evaluate a revision of the Beck Depression Inventory (BDI-II), significant others and staff members were asked to assess the level of depression of a patient by using a version of the BDI-II that was dubbed the BDI-O (for Beck Depression Inventory–Other Report). The BDI-O was reworded in the third person so, for example, "I feel sad" was rewritten as "He/she feels sad." The results of the BDI-O were then correlated with the patient self-report. Significant differences were found between patient and staff and within the staff. One problem was that the staff differed in experience, modality of treatment, and training and standardization of the BDI-O. Informants may shed light on some aspect of a problem behavior about which the individual does not have access or awareness. An important byproduct of this process is that it may ultimately help those in the individual's environment learn how their own behavior may be intimately tied to maintaining the problem behavior of the individual.

Reports of Other Professionals

We view it as essential to obtain reports from any and all previous treating therapists. It would be important to factor in the view of these other therapists along with what was tried in therapy, what worked, and what did not work with a particular patient. Even if the previous therapists had very different theoretical orientations, their view of the clinical behavior can be useful. Reports from occupational therapists, physical therapists, or any other treating professional can all shed light on the patient's style of responding, motivation, and adherence to therapeutic regimen.

Medical History

Data that can be garnered from a patient's medical provider can shed light on medical, physiological, organic, or medication-related problems. If a patient has not seen his or her medical provider within the last 6 moths prior to the evaluation, a check-up with full blood work is essential; for example, a slow and fatigued presentation may be the result of depression or of hypothyroidism. In the event of the latter, it would be better treated medically.

Case Conceptualization

To develop accurate assessment plans, case conceptualizations, and, ultimately, effective treatment plans, cognitive–behavioral therapists must carefully assess the features, context, and manner in which an individual's difficulties develop (Thorpe & Olson, 1997). Persons (1989) and Needleman (1999) offer clinically useful models. The case conceptualization is a template for understanding individuals. The conceptualization can be tested: Does the conceptualization account for the patient's past behaviors, explain the patient's current behaviors, and predict the individual's future behavior (Needleman, 1999)? The formulation is directly linked to the assessment. As a higher order process, case conceptualization firmly rests upon the careful collection, evaluation, and interpretation of valid and reliable behavioral assessment data. The quality of behavioral assessment data directly affects the quality of the formulation. A poorly conceived and implemented behavioral assessment plan could misinform the conceptualization process and ultimately undermine treatment.

The assessment data fuel the case conceptualization process by providing clinically relevant information that helps clinicians and the patients understand their problems more fully. These data are integrated and synthesized with other relevant information about the individual and form a solid foundation for the selection of specific treatment protocols. Finally, this information is helpful in predicting barriers to treatment.

Treatment Planning

Treatment planning and implementation are critical to successful therapy. Both are linked to the therapist's ability to generate clinical hypotheses and develop, refine, and tailor treatment to the individual's needs. Behavioral assessment helps the clinician formulate case-specific treatment plans that are of direct relevance to the individual's treatment (Persons, 1989; Needleman, 1999). Assessment allows the clinician to reduce target problems into observable and measurable units. It also informs the treatment process in an ongoing manner; for example, baseline data provide the clinician with important information about the state of the individual's problem before an intervention has been made. During the course of treatment, the clinician expects that if the treatment is appropriate then changes in the critical aspects of the problem will occur in the desired direction.

Assessment of Malingering and Deception

An important assessment goal of evaluating returning military is the differentiation and identification of malingering. When malingering is unidentified or misunderstood, the label can be inappropriately applied to the veteran and used as an excuse for refusing that individual's services. When overused or misapplied, the label can lead to the reinforcement of symptoms for some gain. Often referred to as *compensation neurosis*, malingering is "the intentional production of false or grossly exaggerated physical or psychological symptoms, motivated by external incentives such as avoiding military duty, avoiding work, obtaining financial compensation, evading criminal prosecution, or obtaining drugs" (APA, 2000, p. 683). In addition to using the label of malingering to avoid military duty, returning service members may use malingering to avoid many other life experiences and perceived stressors. The assessment must differentiate between two other disorders that share some common factors with malingering. These would include factitious and somatoform disorders. The defining factor, according to DSM-IV-TR, is that in factitious disorder "external incentives are absent" and the demonstrated behavior is the result of an "intrapsychic need to maintain the sick role" (p. 298). Malingering is differentiated from somatoform disorders by the intentionality of the symptom production in malingering, whereas in somatoform disorders the symptom production appears to be out of awareness or unconscious. Factors to consider in assessing for malingering include

- The person is being evaluated upon the referral of an attorney.
- Objective findings differ markedly from the level of stress or disability alleged by the patient.
- The individual is not cooperating with the evaluation, diagnosis, and treatment process.
- The individual has been diagnosed with antisocial personality disorder.

- The individual exaggerates his or her symptoms beyond what is realistic (i.e., that seen in population base rates).
- The individual has a history of irregular employment after military service.
- The individual expresses chronic dissatisfaction with his or her job.
- The individual has a history of claims for service-related injury.
- The individual is able to recreate, shop, and socialize but not work.
- The individual describes nightmares that *exactly* replicate the original trauma.
- The individual makes evasive or contradictory statements.
- Psychological testing calls into question the individual's truthfulness.
- The individual displays a flawed understanding and reporting of anatomy (i.e., the patient reports symptoms that are physically unlikely or anatomically unrealistic).

The MMPI-2 is an excellent test for identifying the factor of "faking bad" (i.e., trying to make the symptom picture and problems even worse that they are).

Assessment of Posttraumatic Stress Disorder

The topic of posttraumatic stress disorder (PTSD) is dealt with in greater detail elsewhere in this book, so we will limit our discussion here to some basic questions and issues that the evaluator examining the possibility of PTSD must address. The patient's pretrauma functioning serves as a baseline. This baseline must come from the patient and be corroborated by other people or an objective source (e.g., school records, psychological reports, or psychiatric report). Other issues include the individual's trauma exposure, length of exposure, symptoms of acute stress, manifestations of the PTSD over time, current social supports, any evidence of comorbidity with other disorders (e.g., borderline personality disorder), evidence of malingering, and secondary gain or financial profit from the disorder.

The elements for the diagnosis of PTSD include

- The patient has been exposed to a traumatic event involving intense fear, helplessness, or horror *and* the patient experienced or witnessed an event or events that involved actual death, threatened death, serious injury, or a threat to the person's physical integrity or to the physical integrity of others.
- The traumatic event is persistently reexperienced through intrusive recollections, distressing dreams, or acting or feeling as if the traumatic event were recurring.
- The patient demonstrates persistent avoidance of stimuli associated with the trauma.
- The patient demonstrates numbing of general responsiveness.
- The patient demonstrates increased arousal not present before the trauma.

- The above-noted symptoms have a duration of more than a month.
- The patient demonstrates impaired social, occupational, relationship, or other areas of functioning that were generally acceptable before the trauma.
- The symptoms can be acute (<3 months) or chronic (≥3 months).
- The symptoms may appear after a period of 6 months from the stressor and is noted as delayed onset.

The MMPI or MMPI-2, the Trauma Symptom Inventory (Briere, 1996), or the Posttraumatic Stress Diagnostic Scale (Foa, 1995) are all useful diagnostic tools.

Summary

Effective treatment of the broad range of psychological disorders must begin with a complete and focused assessment. For the assessment to be effective and useful it must be data based and the result of the evaluator using a number of different tools for data collection. These include a personal interview; meeting and interviewing family and significant others; seeking corroborative data from school, military, and medical reports; psychological testing; and a review of the patient's life experience. The experienced clinician will always have a gut response to the presenting data, response style, and appearance of the patient. These gut reactions should be treated as reasonable hypotheses and confirmed through some more data-oriented methodology. It must be noted that all diagnoses are tentative or provisional and can (and should) be changed when new data are collected or collected data are modified. Assessment terms such as "deferred" or "rule out" are frequently used but are often of little value to the treating clinician because it then places the burden for diagnostic clarification on the clinician. As much as possible, the evaluator needs to be able to answer the diagnostic questions. The data from the assessment are the clay from which the diagnostic and treatment conceptualization is molded. The conceptualization is the foundation for the treatment. Thorough assessment will lead to more discrete and focused treatment planning. The comprehensive and thorough treatment plan will allow for more focused therapy, and the more focused the therapy, the better the treatment result.

References

APA. (2000). *Diagnostic and statistical manual of mental disorders* (4th ed., text revision). Washington, D.C.: American Psychiatric Association.

Bailey, P. (1929). *Neuropsychiatry in the American expeditionary forces: Detection and elimination of individuals with nervous or mental disease.* Washington, D.C.: U.S. Government Printing Office.

Beck, A. T. et al. (1979). *Cognitive therapy of depression.* New York: Guilford.

Bellack, A. S., & Hersen, M. (1998). *Behavioral assessment: A practical guide.* Boston: Allyn & Bacon.

Bender, L. (1938). *A visual motor Gestalt test and its clinical use,* Research Monograph No. 3. New York: American Orthopsychiatric Association.

Briere, J. (1996). Psychometric review of the Trauma Symptom Checklist-40. In B. H. Stamm (Ed.), *Measurement of stress, trauma, and adaptation* (pp. 381–383). Lutherville, MD: Sidrian Press.

Cone, J. D. (1997). Issues in functional analysis in behavioral assessment. *Behavior Research and Therapy*, 35, 259–279.

DiTomasso, R. A., & Colameco, S. (1982). Patient self-monitoring of behavior. *The Journal of Family Practice*, 15(1), 79–83.

Exner, J. E. (1997). The future of Rorschach in personality assessment. *Journal of Personality Assessment*, 68(1), 37–46.

Exner, J. E. (2001). A comment on "The misperception of psychopathology: Problems with the norms of the Comprehensive System for the Rorschach." *Clinical Psychology: Science and Practice*, 8, 386–388.

Exner, J. E. (2003). *The Rorschach: A comprehensive system*. Vol. 1. *Basic foundations and principles of interpretation* (4th ed.). Hoboken, NJ: Wiley.

Foa, E. B. (1995). *Posttraumatic stress diagnostic scale*. Minneapolis, MN: NCS Pearson.

Freeman, A. (1989). The development of treatment conceptualization in cognitive therapy. In A. Freeman and F. M. Dattilio (Eds.), *Comprehensive casebook of cognitive therapy* (pp. 13–26). New York: Plenum.

Freeman, A., Pretzer, J., Fleming, B., & Simon, K. M. (2004). *Clinical applications of cognitive therapy* (2nd ed.). New York: Kluwer Academic.

Hammer, E. (1958). *The clinical application of projective drawings*. Springfield, IL: Charles C Thomas.

Haynes, S. N., & Williams, A. E. (2003). Case formulation and the design of behavioral treatment programs: matching treatment mechanisms to causal variables for behavior problems. *European Journal of Psychological Assessment*, 19(3), 164.

Haynes, S. N. et al. (1997). Design of individualized behavioral treatment programs using functional analytical clinical case models. *Psychological Assessment*, 9(4), 334.

Klopfer, B. (1946). *The rorschach technique: A manual for a projective method of personality diagnosis*. Yonkers-on-Hudson, NY: World Book.

Morgillo Freeman, S., & Freeman, A. (Eds.) (2005). *Cognitive behavior therapy in nursing practice*. New York: Springer.

Needleman, L. D. (1999). *Cognitive case conceptualization: A guidebook for practitioners*. Mahwah, NJ: Lawrence Erlbaum Associates.

Persons, J. B. (1989). *Cognitive therapy in practice: A case formulation approach*. New York: Norton.

Piaget, J. (1955). *The child's construction of reality*. London: Routledge.

Rosen, R. (1985). *Anticipatory systems: Philosophical, mathematical and methodological foundations*. London: Pergamon Press.

Taylor, S. (1999). Behavioral assessment: Review and prospect. *Behavior Research and Therapy*, 37(5), 475–482.

Thorpe, G. L. and Olson, S. L. (1997). *Behavior therapy: Concepts, procedures, and applications*. Boston: Allyn & Bacon.

Wolf, T. H. (1961). An individual who made a difference. *American Psychologist*, 16, 245–248.

Wolpe, J. (1973). *The practice of behavior therapy*. New York: Pergamon Press.

Resources

Assessment resources for postdeployment military include these websites:

http://www.behavioralhealth.army.mil/post-deploy/index.html
http://www.ncptsd.va.gov/ncmain/providers/

III
The Individual Service Member—Intervention

10

Theoretical Base for Treatment of Military Personnel

ARTHUR FREEMAN and BRET A. MOORE

Contents

Introduction

Service members, of whatever service branch, begin life as civilians. They are born, develop, and educated, most likely within a civilian milieu. At some point in life they make a decision to enter the military. The reasons for this are personal to the individual; however, when they are sworn into the military they become integral parts of a team. For some, teamwork is something they have experienced throughout life; they have been part of a family team, a sibling team, a team of friends, sports or academic teams, and now a military team. For others, being part of a team is a new and unusual experience. For experienced team members, team rules and structure are familiar friends. For novices, a great deal of education must occur to help them attain team awareness, team familiarity, and finally "teamhood" (i.e., membership in the group). This is accomplished through athletic exercises, competitions between individuals and groups, direct instruction, or group encounters and learning to overcome stressful encounters. It is also inculcated via aphorisms, mottos, and calls to tradition; for example, Army recruits are taught to recite the *Warrior's Creed*:

> I am an American Soldier.
>
> I am a Warrior and a member of a team. I serve the people of the United States and live the Army values.
>
> I will always place the mission first.
>
> I will never accept defeat.
>
> I will never quit.
>
> I will never leave a fallen comrade.
>
> I am disciplined, physically and mentally tough, trained, and proficient in my warrior tasks and drills.
>
> I always maintain my arms, my equipment, and myself.
>
> I am an expert and I am a professional.
>
> I stand ready to deploy, engage, and destroy the enemies of the United States of America in close combat.
>
> I am a guardian of freedom and the American way of life.
>
> I am an American Soldier.

The values, training, culture, and ethos of the military are well described throughout this volume. Our goal in this chapter is to describe a specific cognitive–behavioral psychotherapeutic treatment model that has been developed and used with a broad range of clinical populations and clinical problems and in a range of settings and situations. We first offer a brief introduction to cognitive–behavioral therapy (CBT) and then discuss assessment, treatment conceptualization, and treatment planning. The next part addresses specific cognitive, affective, behavioral, and situational factors. Finally, we offer a clinical example and a summary of the multiple problems encountered by returning service members.

Cognitive–Behavioral Therapy

Many current and former military patients are initially referred because of problems with depression or anxiety. Often lost in the process are the more complex comorbid issues of posttraumatic stress disorder (PTSD) and personality disorders. Patients may seek therapy with a mélange of problems and so many symptoms that the therapist is often at a loss to know where to begin. Freeman (2005) has termed this *symptom profusion*. These individuals may come to therapy under coercion, threat, or simply to get others (family, friends, senior officers) off of their backs. Some who have sought help from primary-care medical providers are referred for therapy when primary-care providers are overwhelmed and need help with comprehensive treatment of these service members. No one treatment can possibly meet the needs of every patient, nor is it likely that a single therapist can be a master of many different treatment models. Finding the appropriate therapist–patient match is essential. This can be done within the broad spectrum of CBT.

Below we describe numerous attributes of CBT. By essentially mixing and matching these attributes, the therapy can be structured to best meet the needs of the patient at a particular time and in any specific circumstance. CBT is active, motivational, directive, collaborative, dynamic, problem oriented, and solution focused, and it has a here-and-now emphasis. It is psychoeducational, structured, time limited, prescriptive, culturally relevant and informed, empirically supported, cognitively appropriate, and integrative, and it uses a one-session model for treatment.

Active

The CBT model requires that the patient be involved in the treatment process. The degree of involvement is negotiable and often dependent on the patient's perception of the therapy, the therapist, and the therapeutic process. The patient who enters therapy to be "worked on" will not fully profit as it means that only half of the therapy dyad is working on the referral issues. Therapists often ask, "What do I do with a patient who has no interest in therapy, refuses to participate, refuses to do homework, denies any and all problems, and consistently misses appointments?" The answer is that perhaps little can be done until the next point is addressed.

Motivational

Many individuals, both military and nonmilitary, must be motivated to be involved in the complex and sometime frightening process of psychotherapy. Through the use of motivational interviewing (Arkowitz, Westra, Miller, &

Rollnick, 2007; Murphy, 2007). If the therapist can assist the patient in increasing his or her motivation for change, then the patient's activity and involvement in the therapy will, ideally, also increase.

Directive

The therapist must be the director, coach, and choreographer of the therapeutic process. The nondirective stance espoused by Rogerians (though not Rogers, in his later work) will be of limited value. If the therapist is in his or her consulting room with a patient, one of them needs to have an idea as to where the process is going or needs to go. The idea of following the patient because only the patient knows where he or she is headed is a questionable thesis. We would submit that if patients knew where to go, they likely would not go for therapy. Another factor to consider is that military or former military are quite used to operating in a hierarchical system of ranks. This can be used to improve both motivation and activity level.

Collaborative

Collaboration is not always 50/50. A well-used aphorism in therapy is that the therapist should not work harder than the patient. The patient must do the bulk of the work. Again, this is a questionable thesis. Initially, the collaboration may be 90/10, with the therapist providing 90% of the therapy collaboration. As the patient's motivation increases, the collaboration may shift to 80/20. The result of successful motivational work may eventually bring the collaboration to 60/40. The patient may never top 40% of the work. This is, in fact, why certain patients or patient groups (e.g., adolescents) require so much more energy and can lead to a greater frequency of therapist burnout.

Dynamic

The dynamic level of CBT involves an understanding of the patient's personal, cultural, social, family, gender, and age-related rules, termed *schema*. Schema are the templates that all persons build to help them understand their world. In all cases, successful service members will have inculcated the basic schema of their service: "Once a Marine, always a Marine" and *semper fi* ("always faithful"). From books and movies, many nonmilitary individuals are familiar with the motto of West Point: "Duty, Honor, Country." These mottos, official and unofficial, are partly what draws individuals into military service and builds the *esprit de corps* of the unit, the larger military group, and the particular branch of the service in general.

Problem Oriented

Therapists often confuse problems with complaints. Complaints may bring an individual into therapy (e.g., "I'm depressed"), but it is the therapist's job to take the complaints and turn them into workable problem lists. Being depressed, for example, may involve low libido, problems with making decisions, negative self-image, thoughts of hopelessness and suicide, sleep or eating disturbances, and a loss of humor. The particular symptoms and their degree of severity are the problems that must be addressed in therapy. Depression is too vague a term and cannot be treated except as a constellation of contributing factors or symptoms.

Solution Focused

Insight, although interesting and important, is not sufficient to bring about change. The goals of therapy must be to assist patients to develop solutions for their problems. These solutions are designed not for cure (a questionable concept) but to move the patient toward better coping. Most important, however, is to help the individual develop a problem-solving strategy for more effective coping. Often, individuals can be superb copers in one arena but somehow fail to apply those same problem-solving skills in other settings and circumstances.

Here and Now

The issues for the CBT therapist are not so much *why* someone is having difficulty but rather *what* maintains their difficulty and *how* change be managed. What rules any individual learns throughout life (schema) will be present and obvious in the here and now. The therapist can deal with issues currently confronting the patient by addressing the present tense and not spending therapy time on the constant or frequent review of what was. The schematic roots of these issues will be found in the schema developed in the person's family of origin.

Psychoeducational

We cannot make the assumption that all persons have the same physical skills nor can we assume that all persons have the same visual scanning ability. A major part of what creates problems for many individuals are general or specific skill deficits in one or many areas; for example, children may have poor attentional skills. Ideally, part of the schooling process is to learn to focus one's attention, to pick out salient details from a Gestalt, to learn to survey one's surroundings, to identify familiar and unfamiliar objects, and to be able to shift one's attention, as needed. The same may be said of social skills. An individual may have the skills to relate to first-degree relatives at home but never master the more complex skills of establishing contact with a stranger, initiating a conversation, maintaining a connection, and then learning how to successfully end the encounter. CBT is a skill-building model, where the behavioral work will focus on skill identification, skill acquisition, skill rehearsal, and evaluation of skills within and without the therapy session.

Structured

For individuals who have been trained to be mission oriented and have learned about the importance of structure as a survival mechanism, a structured element in therapy is important. In fact, free-floating and unstructured therapy may be contraindicated. The therapy must be structured in two ways. First, the overall goal of therapy, addressing the problem list, must remain in constant focus. The therapist's job, in this regard, is to maintain the agreed-upon emphasis throughout therapy. The addition of new issues or items should be negotiated and problem solved with the patient. This provides a focused approach to dealing with life issues. The second type of structuring would be the session itself. At the outset of the session, the therapist and patient should agree on the issues, problems, difficulties, or tribulations to be addressed in that session. The therapist's role is to work at setting an agenda that is broad enough to gain movement in the therapy and narrow enough for the therapy to be successful.

Time-Limited

Time-limited therapy is not necessarily a fixed number of sessions; rather, it is a way of thinking about doing therapy. As an example, the therapy can be organized in modular form, such that a certain number of sessions will focus on a particular area of concern. Modules can be anywhere from 1 to 10 sessions. Advantages of the modular system are that it gives the therapist a way to plan the therapy and the patient does not become overwhelmed by therapy *ad infinitum.*

Prescriptive

The cognitive–behavioral therapist literally writes a prescription or treatment plan addressing how much of the therapy work will be cognitive, how much behavioral, how much focused on the family or interpersonal issues, what intrapersonal issues would need to be addressed, and what emotional issues will be addressed. The prescription is shared with the patient so the patient can become a stronger collaborator who is more motivated and more solution focused. The particular mix of techniques will change with every patient according to the patient's particular needs.

Culturally Relevant and Informed

By focusing on the individual's schema, CBT maintains a strong emphasis on the individual's culture. The therapist must be respectful of differences but cannot be knowledgeable about the broad range of cultures, subcultures, religions, or geographic identifications presented in today's military. By addressing the schema, the therapist becomes aware of the rules that a particular patient lives by and operates under, not the least of which is the military culture.

Empirically Supported

Among the most important and impressive attributes of CBT are the ongoing empirical studies that support the various applications. These studies do not validate the model but undergird and support the model. The history of psychotherapy has been filled with case reports of successful treatment of various disorders, often by charismatic therapists; these reports are then used to justify or support the use of a particular model or technique. Similarly, studies in one area, such as depression, that demonstrate empirical support are then seen to justify a particular model in the treatment of many other related and nonrelated emotional disorders.

Cognitively Appropriate

A common expectation is that adults have moved into what Piaget had theorized was the stage of formal operations. During this stage, the individual would be able to conceptualize and make use of abstractions, (e.g., therapeutic interpretations). Studies by Nigro (2007) have shown that many adults are either still in the stage of concrete operations or on the cusp of achieving formal operational thinking. CBT can easily be structured to direct the therapeutic work to the cognitive level of the individual. Clearly, behavioral work would ft within the concrete operational level.

Integrative

Cognitive–behavioral therapists utilize a broad range of strategies and interventions under the broad umbrella of a cognitive behavioral framework. In-office individual sessions can be combined with group therapy, family therapy, couples therapy, inpatient treatment for respite or detoxification, and psychopharmacological treatment. The various treatment elements are not used in an eclectic mode (i.e., everything works), but rather in an integrative mode where each element fulfills some part of the treatment plan. The CBT therapist will then work collaboratively with other therapists and with medical providers.

Single-Session Treatment

The final attribute of CBT is its emphasis on single-session treatment. This is not meant to suggest that the patient will only be seen for a single session but rather that each session will have a beginning, middle, and end. If the patient does not return for another session, at least there will be closure for the portion in which the patient was involved.

Assessment

The primary step in CBT is assessment. The collection of data begins with the first contact. It is at this point that the therapist begins to internally generate hypotheses regarding the patient. The data will help to confirm or disconfirm

Table 10.1 Deriving CBT Treatment from the Problem

Problem	Treatment
Compulsive/limited behaviors	Build behavioral repertoire.
Inflexible	Increase range of movement.
Thoughtless	Work to become more thoughtful and increase awareness.
Highly noticeable	Work to become less noticeable.
Negative	Work to become less negative.
Maladaptive	Work to become more adaptive.
At odds with general community	Build greater consonance.
Energy consuming	Reduce energy spent.
Egosyntonic	Increase empathy.
Interpersonally conflictual	Build skills and decrease conflict.

any hypotheses. Data can be garnered from the individual, their documented history, reports of others who have treated them, statements and reports from significant others to corroborate data, personal interviews, and psychological testing. Therapists use these data to build their conceptualizations and to structure treatment programs. Often the treatment in CBT is derived directly from the problem; for examples, see Table 10.1. The construction of a conceptualization is, perhaps, the most sophisticated aspect of psychotherapeutic treatment. It is model building at its highest level. Although similar behaviors can be conceptualized in a similar manner, it is essential for therapists to allow that two individuals with apparently the same behavior may have come to that behavior for very different reasons. Similarly, an individual may at different times exhibit similar behaviors but for different reasons.

The Conceptualization Process

The conceptualization is a dynamic formulation. The process will, of necessity, go through several incarnations. The initial conceptualization is followed by all of the requisite revisions and emendations as additional data are collected. The data may come from patient reports or *in vivo* experiences of the therapeutic collaboration. When a conceptualization is cast in granite, the therapy will be limited and rigid. Static conceptualization will limit the therapist's ability to account for actions, thoughts, and behavior that emerge in the therapy. The rigid conceptualization can be characterized as the therapist building a hammer and then seeing everything in sight as a nail to be hit.

One way to better understand the conceptualization process is to think of a cooking metaphor. To be an effective and successful cook, we must start with an idea of where we are going (e.g., "What's for dinner?"). We must have a cognitive representation of the end-product being sought. Let us suppose, for example, that our goal is to make a dish with beef. Even before purchasing the necessary ingredients, we must have an idea of how the finished dish will look, taste, or smell. Once the model is accepted, we must assess our ability to create this dish. We must also decide how much to make and when it has to be ready. We can then assemble the necessary ingredients and be sure that the necessary equipment is in place. We can now begin cooking. If we are missing an ingredient, we could probably substitute another one consistent with the desired goal (e.g., use a different sauce). At other times, we might want to alter the recipe with a specific, desired effect in mind. In the final evaluation, it is clear that the cook must keep in mind a working conceptualization of the desired outcome. That conceptualization includes the desired taste, presentation, and appearance. By having an overall conceptualization of the final goal, the ingredients available, and the limits in place (e.g., time), a cook can prepare dishes that will be memorable for years.

As therapists, we need to have effective conceptualizations of our patients' problems that are in agreement with the goals, time available, and other parameters of therapy while taking into account each patient's skills, our skills, the patient's response to previous therapies, and the extent of the pathology. The core of the conceptualization is developing an understanding of the patient's schema. This is followed by a decision as to the best possible points of entry into the system. Many of these hypotheses will be shared with the patient so the patient can help with the process. The ability to develop conceptualizations and to transform them into therapeutic interventions can be viewed in a hierarchical manner (Freeman, Pretzer, Fleming, & Simon, 1990; Freeman, Mahoney, DeVito, & Martin, 2004). At the lowest skill level, the therapist does not appear to have developed a conceptualization of the case. The session would have no theme or focus, no agenda or structure, and the direction would be dictated by the patient's last statement. If the therapist has begun to develop a conceptualization, the interventions will have a greater focus and direction. At the highest level, a clear treatment conceptualization results in a series of organized and focused treatment interventions.

The therapist uses the conceptual framework to elicit specific thoughts, assumptions, images, meanings, or beliefs. It assists the therapist to develop questions for the guided discovery process. Without the treatment conceptualization, the focus of the CBT work might be vague or irrelevant, even when the basic tools of cognitive therapy are used, such as the dysfunctional thought record or the simpler double or triple column technique to collect thoughts. Newman and Beck (1990) identified eight steps for establishing a treatment plan: (1) problem conceptualization, (2) development of a collaborative relationship, (3) motivation for treatment, (4) patient formulation of the problem, (5) goal setting, (6) patient socialization into the cognitive model, (7) cognitive–behavioral interventions, and (8) relapse prevention.

Understanding Schema

Particular schema may engender a great deal of emotion and be emotionally bound by both the individual's past experience and the sheer weight of the time for which the belief has been held. In addition, the perceived importance and credibility of the source from whom the schema were acquired must also be addressed. A cognitive element of the schema pervades the individual's thoughts and images. With the proper training, individuals can be helped to describe schema in great detail or to deduce them from behavior or automatic thoughts. A behavioral component involves the way the belief system governs the individual's responses to a particular stimulus or set of stimuli. For some patients, the therapeutic sequence may be cognitive→affective→behavioral; for other individuals, it would be behavioral→cognitive→affective. Schema can be

- *Cognitive*—Coded as ideas, beliefs, thoughts
- *Affective*—Coded as emotional stimulants
- *Behavioral*—Propelling the individual to action
- *Problem-solving*—Dealing with new situations
- *Excitatory*—Generating action or arousal

Because the schema determine, in effect, how one defines oneself, both individually and as part of a group, they are important to understand and factor into the treatment conceptualization. They may be said to be unconscious, using a definition of the unconscious as "ideas we are unaware of" or "ideas that we are not aware of simply because they are not in the focus of attention but in the fringe of consciousness" (Campbell, 1989).

Inactive schema are called into play to control behavior in times of stress. Stressors will evoke the inactive schema, which become active and govern behavior. When the stressor is removed, the schema will return to their previously dormant state. Conceptually, understanding the activity–inactivity continuum explains two related clinical phenomena. The first involves the rapid, although transient, positive changes often evidenced in therapy, the so-called *transference cure* or *flight into health*. The result of this phenomenon is that a patient with clinical symptoms seems to have a partial or full recovery in a surprisingly brief period of time and will then seek to terminate therapy as no longer needed. What we may see is a patient, under stress, responding to several dormant schema made active. They may direct the patient's behavior in a number of dysfunctional and self-destructive ways. When the patient enters into therapy and experiences the acceptance and support of the therapist, the stress may be reduced, and the individual will once again operate on the more functionally active schema. If, however, the patient leaves therapy without gaining the skills necessary to cope with the stress, the problem may emerge again when and if the stressor returns. Individuals who develop effective coping strategies may be able to deal with stress throughout their lives and rarely experience activation of the dormant schema.

Another clinical phenomenon is arousal of dormant schema that cause the patient to appear, upon intake or admission to the hospital, to be extremely disturbed. After brief psychotherapy, pharmacotherapy, or combination, the patient appears far better integrated, far more attentive, and indeed far healthier; in fact, after the stress is removed, the therapist may question the existence of any psychopathology.

How much of an effect schema have on an individual's life depends on several factors. The first issue is to evaluate how strongly the schema are held. Schema that are part of the patient's view of self are likely to be more strongly held. These schema are far more difficult to change as they affect how individuals define themselves and see themselves in the context of their world. These compelling schema are often easily seen by the therapist but may be the most difficult to address and change. After all, individuals who hold onto their beliefs even when death threatens are labeled as heroes or martyrs. A related issue is how essential individuals consider particular schema to be to their safety, well-being, or existence. Does the patient engage in any disputation of his or her thoughts, actions, or feelings when a particular schema is activated? This is related to the individual's previous learning *vis-à-vis* the importance and essential nature of a particular schema. Probably the most important issue is related to how early a particular schema was internalized. The earlier the schema was acquired, the more powerful it will be and the greater affect it will have on the individual's behavior. Layden et al. (1993) referred to these schema as being "in a cloud." They influence behavior, but the patient may not be aware of where the direction and power are coming from.

Changing Schema

Because schema define the individual, they are often difficult to change. The greater the survival value of a schema, the greater valence it has for the individual and the greater the difficulty in altering the schema. A warrior returning from a combat zone may visibly flinch upon hearing a loud noise; this is not necessarily a manifestation of PTSD flashbacks but rather a conditioned response to potentially life-threatening circumstances.

Changes in schema can be viewed on a continuum (Table 10.2). The therapeutic goal is to move the individual to a more centrist (flexible) position. In many cases, an individual's schema were more flexible before their military experience. As a result of training and living in harm's way, schema may have become far more rigid as a function of survival. For example, my belief that people in my neighborhood can be trusted and that I will be safe as I stroll down the street would be altered by necessity if I find myself surrounded by armed and motivated enemies who would kill me if they had the opportunity. For some individuals, the survival valence will be difficult to shift.

Table 10.2 Schematic Change Potential

Schematic Paralysis	Schematic Rigidity	Schematic Stability	Schematic Flexibility	Schematic Instability
Ossified	*Dogmatic*	*Steady*	*Creative*	*Chaotic*
Nothing will bring about change	Change may come but with difficulty	Internal rules are clear and predictable	Rules can be bent or changed, as needed	Rules shift and change without warning

Changes or shifts in schema can be categorized as

- *Reconstruction*, which involves slowly dissembling an existing schema and building a more adaptive schema in its place
- *Construction*, which requires that the individual construct rules or schema when none exists to deal with a particular situation
- *Modification*, which involves working within the individual's schematic framework and making small but visible changes to their schema
- *Reinterpretation*, which is when the individual does not change the schema but now uses them in a more adaptive and prosocial manner
- *Camouflage*, which is when the schema are recognized and kept out of the view of most others

Cognitive–Behavioral Treatment

Cognitive–behavioral therapy draws from two basic menus of techniques: cognitive and behavioral. Inherent in both sets of techniques are the emotional, interpersonal, and situational applications.

Cognitive Interventions

Idiosyncratic Meaning   A term or statement used by a patient is not completely understood by the therapist until the patient is asked for the meaning and for clarification of how that idea is relevant for the patient. It is essential to question the patient directly with regard to the meanings of verbalizations. This also models for the patient active listening skills, increased communication, and a means for checking out assumptions.

Questioning the Evidence   It is essential to teach patients to question the evidence that they are using to maintain and strengthen an idea or belief. Questioning the evidence also requires examining the source of data or evidence. Many patients are able to ignore major pieces of data and focus on the few pieces of data that support their preexisting dysfunctional view.

Reattribution   A common statement made by patients is, "It's all my fault." This is especially true when a life has been put at risk or lives have been lost. Although one cannot dismiss the individual's view out of hand, it is unlikely that a single person is totally responsible for everything that has gone wrong within a situation. The therapist can help the patient distribute responsibility among all of the relevant parties. The therapist must not take a position of total support by saying, for example, "It wasn't your fault" or "Don't blame yourself" or "You need to let it go." In this case, the therapist ends up echoing friends and family that the patient has already dismissed as being an unsympathetic (although supportive) cheering squad. By taking a middle ground, the therapist can help the patient reassign responsibility and not assume all of the blame or unrealistically shift all blame to others.

Examining Options and Alternatives   Many individuals may think they have lost all options and become hopeless. Perhaps the prime examples of this feeling of lacking options are suicidal patients, who see their options and alternatives as so limited that among their few choices death might be the easiest or simplest one. The cognitive strategy in this case would require working with the patient to generate additional options.

Decatastrophizing   Also called the "what if" technique, decatastrophizing involves helping patients evaluate whether or not they are overestimating the catastrophic nature of their situation. Questions that might be asked include, "What is the worst thing that can happen?" or "And if it does occur, what would be so terrible?" This technique has the therapist working against a "Chicken Little" style of thinking. If the patient sees an experience (or life itself) as a series of catastrophes and problems, the therapist can work toward greater reality testing. For many service members, the return from a truly catastrophic environment in which life, limb, friendship, and relationships have all been injured or threatened, the therapist cannot remove the past but can focus on the present and the future so the catastrophes of combat do not sear themselves indelibly into the individual's future.

Fantasized Consequences   In this technique, individuals are asked to fantasize situations and describe their images and attendant concerns. Often, when verbalizing their concerns, patients can see the irrationality of their ideas. If the fantasized consequences are realistic, the therapist can work with the patient to assess the actual danger and develop coping strategies. This technique allows patients to bring imaged events, situations, or interactions that have happened previously into the consulting room.

Advantages and Disadvantages   By asking patients to examine both the advantages and disadvantages of an issue, a broader perspective can be achieved. Although patients will often claim that they cannot control their feelings, actions, and thoughts, it is precisely the development of this control that is the strength of CBT.

Turning Adversity to Advantage   At times a crisis or catastrophic experience can be used to advantage. Losing one's job can be a disaster but may, in some cases, lead to a better job or even a new career. It is not simply a matter of "if life gives you lemons, make lemonade." Service members have been to hell and back. They have seen things and experienced or done things that they never thought that they would or could do. The therapist must lead the patient toward an understanding of how the experience of being in harm's way can be used for the good of the individual and society more generally.

Guided Association/Discovery   Through simple questions such as "Then what?" or "What would that mean?" or "What would happen then?" the therapist can help the individual explore the significance of events. The therapist provides the conjunctions to the patient's verbalizations. By asking "And then what?" or "What evidence do we have that this is true?" the therapist can guide the patient along various therapeutic paths, depending on the conceptualization and therapeutic goals.

Use of Exaggeration or Paradox   By taking an idea to its extreme, the therapist can often help to move the individual to a more central position *vis-à-vis* a particular belief. Care must be taken not to insult, ridicule, or embarrass the patient. Patients who are hypersensitive to criticism and ridicule may view a therapist who uses paradoxical strategies as making light of their problems. The therapist who chooses to use the paradoxical or exaggeration techniques must have: (1) a strong working relationship with the patient, (2) good timing, and (3) the good sense to know when to back away from the technique.

Scaling   For those patients who see things as "all or nothing," the technique of scaling or seeing things as existing on a continuum can be very helpful. The scaling of a feeling can force patients to utilize the strategy of gaining distance and perspective. Because patients may be at a point of extreme thoughts and extreme behaviors, any movement toward a midpoint is helpful.

Direct Disputation   Although we do not advocate arguing with a patient, at times direct disputation is necessary. A major guideline for its necessity is a potentially imminent suicide attempt. The therapist must directly and quickly work to challenge hopelessness. Disputation might appear to be the treatment technique of choice, but the therapist risks becoming embroiled in a power struggle or argument with the patient. Disputation coming from outside the patient may, in fact, engender passive resistance and a passive–aggressive response that might include suicide. Disputation, argument, and debate are potentially dangerous tools that must be used carefully, judiciously, and with skill. If the therapist becomes one more harping contact, the patient may turn the therapist off completely.

Behavioral Techniques

The goals of using behavioral techniques within the context of cognitive–behavioral therapy are manifold. The first goal is to utilize direct behavioral strategies and techniques to test dysfunctional thoughts and behaviors. By having the patient try feared or avoided behaviors, old ideas can be directly challenged. A second use of behavioral techniques is to practice new behaviors as homework. Certain behaviors can be practiced in the office and then practiced at home. Homework can range from acting differently, practicing active listening, being verbally or physically affectionate, or doing things in a new way.

Activity Scheduling   The activity schedule is perhaps the most ubiquitous form in the therapist's arsenal. For patients who are feeling overwhelmed, the activity schedule can be used to plan more effective time use. The activity schedule serves as both a retrospective tool to assess past time utilization and a prospective tool to help plan better time use.

Mastery and Pleasure Ratings   The activity schedule can also be used to assess and plan activities that offer patients a sense of personal efficacy (mastery) and pleasure. The greater the mastery and pleasure, the lower the rates of anxiety and depression. By discovering the low- or high-anxiety activities, the therapist and patient can plan to increase the former and decrease the latter.

Social Skills Training   If reality testing is good and a patient lacks specific skills, it is incumbent upon the therapist to either help the patient acquire the necessary skills or make a referral for skills training. Skills acquisition may involve anything from teaching patients how to properly shake hands to practicing conversational skills.

Assertiveness Training   As with social skills training, assertiveness training may be an essential part of the therapy. Patients who are socially anxious can be helped to develop responsible assertive skills.

Bibliotherapy   Several excellent books can be assigned as readings for homework. These books can be used to socialize or educate patients with regard to the basic cognitive therapy model, emphasize specific points made in the session, or introduce new ideas for discussion at future sessions. Some helpful patient resources are listed at the end of this chapter.

Graded Task Assignments   Graded task assignments involve a shaping procedure of small sequential steps that lead to the desired goal. By setting out a task and then arranging the necessary steps in a hierarchy, patients can be helped to make reasonable progress with a minimum of stress. As patients attempt each step, the therapist can be available for support and guidance.

Behavioral Rehearsal/Role Playing   The therapy session is the ideal place to practice many behaviors. The therapist can serve as teacher and guide by offering direct feedback on performance. The therapist can monitor the patient's performance, offer suggestions for improvement, and model new behaviors. In addition, anticipated and actual road blocks can be identified and worked on in the session. Extensive rehearsal can precede the patient's attempting the behavior *in vivo.*

In Vivo *Exposure*   Sometimes the consulting room must be expanded into the outside world. Therapists can accompany patients into feared situations. They can drive with patients across a feared bridge, go to a feared shopping mall, or travel on a feared bus. The *in vivo* exposure can complement an office-based practice and patient-generated homework.

Relaxation Training   The anxious patient can profit from relaxation training because the anxiety response and the relaxation response are mutually exclusive. Relaxation training can be taught in the office and then practiced by the patient as homework. Ready-made relaxation tapes can be purchased or the therapist can easily tailor a tape for a particular patient. A tape made by the therapist can include the patient's name and can focus on the patient's particular symptoms. The tape can then be modified, as needed.

Homework   Homework, or self-help work, is an essential part of cognitive behavioral work. Simply, when therapy ends, regardless of the number of sessions, focus of the sessions, or goals of the therapy, everything that the patient does must be viewed as homework. If patients learn in therapy that they must depend on their therapists and the therapy sessions to cope effectively without acquiring new and specific skills, then their therapy would have to continue *ad infinitum.* If, however, patients have acquired the basic and necessary skills required for effective coping, then they will most likely be able to cope more effectively. Homework is a key ingredient in therapeutic change and has several positive effects on psychotherapy:

- Homework extends the therapy contact. Rather than therapy being one or two hours per week, out-of-session activity allows patients to be involved in therapeutic activities for as many hours of the day or days of the week they choose.

- Homework offers the therapist an opportunity to gauge the patient's level of motivation for therapy. If a patient is motivated for change, then the homework provides an arena for action.
- Self-help work gives patients an opportunity to practice skills learned in the sessions. The session may be viewed as a lecture in a college course, while the homework is the laboratory for testing out new ideas, behaviors, or emotions.
- Homework can be used to set up situations that allow patients to test assumptions identified during their therapy sessions. Homework offers an opportunity to collect data relevant to insights gained in the sessions.
- Homework offers a continuity between the sessions so therapy is not a series of discrete moments in time but a contiguous set of experiences. By reviewing the patient's homework, the therapist demonstrates the importance of the homework and can show how it provides value to the ongoing therapy.
- Homework can allow the involvement of significant others in the therapy. Some homework may involve how others respond to the patient. The accumulation of these views can be useful.
- Homework is essential for preventing relapse. After all, when therapy ends, everything the patient does is homework or self-help.

Homework must be introduced very early in the therapeutic collaboration. It is introduced not so much as one additional trial to endure but rather as a course of action that will make the therapy work faster and better, even to the point of having the patient come to therapy for a shorter time. Paradoxically, the problematic affect and behaviors that distress and impair the patients in daily life and impel them to seek treatment often become the very obstacles to complying with the treatment interventions that might relieve their suffering. Homework noncompliance correlates with premature treatment termination (Burns & Nolen-Hoeksema, 1991).

Patients who are the most willing to attempt new strategies for coping with personal problems before beginning treatment engage more actively in the treatment process, carry the therapy outside of the treatment room, and experience relatively greater clinical improvement. Surprisingly, although willingness was found to be only marginally correlated with actual homework compliance between sessions, both willingness and homework compliance apparently make separate and additive contributions to mitigating depression (Burns & Nolen-Hoeksema, 1991; Persons, Burns, & Perloff, 1988). Burns and Nolen-Hoeksema (1991) hypothesized that high willingness may reflect increased motivation and a general cooperative response set indicating that these patients would be more willing to express their feelings openly, take medications consistently, or engage in introspection.

Homework provides a means for the therapist and patient to collaboratively achieve the patient's most desired treatment goals and increase the patient's motivation to actively participate in the process. Thus, it becomes vital to assess task-interfering beliefs when patients fail to complete homework assignments or express low willingness to try new skills. By presenting a cogent rationale for how each homework activity is relevant to the patient's treatment goals, the therapist can increase both the salience of the goals and the probability of homework adherence (Beck, Brown, Berchick, & Stewart, 1979). It is vital to reframe each collaboratively generated homework assignment as a positive incremental step on the road to improving social skills, mood elevation, anxiety reduction, and other desirable goals. Consequently, teaching patients how to deal with anxiety and tolerate frustration (McMullin, 2000) should reduce discomfort, provide generalizable skills, and prevent premature termination (Freeman & Simon, 1989).

To further subvert homework nonadherence, the therapist must identify the patient's cognitions regarding the homework assignment and the consequences of change (Beck, Brown, Berchick, & Stewart, 1990; Beck, Freeman, & Davis, 2004; Freeman et al., 1990, 2004). Cognitive restructuring and graded task assignments can help to reduce distorted task-interfering cognitions. Increasing emotional tolerance and social skills training can further reduce avoidance (Beck et al., 1990, 2004; Linehan, 1993).

When selecting homework activities the therapist must consider the patient's current abilities and readiness to change; for example, the patient's perceptions of the problem and their motivation, distress tolerance, communication skills, and even reading or writing ability must be considered (Beck et al., 1979). Allowing patients to select among homework activities that are equivalent to their own abilities can further increase commitment, responsibility, and homework adherence. Merely offering the patient a choice of ways and times to complete the assignments can have similar effects (Beck et al., 1979).

When a homework assignment is selected, in-session cognitive or behavioral rehearsal can help to shape adaptive skills, increase the probability of homework adherence, and illuminate any potential obstacles to completing the task (Freeman et al., 1990, 2004; Beck et al., 1979). In addition, the therapist may ask patients how confident they are that they will be able to complete the homework assignment on a scale of 1 to 100. If the patient indicates low levels of confidence, the therapist can solicit expected obstacles, which can then be directly addressed. These obstacles may include distressing beliefs, cognitive and emotional avoidance, and real-world problems such as scheduling, inclement weather, and lack of transportation, among others. Solving each obstacle can be the most valuable grist for the therapeutic mill.

To increase motivation and reduce perfectionism, each task may be framed as an experiment to test beliefs, practice skills, or find cognitive and other obstacles to goal attainment. It is a no-lose, win–win proposition. Either patients return to the next session having completed their homework or they have identified the obstacles that are impeding their progress (Beck et al., 1979; Freeman et al., 1990, 2004).

Regardless of the task, homework assignments should be concrete teaching, hypothesis-testing skills. The essence of homework is to have patients examine specific beliefs, conduct direct tests of these beliefs, and practice hypothesis-testing skills in specific in-session and between-session exercises (Beck et al., 1979; DeRubeis & Feely, 1990). Belief testing also produces more improvement in depressive symptoms after treatment than less cognitive tasks. Patient workbooks with accompanying therapist manuals offer the therapist and the patient structure and guidance in assigning homework (Freeman & Fusco, 2003; Fusco & Freeman, 2003; Greenberg & Padesky, 1995; Linehan, 1993).

The therapist can also provide corrective feedback to shape more adaptive behavior and cognitions (Meichenbaum & Turk, 1987). Before attempting any intervention, the therapist must explain the rationale, objective, and potential benefits to be gained. Homework must be framed as a win–win situation (Linehan, 1993). The patient learns adaptive new skills, solves vexing problems, or discovers the obstacles that then become the focus of the next session (Beck et al., 1979).

It is wise to recruit social support from significant others in the patient's own environment (Meichenbaum & Turk, 1987). This support should increase public commitment, promote between-session practice, assist in contingency management/positive reinforcement regimes, and prevent inadvertent sabotage of patients' efforts. Patients also may be encouraged to use positive labels to produce similarly salubrious expectations and consonant behavior; for example, after a successful impulse control homework assignment, borderline personality disorder (BPD) patients may label themselves as "thoughtful."

Optimum behavioral rehearsal and skills training occur in situations similar to those in which the skills will be used. Both *behavioral rehearsal* and *covert rehearsal* are elegant means of increasing motivation in that they illuminate and reduce obstacles to change and provide an opportunity for the therapist to shape desired behavior. Therapist reinforcement can be particularly powerful in teaching, shaping, and strengthening adaptive behavior for individuals who have been nurtured in punishing, negative environments, such as BPD patients (Linehan, 1993). Patient perceptions of therapist empathy correlate with patient motivation and homework compliance (Burns & Nolen-Hoeksema, 1991).

The Dysfunctional Thought Record (Beck et al., 1979) and downward arrow technique (Burns, 1980) are also useful homework assignments that can be used in identifying and modifying activated schema that generate negative cognitions and high levels of affect and blame. This makes homework an ideal method to generate and test schema-related hypotheses. Actual skills deficits should be addressed with specific skills training (Beck et al., 2004; Linehan, 1993).

Clinical Example

Tamara, age 40, was referred by her primary care provider because of "anger" issues. She reported that she and her husband Steve, age 43, would have frequent arguments that ended up with both of them yelling and the slamming of doors. They had been married 7 years, and it was a second marriage for both, as their first marriages ended in divorce. Tamara was previously married for 5 years, Steve for 8 years. Neither had children from their previous marriages. They had two children together, ages 5 (son) and 4 (daughter). Steve had an independent practice as an accountant, and Tamara was a social worker.

As was policy, halfway through the initial intake, the intake worker consulted with the clinical director regarding diagnosis and potential treatment recommendations. The intake worker offered a tentative diagnosis of paranoid personality disorder with borderline features. When asked what the basis was for this rather substantial diagnosis, the intake worker detailed the following:

- Tamara was constantly scanning for imminent attacks and would often work to prevent the attacks from her husband and others preemptively.
- Tamara was hypervigilant.
- She questioned the interviewer intensively about confidentiality, access to her records, who could obtain records and reports, what the interviewer was writing down, whether she could see the interview notes, and why the interviewer needed to speak to the clinical director.
- She refused to answer several questions.
- She refused to complete the Beck Depression Inventory.
- She refused to even discuss joint sessions with Steve.
- She was concerned that reports of her therapy would go to Child Protective Services and that they might take her children.
- She said that her coworkers were out to get her.
- She commented on the lack of fairness all around her.

When the clinical director, as was the intake protocol, met with Tamara, several things were noted and discussed. First, Tamara had a distinct accent. When asked where she originally came from, she reported Israel. As the clinical director continued to question Tamara, several issues were confirmed. She served in the Israeli Defense Force and had been in a unit that was under frequent assault by mortars and rocket propelled grenades. She grew up in a *kibbutz* on the border with Lebanon and heard almost daily air raid sirens alerting the *kibbutzniks* of imminent attack. The siren meant that children had to seek shelter in bunkers while adults worked to repel the attack. On many occasions, the *kibbutz* was attacked by guerillas with automatic weapons. Although she had never been injured as a child or as a soldier, she was hypervigilant to any possible attack.

She met Steve after she had moved to the United States at age 30. Although Steve had never attacked her physically, their verbal altercations were significant even at the beginning of the relationship. He claimed that her "spunk" and "no-bullshit" attitude were two of the things that attracted him to her.

It was clear that Tamara's early schema regarding potential danger, imminent attack, helplessness as a child against these attacks, the need for self-sufficiency and maintaining a constant survival mode of thinking, and sensitivity to any criticism all now served to keep her safe but also isolated from her husband, children, and coworkers and made it difficult to make and keep friends.

The therapy focused on explicating her schema and testing the need for the schema in the extreme form that they now took. Schematic reconstruction and modification were used to move Tamara from "The world is a dangerous place and one must attack before one is attacked" to "The world can be a dangerous place, and one must be prepared as best as one can." This was a position that discomforted Tamara, but after performing a cost–benefit analysis (advantages vs. disadvantages) she accepted that it made sense and would reduce conflict at home. She further agreed to couples therapy after 21 sessions of individual CBT.

Summary

Cognitive–behavior therapy (CBT) is a multifaceted treatment model for a broad range of emotional disorders that is active, motivational, directive, collaborative, dynamic, problem oriented, and solution focused. It has a here-and-now emphasis and is psychoeducational, structured, time limited, prescriptive, culturally relevant and informed, empirically supported, cognitively appropriate, integrative, and based on a one-session model for treatment. By virtue of these attributes, CBT has emerged as the treatment of choice within a broad range of psychological disorders and populations of patients. It is practiced in all clinical settings and applied to every delivery option for therapy. Timely help for pre- and postdeployment service members can have a positive impact on these individuals, their families, their significant others, and our society as a whole. They have already paid a steep price, they should not have to pay any further.

References

Arkowitz, H., Westra, H. A., Miller, W. R., & Rollnick, S. (2007). *Motivational interviewing in the treatment of psychological problems.* New York: Guilford Press.

Beck, A. T., Kovacs, M., & Weissman, A. (1979). Assessment of suicidal intention: The scale for suicide ideation. *Journal of Consulting and Clinical Psychology*, 47(2), 343–352.

Beck, A. T., Brown, G., Berchick, R., & Stewart, B. (1990). Relationship between hopelessness and ultimate suicide: A replication with psychiatric outpatients. *American Journal of Psychiatry*, 147(2), 190–195.

Beck, A. T., Emery, G., & Greenberg, R. (1996). Cognitive therapy for evaluation anxieties. In C. G. Lindemann (Ed.), *Handbook of the treatment of the anxiety disorders* (2nd ed., pp. 235–260). Lanham, MD: Jason Aronson.

Beck, A. T., Freeman, A., & Davis, D. D. (2004). *Cognitive therapy of personality disorders* (2nd ed.). New York: Guilford Press.

Burns, D. (1980). *Feeling good.* New York: Morrow.

Burns, D., & Nolen-Hoeksema, S. (1991). Coping styles, homework compliance, and the effectiveness of cognitive–behavioral therapy. *Journal of Consulting and Clinical Psychology,* 59(2), 305–311.

Campbell, D. (1989). Being mechanistic/materialistic/realistic about the process of knowing. *Canadian Psychology/Psychologie Canadienne,* 30(2), 184–185.

DeRubeis, R. J., & Feely, M. (1990). Determinants of change in cognitive therapy for depression. *Cognitive Therapy and Research,* 14(5), 469–482.

Freeman, A., & Fusco, G. M. (2003). *Borderline personality disorder: A therapist's manual for taking control.* New York: W.W. Norton.

Freeman, A., & Simon, K. (1989). Cognitive therapy of anxiety. In: H. Arkowitz, L. E. Beutler, A. Freeman, & K. M. Simon (Eds.), *Comprehensive handbook of cognitive therapy* (pp. 347–365). New York: Plenum Press.

Freeman, A., Pretzer, J., Fleming, B., & Simon, K. (1990). *Clinical applications of cognitive therapy.* New York: Plenum Press.

Freeman, A., Mahoney, M., DeVito, P., & Martin, D. (2004). *Cognition and psychotherapy* (2nd ed.). New York: Springer.

Freeman, A., Stone, M., & Martin, D. (2005). Similarities and differences in treatment modalities. In A. Freeman, M. H. Stone, & D. Martin (Eds.), *Comparative treatments for borderline personality disorder* (pp. 259–287). New York: Springer.

Fusco, G. M., & Freeman, A. (2003). *Borderline personality disorder: A patient's workbook for taking control.* New York: W.W. Norton.

Greenberg, D., & Padesky, C. (1995). *Mind over mood.* New York: Gulford Press.

Kazantzis, N. Deane, F. P., Ronan, K. R., & L'Abate. (2005). *Using homework assignments in cognitive–behavior therapy.* New York: Routledge.

Layden, M., Newman, C., Freeman, A., & Morse, S. (1993). *Cognitive therapy of borderline personality disorder.* Needham Heights, MA: Allyn & Bacon.

Linehan, M. (1993). *Skills training manual for treating borderline personality disorder.* New York: Guilford Press.

McMullin, R. E. (2000). *The new handbook of cognitive therapy techniques.* New York: W.W. Norton.

Meichenbaum, D. and Turk, D. C. (1987). *Facilitating treatment adherence: a practitioner's guidebook.* New York: Plenum Publishers.

Murphy, R. T. (2007). Enhancing combat veterans' motivation to change posttraumatic stress disorder symptoms and other problem behaviors. In H. Arkowitz, H. A. Westra, W. R. Miller, & S. Rollnick (Eds.), *Motivational interviewing in the treatment of psychological problems* (pp. 57–84). New York: Guilford Press.

Newman, C., & Beck, A. (1990). Cognitive therapy of affective disorders. In B. B. Wolman & G. Stricker (Eds.), *Depressive disorders: facts, theories, and treatment methods* (pp. 343–367). New York: John Wiley & Sons.

Nigro, C. (2007). Formal operational thinking with adults: testing the Piagetian model (Jean Piaget). *Dissertation Abstracts International: Section B: The Sciences and Engineering,* 67(9-B), 5446.

Persons, J., Burns, D., & Perloff, J. (1988). Predictors of dropout and outcome in cognitive therapy for depression in a private practice setting. *Cognitive Therapy and Research,* 12(6), 557–575.

11
Core Psychotherapeutic Tasks With Returning Soldiers: A Case Conceptualization Approach

DONALD MEICHENBAUM

Contents

I spend a good deal of time consulting at psychiatric hospitals, including Veterans Affairs hospitals. In each setting, the staff identify the most challenging or difficult client or group of clients, and they then ask me to interview the clients behind a one-way mirror. Before I see a client, a case conference is held where all members of the mental-health treatment team present relevant case material.

I quickly discovered that I needed some sort of framework to integrate this diverse current and developmental information about past and present risk and protective factors before seeing a client. As Taylor (2006) has noted, it is necessary to identify the predisposing, precipitating, perpetuating, and protective factors that contribute to a client's problems.

In this chapter, I describe a Case Conceptualization Model (CCM) that I have developed. A well-formulated CCM can help give direction to the assessment and treatment plans that are designed to match the client's needs and situation. Such a case formulation is both a descriptive and explanatory summary of the client's problems, as well as the client's strengths, and reflects a set of testable hypotheses of the probable contributory and maintaining factors. A well-formulated case conceptualization also includes a collaboratively established set of treatment goals and a consideration of possible barriers and provides a means for evaluating progress.

As Beck and his colleagues (1993, p. 80) observed: "Without a case formulation, the therapist is proceeding like a ship without a rudder, drifting aimlessly through the session." A similar sentiment has been voiced by Persons (1989) and Zayfert and Becker (2007). A well-formulated CCM should be collaboratively generated with the client (and significant others). The review of the CCM with the client and significant others (see Table 11.2) serves a valuable psychotherapeutic purpose, especially when the CCM covers the "rest of the client's story" or the client's strengths and resilience, thus nurturing hope by means of collaborative goal setting. But, such treatment goal setting must be tempered by an appreciation of possible barriers and potential roadblocks and a discussion of how to anticipate and address such potential obstacles.

The CCM presented here covers the following sources of information:

- Background demographic and military relevant information, as well as reasons for referral
- Presenting problems and their history, especially in terms of current risk assessment to self and to others and the impact of symptoms on the level of the client's daily functioning and role performance
- Presence and impact of comorbid disorders
- Current transitional and developmental stressors
- Current and past treatments received in terms of efficacy, treatment adherence, and client satisfaction
- Signs of resilience and evidence of individual, social, and cultural or systemic strengths
- Summary of risk and protective factors
- List of doable and measurable short-term, intermediate, and long-term client goals and subgoals and a multifaceted individualized treatment game plan for achieving them
- Individual (motivational), social, and systemic barriers or possible roadblocks to client participation and treatment adherence and maintaining improvement

The Flowchart of Nine Boxes shown in Table 11.1 diagrammatically represents the CCM; breakout information to be considered under each category is provided at the end of the chapter. This information should be shared with the client and with the treatment team using the Feedback Sheet format (Table 11.2).

Table 11.1 Generic Case Conceptualization Model Tailored to Returning Soldiers (Multiple-Focused Assessment Strategy)

	1A. Background Information 1B. Military History (Predeployment, Deployment, Postdeployment) 1C. Reasons for Referral

↓

9. Barriers 9A. Individual 9B. Social 9C. Systemic	2A. Presenting Problems (Symptomatic Functioning) 2B. Risk Assessment Toward Self and Toward Others 2C. Level of Functioning (Interpersonal Problems, Social Role Performance)

↑ ↓

8. Outcomes (Goal Attainment Scaling) 8A. Short-Term 8B. Intermediate 8C. Long-Term	3. Comorbidity (TBI Involvement?) 3A. Axis I 3B. Axis II 3C. Axis III 3D. Impact of Comorbidity

↑ ↓

7. Summary of Risk and Protective Factors	4. Stressors (Present and Past) 4A. Current 4B. Ecological 4C. Developmental 4D. Familial

↑ ↓

6. Strengths 6A. Individual 6B. Social 6C. Systemic	5. Treatments Received (Current, Past) 5A. Efficacy 5B. Adherence 5C. Satisfaction

Table 11.2 Feedback Sheet on Case Conceptualization

"Let me see if I understand ..."

9. Barriers	1 and 2. Military Experiences, Referral Sources, and Presenting Problems
"Let me raise one last question, if I may. Can you envision, can you foresee, anything that might get in the way? Any possible obstacles or barriers to your achieving your treatment goals?" (Consider with the patient possible and systemic barriers. Do not address the potential barriers until some hope and resources have been addressed and documented.) "Let's consider how we can anticipate, plan for, and address these barriers."	"What brings you here?" "And this is particularly bad when ... ." "But, it tends to improve when you ... ." "How is it affecting you in terms of relationships, work, etc.?" "Would it be okay if we discuss how your current reactions are tied into your past military experiences?" (Review information from 1B.)

↑   ↓

8. Outcomes (Goal Attainment Scaling)	3. Comorbidity (TBI Involvement?)
"Let's consider your expectations for this treatment. As a result of us working together, what would you like to see change in the short term?" "How are things now in your life? How would you like them to be?" "How can we work together to help you achieve these short-term, intermediate, and long-term goals?" "What has worked for you in the past?" "How can our current efforts be informed by your past experiences?" "If you achieve your goals, what would you see change?" "Who else would notice these changes?"	"In addition, you are also experiencing or struggling with ... ." "And the impact of this in terms of your day-to-day experience is ... ."

↑   ↓

Observations About the Heuristic Value of the Case Conceptualization Model

- The information included under each category in Table 11.1 has been gleaned from research and findings on combat veterans summarized by Castro, Adler, and Britt (2006); Friedman, Keane, and Resick (2007); Hoge and McGunk (2008); Hopewell and Christopher (2007); Hosek, Kavanaugh, and Miller (2006); King, King, Foy, and Gudanowski (1996); McNally (2003); and Meichenbaum (1994). A detailed summary of findings regarding the impact of predeployment, deployment, and postdeployment on soldiers can be found on the Melissa Institute website (www.melissainstitute.org).

Table 11.2 (cont.) Feedback Sheet on Case Conceptualization

7. Summary of Risk and Protective Factors	4. Stressors (Present and Past)
"Have I captured what you were saying?" (Summarize risk and protective factors.) "Of these different areas, where do you think we should begin?" (Collaborate and negotiate a treatment plan with the patient. Do not become a surrogate "frontal lobe" for the patient.)	"Some of the factors (stressors) that you are currently experiencing that seem to maintain your problems are … ." "Ones that seem to exacerbate your problems are … (current and ecological stressors)." "And it's not only now but this has been going on for some time, as evidenced by … ." (Review impact of military experiences and developmental stressors.) "It's not something that only you are experiencing; your family members have also been struggling with … ." "The impact on you has been … (familial stressors and family psychopathology)."

↑	↓
6. Strengths	**5. Treatments Received (Current and Past)**
"But, in spite of … you have been able to … ." "Some of the strengths (signs of resilience) that you have evidenced or that you bring to the current situation are … ." "Moreover, some of the people (resources) you can call upon (access) are … and they can be helpful by doing … (social supports)." "And some of the services you can access are … (systemic resources).	"For these problems the treatments that you have received were … ." (Note type, time, by whom.) "The most effective treatment was … as evidenced by … ." "But, you had difficulty following through with the treatment, as evidenced by … ." "Some of the difficulties you encountered during the treatment were … ." "But, you were specifically satisfied with … and would recommend or consider … ."

"Let's review once again." Go back over the case conceptualization and have the client put the treatment plan into his or her own words. Involve significant others in the case conceptualization and treatment plan. Solicit their input and feedback. Reassess the treatment plan with the client throughout treatment. Keep track of treatment interventions using the coded activities (2A, 3B, 5B, 4C, 6B, etc.). Maintain progress notes and share these with the client and with other members of the treatment team.

- When I present the CCM model to mental health workers, I highlight that I have reduced their entire professional activities to one page and to nine boxes. Every single thing they do in their clinical practice can be coded in this fashion. As an example, after working with the client, the therapist may specify that the treatment plan consists of 2A (e.g., attempt to help the client reduce presenting problems of intrusive ideation, nightmares, and avoidance behavior). This will be accompanied by 5B (e.g., an adherence history) and 6B (e.g., support of significant others). As therapy evolves, integrative treatment á la the work of Mueser, Noordsby, Drake, and Fox (2003) and Najavits (2006) will be offered to address 3A (e.g., issue of comorbidity of substance abuse). The treatment will be strengths based (6A) by helping the client construct meaning from his or her combat experience consistent with the work of Baumeister (1991), Frankl (1984), Park and Folkman (1997), and Schok, Kleber, Elands, and Weerts (2008). With some stability in the client's adjustment, the attention of the treatment will be turned toward (4C) or the impact of developmental traumas á la exposure-based interventions (Riggs, Cahill, & Foa, 2006; Taylor, 2006; Zayfert & Becker, 2007). Thus, the therapist can summarize for the client and the treatment team the individualized therapy plan that includes 2A (symptom management), 5B (assessment of the history of treatment nonadherence), 6B (involvement of social supports), 6A (bolstering of the client's individual strengths in the form of coconstructing personal meaning), and 9A (bolstering motivation by means of motivational interviewing procedures).

- The treatment team is asked to test out the usefulness of this CCM by coding their client's progress notes in terms of the CCM categories. Is there anything that they and their client have worked on that cannot be coded?

- Another setting in which the treatment team can appreciate the potential usefulness of the CCM is when they take the Flow Chart of Nine Boxes to a case conference and listen to their colleagues present client information. Did they cover information in all of the boxes? Did they include information about Box 6 (client's strengths)? Did they assess for potential barriers (Box 9) and present a game plan for addressing these obstacles (e.g., use of relapse prevention procedures)?

- The CCM model can be applied to a variety of clinical populations, including juvenile offenders, individuals with traumatic brain injuries, depressed and victimized children, and other clinical populations (Meichenbaum, 2004). The CCM and the accompanying Feedback Sheet can be readily adapted to a computer-generated report and to progress notekeeping, as the therapist incorporates information from the various Nine Boxes.

- A caveat should be noted. Therapists will never have all of the information described in the Breakout Boxes shown in Table 11.1, but nevertheless they can proceed in an informed collaborative manner with their clients. All along the way, therapists should provide clients with information so the clients have a "big picture" of where they have been and where they are going in treatment.

- Finally, the CCM lends itself to collecting in-house data on the effectiveness of the respective treatment options. Imagine that you have asked me to pick up a case of yours and all I had access to was your progress notes. Would I be able to tell what you have been working on with your client (where you have been and where you plan to go); moreover, if I asked your client would I receive a similar account? If you use the CCM, I could open your folder and see the activities you have coded (2A, 5B, 6B, 3A, 4B, and 9A) and know how you spent your time with your client. Not only that, we could begin to collect data on just how effective such a combination of interventions has proven to be in achieving the short-term, intermediate, and long-term treatment goals (8A, 8B, and 8C).

Because I consult with many different clinical groups, I have adopted the motto "Have boxes, will consult!" (Remember the old television program where the character Paladin, played by Richard Boone, would say, "Have gun, will travel"?)

Breakout Box Information for Case Conceptualization Model

1. Background Information and Reason for Referral

1A. Background Information
- Age of client
- Minority status, ethnic identity and degree of acculturation and adherence to ancestral traditions and values, language preference, immigrant status
- Gender, sexual orientation, marital status
- Spouse in military
- Children and childcare status (e.g., single mother in military with children)
- Living conditions (homeless, living on base, living with others)
- Level of education; evidence of level of IQ and cognitive competence
- Current socioeconomic status
- Employment history and work-related difficulties, first job, time to acquire first job, how long job was kept, reasons for ending jobs
- Financial situation
- Life satisfaction
- Nature of daily routine (purposeful activities, degree of social isolation, alienation, degree of social integration)

1B. Military History
- Current rank, service branch, active duty vs. reservist, job in military (e.g., consider high-risk nature of job such as body recovery and identification, bomb removal, Special Forces, providing medical

and psychiatric care), station and length of station, length of service, number of moves, prior deployments, degree of leadership evident in service

- Predeployment information (age of entry into military, reasons for joining the military)
- Predeployment stressors such as family stressors, issues of child care, degree of preparation for deployment by family members

Deployment information and war-zone experiences (*administer gender-specific combat experience scales for both male and female veterans*):

- Number and length of deployments in the last 5 years
- Degree of perceived threat to one's life
- Nature and degree of injuries and nature of treatment required
- Race-related military stress experienced
- Disproportinate level of combat stress experienced
- Near-death experiences ("alive day" memories—date, what happened, how client survived)
- Exposure to gruesome, macabre deaths; human atrocities; human suffering
- Death of a buddy
- Number of injuries and deaths among friends and members of client's unit
- Exposure to extreme environmental stressors such as toxins
- Combat stress reactions or acute stress reactions experienced and what battlefield treatments were applied
- Being a prisoner of war, experiencing torture, loss of 35% of body weight
- During deployment was assaulted by others, experienced sexual harrassment and sexual abuse (obtain details and resultant consequences and any actions taken, shared experiences with supportive individuals)
- Perceptions of military leadership and attitudes toward military mission (then and now)
- What lingers from deployment
- "Best" and "worst" parts of deployment
- Miss anything about deployment
- How client survived
- Lessons client learned

Postdeployment information (*administer homecoming stress scale*):

- Time period since return from deployment
- Nature of adjustment difficulties involving family, children, work, finding purposeful activities

- Rehabilitation history since leaving combat zone
- Pending compensation issues
- Experiences with VA system
- Services received and level of satisfaction
- Decision to leave military and attendant stressors
- Future plans

1C. Reasons for Referral
- Time period since being deployed to point of referral
- Referred by others (mandated treatments) or self-referral (perception of having "problems)
- Reasons for not seeking treatment (perceived)—stigmitization, fear, embarrassment, professional career concerns
- Applying for or currently receiving compensation for service-related disability
- Motivation for treatment and change

2. Presenting Problems and Symptomatic Functioning

Administer the Post-Deployment Health Reassessment (PDHRA) and PTSD assessment and comorbidity measures.

- Check for partial PTSD and conduct a history of waxing and waning of PTSD and partial PTSD and related symptoms.
- Focus on examples of emotional dysregulation: impulse control problems; heightened arousal; being edgy, tense, or easily startled; irritability; restlessness; overresponding emotionally and not "letting it go"; anger, aggressive behavior, hostility, and carrying a chip on one's shoulder; anxiety; depression; emotional numbing; complicated grief; guilt and shame reactions.
- Look for cognitive alterations of memory and attentional problems, concentration difficulties, hypervigilance, rumination, cognitive distortions, and self-blame.
- Look for relationship difficulties with significant others, especially family members; feelings of loneliness, alienation, and being forgotten or misunderstood; avoidance, disengaging, or withdrawing behavior; difficulties with intimacy; distrust.
- Look for sleep disturbance (terrifying nightmares, insomnia); increased risk-taking behaviors (conduct risk assessment to self and others); spiritual and religious concerns and moral distress; and impact of belief system concerning self, others, and the future.
- Identify target symptoms, especially presence of avoidance, numbing, and dissociative behaviors.
- Indicate psychiatric diagnosis ("caseness").

- Be sensitive to culturally based syndromes (e.g., *ataques de nervios* for Caribbean clients) and how culture influences the expression of symptoms.

2C. Level of Functioning

Administer social adjustment and quality of life measure to determine interpersonal problems and ability to fulfill social role performance.

- Consider problems in living and lingering impact of war experiences on life satisfaction and on level of ability to function (e.g., meet personal goals).
- Indicate scores in DSM Axes IV and V.

3. Presence of Comorbid Disorders

Assess for the presence of comorbid psychiatric and medical disorders covering DSM Axes I, II, and III (3A, 3B, 3C). Assess for presence of traumatic brain injuries.

- Consider current *and* developmental evidence of comorbid disorders.
- Obtain a lifespan perspective and consider the sequence of comorbid disorders, including premilitary experiences.
- Use timelines to demarcate the developmental sequence of comorbid disorders, stressful events, and intervention history.
- Assess for common comorbid disorders: PTSD and substance abuse disorder, PTSD and anger-related problems, PTSD and depression and anxiety disorders, PTSD and borderline personality disorder and antisocial personality disorders, and acute stress disorder (combat stress reactions) and PTSD.
- Considered the presence of traumatic brain injuries and concussive disorders, as well as any evidence of a developmental history of "soft neurological signs." The second class of comorbid disorders is the presence of medical disorders and somatoform disorders and a history of accompanying high medical utilization.

3D. Impact of Comorbid Disorders on Life Functioning

Assess for the impact on self and others resulting from the presence of comorbid disorders and interpersonal and work-related difficulties.

4. Stressors (Present/Past)

4A. Current Stressors

- Address stressors tied to predeployment, deployment, and postdeployment phases. In terms of postdeployment, administer the homecoming stress scale. Address issues of relocation, reintegration into new roles, reconnection and intimacy with others, and issues related to family adjustment.

- Other stressors may be related to health, financial, compensation, and legal areas; ways to handle reminders of trauma; sleep disturbances; presence of comorbid disorders such as substance abuse; cumulation of day-to-day irritations and frustrations.
- If the client is leaving the military, address the impact of stressors such as job change, relocation, change in peer group, loss of access to military services, and loss of order and a disciplined environment.

4B. Ecological Stressors

- Ecological stressors reflect the impact of the social milieu (e.g., community attitude toward and resources for returning veterans).
- Is the veteran likely to be shunned, ignored, rejected or unwelcomed or experience a lack of recognition, a discriminatory environment, unavailability of treatment, and supportive services?
- The social milieu may be compounded by media exaggerating descriptions of so-called "dysfunctional aggressive-crazed veterans." It is difficult to combat stereotypes.
- Trauma events may result in the loss of social and material resources.

4C. Developmental Stressors

Administer the antecedent trauma questionnaire.

- Premilitary stressors are cumulative in nature and may include instances of prior victimization, adverse effects of emotional abuse in childhood, family instability, distress and psychopathology, childhood behavior problems, antisocial behavioral history, prior psychiatric and substance abuse history, exposure to poverty or racism, and other forms of discrimination.
- Ascertain if any treatment services were provided for the impact of developmental stressors.
- Consider the lingering impact of experiences that were beyond "normal" combat, especially if the client is younger than 24 years of age.

4D. Familial Stressors (Present/Past)

- Consider the role of developmental psychopathology and distress in the family with regard to creating a vulnerability (diathesis) for response to traumatic events.
- Consider the role of current ongoing familial stress in the form of intimate partner violence and the experience of a rejecting, highly expressed emotional family environment (critical, intrusive, unsupportive).

- Consider issues of marital infidelity, marital distress, and parent–child conflict.
- Comment on changes in parenting roles and the ability to engage in emotional expressiveness.
- Provide supervision and proper monitoring, and foster family cohesion.
- Comment on the lingering impact of family instability, emotional dysfunction and conflict, and family psychopathology.
- Identify the degree of cultural distress beyond the family (e.g., incidence of substance abuse, violence, ongoing racism and discrimination).
- Consider the impact of intergenerational transmission of traumatic reactions (e.g., offspring of Holocaust-surviving parents, parents who were victimized and experienced PTSD and related problems), the transmission of racism and discrimination, and a multigenerational trauma history.

5. Treatments (Current/Past)

5A. Efficacy

- Enumerate various forms of psychosocial treatment and pharmacotherapy over the lifespan for the client and family members. Use Timeline 1 to provide as much information as possible (e.g., dates, onset and duration of each treatment, type of treatment, who administered the treatment, any evidence of efficacy or outcomes).
- Include treatments the client received while in the military during and after deployment and active case management; for example, include information on such treatments as antimalaria drugs and treatment of combat stress reactions (peritraumaric distress, intense emotional and arousal level during and after traumatic event).
- Include the use of community-based services, such as clergy, spiritual healers, traditional healers, elders, and alternative forms of treatment.

5B. Treatment Adherence/Compliance Information

- Address conditions under which treatment interventions were provided—namely, mandated treatments vs. self-referral vs. referred by others.
- Obtain adherence/compliance history for attending treatment, taking medication, and actively participating in treatment.
- Discuss individual, social, and systemic barriers that interfered with client attendance and participation in various treatment interventions.

- Consider the need for motivational interviewing procedures to enhance participation in treatment.

5C. Patient Satisfaction with Prior Treatment

- Prior satisfaction is a good predictor of future participation, engagement, and outcome.
- Assess the client's reactions to previous treatments and current views (theories regarding his or her distress and possible forms of intervention).
- Consider attitude and reactions toward exposure-based interventions to address the lingering effects of trauma experiences.

6. Strengths and Signs of Resilience

6A. Individual Strengths

Administer scales and interviews designed to tap resilience and posttraumatic growth.

- Consider ways to nurture and fortify resilience.
- Assess capacity for emotion-regulation skills.
- Identify preexisting spiritual and religious beliefs.
- Evaluate the client's pride in military accomplishments.
- Assess the client's coping skills (e.g., how he or she got through deployment), allegiance to military branch and devotion to country (e.g., client's belief that sacrifice was worth it), self-reliance, levels of self-efficacy, expression of inner strengths, positive reframing, positive social comparisons, and optimism (e.g., realistic?).
- Does the client engage in goal setting?
- Does the client have a "warrior" mentality and a sense of honor?
- Does the client have a proactive orientation and active (vs. avoidant) coping style?
- Is the client able to compartmentalize and not be blindsided by reminders of emotions and social conflicts?
- Evaluate the client's willingness to seek help.
- Determine the client's reasons for living; does the client construct meaning from his or her war experience as reflected in a feeling of greater independence?
- Has the client developed coping skills, a sense of self-improvement, and a greater purpose in life (a form of benefit finding) from his or her military participation?
- Identify the client's relapse-prevention skills.
- Do the client's academic achievements reflect a higher IQ and signs of cognitive functioning?

- Encourage the client not to engage in negative-thinking traps, self-blame, and avoidant behaviors; the client should not view failures as endpoints.
- What financial and human resources does the client have?

6B. Presence of Social Supports (Past/Present)

- Evaluate current social supports and the density of the client's social network of family, friends, coworkers, military buddies, and health-care providers.
- Determine the client's sense of commitment and responsibility to his or her "band of brothers" and "comrades in arms"; the client's camaraderie, sense of fellowship, and sense of military community; and how the client faces common challenges.
- Can the client share his or her experience to help others?
- Does the client have the interpersonal skills necessasry to access and use social supports?
- Can the client benefit from the military culture of order and discipline?
- Look for signs of cultural resilience. (What did the client's forefathers do to survive traumatic events; what lessons did they pass down?)
- Does the client have access to religious or spiritual traditions and groups?

6C. Systemic Strengths

- What services and supportive healthcare systems are available? What active outreach and follow-through programs?
- Culturally sensitive interventions should be family friendly.
- Did the client display positive and effective leadership while deployed?
- Does the client believe in the military mission and project a sense of honor?
- Is the shared meaning of sacrifice honored and respected by the community at large?

7. Summary of Risk and Protective Factors

- Empathize, validate, and normalize reactions to risk factors but also highlight the "rest of the client's story" that reflects strengths that the client was able to demonstrate in spite of the presence of risk factors.
- Collaborate with the client in enumerating the presence of protective factors (draw a timeline from birth to present that indicates the presence of resilience and signs of cultural resilience that occurred before the client was born).

8. Possible Outcomes

- Use goal attainment scaling (GAS) to identify doable, measurable short-term (8A), intermediate (8B), and long-term (8C) goals.
- Identify at least three treatment goals, one of which should be working toward a positive purposeful objective.
- Collaboratively agree on what would be a demonstration of 100% improvement, 50% improvement, or 0% improvement for each target goal.
- Indicate objectives within a specified timeframe.

9. Barriers or Possible Roadblocks to Maintaining Improvement

9A. Individual Barriers

- Presence of severity and chronicity of disorders
- Presence of comorbid psychiatric and medical disorders
- Neuropsychological impairment
- "Paralysis of will" (i.e., level of depression)
- Presence of risk factors to self and others
- Fear about treatment
- Lack of interest or motivation to change (client does not see that there is a problem)
- Concerns about stigmatization and the impact of receiving treatment on career
- Defensiveness
- Hostility and paranoid features (i.e., distrust and animosity toward healthcare providers and authorities)
- Perceived failure of prior treatment and concerns about side-effects of treatment

9B. Social Barriers

- Absence of supportive social supports
- Significant others who undermine treatment (e.g., codependent substance-abusing spouse and friends) and are an additional source of stress (spousal conflict, highly emotional environment)
- Degree of family members' psychopathology
- Culturally based prohibitions against seeking help or displaying emotions
- Types of trauma that should not be disclosed and discussed, such as incest
- Culturally based explanations (e.g., devil's work) that may undermine the use of exposure-based interventions

9C. Systemic Barriers and Roadblocks

Potential treatment-related barriers may include:

- Long waiting lists and an absence of specifically well-trained providers
- Lack of insurance coverage
- Lack of and delay in receiving combat-related compensation to cover treatment
- Absence of gender-specific and culturally sensitive interventions and, where indicated, a racially culturally matched therapist
- Absence of spiritually oriented treatment that can be integrated with psychotherapeutic procedures
- Lack of integrative treatment (e.g., available treatment provides only sequential or parallel ways of conducting treatment and no integrative services for PTSD and substance abuse disorders)
- No active follow-through interventions
- No treatment services for those who are leaving the military
- Little or no focus on the maintenance and generalization of treatment changes
- Limited length of treatment sessions
- Desire to avoid group treatment

Practical barriers may include:

- Absence of transportation to treatment center
- No childcare services
- Inflexible meeting times for treatment sessions
- Living in a remote rural area with few services
- Other individual, social, and systemic barriers, which should be discussed with the individual

References

Baumeister, R. F. (1991). *Meanings of life*. New York: Guilford Press.

Beck, A. T., Wright, F. D., Newman, C. E., & Liese, B. S. (1993). *Cognitive therapy of substance abuse*. New York: Gulford Press.

Castro, C. A., Adler, A. B., & Britt, C. A. (Eds.). (2006). *Military life: The psychology of serving in peace and combat*. Bridgeport, CT: Praeger.

Frankl, V. E. (1984). *Man's search for meaning* (3rd ed.). New York: Touchstone.

Friedman, M. J., Keane, T. M., & Resick, P. A. (Eds.). (2007). *Handbook of PTSD: Science and practice*. New York: Guilford Press.

Hoge, C. W., & McGunk, D. (2008). Mild traumatic brain injury in U.S. soldiers returning from Iraq. *New England Journal of Medicine*, 358, 453–463.

Hopewell, C. A., & Christopher, R. (2007). *Military personnel and combat trauma*. Sparks, NV: Sparks Publishers.

Hosek, J., Kavanaugh, J., & Miller, L. (2006). *How deployments affect service members*. Santa Monica, CA: The Rand Corporation.

Keane, J. M., Marshall, M., & Taft, C. (2006). Posttraumatic stress disorder: Epidemiology, etiology, and treatment outcome. *Annual Review of Clinical Psychology*, 2, 161–197.

King, D. W., King, L. A., Foy, D. W., & Gudanowski, D. M. (1996). Prewar factors in combat-related PTSD. *Journal of Consulting and Clinical Psychology*, 64, 520–531.

McNally, R. J. (2003). Progress and controversy in the study of posttraumatic stress disorder. *Annual Review of Psychology*, 54, 229–242.

Meichenbaum, D. (1994). *Treatment of adults with posttraumatic stress disorder: A clinical handbook.* Clearwater, FL: Institute Press.

Meichenbaum, D. (2004). *Treatment of individuals with anger-control problems and aggressive behavior.* Clearwater, FL: Institute Press.

Mueser, K. T., Noordsby, D. L., Drake, R. E., & Fox, L. (2003). *Integrated treatment for dual disorders.* New York: Guilford Press.

Murphy, R.T. (2008). Enhancing combat veterans' motivation to change posttraumatic stress disorder symptoms and other problem behaviors. In H. Arkowitz, H. A. Westra, W. R. Miller, and S. Rollnick (Eds.), *Motivational interviewing in the treatment of psychological problems* (pp. 57–84). New York: Guilford Press.

Najavits, L. M. (2006). Seeking safety: therapy for posttraumatic stress disorder and substance use disorder. In V. M. Follette and J. I. Ruzek (Eds.), *Cognitive–behavioral therapies for trauma* (2nd ed., pp. 228–257). New York: Guilford Press.

Park, C. L., & Folkman, S. (1997). Meaning in the context of stress and coping. *Review of General Psychology*, 1, 115–144.

Persons, J. B. (1989). *Cognitive therapy in practice: A case formulation approach.* New York: W.W. Norton.

Riggs, D. S., Cahill, S. P., & Foa, E. (2006). Prolonged exposure treatment of posttraumatic stress disorder. In V. M. Follette and J. I. Ruzek (Eds.), *Cognitive–behavioral therapies for trauma* (2nd ed., pp. 65–95). New York: Guilford Press.

Schok, M. L., Kleber, R. J., Elands, M., & Weerts, J. M. (2008). Meaning as a mission: A review of empirical studies on appraisals of war and peacekeeping experiences. *Clinical Psychology Review*, 28, 357–365.

Taylor, S. (2006). *Clinician's guide to PTSD: A cognitive–behavioral approach.* New York: Guilford Press.

Zayfert, C., & Becker, C. B. (2007). *Cognitive–behavioral therapy for PTSD: A case formulation approach.* New York: Guilford Press.

12
Treatment of Anxiety Disorders

DAVID S. RIGGS

Contents

Treating anxiety symptoms and disorders among service members returning from combat requires considerable understanding about the role that exposure to combat stress plays in the development of such problems. Exposure to a traumatic event such as combat is necessary, though not sufficient, for the development of posttraumatic stress disorder (PTSD), but such exposure can also cause a variety of other psychological difficulties, including depression and other anxiety disorders. The focus of this chapter is on the treatment of anxiety symptoms and disorders, and we discuss issues as they apply to a variety of disorders; however, the most salient of these for combat veterans are those that are specifically tied to traumatic events: PTSD and acute stress disorder (ASD) (APA, 1994).

The treatment of anxiety is relatively straightforward. Indeed, in terms of psychotherapeutic interventions anxiety disorders may be the most clearly amenable to treatment; however, it is important to recognize that not all anxiety or fear is in need of treatment, so it is important to accurately distinguish individuals who may require treatment from those who do not. This can be a tricky proposition, particularly in deployed settings where feelings of fear and anxiety will be relatively common. This discussion can be complicated by other issues as well, such as definitional issues including definitions of such basic concepts as fear and anxiety.

Many use the terms *fear* and *anxiety* interchangeably, and those who distinguish between them often do so in ways that appear inconsistent. For the purposes of this chapter, we have adopted the definition of anxiety used in the fourth edition of the *Diagnostic and Statistical Manual of Mental Disorders* (DSM-IV), where anxiety is defined as an "apprehensive anticipation of future danger or misfortune accompanied by a feeling of dysphoria or somatic symptoms of tension" (APA, 2000, p. 820). This is distinct from the basic emotion of fear, which is seen as a reaction to perceived danger in a person's immediate environment. In essence, then, anxiety is seen as a response to the anticipation of bad events, including potential threats, whereas fear is seen as a response to stimuli in the current environment that are thought to pose a threat to the person (Antony & Barlow, 1996). Another distinction between anxiety and fear suggested by Antony and Barlow rests on the function of the two emotional states. Fear, as a basic emotional response to threat, is seen as an alarm reaction that potentiates responses that lead to escape from the perceived danger and therefore may be adaptive. In contrast, anxiety is thought to contribute to states of worry and negative affect that are generally thought to be maladaptive.

This becomes important when discussing the treatment of anxiety disorders, particularly among service members faced with combat deployments. The combat environment is full of true threats to the lives and physical well-being of the individuals who must function in that environment. It is not surprising that individuals placed into combat situations will experience significant levels of fear. Indeed, it has been said that fear is ubiquitous in combat, and a good deal of military training is designed to help combatants overcome the normal fear response that occurs in combat situations (Grossman & Christensen, 2004). Similarly, in anticipation of a combat deployment or a specific action on the battlefield one would expect service members to experience elevated levels of anxiety. Unless these feelings of fear and anxiety on the battlefield or in anticipation of battle become excessive and begin to interfere significantly with other aspects of their lives, they would not be considered problematic, pathological, or in need of treatment.

In addition to fear and anxiety associated with anticipated or actual combat, service members who experience combat trauma also are likely to have emotional reactions that persist beyond the time of combat itself. These reactions may persist for hours, days, weeks, or longer. Again, the presence of emotional reactions following a combat experience is not in itself pathological. Indeed, within the DSM-IV classification system, there is no trauma-related diagnosis for reactions that last less than 2 days following the event (APA, 1994). The ICD-10 classification system includes a category of acute stress reactions for extreme responses during the first hours after an event, but these reactions are seen to dissipate rapidly (within hours) if the stressor is removed and within 1 to 2 days when the stressor persists.

Because intense emotional reactions are expected following intense, traumatic events such as combat missions, they should probably be viewed as normative or common reactions. This is true whether we are discussing reactions that occur in the immediate aftermath of combat while service members are still deployed or reactions that continue after they have returned home. They become "symptoms" only as they persist over time or begin to interfere with an individual's ability to function effectively. Importantly, the definition of effective functioning can differ dramatically depending on the setting in which the evaluation takes place. Thus, the requirements for effective functioning in a deployed combat environment are likely quite different than the determinants of effective functioning in the home environment. As a result, determining whether a person's feelings of fear or anxiety constitute "symptoms" that should be treated will depend, at least in part, on when and where they are evaluated.

It is important to note that the distinction between anxiety and fear discussed above does not rest on the accuracy or inaccuracy of the perception of danger; that is, one may be anxious in anticipation of an event that is truly dangerous (e.g., a soldier preparing to enter battle) or in anticipation of an event that carries little true threat (e.g., going to the grocery store). Similarly, I can experience fear in response to a true threat in the environment (e.g., someone pointing a weapon at me) or to something that I perceive as dangerous but is not (e.g., entering a crowded room). This distinction is potentially important when examining the question of when to provide or encourage treatment of anxiety symptoms. For the most part, such treatment becomes necessary when the anxiety or fear is interfering with appropriate and adaptive functioning. This can happen in a number of ways, but they all boil down to two basic premises. Fear or anxiety associated with events or stimuli that are not truly dangerous will likely interfere with functional behavior. Alternatively, extreme anxiety in anticipation of events, even truly dangerous events, can interfere with functioning prior to the actual event.

Assessment of Anxiety Symptoms and Disorders

The discussion above makes it clear that a careful assessment of anxiety and its functional impact is necessary prior to engaging in treatment. This assessment should include an evaluation of a number of issues beyond the mere presence and severity of symptoms to include medical conditions and skill deficits that may contribute to (or be exacerbated by) the anxiety or limit the treatment options that can be used to treat the individual. Other issues such as the course of symptoms, family history, prior treatment, and comorbid problems may also be important assessment questions in certain settings (Antony, 2001). Each of these areas will be more or less important in different settings. Regardless of emphasis, though, the core question of the functional impact of the anxiety and fear in the current environment will be critical to determining whether, when, and how to implement treatment.

Assessment Issues Related to Specific Disorders

Trauma-Related Diagnoses

Within the DSM-IV classification system (APA, 1994), two disorders are specifically tied to the experience of a traumatic event: acute stress disorder (ASD) and posttraumatic stress disorder (PTSD). The World Health Organization (WHO) *International Classification of Diseases*, 10th revision (ICD-10), includes a diagnosis of acute stress reaction (ASR) for brief, transient, but extreme reactions to events. Some slight differences in symptoms and presentation exist among these disorders, but the primary distinction is one of duration. Symptoms of ASR are expected to begin dissipating almost immediately after the stressor is removed and to be largely remitted within 48 hours. ASD cannot be diagnosed unless the reactions persist for at least 2 days, nor can ASD be diagnosed more than 1 month post-trauma. In comparison, PTSD can only be diagnosed if the symptoms persist for at least 1 month.

The symptoms of ASD and PTSD are clearly overlapping, with both diagnoses including symptoms of reexperiencing the trauma (e.g., intrusive thoughts, nightmares, flashbacks), avoidance (e.g., avoiding reminders, emotional numbing, detachment from others), and arousal (e.g., hypervigilance, sleep disturbance, irritability). The primary distinction between the two diagnoses is a greater emphasis on the presence of dissociative symptoms in the ASD criteria. In addition to the overlap between these two trauma-related diagnoses, symptoms of PTSD and ASD are shared with diagnoses of depression, other anxiety disorders, dissociative disorders, and, at least in the case of chronic PTSD, personality disorders (Riggs & Keane, 2006). Differentiating among these various disorders typically depends on establishing a link between the traumatic event and the onset of symptoms. This link is often established by the presence of reexperiencing symptoms that directly relate to an identified trauma but also relies on the temporal link between the traumatic experience and the onset of symptoms. It should be noted, however, that other disorders including panic disorder, phobias, depression, substance abuse/dependence, and perhaps even obsessive–compulsive disorder may begin in the aftermath of traumatic exposure (Riggs & Keane, 2006).

The distinct delimitation of the duration of symptoms in ASD and PTSD substantially reduces diagnostic confusion between these disorders, but some difficult issues may still arise. Perhaps the most difficult issue has to do with determination of when the "clock starts ticking." Both ASD and PTSD are conceptualized as disorders that manifest after the traumatic stressor is over. In cases of single traumatic events such as an assault or an automobile accident, it is relatively straightforward to determine when the event happened. Assuming that the posttraumatic symptoms manifest immediately after the event, as is typical, then determining the duration of the symptoms (and therefore the appropriate diagnosis) is fairly simple. Complications arise in situations where the traumatic event is repeated, such as in combat or abusive relationships, and when the offset of the stressor is not clear, such as natural disasters, where damage to the infrastructure can lead to ongoing stress.

Implied in the definitions of both ASD and PTSD is the assumption that the symptoms extend beyond the time of the true threat inherent in the trauma. Determining whether to diagnose PTSD while a threat continues can be complicated. To illustrate the difficulties that may arise in cases of chronic or repeated trauma, let's examine a hypothetical case of a soldier deployed into a combat situation.

Shortly after arriving in the combat theater, his unit comes under enemy fire and several of his comrades are seriously injured. This soldier is clearly emotionally distraught following this experience and remains upset for several weeks. The quality of his work begins to deteriorate, although he continues to function as a member of his unit. Over the course of the next several months, this soldier continues to participate in combat operations involving multiple firefights and culminating in a major operation extending over several days of combat during which three members of the unit are killed. At an evaluation conducted when he returns home from his deployment two months after this large operation, our soldier exhibits symptoms consistent with PTSD, including intrusive memories and nightmares about his first combat experience. Upon questioning from the evaluator, the soldier reports that he first began having problems shortly after arriving in the combat environment, about the time of his initial combat exposure, but that his emotions really began to create problems for him in the last month of his deployment, after the members of his unit were killed.

Recognizing that the symptoms of ASD cannot persist for more than a month and symptoms must be present for at least a month after the trauma to warrant a diagnosis of PTSD, this scenario raises questions as to the appropriate diagnosis for this soldier. If we date the onset of symptoms to his first combat experience, then we would conclude that he has chronic PTSD (lasting more than 3 months) and that it began while he was deployed. If the symptoms and the PTSD diagnosis are dated to 1 month after his last combat exposure, corresponding to the deaths of comrades, then he would be diagnosed with acute PTSD (lasting 1 to 3 months). However, if the onset of symptoms is tied to his departure from the war zone, then a diagnosis of ASD may be more appropriate than PTSD. This question is further complicated by the knowledge that many of the symptoms of PTSD (e.g., sleep problems, elevated startle and vigilance, irritability, increased arousal) are common among soldiers in deployed settings due to the true threats that exist in such environments.

Panic Disorder

Panic disorder (PD) can develop following trauma exposure. This disorder is characterized by repeated and unexpected panic attacks and anxiety over the possibility of future attacks or the consequences of the attacks. Panic attacks typically include intense physiological reactions, such as increased heart rate, respiration rate, muscle tension, and perspiration. Individuals may also experience feelings of dizziness, nausea, and tingling in their extremities during an attack. In addition to physiological symptoms, individuals with PD typically have distortions in their thoughts about the attacks or their consequences as well as biases in the processing of information (Khawaja & Oei, 1998). Often, there are associated behavioral changes to reduce the probability or severity of attacks such as avoiding certain places or not going out alone. When this avoidance extends to include a wide range of situations, the individual may experience substantial functional impairment and be diagnosed with agoraphobia in conjunction with the PD. It is equally important, however, for understanding an individual's PD symptoms to identify subtle forms of avoidance such as only going places when accompanied by a trusted person or not climbing a flight of stairs as well as other behaviors designed to ensure safety, such as always carrying medication. Although such avoidance may not create the significant functional impairment associated with agoraphobia, they do serve to maintain the panic symptoms and may reduce the efficacy of treatment.

It is important to note that panic or anxiety attacks can be associated with many causes, and the presence of such attacks does not necessarily mean that the individual has a panic disorder. Symptoms and reactions that may be mistaken for panic can result from medical conditions (e.g., hyperthyroidism) as well as the use of substances such as cocaine or caffeine or withdrawal from substances such as alcohol or pain medication. Prior to diagnosing PD, it is necessary to rule out these other possibilities.

Symptoms that mimic panic attacks are also present in a number of other anxiety disorders such as PTSD and social phobia. One way to differentiate panic attacks related to PD from those in the context of other anxiety disorders is by the content of the fear associated with the attacks. In most cases, individuals with PD are fearful of the attacks themselves (e.g., "I am having a heart attack," "I am going crazy") or the consequences of the attack (e.g., "Someone will notice my panic"). In contrast, individuals with other anxiety disorders tend to fear the consequences of the situation that cues the attack (e.g., "People will laugh at me," "I will remember my trauma experience"). Another means for differentiating PD from other disorders is the fact that the attacks associated with PD are typically described as occurring "out of the blue" in contrast to those associated with other disorders that typically occur in the presence of environmental cues (e.g., trauma reminders). As a result, one means to differentiate PD from panic associated with other disorders is the absence of identifiable cues for the panic attacks. One must remember, however, that some individuals with PTSD and other disorders will fail to identify cues, and some with PD will identify certain locations (typically ones that are difficult to leave) or internal sensations (e.g., dizziness, shortness of breath) as cues for their attacks.

Social Phobia

Although not explicitly tied to trauma exposure, social phobia (SP) may develop in individuals exposed to trauma, particularly interpersonal trauma. This disorder is characterized by significant fear associated with situations that involve social interactions or performance situations, particularly where scrutiny or embarrassment could occur. In addition, many individuals with SP experience significant anxiety in anticipation of social situations. Although SP may lead to significant impairment, people with SP often do not seek treatment (Magee, Eaton, Wittchen, McGonagle, & Kessler, 1996). In an effort to manage their fear and anxiety, individuals with SP will often attempt to avoid situations that elicit their fear. When it is not possible to avoid these situations, people with SP may simply endure their distress or they may engage in behaviors designed to manage their fear. The fear experienced in social situations by those suffering with SP can be very intense and, as mentioned above, may manifest as panic attacks that can make the differential diagnosis of SP and panic disorder difficult.

In addition to social phobia, fear and anxiety associated with social situations are also associated with a number of other anxiety disorders, including PTSD, panic disorder, and obsessive–compulsive disorder. Similarly, behaviors used to reduce feelings of anxiety in social situations (e.g., placing one's hands in pockets to avoid the appearance of trembling) can also mimic avoidance behaviors in other disorders such as OCD. As in other cases where the presentation of anxiety disorders overlaps, one means for differentiating among them is to examine the focus of the consequences associated with the fears. Typically, individuals with social phobia fear the possibility that they will do something that will lead others to judge them negatively. As a comparison, people with PTSD tend to fear interactions with other people because of the possibility that someone will attack them and people with OCD may fear becoming contaminated by germs when interacting with others.

Obsessive–Compulsive Disorder

Several cases have been described in the literature where symptoms of obsessive–compulsive disorder (OCD) appeared to develop in the aftermath of traumatic exposure (de Silva & Marks, 1999; Gershuny, Baer, Radomsky, Wilson, & Jenike, 2003; Pitman, 1993), and recently a number of articles have posited a link between OCD and PTSD (de Silva & Marks, 2001; Huppert et al., 2005). OCD is characterized by: (1) intrusive, distressing thoughts, images, or urges, and (2) repetitive behaviors or thoughts that serve to reduce obsessional distress (APA, 1994). As with other anxiety disorders, OCD can include substantial avoidance of situations and stimuli that increase obsessional fears. A number of other disorders also include symptoms very similar to those of OCD; for example, individuals with hypochondriasis and body dysmorphic disorder (BDD) have obsessional worries about their physical health or a physical characteristic and can also manifest rituals (typically checking) similar to OCD. Because of the overlap in symptoms, some have suggested conceptualizing these additional disorders as "OCD spectrum disorders."

Symptoms of a number of other disorders can mimic the obsessions and, to a lesser extent, the compulsions of OCD; for example, depression and generalized anxiety disorder are associated with worry and ruminations that may appear obsessional. In contrast to the obsessions associated with OCD, however, the ruminations and worries associated with depression tend to be mood congruent. PTSD symptoms also include intrusive and distressing thoughts similar to obsessions, although the focus is typically different. In the case of PTSD, the intrusive thoughts typically revolve around recollections of past traumatic events, whereas OCD obsessions tend to focus on the future. Individuals with PTSD can also display repetitive or ritualized behaviors, such as checking doors and windows to ensure they are locked, that can appear quite compulsive.

Treatments for Anxiety Disorders

As mentioned above, anxiety symptoms and disorders are among the most amenable to psychotherapy and psychotropic medication, and a number of different psychological and psychopharmacological approaches are available for treating pathological anxiety. Generally, medication and cognitive–behavioral therapies represent the first-line approaches to treatment, and they are frequently used together. Several classes of medication have been found effective in managing symptoms of anxiety, including selective serotonin reuptake inhibitors (SSRIs), tricyclic antidepressants, benzodiazepines, and (particularly to manage physiological symptoms of anxiety) beta-blockers. The most effective psychotherapeutic approaches to treating anxiety are drawn from the cognitive–behavioral tradition and are based on techniques including exposure therapy, anxiety management strategies, and cognitive restructuring. Specific therapeutic programs have been developed for the various anxiety disorders, most of which combine elements of all of these cognitive–behavioral approaches. Although these treatment programs have many common elements, they tend to differ in the degree to which they emphasize one or more of the cognitive–behavioral techniques. Rather than trying to review all of the treatments that have been used or suggested for anxiety disorders, this chapter discusses treatments that have received empirical support for particular disorders. Furthermore, because of the nature of this volume, this chapter focuses on the treatment of the trauma-related disorders acute stress disorder and posttraumatic stress disorder, although we also discuss treatments for other anxiety disorders that may arise in the aftermath of combat such as panic disorder, social phobia, and obsessive–compulsive disorder.

Psychotherapeutic Treatments for Trauma-Related Disorders

Most of the treatments with established efficacy are based on techniques of cognitive–behavioral therapy (CBT), and programs typically include elements of exposure therapy, cognitive therapy, and anxiety management techniques (Foa & Meadows, 1997; Foa & Rothbaum, 1998; Foa, Rothbaum, & Furr, 2003). These treatment programs tend to share many characteristics, with most variations in procedures reflecting differences in emphasis rather than dramatically different approaches to treatment. Treatments for posttrauma symptoms have been developed to treat chronic PTSD; some recent studies have examined the utility of these same techniques for treating trauma survivors more acutely, particularly those suffering with acute stress disorder.

Treatment of Acute Stress Disorder

Because most trauma survivors will recover naturally over the first weeks or months after the event (Blanchard et al., 1996; Koren, Arnon, & Klein, 1999; Riggs, Rothbaum, & Foa, 1995; Rothbaum, Foa, Riggs, Murdock, & Walsh, 1992), effective treatment for most produce more improvement than what occurs naturally. Because of this natural recovery and the time-limited nature of the ASD diagnosis, the outcome measures of interest in most existing studies has been whether or not the treated individuals developed more chronic PTSD symptoms. Alternatively, some studies have examined whether early intervention can decrease the length of time that it takes trauma-exposed individual to recover. As a result, treatments delivered during the first few weeks following a trauma might be better described as secondary prevention programs for PTSD rather than treatments for ASD *per se.*

Relatively less research has been conducted to examine the efficacy of treatments for trauma-related symptoms shortly after trauma exposure than for chronic PTSD (Riggs & Foa, 2008); however, the intervention programs that have been found to be somewhat effective when delivered within the first few weeks of trauma exposure have utilized CBT techniques that have been found effective in treating more chronic PTSD symptoms.

Most of the research on early intervention programs specifically targeting individuals who meet diagnostic criteria for ASD has been conducted by Bryant and colleagues (1998, 1999, 2003c, 2005). These researchers developed a brief (five-session) intervention that incorporated a number of different CBT techniques (e.g., exposure, cognitive therapy, anxiety management skills) with treatment initiated 2 to 4 weeks after the traumatic event (typically a motor vehicle accident). Across several studies this program has proven effective in reducing acute symptom severity and the number of individuals who develop chronic PTSD (Bryant, 2006); for example, in one study only 17% of individuals treated with this CBT program were diagnosed with PTSD 6 months after treatment compared to 67% of individuals in a supportive/problem-solving treatment (Bryant, Harvey, Sackville, Dang, & Basten, 1998). Similar results were found when the anxiety management component of the CBT protocol was removed (Bryant, Sackville, Dangh, Moulds, & Guthrie, 1999) and when hypnosis was added to the CBT program (Bryant, Moulds, Guthrie, & Nixon, 2005). Follow-up data collected 4 years after treatment for ASD indicate that the early treatment programs produce gains that are maintained over at least 4 years (Bryant, Moulds, & Nixon, 2003a). Finally, in a study that may have particular salience for Operation Enduring Freedom (OEF)/Operation Iraqi Freedom (OIF) veterans, Bryant, Moulds, Guthrie, and Nixon (2003c) examined the effects of brief, early treatment for individuals with ASD and mild traumatic brain injury (i.e., lost consciousness at the time of the trauma). Results from this study indicate that such early intervention can be effective even with this complicated group of survivors; in this case, only 8% of the group treated with CBT had PTSD 6 months following treatment compared to 58% of those treated with supportive counseling.

Similar CBT programs designed to be delivered shortly after a trauma have been examined by Foa and colleagues (1995, 2006). These studies have not targeted ASD; rather, they have targeted assault survivors with significant symptoms of PTSD within 2 to 3 weeks of the trauma. The treatment program developed by Foa and colleagues is similar to the one Bryant's group tested in that it includes elements of exposure, cognitive restructuring, education, and anxiety management (relaxation) training. In the first study of this program, Foa et al. (1995) found that a four-session treatment reduced the rate of PTSD diagnosed approximately 8 weeks after a sexual assault from 70% for a group that completed weekly assessments to 10% in the group treated with CBT. At the time of a follow-up assessment conducted 6 months after treatment, the two groups did not differ in the rate of diagnosable PTSD (11% vs. 22%, respectively, for the treated and untreated groups). In a second study, Foa et al. (2006) compared recent assault survivors with significant PTSD symptoms who were treated in the CBT program to another group that received supportive counseling and a third group that completed assessments but no treatment. At the end of treatment, all three groups showed improvement in PTSD symptoms, with those in the CBT group reporting less severe symptoms than those treated with supportive counseling; the assessment-only group did not differ from either treated group. As was the case in the earlier study of this treatment program, at the time of the follow-up assessment there were no differences among the treatments. The authors concluded that early treatment was effective at accelerating recovery following a trauma but the natural recovery process will tend to wash out these differences over time (Foa, Zoellner, & Feeny, 2006).

Treatment of Chronic PTSD

Several CBT treatment programs have been shown to significantly reduce symptoms of chronic PTSD (Foa & Meadows, 1997; Foa & Rothbaum, 1998; Foa et al., 2003; Riggs & Foa, 2008). The Department of Defense/ Department of Veterans Affairs practice guidelines list four of these programs as having substantial and strong empirical support that they are effective in treating chronic PTSD. These four programs are (1) exposure therapy, (2) cognitive therapy, (3) stress inoculation therapy, and (4) eye movement desensitization and reprocessing.

Exposure Therapy

The details of how exposure therapy is implemented differ across specific programs, but most programs for PTSD ask the trauma survivor to repeatedly recall their memories of the trauma (commonly referred to as *imaginal exposure*). Programs may also include *in vivo* exercises to promote exposure to situations, objects, or other environmental stimuli that trigger distress or memories of the trauma (Foa et al., 1999, 2005; Resick, Nishith, Weaver, Astin, & Feuer, 2002). Typically, components of the treatment programs will focus on educating clients about trauma and posttraumatic reactions and providing them with skills for managing their distress (Foa et al., 1999, 2005; Keane, Fairbank, Caddell, & Zimering, 1989). Some programs also include other elements such as cognitive restructuring and affect management techniques (Blanchard et al., 2003; Cloitre, Koenen, Cohen, & Han, 2002). Across these variations, exposure therapy has been found effective in many studies with survivors of a wide range of traumatic events (Riggs & Foa, 2008).

One of the most studied versions of exposure therapy is prolonged exposure (PE), developed by Foa and colleagues (2007). This program includes both imaginal and *in vivo* exposure exercises, psychoeducation, and minimal anxiety management training with a clear emphasis on the exposure elements. Studies have found PE more effective in treating PTSD compared to supportive counseling (Foa, Rothbaum, Riggs, & Murdock, 1991), a waitlist condition (Foa et al., 1991, 1999, 2005; Resick et al., 2002), and problem-focused therapy (Schnurr et al., 2007). Treatment gains associated with PE are comparable to those obtained with other CBT programs, including stress inoculation therapy (SIT) (Foa et al., 1991, 1999), cognitive processing therapy (CPT) (Resick et al., 2002), and eye movement desensitization and reprocessing (EMDR) (Rothbaum, Astin, & Marsteller, 2005), and they are generally maintained over time. In studies where symptoms of depression and anxiety are assessed, PE has been shown to reduce these symptoms as well (Foa et al., 2005; Resick et al., 2002; Schnurr et al., 2007). Despite the excellent outcome following PE treatment, several researchers have attempted to improve outcome by adding other treatment techniques to the standard PE program. In general, these studies have found no particular benefit for adding SIT (Foa et al., 1999) or cognitive restructuring (Foa et al., 2005) to the PE program.

Programs other than PE that include significant exposure components are also effective in treating PTSD symptoms; for example, several trials have examined a program developed by Marks and colleagues that was very similar to the PE program, except that the exposure exercises were delivered sequentially with imaginal exposure exercises presented during the first several sessions and *in vivo* exercises conducted through the final sessions. This program has been found to be more effective than relaxation training (Marks et al., 1998; Taylor et al., 2003) and equally effective as cognitive restructuring (Marks et al., 1998) and EMDR (Power et al., 2002; Taylor et al., 2003). Blanchard et al. (2003) compared a CBT program that included writing and rereading a description of the trauma as well as *in vivo* exposure and additional techniques to supportive counseling and a waitlist; the CBT group showed greater improvement than those in the other conditions. Bryant and colleagues (2003b) examined the efficacy of a program that used only imaginal exposure to the combination of imaginal exposure and cognitive therapy. Both groups showed greater reduction in symptoms than persons in a waitlist condition, although the combined exposure and cognitive therapy group improved more than the imaginal-exposure-only group on measures of trauma-related cognitions and reexperiencing (Bryant, Moulds, Guthrie, Dang, & Nixon, 2003b).

Cognitive Therapy

As with exposure therapy, several cognitive therapy programs have been shown to be effective in reducing PTSD symptoms. These treatments share many characteristics with one another, although certain details differ across programs. These treatments focus on identifying and modifying dysfunctional trauma-related cognitions; these cognitions typically involve inaccurate or exaggerated beliefs about the traumatic event, the person, or the world. Precisely which cognitions are identified differs across treatment programs, with some focused on thoughts related to current functioning (Blanchard et al., 2003), others focused more directly on thoughts related to the traumatic event (Resick et al., 2002), and still others attending more to cognitions about trauma-related reactions (Ehlers, Clark, Hackmann, McManus, & Fennell, 2005). When the dysfunctional thoughts have been identified, the therapist works with the client to shift the cognitions to replace them with more functional cognitions using a variety of techniques such as Socratic questioning (Ehlers et al., 2005; Foa et al., 2005; Resick et al., 2002). Most cognitive therapy programs include elements of exposure therapy such as writing, imagining the trauma, or *in vivo* exercises; however, the rationale for including these exercises is typically different than that for exposure therapy, so the exposure exercises incorporated into these cognitive programs are usually briefer and less frequent than those used in the exposure-focused programs.

Cognitive processing therapy (CPT) (Resick & Schnicke, 1992) is the most extensively studied cognitive therapy programs for PTSD. In addition to the core cognitive techniques focused on addressing specific themes (e.g., safety, esteem, intimacy) that are theoretically linked to trauma, CPT includes a writing/reading exposure component and education about trauma reactions. Three studies have found CPT to be more effective in reducing PTSD symptoms compared to waitlist comparisons (Chard, 2005; Monson et al., 2006; Resick et al., 2002). As with PE, treatment with CPT appears to reduce depression and anxiety as well as the targeted PTSD symptoms, and treatment gains are maintained after therapy is ended (Resick et al., 2002). Monson et al. (2006) found that the CPT protocol was effective in reducing symptoms of PTSD and anxiety in a sample of veterans, but the CPT group did not differ from the waitlist group on measures of hyperarousal and behavioral avoidance. A modified version of CPT that included additional sessions (some conducted as groups) and incorporated training in a number of interpersonal skills was effective in reducing PTSD and depression symptoms in a group of adult survivors of childhood sexual abuse (Chard, 2005).

A second cognitive therapy approach that appears effective in treating PTSD has been developed by Ehlers and colleagues (2003, 2005). This approach differs somewhat from CPT in that it focuses on modifying negative appraisals, changing the nature of the traumatic memory, and reducing reliance on short-term distress management strategies (Ehlers et al., 2005). In addition to the cognitive interventions, this program includes a number of exposure techniques such as writing about the event and visiting the location of the trauma. Studies indicated that this treatment was effective in reducing PTSD, depression, and anxiety symptoms compared to a waitlist condition (Ehlers et al., 2003, 2005).

A third cognitive approach focuses on changing cognitions associated with feelings of guilt associated with the traumatic exposure (Kubany, Hill, & Owens, 2003; Kubany et al., 2004). As with other cognitive therapies, this treatment program includes a limited exposure component. The program also incorporates assertiveness training, problem-solving, and anger-management skills training. This program has not been studied as extensively as the other cognitive therapy programs discussed previously, but the available data suggest that the treatment can be effective at reducing the severity of PTSD, depression, guilt, and low self-esteem in female victims of domestic violence (Kubany et al., 2003, 2004).

Additional evidence that cognitive therapy can be effective in reducing PTSD symptoms comes from studies comparing different treatments for PTSD; for example, Resick et al. (2002) found that CPT and PE were equally effective in treating PTSD symptoms and that both were more effective than a waitlist condition. Similarly, Marks et al. (1998) directly compared PE and a cognitive restructuring condition to a waitlist condition. Results indicated that both treatments reduced symptoms of PTSD and depression compared to the waitlist and that the two active treatments did not differ from one another. Additional studies have examined the effects of adding cognitive therapy elements to exposure therapy programs. The resulting combined treatment programs are similar to the cognitive therapy programs described above, although they include a heavier dose of exposure. Several studies have found these combined treatments to be more effective than waitlist and equally effective as exposure therapy (Foa et al., 2005; Marks, Lovell, Noshirvani, Livanou, & Thrasher, 1998; Paunovic & Ost, 2003).

Anxiety Management

Some treatments used for PTSD focus on teaching trauma survivors to control anxiety symptoms and manage ongoing stress; for example, Veronen and Kilpatrick (1983) adapted stress inoculation training (SIT) (Meichenbaum, 1974) to treat rape survivors. SIT programs typically include a number of different techniques focused on managing distress and anxiety such as breathing and relaxation training, guided self-talk, assertiveness training, role playing, covert modeling, and thought stopping. Typically, these programs also include components that are included in other treatment approaches such as education about trauma-related reactions and cognitive restructuring. Some SIT programs also have explicit exposure elements (Veronen & Kilpatrick, 1983), whereas others do not (Foa et al., 1991, 1999). In addition, a number of treatment programs that have been found effective in treating PTSD include anxiety management (Blanchard et al., 2003) or emotion regulation (Cloitre et al., 2002) components.

Anxiety management programs such as SIT as treatments for PTSD have not been the focus of much systematic research; however, several well-controlled studies have included these approaches in studies more focused on testing the efficacy of exposure and cognitive therapies. In particular, two studies conducted by Foa and colleagues (1991, 1999) were designed such that the effects of SIT could be compared to a waitlist condition. SIT produced significantly greater reduction in PTSD symptoms than in the waitlist condition. Further, these gains were maintained through follow-up assessments (Foa et al., 1991, 1999). Treatment gains with SIT were comparable to PE, though the effect sizes associated with PE were somewhat larger than those in SIT (Foa et al., 1999).

Eye Movement Desensitization and Reprocessing

Eye movement desensitization and reprocessing (EMDR) (Shapiro, 1989, 2001) was not originally developed as a CBT program, but it shares many characteristics with exposure and cognitive therapy treatments for PTSD. EMDR treatment programs ask clients to focus on their mental image of their trauma paired with any negative thoughts or feelings associated with the memory. Concurrent with this, patients engage in alternating bilateral stimulation such as rapid eye movements. As treatment progresses, the client is instructed to introduce alternative appraisals about the trauma and then repeat the eye movements.

Numerous studies have examined the efficacy of EMDR. The results of these studies suggest that EMDR is more effective than waitlist control conditions in reducing PTSD symptoms. Results from one study found significantly more improvement in PTSD symptoms following nine EMDR sessions than in a waitlist condition (Rothbaum et al., 2005). A second study found that EMDR treatment was more effective than standard psychiatric care for reducing PTSD symptoms (Marcus et al., 1997).

Studies that directly compare EMDR to CBT programs produce mixed results. Some studies suggest that EMDR may be somewhat superior to other treatment programs; for example, in a study comparing seven sessions of EMDR to seven sessions of PE combined with SIT, Lee and colleagues (2002) found comparable reductions in PTSD symptoms at posttreatment assessment. Three months after treatment, the EMDR group had lower symptom scores than the PE–SIT group, particularly for intrusion symptoms. Similarly, Power et al. (2002) reported larger effect sizes associated with EMDR than with a combined exposure and cognitive restructuring treatment. In contrast, other studies found that EMDR produces smaller changes than exposure therapy. In one study, assessments at posttreatment and 3 months' follow-up indicated that a program that combined exposure therapy, cognitive restructuring, and SIT produced greater results than EMDR (Devilly & Spence, 1999). This was particularly the case at the follow-up assessment due to some return of symptoms in the EMDR group. In a second study, Taylor et al. (2003) compared exposure therapy, EMDR, and relaxation training; in this study, exposure therapy produced greater improvement in PTSD symptoms than relaxation, but EMDR did not. Some studies have found generally comparable results for treatment with exposure and EMDR, with both being more effective than a waitlist condition (Rothbaum et al., 2005).

Pharmacological Treatment of PTSD

Considerable research has examined the potential of medications to effectively treat chronic PTSD (Bisson, 2007). Although there are considerable limitations to this literature (e.g., small sample sizes, limited replication studies), the research does indicate that certain medications, particularly selective serotonin reuptake inhibitors (SSRIs) can help to reduce PTSD symptoms. Research on other classes of medication are considerably more limited (typically only a single trial is available), although there are indications that certain tricyclic antidepressants (e.g., amitriptyline) and monoamine oxidase inhibitors (e.g., phenelzine) may also be helpful in treating PTSD, as might mirtazapine (Bisson, 2007).

Most of the pharmacological research on the treatment of PTSD has focused on SSRIs, particularly paroxetine, and sertraline. Results for paroxetine are generally positive, with significantly more reduction in symptoms associated with the medication as compared to placebo treatment (Marshall et al., 2001; Tucker et al., 2001), but the clinical significance of this improvement was limited (Bisson, 2007). Data on sertraline are more ambiguous. Results from four published studies suggest that sertraline can be effective in reducing PTSD symptoms (Brady et al., 2000; Davidson et al., 2001; Zohar, 2002); however, the National Institute for Health and Clinical Excellence (NICE) clinical guidelines development group in the United Kingdom did not find an overall significant effect of sertraline over placebo (National Collaborating Center for Mental Health, 2005). This was in part because of the inclusion of two unpublished studies in their review (Bisson, 2007).

Treatment of Panic Disorder

Both CBT and medications have been shown effective in reducing the symptoms of panic disorder (Allen & Barlow, 2006), and CBT is considered to be the gold-standard treatment (Craske & Barlow, 2001) because of its efficacy and the tendency toward relapse when medication is discontinued (Noyes et al., 1991; Rapaport et al., 2001). Combined CBT and pharmacological treatments have also been examined with mixed results depending on when participants are evaluated (Barlow, Gorman, Shear, & Woods, 2000).

Cognitive–Behavioral Therapy

The most well-researched CBT approach to treating panic disorder is based on the work of Barlow and colleagues (Craske, Barlow, & Meadows, 2000). This approach, sometimes labeled *panic control therapy*, combines the basic elements of CBT for anxiety disorders (psychoeducation, exposure to feared stimuli, cognitive restructuring) to address fear associated with specific locations/situation and the physiological sensations that signal a panic attack. In this 12-session program, the initial session is used to provide an overview of the treatment program, develop a working model of the client's current problems, and educate the client about the conceptual model underlying the treatment. The next sessions focus on exposure to interoceptive cues associated with panic attacks such as dizziness, shortness of breath, hyperventilation, and elevated heart rate. Clients work with the therapist to identify exercises that elicit physiological responses that mimic their own panic experiences and then repeatedly engage in those exercises to reduce the intensity of their emotional reactions to these physiological states. Subsequent sessions focus on cognitive restructuring focused on the client's cognitive misappraisal of the panic symptoms. The last several sessions shift the focus to developing *in vivo* exposure exercises to allow clients to habituate their anxiety to various environmental and interoceptive cues (Craske & Barlow, 2001).

A second CBT approach that has proven effective in treating panic disorder is based on the work of Clark and colleagues (Clark, 1986; Salkovskis & Clark, 1991). This approach shares many characteristics with the Barlow program, but it places much more emphasis on cognitive restructuring and less on interoceptive exposure than does the Barlow treatment program. In this treatment, sessions focus on challenging clients' catastrophic interpretations of their bodily sensations and replacing them with more realistic interpretations. In addition to some limited interoceptive exposure, this program also includes behavioral experiments that are designed not to promote habituation as in the Barlow program but to allow clients to challenge their interpretations and the probability that their feared consequences will occur. As with the *in vivo* exposure exercises included in Barlow's treatment, clients are asked to engage in these experiments without engaging in safety behaviors (e.g., holding onto something when feeling dizzy), but rather than emphasizing the potential of such behaviors to interfere with habituation it is thought that this allows the client to disconfirm the feared consequences of the symptoms (Clark et al., 1999).

Numerous studies using variations on these protocols have supported the efficacy of CBT for panic disorder. Several metaanalyses (Chambless & Gillis, 1993; Gould et al., 1997; Westen & Morrison, 2001) of treatments for panic disorder have found a consistent advantage for cognitive–behavioral treatments over waitlist or various comparison treatments (e.g., placebo therapy, supportive therapy, pill placebo). Across the studies, over 70% of those treated with CBT showed improvement compared to about 25% of those in the comparison conditions, and the effect sizes associated with CBT that combined exposure and cognitive restructuring were very large. Although few studies reported extensive follow-up data, there are indications that the gains made with CBT are maintained over time (Westen & Morrison, 2001).

Pharmacological Treatment of Panic Disorder

Several classes of medication, including SSRIs, tricyclics, and benzodiazepines, have been found effective in reducing panic attacks and improving the level of functioning of those treated (Bakker, van Balkom, & Spinhoven, 2002; Mitte, 2005; Otto, Tuby, Gould, McLean, & Pollack, 2001; Rickels & Schweizer, 1998). At this time, the SSRIs are typically the first-line treatment because they have fewer adverse side effects than tricyclics (Bakker et al., 2002) and fewer issues related to dependency and withdrawal than the benzodiazepines (Gorman et al., 1998). Reactions upon discontinuation of benzodiazepines may be particularly problematic for panic patients because they are sensitive to the physiological arousal that can occur (Otto, Hong, & Safren, 2002). Similarly, because physiological arousal can also occur with the introduction of some psychotropic agents (e.g., SSRIs), it is important to monitor these reactions and treatment compliance when medications are initiated or dosages are increased. Some have suggested combining a benzodiazepine with an SSRI initially to reduce these problems and then continuing with SSRI monotherapy (Starcevic, 2004). This approach may have the added advantage of having a faster onset than SSRI monotherapy (Goddard et al., 2001).

One large multisite study compared the efficacy of CBT, medication (imipramine), and a combination of the two treatments (Barlow et al., 2000). The results immediately after treatment was completed indicated that all active treatments (CBT alone, imipramine alone, CBT combined with imipramine, and CBT with a pill placebo) were more effective than placebo alone. No differences were found among the active treatments. Following a maintenance phase of the trial, during which treatment continued with monthly meetings, the combination treatment (CBT plus imipramine) demonstrated greater improvement than the other treatment groups (although the differences were relatively small). Six months after the termination of treatment the only groups that continued to show greater improvement than the placebo group were those that received CBT without medication (CBT alone and CBT plus placebo). The groups treated with medication, whether alone or in combination with CBT, relapsed when the medication was removed. Thus, it appeared that CBT without medication appeared to produce a more lasting improvement than the combination of CBT and medication (Allen & Barlow, 2006).

Treatment of Social Phobia

As is the case with other anxiety disorders, both medications and psychotherapy have been found effective in reducing the symptoms of social phobia, and the most studied psychotherapies are cognitive–behavioral in nature (Davidson, 2006).

Cognitive–Behavioral Therapy

Cognitive–behavioral approaches to treating social phobia tend to incorporate elements of exposure and cognitive restructuring techniques much as do the treatments of other anxiety disorders. In some cases, they also include a skills training component designed to address identified social skills deficits. In a review of the literature on psychosocial treatments of social phobia, Taylor (1996) determined that exposure, cognitive restructuring, social skills training, and a combination of these treatments were all superior to a waitlist condition when assessments were conducted immediately following treatment or at follow-up. One metaanalysis of the treatment literature indicated that treatments that included exposure (alone or in combination with cognitive restructuring) was more effective than cognitive therapy or social skills training alone (Gould et al., 1997). However, a later metaanalysis found all forms of CBT to be moderately effective, with none of the approaches being more or less effective than the others (Federoff & Taylor, 2001). In sum, CBT treatments appear more effective than no treatment for social phobia, although no particular treatment has been found clearly superior to others. Furthermore, a substantial number of people with social phobia do not respond to CBT, and even among those who do improve significant residual symptoms persist (Davidson, 2006).

Pharmacological Treatment of Social Phobia

A number of medications, including SSRIs and benzodiazepines, have been found to be effective in reducing symptoms of social phobia. Additional medications that appear somewhat effective in reducing social phobia symptoms include some monoamine oxidase inhibitors (MAOIs) and anticonvulsants (Davidson, 2006). A review treatments for social phobia found that benzodazepines were associated with the largest effect sizes, followed by MAOIs, SSRIs, and anticonvulsants (Hildago, Barnett, & Davidson, 2001). As is the case with many pharmacological treatments of anxiety disorders, it may be necessary to maintain patients on medication, as discontinuing the medicine is associated with a return of symptoms (Davidson, 2006).

Several studies of benzodiazepine medications provide consistent support for their efficacy in reducing social phobia symptoms, and Davidson (2006) concluded that the rapid onset and high efficacy suggest that these medications may serve as the first line of pharmacological treatment for the disorder. This approach, however, should be tempered by the acknowledged difficulties with misuse of these medications, difficulties upon discontinuation, and their limited effect on comorbid problems such as depression and PTSD (Davidson, 2006).

Davidson (2006) reviewed eight studies examining various SSRIs (fluvoxamine, paroxetine, sertraline) that have demonstrated consistently better response rates (from 40 to 70%) than placebo controls (from 7 to 32%); however, Davidson (2006) also identified two trials of the medication fluoxetine that failed to show greater efficacy than placebo, so it should not be assumed that any SSRI will work to reduce social phobia. As with the SSRIs, several MAOIs have been shown consistently to improve social phobia across multiple studies (Davidson, 2006), but their use in general clinical practice is limited because many individuals are unable to tolerate the side effects and food restrictions that are required when taking these medications.

Treatment of Obsessive–Compulsive Disorder

Exposure and Ritual Prevention

Exposure and response (ritual) prevention (EX/RP) is effective at reducing obsessive–compulsive (OCD) symptoms, both obsessions and the urges to complete compulsive rituals (for a comprehensive review, see Franklin & Foa, 2002). The treatment encourages clients to confront stimuli that are associated with obsessive fear through imaginal and *in vivo* exercises. In contrast to the imaginal exposure used with PTSD patients, where the focus is on exposure to traumatic memories, imaginal exposure for OCD tends to focus on the feared consequences of failing to engage in compulsive rituals. Through exposure and abstinence from rituals, it is thought that these clients will habituate to feared situations and disconfirm inaccurate beliefs about the danger of certain situation and the safety offered by the completion of rituals (Foa & Kozak, 1986). In these treatments, the therapist does not actually prevent the client from completing rituals; rather, the client is asked and encouraged to refrain voluntarily (Riggs & Foa, 2006).

Many controlled studies provide empirical support for the efficacy of EX/RP for OCD over waitlist or limited contact control groups (Franklin & Foa, 2002). Additionally, several randomized controlled trials have found EX/RP to be more effective than a variety of control treatments such as relaxation (Fals-Stewart et al., 1993; Marks et al., 1975) and anxiety management training (Lindsay et al., 1997). Metaanalyses of treatment outcome studies (Abramowitz, 1996; Cox et al., 1993; van Balkom et al., 1998) indicate that, across studies, EX/RP is associated with large effect sizes (≥ 1.0). Further, the effects of EX/RP treatment appear to be maintained over time (Foa & Kozak, 1996). The effects of EX/RP appear robust and resilient to procedural variations (Riggs & Foa, 2006); however, the results of a metaanalysis of treatment studies suggest that strict ritual prevention instructions worked better than instructions to merely limit rituals or no ritual prevention instructions at all (Abramowitz, 1996). Based on the same review, Abramowitz (1996) concluded that the combination of imaginal and *in vivo* exposure performed better than *in vivo* exposure alone.

Pharmacological Treatment of Obsessive–Compulsive Disorder

Several comprehensive reviews of pharmacotherapy for OCD are available; see, for example, Dougherty, Rauch, and Jenike (2002) for adult OCD and March, Franklin, Nelson, and Foa (2001) and Thomsen (2002) for pediatric OCD. Based on the available literature, it seems clear that several SSRIs including fluoxetine (Geller et al., 2001; Tollefson et al., 1994), sertraline (Greist, Jefferson, Kobak, Katzelnick, & Serlin, 1995; March et al., 1998), fluvoxamine (Goodman et al., 1989; Riddle et al., 2001), and paroxetine (Wheadon, Bushnell, & Steiner, 1993) can effectively reduce OCD symptoms. In addition, the tricyclic antidepressant clomipramine appears effective for OCD (Clomipramine Collaborative Group, 1991; DeVeaugh-Geiss et al., 1992).

Summary

In sum, anxiety symptoms and disorders are quite amenable to treatment; for most disorders there are both pharmacological and psychotherapeutic (primarily cognitive–behavioral) treatments with established efficacy. The identification of anxiety or fear reactions as "symptoms" to be treated must take into account the functional impact of those reactions and the setting in which they occur. This is particularly important in the case of service members deployed to combat or returning from such deployment. Fear and anxiety are appropriate reactions when faced with real dangers and should be considered pathological only when they interfere with appropriate and necessary functioning. Similarly, reactions that may represent symptoms of posttraumatic reactions (e.g., sleep disturbance, hypervigilance) are also expected and are common reactions to traumatic exposure that will likely reduce when the exposure ceases. Thus, identifying these reactions as "symptoms" may be difficult when the individual continues to be exposed to traumatic or potentially traumatic events such as during combat deployments. Again, a useful rule of thumb for determining the pathology of these reactions is to evaluate the functional impairment associated with them.

References

Abramowitz, J. S. (1996). Variants of exposure and response prevention in the treatment of obsessive-compulsive disorder: A meta-analysis. *Behavior Therapy, 27,* 583–600.

Allen, L. B., & Barlow, D. H. (2006). The treatment of panic disorder: Outcomes and basic processes. In B. Rothbaum (Ed.), *Pathological anxiety: Emotional processing in etiology and treatment* (pp. 166–180). New York: Guilford.

APA. (1994). *Diagnostic and statistical manual of mental disorders* (4th ed.). Washington, D.C.: American Psychiatric Association.

APA. (2000). *Diagnostic and statistical manual of mental disorders* (4th ed., text revision). Washington, D.C.: American Psychiatric Association.

Antony, M. M. (2001). Assessment of anxiety and the anxiety disorders: An overview. In M. M. Antony, S. M. Orsillo, & L. Roemer (Eds.), *Practitioner's guide to empirically based measures of anxiety* (pp 9–17). New York: Kluwer Academic/Plenum.

Antony, M. M., & Barlow, D. H. (1996). Social and specific phobias. In D. H. Taylor & A. Tasman (Eds.), *Psychiatry: Self-assessment and review* (pp. 196–241). New York: Guilford.

Bakker, A., van Balkom, A. J., & Spinhoven, P. (2002). SSRIs vs. TCAs in the treatment of panic disorder: A meta-analysis. *Acta Psychiatrica Scandinavia, 106,* 163–167.

Barlow, D. H., Gorman, J. M., Shear, M. K., & Woods, S. W. (2000). Cognitive–behavioral therapy, imipramine or their combination for panic disorder: A randomized controlled trial. *Journal of the American Medical Association, 283,* 2529–2536.

Bisson, J. I. (2007). Pharmacological treatment of post-traumatic stress disorder. *Advances in Psychiatric Treatment, 13,* 119–126.

Blanchard, E. B., Hickling, E. J., Barton, K. A., Taylor, A. E., Loos, W. R., & Jones-Alexander, J. (1996). One-year prospective follow-up of motor vehicle accident victims. *Behaviour Research and Therapy, 34,* 775–786.

Blanchard, E. B., Hickling, E. J., Devineni, T., Veazey, C. H., Galovski, T. E., Mundy, E. et al. (2003). A controlled evaluation of cognitive behavioral therapy for posttraumatic stress in motor vehicle accident survivors. *Behavior Research and Therapy, 41*(1), 79–96.

Brady, K., Pearlstein, T., Asnis, G. et al. (2000). Efficacy and safety of sertraline treatment of posttraumatic stress disorder: A randomized controlled trial. *Journal of the American Medical Association, 283,* 1837–1844.

Bryant, R. A. (2006). Cognitive–behavioral therapy for acute stress disorder. In A. M. Follette & J. I. Ruzek (Eds.), *Cognitive–behavioral therapies for trauma* (2nd ed., pp. 201–227). New York: Guilford.

Bryant, R. A., Harvey, A. G., Sackville, T., Dang, S. T., & Basten, C. (1998). Treatment of acute stress disorder: A comparison between cognitive–behavioral therapy and supportive counseling. *Journal of Consulting and Clinical Psychology*, 66, 862–866.

Bryant, R. A., Sackville, T., Dangh, S. T., Moulds, M., & Guthrie, R. (1999). Treating acute stress disorder: An evaluation of cognitive behavior therapy and supportive counseling techniques. *American Journal of Psychiatry*, 156, 1780–1786.

Bryant, R. A., Moulds, M. L., & Nixon, R. D. V. (2003a). Cognitive behaviour therapy of acute stress disorder: A four-year follow-up. *Behaviour Research and Therapy*, 41, 489–494.

Bryant, R. A., Moulds, M. L., Guthrie, R. M., Dang, S. T., & Nixon, R.D. V. (2003b). Imaginal exposure alone and imaginal exposure with cognitive restructuring in treatment of posttraumatic stress disorder. *Journal of Consulting and Clinical Psychology*, 71, 706–712.

Bryant, R. A., Moulds, M., Guthrie, R., & Nixon, R. D. V. (2003c). Treating acute stress disorder following mild traumatic brain injury. *American Journal of Psychiatry*, 160(3), 585–587.

Bryant, R. A., Moulds, M., Guthrie, R., & Nixon, R. D. V. (2005). The additive benefit of hypnotherapy and cognitive behavior therapy in treating acute stress disorder. *Journal of Consulting and Clinical Psychology*, 73, 334–340.

Chambless, D. L., & Gillis, M. M. (1993). Cognitive therapy for the anxiety disorders. *Journal of Consulting and Clinical Psychology*, 61, 248–260.

Chard, K. M. (2005). An evaluation of cognitive processing therapy for the treatment of posttraumatic stress disorder related to childhood sexual abuse. *Journal of Consulting and Clinical Psychology*, 73, 965–971.

Clark, D. M. (1986). A cognitive approach to panic. *Behaviour Research and Therapy*, 24, 461–470.

Clark, D. M., Salkovskis P. M., Hackmann A. et al. (1999). Brief cognitive therapy for panic disorder: A randomized controlled trial. *Journal of Consulting and Clinical Psychology*, 67, 583–589.

Cloitre, M., Koenen, K. C., Cohen, L. R., & Han, H. (2002). Skills training in affective and interpersonal regulation followed by exposure: A phase-based treatment for PTSD related to childhood abuse. *Journal of Consulting and Clinical Psychology*, 70(5), 1067–1074.

Clomipramine Collaborative Group. (1991). Clomipramine in the treatment of patients with obsessive–compulsive disorder. *Archives of General Psychiatry*, 48, 730–738.

Cox, B. J., Swinson, R. P., Morrison, B., & Lee, P. S. (1993). Clomipramine, fluoxetine, and behavior therapy in the treatment of obsessive–compulsive disorder: A meta-analysis. *Journal of Behavior Therapy and Experimental Psychiatry*, 24(2), 149–153.

Craske, M. G., & Barlow, D. H. (2001). Panic disorder and agoraphobia. In D. H. Barlow (Ed.), *Clinical handbook of psychological disorders*, 3rd ed. (pp. 1–59). New York: Guilford.

Craske, M. G., Barlow, D. H., & Meadows, E. A. (2000). *Mastery of your anxiety and panic: Therapist guide for anxiety, panic, and agoraphobia* (3rd ed.). Boulder, CO: Graywind.

Davidson, J. R. T. (2006). Social phobia then, now, the future. In B. Rothbaum (Ed.), *Pathological anxiety: Emotional processing in etiology and treatment* (pp. 115–131). New York: Guilford.

Davidson, J. R. T., Rothbaum, B. O., van der Kolk, B. A. et al. (2001). Multicenter, double-blind comparison of sertraline and placebo in the treatment of posttraumatic stress disorder. *Archives of General Psychiatry*, 58, 485–492.

de Silva, P., & Marks, M. (1999). The role of traumatic experiences in the genesis of obsessive compulsive disorder. *Behaviour Research and Therapy*, 37, 941–951.

de Silva, P., & Marks, M. (2001). Traumatic experiences, post-traumatic stress disorder and obsessive–compulsive disorder. *International Review of Psychiatry*, 13, 172–180.

DeVeaugh-Geis, J., Moroz, G., Biederman, J., Cantwell, D. P. et al. (1992). Clomipramine hydrochloride in childhood and adolescent obsessive-compulsive disorder: A multicenter trial. *Journal of the American Academy of Child and Adolescent Psychiatry*, 31(1), 45–49.

Devilly, G. J., & Spence, S. H. (1999). The relative efficacy and treatment distress of EMDR and a cognitive–behavior trauma treatment protocol in the amelioration of posttraumatic stress disorder. *Journal of Anxiety Disorders*, 13(1–2), 131–157.

Dougherty, D. D., Rauch, S. L., & Jenike, M. A. (2002). Pharmacological treatments for obsessive–compulsive disorder. In P. E. Nathan & J. M. Gorman (Eds.), *A guide to treatments that work* (2nd ed., pp. 387–410). London: Oxford University Press.

Ehlers, A., Clark, D.M., Hackmann, A. H., McManus F., Fennell, M., Herbert, C., & Mayou, R. (2003). A randomized controlled trial of cognitive therapy, self-help booklet, and repeated assessment as early interventions for PTSD. *Archives of General Psychiatry*, 60, 1024–1032.

Ehlers, A., Clark, D. M., Hackmann, A. H., McManus, F., & Fennell, M. (2005). Cognitive therapy for post-traumatic stress disorder: Development and evaluation. *Behaviour Research and Therapy*, 43, 413–431.

Fals-Stewart, W., Marks, A. P., & Schafer, J. (1993). A comparison of behavioral group therapy and individual behavior therapy in treating obsessive-compulsive disorder. *Journal of Nervous and Mental Disease*, 181(3), 189–193.

Federoff, I. C., & Taylor, S. (2001). Psychological and pharmacological treatments of social phobia: A meta-analysis. *Journal of Clinical Psychopharmacology*, 21, 311–324.

Foa, E. B., & Kozak, M. J. (1986). Emotional processing of fear: Exposure to corrective information. *Psychological Bulletin*, 99, 20–35.

Foa, E. B., & Kozak, M. J. (1996). Psychological treatment for obsessive-compulsive disorder. In M. R. Mavissakalian & R. F. Prien (Eds.), *Long-term treatments of anxiety disorders* (pp. 285–309). Washington, D.C.: American Psychiatric Press.

Foa, E. B., & Meadows, E. A. (1997). Psychosocial treatments for post-traumatic stress disorder: A critical review. *Annual Review of Psychology*, 48, 449–480.

Foa, E. B., & Rothbaum, B. O. (1998). *Treating the trauma of rape: Cognitive–behavioral therapy for PTSD*. New York: Guilford.

Foa, E. B., Rothbaum, B. O., Riggs, D. S., & Murdock, T. B. (1991). Treatment of post-traumatic stress disorder in rape victims: A comparison between cognitive–behavioral procedures and counseling. *Journal of Consulting and Clinical Psychology*, 59, 715–723.

Foa, E. B., Hearst-Ikeda, D., & Perry, K. J. (1995). Evaluation of a brief cognitive–behavior program for the prevention of chronic PTSD in recent assault victims. *Journal of Consulting and Clinical Psychology*, 63, 948–955.

Foa, E. B., Dancu, C. V., Hembree, E. A., Jaycox, L. H., Meadows, E. A., & Street, G. (1999). The efficacy of exposure therapy, stress inoculation training and their combination in ameliorating PTSD for female victims of assault. *Journal of Consulting and Clinical Psychology*, 67, 194–200.

Foa, E. B., Rothbaum, B. O., & Furr, J. M. (2003). Augmenting exposure therapy with other CBT procedures. *Psychiatric Annals, 33*, 47–53.

Foa, E. B., Hembree, E. A., Cahill, S. P., Rauch, S. A., Riggs, D. S., Feeney, N. C., & Yadin, E. (2005). Randomized trial of prolonged exposure for PTSD with and without cognitive restructuring: Outcome at academic and community clinics. *Journal of Consulting and Clinical Psychology, 73*, 953–964.

Foa, E. B., Zoellner, L. A., & Feeny, N. C. (2006). An evaluation of three brief programs for facilitating recovery after assault. *Journal of Traumatic Stress, 19*(1), 29–43.

Foa, E. B., Hembree, E. A., & Rothbaum, B. O. (2007). *Prolonged exposure therapy for PTSD: Emotional processing of traumatic experiences.* New York: Oxford.

Franklin, M. E., & Foa, E. B. (2002). Cognitive behavioral treatments for obsessive compulsive disorder. In P. E. Nathan & J. M. Gorman (Eds.), *A Guide to Treatments That Work* (2nd ed., pp. 387–410). London: Oxford University Press.

Geller, D. A., Hoog, S. L., Heiligenstein, J. H., Ricardi, R. K., Tamura, R., Kluszynski, S., & Jacobson, J. G. (2001). Fluoxetine treatment for obsessive-compulsive disorder in children and adolescents: A placebo-controlled clinical trial. *Journal of the American Academy of Child and Adolescent Psychiatry, 40*(7), 773–779.

Gershuny, B. S., Baer, L., Radomsky, A. S., Wilson, K. A., & Jenike, M. A. (2003). Connections among symptoms of obsessive–compulsive disorder and post-traumatic stress disorder: A case series. *Behaviour Research and Therapy, 41*, 1029–1041.

Goddard, A. W., Brouette, T., Almai, A. et al. (2001). Early coadministration of clonaze-pam with sertraline for panic disorder. *Archives of General Psychiatry, 58*, 681–686.

Goodman, W. K., Price, L. H., Rasmussen, S. A., Delgado, P. L., Heninger, G. R., & Charney, D. S. (1989). Efficacy of fluvoxamine in obsessive-compulsive disorder: A double-blind comparison with placebo. *Archives of General Psychiatry, 46*, 36–44.

Gorman, J., Shear, K., Cowley, D. S. et al. (1998). Practice guideline for the treatment of patients with panic disorder. In J. McIntyre & S. Charles (Eds.), *American Psychiatric Association practice guidelines for the treatment of psychiatric disorders* (pp. 635–696). Washington, D.C.: American Psychiatric Association.

Gould, R. A., Otto, M. W., & Pollack, M. H. (1995). A meta-analysis of treatment outcome for panic disorder. *Clinical Psychology Review, 15*, 819–844.

Gould, R. A., Buckminster, S. Pollack, M. H., Otto, M. W., & Yap, L. (1997). Cognitive–behavioral and pharmacological treatment for social phobia: A meta-analysis. *Clinical Psychology, Science and Practice, 4*, 291–306.

Greist, J. H., Jefferson, J. W., Kobak, K. A., Katzelnick, D. J., & Serlin, R. C. (1995). Efficacy and tolerability of serotonin transport inhibitors in obsessive-compulsive disorder: A meta-analysis. *Archives of General Psychiatry, 52*, 53–60.

Grossman, D., & Christensen, L. W. (2004). *On combat.* Bellville, IL: PPCT Research Publications.

Hildago, R. B., Barnett, S. D., & Davidson, J. R. T. (2001). Social anxiety disorder in review: Two decades of progress. *International Journal of Neuropsychopharmacology, 4*, 279–298.

Huppert, J. D., Moser, J. S., Gershuny, B. S., Riggs, D. S., Spokas, M., Filip, J., Hajcak, G., Parker, H. A., Baer, L., & Foa, E. B. (2005). The relationship between obsessive–compulsive and posttraumatic stress symptoms in clinical and non-clinical samples. *Journal of Anxiety Disorders, 19*, 127–136.

Keane, T. M., Fairbank, J. A., Caddell, J. M., & Zimering, R. T. (1989). Implosive (flooding) therapy reduces symptoms of PTSD in Vietnam combat veterans. *Behavior Therapy, 20*, 245–260.

Khawaja, N. G., & Oei, T. P. S. (1998). Catastrophic cognitions in panic disorder with and without agoraphobia. *Clinical Psychology Review*, 18, 341–365.

Koren, D., Arnon, I., & Klein, E. (1999). Acute stress response and posttraumatic stress disorder in traffic accident victims: A one-year prospective, follow-up study. *American Journal of Psychiatry*, 156, 369–373.

Kubany, E.S., Hill, E. E., & Owens, J.A. (2003). Cognitive trauma therapy for battered women with PTSD: Preliminary findings. *Journal of Traumatic Stress*, 16(1), 81–91.

Kubany, E. S., Hill, E. E., Owens, J. A., Iannce-Spencer, C., McCaig, M. A., Tremayne, K. J. et al. (2004). Cognitive trauma therapy for battered women with PTSD (CTT-BW). *Journal of Consulting and Clinical Psychology*, 72(1), 3–18.

Lee, C., Gavriel, H., Drummond, P., Richards, J., & Greenwald, R. (2002). Treatment of PTSD: Stress inoculation training with prolonged exposure compared to EMDR. *Journal of Clinical Psychology*, 58, 1071–1089.

Lindsay, M., Crino, R., & Andrews, G. (1997). Controlled trial of exposure and response prevention in obsessive-compulsive disorder. *British Journal of Psychiatry*, 171, 135–139.

Magee, W. J., Eaton, W. W., Wittchen, H. U., McGonagle, K. A., & Kessler, R. C. (1996). Agoraphobia, simple phobia, and social phobia in the National Comorbidity Survey. *Archives of General Psychiatry*, 53, 159–168.

March, J. S., Biederman, J., Wolkow, R., Safferman, A., Mardekian, J., Cook, E. H., Cutler, N. R., Dominguez, R., Ferguson, J., Muller, B., Riesenberg, R., Rosenthal, M., Sallee, F. R., & Wagner, K. D. (1998). Setraline in children and adolescents with obsessive-compulsive disorder: A multicenter randomized controlled trial. *Journal of the American Medical Association*, 280(20), 1752–1756.

March, J. S., Franklin, M. E., Nelson, A. H., & Foa, E. B. (2001). Cognitive–behavioral psychotherapy for pediatric obsessive–compulsive disorder. *Journal of Clinical Child Psychology*, 30, 8–18.

Marcus, S., Marquis, P., & Sakai, C. (1997). Controlled study of treatment of PTSD using EMDR in an HMO setting. *Psychotherapy*, 34, 307–315.

Marks, I., Hodgson, R., & Rachman, S. (1975). Treatment of chronic obsessive-compulsive neurosis by *in vivo* exposure. *British Journal of Psychiatry*, 127, 349–364.

Marks, I., Lovell, K., Noshirvani, H., Livanou, M., & Thrasher, S. (1998). Treatment of posttraumatic stress disorder by exposure and/or cognitive restructuring. *Archives of General Psychiatry*, 55, 317–325.

Marshall, R. D., Beebe, K. L., Oldham, M. et al. (2001). Efficacy and safety of paroxatine treatment for chronic PTSD: A fixed dose, placebo controlled study. *American Journal of Psychiatry*, 158, 1982–1988.

Meichenbaum, D. (1974). *Cognitive behavior modification*. Morristown, NJ: General Learning Press.

Mitte, K. (2005). A meta-analysis of the efficacy of psycho- and pharmacotherapy in panic disorder with and without agoraphobia. *Journal of Affective Disorders*, 88, 27–45.

Monson, C. M., Schnurr, P. P., Resick, P. A., Friedman, M. J., Young-Xu, Y., & Stevens, S. P. (2006). Cognitive processing therapy for veterans with military-related posttraumatic stress disorder. *Journal of Consulting and Clinical Psychology*, 74, 898–907.

National Collaborating Centre for Mental Health. (2005). *Posttraumatic stress disorder: The management of PTSD in adults and children in primary and secondary care*, National Clinical Practice Guideline No. 26. London: British Psychological Society and Gaskell.

Noyes, R., Garvey, M. J., Cook, B., & Suelzer, M. (1991). Controlled discontinuation of benzodiazepine treatment for patients with panic disorder. *American Journal of Psychiatry*, 148, 517–523.

Otto, M. W., Tuby, K. S., Gould, R. A., McLean, R. Y., & Pollack, M.H. (2001). An effect-size analysis of the relative efficacy and tolerability of serotonin selective reuptake inhibitors for panic disorder. *American Journal of Psychiatry*, 158, 1989–1992.

Otto, M. W., Hong, J. J., & Safren, S. A. (2002). Benzodiazepine discontinuation difficulties in panic disorder: Conceptual model and outcome for cognitive–behavior therapy. *Current Pharmacological Design*, 8, 75–80.

Paunovic, N., & Ost, L. G. (2001). Cognitive–behavior therapy vs. exposure therapy in the treatment of PTSD in refugees. *Behaviour Research and Therapy*, 39, 1183–1197.

Pitman, R. K. (1993). Posttraumatic obsessive–compulsive disorder: A case study. *Comprehensive Psychiatry*, 34, 102–107.

Power, K., McGoldrick, T., Brown, K., Buchanan, R., Sharp, D., Swanson, V., & Karatzias, A. (2002). A controlled comparison of eye movement desensitization and reprocessing versus exposure plus cognitive restructuring versus waiting list in the treatment of post-traumatic stress disorder. *Clinical Psychology and Psychotherapy*, 9, 299–318.

Rapaport, M. H., Wolkow, R., Rubin, A. et al. (2001). Sertraline treatment of panic disorder: Results of a long-term study. *Acta Psychiatrica Scandinavia*, 104, 289–298.

Resick, P. A., & Schnicke, M. K. (1992). Cognitive processing therapy for sexual assault victims. *Journal of Consulting and Clinical Psychology*, 60(5), 748–756.

Resick, P. A., Nishith, P., Weaver, T. A., Astin, M.C., & Feuer, C. A. (2002). A comparison of cognitive processing therapy with prolonged exposure and a waiting condition for the treatment of posttraumatic stress disorder in female rape victims. *Journal of Consulting and Clinical Psychology*, 70(4), 867–879.

Rickels, K., & Schweizer, E. (1998). Panic disorder: Long-term pharmacotherapy and discontinuation. *Journal of Clinical Psychopharmacology*, 18(Suppl. 2), 12–18.

Riddle, M. A., Reeve, E. A., Yaryura-Tobias, J. A., Yang, H. M., Claghorn, J. L., Gaffney, G., Greist, J. H., Holland, D., McConville, B. J., Pigott, T., & Walkup, J. T. (2001). Fluvoxamine for children and adolescents with obsessive–compulsive disorder: A randomized, controlled, multicenter trial. *Journal of the American Academy of Child and Adolescent Psychiatry*, 40, 222–229.

Riggs, D. S., & Foa, E. B. (2006). Obsessive–compulsive disorder. In F. Andrasik (Ed.), *Comprehensive handbook of personality and psychopathology* (Vol. 2, pp. 169–188). Hoboken, NJ: John Wiley & Sons.

Riggs, D. S., & Foa, E. B. (2008). Psychological treatment of PTSD and other trauma-related disorders. In M. M. Antony & M. B. Stein (Eds.), *Oxford handbook of anxiety and related disorders* (pp. 417–428). New York: Oxford.

Riggs, D. S., & Keane, T. M. (2006). Assessment strategies in the anxiety disorders. In B. Rothbaum (Ed.), *Pathological anxiety: Emotional processing in etiology and treatment* (pp. 91–114). New York: Guilford.

Riggs, D. S., Rothbaum. B. O., & Foa, E. B. (1995). A prospective examination of symptoms of posttraumatic stress disorder in victims of nonsexual assault. *Journal of Interpersonal Violence*, 10, 201–213.

Rothbaum, B. O., Foa, E. B., Riggs, D. S., Murdock, T., & Walsh, W. (1992). A prospective examination of post-traumatic stress disorder in rape victims. *Journal of Traumatic Stress*, 5, 455–475.

Rothbaum, B. O., Astin, M. C., & Marsteller, F. (2005). Prolonged exposure versus eye movement desensitization and reprocessing (EMDR) for PTDS rape victims. *Journal of Traumatic Stress*, 18, 607–616.

Salkovskis, P. M., & Clark, D. M. (1991). Cognitive therapy for panic disorder. *Journal of Cognitive Therapy*, 5, 215–226.

Schnurr, P. P., Friedman, M. J., Engel, C. C., Foa, E. B., Shea, M. T., Chow, B. K. et al. (2007). Cognitive behavioral therapy for posttraumatic stress disorder in women: A randomized controlled trial. *Journal of the American Medical Association*, 297, 820–830.

Shapiro, F. (1989). Efficacy of eye movement desensitization procedure in the treatment of traumatic memories. *Journal of Traumatic Stress*, 2, 199–223.

Shapiro, F. (2001). *Eye movement desensitization and reprocessing: Basic principles, protocols, and procedures* (2nd ed.). New York: Guilford.

Starcevic, V., Linden, M., Uhlenhuth, E. H., Kolar, D., & Latas, M. (2004). Treatment of panic disorder with agoraphobia in an anxiety disorders clinic: Factors influencing psychiatrists' treatment choices. *Psychiatry Research*, 125, 41–52.

Taylor, S. (1996). Meta-analysis of cognitive–behavioral treatment for social phobia. *Journal of Behavior Therapy and Experimental Psychiatry*, 27, 1–9.

Taylor, S., Thordarson, D. S., Maxfield, L., Fedoroff, I. C., Lovell, K., & Ogrodniczuk, J. (2003). Comparative efficacy, speed, and adverse effects of three PTSD treatments: Exposure therapy, EMDR, and relaxation training. *Journal of Consulting and Clinical Psychology*, 71, 330–338.

Thomsen, P. H. (2002). Pharmacological treatment of pediatric obsessive compulsive disorder. *Expert Review of Neurotherapeutics*, 2, 549–554.

Tollefson, G. D., Rampey, A. H., Potvin, J. H., Jenike, M. A., Rush, A. J., Dominguez, R. A., Koran, L. M., Shear, K., Goodman, W. K., & Genduso, L. A. (1994). A multicenter investigation of fixed-dose fluoxetine in the treatment of obsessive–compulsive disorder. *Archives of General Psychiatry*, 51, 559–567.

Tucker, P., Zaninelli, R., Yehuda, R. et al. (2001). Paroxetine in the treatment of chronic posttraumatic stress disorder: Results of a placebo-controlled, flexible dosage trial. *Journal of Clinical Psychiatry*, 62, 860–868.

van Balkom, A., J., L. M., de Haan, E., van Oppen, P., Spinhoven, P., Hoogduin, K. A. L., & van Dyk, R. (1998). Cognitive and behavioral therapies alone versus in combination with fluvoxamine in the treatment of obsessive–compulsive disorder. *Journal of Nervous and Mental Disease*, 186, 492–499.

Veronen, L. J., & Kilpatrick, D. G. (1983). Stress management for rape victims. In D. Meichenbaum & M. E. Jaremko (Eds.), *Stress reduction and prevention* (pp. 341–374). New York: Plenum Press.

Westen, D., & Morrison, K. (2001). A multidimensional meta-analysis of treatments for depression, panic, and generalized anxiety disorder: An empirical examination of the status of empirically supported therapies. *Journal of Consulting and Clinical Psychology*, 69, 875–899.

Wheadon, D. E., Bushnell, W. D., & Steiner, M. (1993). A fixed dose comparison of 20, 40 or 60 mg paroxetine to placebo in the treatment of OCD. Paper presented at the annual meeting of the American College of Neuropsychopharmaclogy, Honolulu, HI.

Zohar, J., Amital, D., Miodownik, C. et al. (2002). Double-blind placebo-controlled pilot study of sertraline in military veterans with posttraumatic stress disorder. *Journal of Clinical Psychopharmacology*, 22, 190–195.

13
Depression and Suicide: A Diathesis–Stress Model for Understanding and Treatment

M. DAVID RUDD

Contents

Understanding the Relationship Between Depression and Suicide

The relationship between depression and suicide is often misunderstood and misquoted, with many citing the suicide prevalence rate for depression as 15%; that is, that 15% of clinically depressed individuals will eventually die by suicide (Guze & Robins, 1970). This figure has come under scrutiny and appropriate criticism, with Blair-West, Mellsop, and Eyeson-Annan (1997, 1999) noting that for the prevalence rate to be 15% the annual suicide rate in the United States would have to be at least fourfold higher. Although the suicide rate for those suffering from depression is higher than that for the general population, it is a simple reality that the vast majority of depressed individuals do *not* attempt suicide or die by suicide, as most experience symptom improvement if not full recovery given a range of available and effective treatments. Even if the suicide prevalence rate for those with depression was 15%, then a full 85% would survive.

One reason why the relationship between depression and suicide is so frequently misunderstood is that greater than 60% of those who eventually die by suicide suffered a depressive spectrum disorder at some point, but not necessarily at the time of death (Lonnqvist, 2000). As Joiner, Van Orden, and Rudd

(2008) have noted, the low base rate of suicide in the general population compounds the problem further, with the suicide prevalence rate for those suffering from depression more likely in the range of 6%, almost two thirds lower than originally thought. The standard mortality ratio, however, still indicates that those with major depression are 20 times more likely to die by suicide than someone in the general population. At the heart of the problem, though, is that even a 20-fold increase in the low base rate for the general population still produces a relatively small overall percentage, making suicide a rare event statistically. Regardless of how these data are interpreted, depression clearly is a significant risk factor for suicidality, including ideation, attempts, and death; however, it is not the only diagnosis and clinical syndrome of concern. Ultimately, though, what receives the greatest public (and sometimes clinical) attention is suicidality. The *Washington Post* (Priest, 2008) recently noted a fivefold increase in the suicide attempt rate and a significant increase in the suicide rate in U.S. Army personnel since the beginning of the Iraq war. Such an increase underscores the need for effective, efficient, and portable treatments. Despite the low base rate nature of suicide, the rates are nonetheless alarming, particularly when compared to other causes of death. And, in light of available and effective treatments (both psychotherapy and medicines) for psychiatric illness, death by suicide is certainly tragic.

Worldwide, there are almost one million suicides each year (DeLao, Bertolote, & Lester, 2002). Accurate data on suicide attempts are far more difficult and complex to accumulate. The best, and most conservative, estimates to date indicate that up to 5% of the U.S. population has made a suicide attempt (approximately 15 million adults), with upwards of 25 attempts for every death (Kessler, Borges, & Walters, 1999). Estimated rates of suicidal ideation vary greatly, depending on the specific population being studied (clinical vs. nonclinical). Linehan (1982) estimated that 31% of the clinical population and 24% of the general population have considered suicide at some point in their lives. The presence or absence of intent to die is unclear in many of these estimates, making the data difficult to interpret and use in a meaningful fashion. Despite the considerable disparity in observable rates, it is reasonable to say that suicidality (broadly defined to include deaths and attempts, both single and multiple) is a serious and persistent public health problem and an area of specific and unique concern for the military. Deployment into a combat zone, specifically, carries many unique demands and makes the job of military psychologists that much more challenging. Having a workable and empirically grounded model for understanding and treating problems such as suicidality is essential. It is also critical to have a model that translates well into nontraditional settings. As will be discussed later, treatment outcome research has allowed us to identify "common elements of treatments that work" with suicidal patients; all are straightforward, flexible, and transferable into a broad range of treatment contexts, with some elements even portable into combat zones.

Having served as an Army psychologist during the Gulf War, I am keenly aware of both the type and nature of mental health decisions presented during wartime, including pre- and postdeployment decisions. Depression and suicidality are certainly two primary concerns that seriously limit the combat effectiveness of any soldier suffering and have broader implications for the safety and effectiveness of the entire unit, both small and large. This chapter offers a broad conceptualization of diathesis–stress approaches to the treatment of depression and suicidality and identifies common elements of treatments that work that can be effectively employed in a range of treatment contexts. Because suicidality is the most extreme manifestation of psychiatric symptomatology, multiply determined (e.g., depression, anxiety, substance abuse), and among the most challenging issues encountered in a clinical environment, it is the central focus of the diathesis–stress model, the discussion of subsequent treatment studies, and implications for day-to-day clinical care.

What is critical to understanding the link between depression and suicide (as well as a host of other psychiatric disorders and suicide), along with the emergence and persistence of suicidality over time, is parsimonious and empirically grounded theory. Useful and empirically supported theory translates to effective treatment paradigms. The most oft-cited approach to understanding, modeling, and treating suicidality is some variant of the diathesis–stress model, with the nature of the diathesis varying somewhat, but the majority revolving around recognition and treatment of cognitive diatheses. This is particularly salient when considering treatments that have been proven effective. It terms of cognitive vulnerabilities for depression and suicide, the list is a relatively long one, including helplessness, hopelessness, cognitive rigidity, attentional bias, overgeneral memory, dysfunctional attitudes, cognitive distortions (negative automatic thoughts, schema, and core beliefs), and ultimately impaired problem solving. In most cognitive-oriented approaches, impaired problem solving is viewed as the eventual and observable consequence of cognitive disruption and dysfluency, something that takes many forms.

Making a Case for the Central Role of Cognitive Diatheses

Schotte and Clum (1982) were among the first to offer empirical support for a diathesis–stress model of suicidality incorporating negative life events, cognitive rigidity, deficient problem solving, and hopelessness. At the heart of the theory is the diathesis of cognitive rigidity, with those evidencing cognitive rigidity experiencing impaired problem solving (i.e., difficulty generating, testing, and modifying potential solutions), along with the increased likelihood of hopelessness and eventually suicidality (to include ideation and attempts). Schotte and Clum's approach is similar in many ways to the hopelessness theory of depression proposed by Abramson, Metalsky, and Alloy (1989). Abramson and her colleagues provided solid empirical support for the critical role of cognitive processes in more severe depression. In particular, they found that the affinity to attribute negative life events to stable and global causes (e.g., "I didn't get the job because I'm dumb") creates considerable vulnerability for *hopeless depression*, a mood state characterized by "sadness, suicidality, low energy, apathy, psychomotor retardation, sleep disturbance, and poor concentration" (Abramson et al., 1989). Rudd et al. (1994, 1996b) and Dixon, Heppner, and Rudd (1991) extended Schotte and Clum's original work using a military population, with results proving identical despite the unique nature of the sample. Since these early contributions, a number of variants of the diathesis–stress model have emerged in the literature, with almost all focusing on similar cognitive mechanisms. All of them depend, at least in part, on the original cognitive model offered by Beck (1967, 1976).

The Importance of Beck's Cognitive Theory: An Essential Foundation for Understanding and Treatment

Central to the notion of a cognitive diathesis in depression and suicide is that the meaning that an individual assigns to various life events and experiences is critical to understanding the affective/emotional response and associated behavior (Beck, 1967). In short, how an individual perceives an event is essential to understanding the emotional and behavioral correlates and provides the foundation for effective treatment. In accordance with cognitive theory, stressors and environmental events alone are not explanatory for depression or suicidality. Although deployment and combat exposure are certainly unusual and unique stressors, the majority of soldiers do not develop psychiatric symptomatology secondary to such exposure (see Chapter 6 in this volume). Individual interpretation and perception are central to understanding emotional impact and behavioral responses, including suicidality, hence the importance of cognitive-based diatheses, such as attribution errors, cognitive distortions, rigidity, attentional bias, overgeneral memory, and impaired problem solving. All soldiers enter combat preloaded with a cognitive perspective, one that has evolved across their particular developmental trajectory and may well have hidden vulnerabilities or resiliencies.

Beck, Brown, Berchick, and Stewart (1990) emphasized cognitive aspects of psychological functioning in understanding depression and suicidal behavior, speculating that the central factor is the emergence of negative automatic thoughts, related schema, core beliefs, and subsequent hopelessness (defined as *negative future expectancies*). Considerable empirical evidence has accumulated over the last several decades supporting the role of hopelessness in cases of suicide and suicide attempts. Of importance, Beck postulated that hopelessness pervades all aspects of the cognitive triad—that is, beliefs about self, others, and the future. When talking about depression and suicide, it is important to recognize that cognitions (including maladaptive ones) are not random but are determined by our individual developmental trajectories, including the unique characteristics of each life, particularly past trauma.

In cases where suicide is an issue, negative schemas and core beliefs are believed to be the central problem. Clark and Beck (1999, p. 79) defined schemas as "relatively enduring internal structures of stored generic or proto-typical features of stimuli, ideas, or experiences that are used to organize new information in a meaningful way thereby determining how phenomena are perceived and conceptualized." Accordingly, schemas are central to understanding how someone *organizes* and perceives the world. A previous history of abuse or trauma, in particular, is likely to result in negative schemas across many fronts, creating a foundation for later difficulty secondary to several identifiable information processing problems noted below. Beck (1996, p. 16) defined core beliefs as "the most fundamental level of belief; they are global, rigid, and overgeneralized." The rigid and inflexible nature of core beliefs compounds the problem, particularly during periods of acute stress or crisis. When someone feels worthless, it makes the challenge of processing and dealing effectively with the unique experience of combat exposure even more difficult. Negative schemas, automatic thoughts, and core beliefs are believed to be secondary to psychological disturbance (e.g., depression and anxiety), with the net result being biased, distorted information processing. In some ways, this hints at the potential for chronic problems. Those most vulnerable for problems during deployment and combat, particularly suicidality, are likely to evidence longstanding psychopathology (Rudd, 2006). Cognitively oriented diathesis–stress models emphasize the need for careful screening and identification of developmental trauma that may be the genesis for later problems. The nature of the distorted information processing takes a number of forms, including attentional bias and overgeneral memory, with eventual impairment in day-to-day problem solving. The observable outcome for mental-health professionals is impairment in psychological functioning and resultant symptoms, characteristic of Axis I and II psychiatric disorders.

More recently, Wenzel et al. (2008) have speculated that there are not just depressive schemas and have proposed the presence of *suicide schemas*—that is, beliefs directly related to exacerbation of hopelessness and emergence of intent to die by suicide. In other words, suicide schemas are the beliefs that provide the motivation or intent to die—for example, "I'm worthless and don't deserve to live. I'm a burden on my family. They'd be better off if I were dead." Similarly, Rudd (2006) proposed the presence of a suicidal belief system, noting that hopelessness is the pervasive, but not the only, feature of impaired information processing. Rudd (2006) identified multiple themes that characterize the suicidal belief system, including helplessness ("I can't fix this problem"), unlovability ("I don't deserve to live"), poor distress tolerance ("I can't stand this pain anymore"), and perceived burdensomeness ("Everyone would be better off if I were dead"). All of these beliefs indicate an underlying cognitive diathesis that must be targeted in treatment.

According to Rudd (2006) and Wenzel et al. (2008), the focus on maladaptive information is at the expense of positive, productive (disconfirming) information that could generate a sense of hope and result in improved problem solving. It is believed that a cascade of information processing problems occurs, with the net result being impairment in day-to-day problem solving. Bias in information processing is observable in two ways in severely depressed and suicidal individuals, including both attention and memory. Wenzel et al. (2008) have speculated about the presence of a *suicide-relevant attentional bias*, which results in the selective processing of depressive and suicide-relevant stimuli; in other words, those at risk ruminate about their reasons for dying rather than reasons for living. It is conceptualized as a very real cognitive impairment secondary to the depression, anxiety, or other active diagnosis. The net impact is exacerbation of hopelessness, reinforcement of active negative schema and core beliefs (about self and others), heightened symptomatology, and increased suicide risk.

Similarly, suicidal individuals also evidence impairment in memory processes, including an overgeneral memory style (Williams, Barnhoffer, Crane, & Duggan, 2006). It is believed that an overgeneral memory style exacerbates hopelessness and impairs problem solving during a suicidal crisis because it is likely that a suicidal person will have trouble recalling specific reasons for living, potential solutions to the conflict, or being hopeful about life. Suicidal individuals have been found to have marked difficulty remembering specific positive experiences that might offer some prophylaxis during periods of acute stress (Williams, 1996). Emerging evidence suggests that the nature of the cognitive diathesis is layered and multiply determined, including problems with attributions; negative underlying automatic thoughts, schemas, and core beliefs; attentional bias; and impaired memory functioning, along with eventual difficulty in application with daily problem solving and coping.

In the last few years, some important changes have occurred in cognitive theory and, accordingly, clinical application in treatment. In some ways, this represents an acknowledgment of recent empirical findings suggesting a multidetermined and multilayered cognitive system with a number of identifiable cognitive diatheses. Beck (1996) modified traditional, linear cognitive theory and proposed the presence of *modes*. Modes are defined as structural/organizational unites that contain schemas. Modes are interconnected networks of cognitive, affective, motivational, physiological, and behavioral schemas that are activated simultaneously by relevant internal and external events. Beck speculated that repeated activation of the modes lowers the threshold for future activation of the mode, thus creating a sensitivity or vulnerability for future problems. Rudd et al. (2004) applied the theory of modes specifically to suicide, proposing a *suicidal mode*, again characterized by hopelessness and summarized as a "suicidal belief system." Rudd (2006b) subsequently extended the notion of the suicidal mode and offered a fluid vulnerability theory (FVT) to explain the persistence of the cognitive diathesis in suicidality. In short, FVT explains the emergence, subsequent resolution, and reemergence of suicidality over extended periods of time. The central focus of FVT is that a range of precipitants, both internal (e.g., thoughts, feelings) and external (e.g., combat exposure) can activate the suicidal mode and associated suicidal belief system, with periods of suicide risk being time limited but vulnerable to activation again, particularly if an individual has a low threshold for activation. Those with the greatest vulnerability are those with traumatic histories and previous suicide attempts (Rudd, 2006b).

Wenzel, Brown, and Beck (2008) recently expanded on the theory of modes in four important ways, with all being important for understanding the nature of the underlying cognitive diathesis. First, they clarified the nature of *suicidal cognitions*, differentiating between cognitions characterized by trait hopelessness and those characterized by unbearability (Joiner, Brown, & Wingate, 2005), recognizing multiple themes of hopelessness consistent with the work of Rudd (2006b). Second, their recent modification has helped clarify the interplay among risk factors, psychiatric illness, activation of suicide cognitions, associated emotional distress, and related self-destructive behavior and has identified variable cognitive pathways to suicidality. More specifically, different suicidal beliefs are associated with different stressors, risk factors, and psychiatric symptoms. In other words, they have retained the notion of *cognitive content specificity*; that is, different beliefs are related to different symptoms, disorders, and behavioral problems. It is possible that those in a military or combat environment have identifiable unique features to their suicide schemas, particularly given the unique nature of the stressor. Research to date has not confirmed this hypothesis, but some themes of hopelessness are potentially unique to the military environment (e.g., survivor's guilt). Of particular importance, the most recent expansion of cognitive theory is much more friendly and amenable to integration of related cognitive diatheses, including attentional bias and overgeneral memory.

One of the unique and attractive elements of Beck's approach and Rudd's subsequent elaboration is the integrative nature of the theory. Although cognitive process (e.g., the cognitive triad, related core beliefs, hopelessness) is central to understanding depression and suicidality, the theory of modes integrates affective, motivational, physiological, and behavioral elements more precisely than some other psychologically based approaches. As a result, the theory is flexible and amenable to integration of social factors, relationships, skills building (e.g., emotion regulation training), and unique stressors such as combat exposure, along with a host of others. Regardless, though, it is hypothesized that cognitive change, not just behavioral change, is central to effective treatment and lasting change. Several theoretical variants worth mentioning, including the work of Linehan (1993), Williams (2001), and Joiner (2005).

Williams' Theory: The Importance of Overgeneral Memory

As mentioned above, Williams et al. (2006) have discussed the critical role of overgeneral memory in the suicidal process. Briefly stated, Williams and his colleagues have provided solid empirical evidence that those at risk for suicidal behavior have impairment in autobiographical memory. In other words, they have a deficit in the specificity of autobiographical memory, making it difficult to recall reasons for living and accessing other cognitive material that facilitates both hope and problem solving (in varied forms) during times of acute and enduring stress. Those at risk exhibit a tendency to ruminate about reasons for dying during periods of stress. This is particularly salient for vulnerable individuals during unique circumstances such as combat. It does not appear that overgeneral memory correlates with the severity of depression or related mood disturbance. It appears to be a relatively independent cognitive and psychological variable that elevates suicide risk, a unique cognitive diathesis with clear implications for treatment. Williams has been unable to identify a specific mechanism of action for overgeneral autobiographical memory but reported convincing empirical evidence that links it to trauma, particularly that which appears during childhood and adolescence. Without question, this has implications for those exposed to combat and the subsequent development of posttraumatic stress disorder (PTSD). Whether or not those with multiple trauma exposure are more vulnerable is yet to be determined. In addition to treatment, this finding has important implications for screening and anticipation of risk in unusual and extreme circumstances such as combat, particularly for those with abuse and trauma histories prior to deployment. Those with a previous history of trauma require careful and thorough scrutiny of previous coping during periods of acute and prolonged stress.

From a survival perspective, it is hypothesized that overgeneral memory helps diffuse the impact (and potential psychological damage) of negative emotions, consistent with an affective gating mechanism. The complication of affective gating, though, is that overgeneral memory contributes to an escalation of risk for a suicidal crisis by limiting problem solving and future oriented thinking (i.e., hope). Although overgeneral memory, according to Williams, blunts immediate emotional distress and dysphoria, it comes at a considerable practical cost. Williams has summarized empirical findings about the consequences of overgeneral memory in relation to suicidality: (1) episodes of emotional upset endure for longer duration secondary to impaired problem solving, (2) overgeneral memory impairs interpersonal problem solving (i.e., it has a social consequence), and (3) it limits an individual's ability to think in future terms and manifest hope. The net outcome is increased risk for suicidality, broadly defined. As is evident, there are also social consequences. This is particularly true for the military environment and has very real implications for unit functioning, an element that must be considered during the assessment process and one that impacts the eventual clinical decision.

Linehan's DBT and Emotion Regulation Theory

Linehan's (1993) dialectical behavior therapy (DBT), perhaps the best known treatment for suicidality, is based on the critical role played by emotion regulation and the pathological variant, emotion dysregulation. Linehan proposed that emotion dysregulation is central to understanding and altering suicidal behavior. Although not traditionally considered a cognitive diathesis, emotion regulation has both cognitive and interpersonal components, with an emphasis on the transactional nature of the relationship. The empirical base for DBT is easily the strongest of any single treatment targeting suicidality, with six randomized, controlled trials documenting treatment efficacy and two of those demonstrating an ability to reduce the frequency of suicide attempts following treatment (Rudd, Joiner, Trotter, Williams, & Cordero, 2008).

Unlike some other psychologically based theories, Linehan (1993) acknowledged the importance of both biological and social factors and proposed the environment–person system (EPS) as critical to understanding suicidality. She emphasized its transactional nature and in this way differentiated her approach from traditional diathesis–stress models. Consistent with the notion of reciprocal determinism, Linehan (1993) differentiated the transactional nature of the model that provides the foundation for DBT as assuming that individual functioning and environmental conditions are "mutually and continuously interactive, reciprocal, and interdependent" (p. 39). At the heart of the theory is that individuals and the environment both continuously adapt to one another and, accordingly, influence and change one another. As with Beck (1967) and Williams et al. (2006), the central diathesis is a cognitive one, with emotion dysregulation eventually impairing problem solving, along with available and accessible social support. The transactional approach is amenable to biological influences as well; that is, despite a supportive and nurturing environment psychopathology can still emerge as the result of underlying genetic influence. Someone with a heavy genetic loading can manifest psychopathology despite a supportive and nurturing environment, although it is certainly less likely.

Linehan's (1993) approach is in sharp contrast to more traditional linear models, particularly more ubiquitous diathesis–stress models. She identified two critical subsystems, including the environmental subsystem and the behavioral subsystem. The environmental subsystem incorporates social support, life change, suicidal models, and suicidal consequences (i.e., reinforcement of suicidal behavior and related contingencies). This has implications for how suicidality is managed within the military system; that is, what type of modeling occurs? This is a particularly salient concern during wartime. Are distressed soldiers removed from the unit? Are they hospitalized? Are they evacuated stateside? The behavioral subsystem includes physiological and affective elements, the overt motor system, and the cognitive system. Of particular relevance, an individual's response to stress and emotion regulation skill are determined, at least partially, by his or her underlying biology and physiology. The individual's underlying biology is characterized by sensitivity to negative stressors and sharp emotional reactions, coupled with poor skills to facilitate emotional recovery. As is evident, the role of preexisting psychopathology (e.g., depression, posttraumatic stress disorder) only further impairs emotion regulation skills and associated problem solving. According to Linehan, suicidal behavior (and related self-injury and mutilation) emerges as an effort to cope or regulate affect, not necessarily because someone is genuinely motivated to die. As is evident, this fundamental assumption drives the nature of treatment, with emotion regulation being an essential skill to target.

Joiner's Theory and Interpersonal Consequences

Joiner (2005) hypothesized that psychological factors converge with interpersonal ones to predict suicide, including both the desire to die and the capability to do so. In many ways, Joiner proposes dual diatheses, much like Linehan (1993), including cognitive and interpersonal ones. The desire to die by suicide is believed to be the function of two interpersonal constructs, perceived burdensomeness (i.e., feeling like a burden to friends and loved ones) and thwarted belongingness (i.e., a sense of low belongingness or social alienation), both of which are cognitive-based diatheses that emerge as the result of interpersonal context. Joiner suggests that thwarted belongingness emerges secondary to an unmet need to belong. Perceived burdensomeness results from the inability to engage in meaningful, reciprocal relationships. The capability for suicide is defined as a fearlessness of pain, injury, and death.

Joiner believes that individuals acquire the capability for suicide through a process of repeatedly experiencing painful and otherwise provocative events, such as self-injury, physical fights, accidental injury secondary to high-risk behavior, and workplace exposure such as that of soldiers and physicians. Joiner's theory, in particular, identifies combat exposure as being especially toxic for those with an underlying diathesis. Soldiers, due to repeated exposure to life-threatening events (and related habituation), are vulnerable to developing the most toxic risk factor, the capability for suicide (i.e., a loss of fearfulness about death). Both direct and vicarious exposure can heighten the capability for suicide, meaning that those in a war zone are vulnerable, regardless of whether or not they have had direct combat exposure. Another way of thinking about this is that those handling and processing the deceased in the morgue are at particular risk. The net outcome of repeated exposure to such events is the loss of the instinctual *fear of death* and what has often been described as the survival instinct or self-preservation motive.

It is the interaction of multiple factors in Joiner's theory that results in death by suicide. Suicidal desire alone is not sufficient. Suicidal desire must be coupled with acquired capability for lethal harm in order for an individual to die by suicide. In short, suicide is the outcome of three intersecting elements: thwarted belongingness, perceived burdensomeness, and acquired capability, with the central diathesis being cognitively based. An impressive growing body of work provides empirical support for Joiner's (2005) theory.

Identifying Treatments That Work

Fifty-three clinical trails targeting suicidality are available in the literature, with the majority (53%) having a cognitive–behavioral therapy (CBT) orientation (Rudd et al., 2008). The CBT orientation is attractive for those in a military setting, particularly during deployments. Some shorter term CBT formats are actually amenable to use in combat settings, requiring very little time away from the unit. With respect to suicidality, though, very short-term interventions are likely to be problematic given that those with a history of suicide attempts, and particularly multiple attempts, are known to struggle with far more severe and enduring psychopathology (Rudd et al., 1996a). For the most part, CBT-based approaches are manualized and portable, a critical consideration in the military environment where broader dissemination and flexible application are critical. It is certainly arguable that there are fewer nuances with CBT interventions, making training and dissemination of identified techniques easier. When considering treatments for suicidality, two questions are of particular importance: First, what treatments actually work? In other words, which ones reduce the rates of suicide attempts following care? Second, what do they have in common that might help us understand why they work? It is also important to consider any implications for clinical practice in atypical settings such as the military, particularly deployments. Although dismantling studies are yet to be conducted that would answer these questions in precise fashion, we do have enough data to engage in informed discussion.

When referring to treatment outcome studies in suicidality, it is important to mention a recent metaanalysis concluding that "results do not provide evidence that additional psychosocial interventions following self-harm have a marked effect on the likelihood of subsequent suicide" (Crawford, Thomas, Khan, & Kulinskaya, 2007). As has been mentioned elsewhere (Rudd, 2007; Rudd et al., 2008), a number of confounds exist that limit the accuracy of their conclusion, but two are of primary concern. First, the interventions included in the metaanalysis were not developed nor intended to reduce suicide rates; rather, they targeted suicide attempts and associated symptoms such as suicidal ideation, hopelessness, and depression. Second, it is arguable that the studies included are not actually amenable to a metaanalytic approach, given the variable inclusion/exclusion criteria and treatment targets, with the net outcome being that the metaanalysis inappropriately assumed intervention and treatment studies were comparable and included samples of highly disparate ages, ranging from 12 to over 50. Actually, several of the interventions included in the metaanalysis were of, at best, questionable value (e.g., random follow-up telephone calls). Rudd (2007) provides a detailed list of identifiable confounds that corrupt the conclusions offered by Crawford et al. (2007).

Previous Reviews of Psychological and Behavioral Treatments

Comprehensive reviews targeting suicidality available in the literature include Gunnell and Frankel (1994), Hepp, Wittman, Schnyder, & Michel (2004), Linehan (1997), and Rudd (2006), as well as several metaanalyses that are methodologically sound (Hawton et al., 1998, 2005; van der Sande et al., 1997). These reviews include in-depth discussions of methodological problems and limitations across studies. The focus of this chapter is quite different. More specifically, the intent here is to identify common elements of treatments that work and explore their potential application in nontraditional clinical settings such as the military. The military environment places unique demands on treatment approaches, including the need to be transferable (across contexts and service providers), understandable, and flexible, particularly in terms of intensity and duration.

Common Elements of Treatments That Work

The best and most accurate marker of lower risk for suicide following treatment is a reduction in suicide attempts during the follow-up period (Rudd, Joiner, & Rajab, 2004). If we focus specifically on what treatments are effective at reducing suicide attempts and also do *not* have serious or disqualifying methodological problems, the list is remarkably short, including only nine studies with a CBT focus or orientation (Brown et al., 2006; Koons et al., 2001; Linehan et al., 1991, 2004, 2006; McLeavey, Daly, Ludgate, & Murray, 1994; Nordentoft et al., 2002; Salkovkis, Atha, & Storer, 1990; van den Bosch, Koeter, Stijnen, Verheul, & van den Brink, 2005). Consistent with the notion of diathesis–stress models, though, a number of other treatments have been demonstrated to reduce associated symptoms such as depression, anxiety, hopelessness, and features of suicidal thinking (e.g., specificity and intensity). As was mentioned before, though, the variable nature of associated symptoms such as suicidal ideation, depression, and hopelessness significantly limits their utility as outcome measures. In other words, it is simply not surprising that individuals who are feeling their worst (common to a suicidal crisis) improve symptomatically after the immediate crisis resolves. Actually, Rudd's (2006) FVT addresses this phenomenon from both theoretical and clinical perspectives. Remarkably, only four studies have adequate follow-up and controls to be considered effective at reducing subsequent suicide attempt rates: Brown et al. (2006), Linehan et al. (1991, 2006), and Salkovkis et al. (1990).

As was mentioned earlier, an important question is whether or not there are identifiable *common elements* across treatments that work at reducing subsequent suicide attempt rates. Among the studies referenced above, a review of these studies supports several conclusions about common elements, techniques, and interventions, all with implications for clinical practice in military settings.

Simple Models as the Basis of Treatment   All of the treatments have simple theoretical models that are embedded in empirical research, are consistent with a diathesis–stress approach, and stress the importance of cognitive processes. They all identify cognitions, emotional processing, and associated behavioral responses as critical to understanding the motivation to die, emotional distress (and symptoms), and ultimately changing the suicidal process. Although somewhat artificial, the models can be translated in linear and visual fashion, and a drawing on a chalkboard or sheet of paper can improve understanding and recognition of treatment targets. These treatments have made it easy to sit down with patients and explain in understandable language why they have tried or are thinking about killing themselves. From an empirical perspective, it would certainly be interesting to look at the relationship between how well a patient understands an intervention model, complies with care (including emergency management), and expresses hopefulness and eventual treatment outcomes. Emerging evidence seems to clearly indicate that simple models work, but exactly why we do not know. This does prompt some important questions about clinical impact. Do simple and straightforward models facilitate hope? Do they improve motivation? Do patients feel more in control and, accordingly, respond with fewer crises? Are patients more compliant with treatment demands? Are they more willing to access available emergency care when needed? Does this lower subsequent suicide attempt rates? From a military perspective, are simple models easier to implement in nontraditional environments? Do simple models make it easier to train nondoctoral providers for clinical delivery, thus improving treatment efficacy and eventual success?

Maintain Treatment Fidelity   Treatment fidelity has proven to be a critical factor. This means that clinicians were trained to a target standard and provided appropriate supervision and oversight, and fidelity was ensured. As mentioned above, though, it is likely that a relationship exists between the complexity of the theoretical model and eventual treatment fidelity. This has yet to be proven empirically but is a reasonable hypothesis. Fidelity, in many ways, is consonant with how well defined the treatment is across many fronts, including its identified component parts (e.g., skills training, crisis management, safety plans), as well as the more nuanced aspects of care. Manualized CBT-based approaches lend themselves to greater fidelity given the nature of the orientation, although it might just be easier to monitor and track targeted treatment interventions. In the military environment, this is particularly important because much of the care is provided by nondoctoral-level individuals. What is also clear is that all four studies conceptualized the treatment of suicidality as requiring unique competencies, consistent with the recent movement to identify core competencies in the assessment, management, and treatment of suicidality (Rudd et al., 2008). The core competency movement requires clinicians target specific markers of suicidality in direct fashion, rather than indirect markers such as depression, anxiety, and the like. Suicidality is conceptualized as distinct and separate from an Axis I or II diagnosis.

Targeting Compliance Is Important   Effective treatments in these studies also targeted treatment compliance in specific and consistent fashion throughout the entirety of treatment. More specifically, all successful treatments had specific interventions and techniques that targeted poor compliance and motivation for care. When compliance problems emerged they were addressed immediately and specifically. Treatment is only effective if the patient is active, involved, and invested. In a military environment, compliance with care is less of a significant issue, but not entirely. Although someone can be assigned to treatment as the primary focus of his or her daily activity, inherent motivation and commitment cannot always be guaranteed. Someone can attend treatment but not fully participate and consequently not benefit in a meaningful way. It is clear from effective treatments that compliance with care must be a central and primary focus, with clear plans about what to do if noncompliance emerges. Just as suicidal behavior must be a primary target, motivation and investment in care are also important. When motivation, investment, and involvement drop, they should become a primary treatment target until effectively resolved. Suicidality revolves, in large part, around personal motivations, motivations about both living and dying.

Targeting Identifiable Skills   Consistent with easy-to-understand theoretical models of suicidality driving the treatment process, effective treatments targeted clearly identifiable skill sets (e.g., emotion regulation, anger management, problem solving, interpersonal relationships, cognitive distortions). In these treatments, patients understood what was wrong and what to do about it to reduce suicidal thinking and behaviors. They also had the opportunity to practice and build skill sets over time. This is entirely consistent with the structured military environment. One issue that would certainly be a concern, though, is how long does it take to establish, refine, and generalize a targeted skill?

Personal Responsibility   Consistent with each of the above points, effective treatments emphasized self-reliance, awareness, and personal responsibility. Effective treatments are clear in the goal that if patients developed appropriate skills, the distress and upset tied to early events would diminish, as would associated suicidal urges. Consistent with this goal, patients assumed a considerable degree of personal responsibility for their care, with a particular emphasis on identifying warning signs and crisis management. Again, this is consistent with the issue of improved compliance and motivation for care. Although numerous models are available for facilitating compliance, and crisis management, I would encourage clinicians to consider use of the *commitment to treatment agreement* (Rudd, Mandrusiak, & Joiner, 2006). It provides an approach and structure consistent with shared responsibility in treatment, clearly articulating the responsibilities of both clinician and patient.

Easy Access to Treatment and Crisis Services   Effective treatments emphasize the importance of crisis management and access to available emergency services during and after treatment, with a clear plan of action being identified. Ideally, patients are taught (again consistent with a diathesis–stress approach) to recognize warning signs that may signal the emergence of a suicidal crisis and intervene as early as possible. Additionally, effective treatments more often than not dedicated time to practicing the skills sets necessary for effective crisis management; among the range of creative techniques used, role playing proved to be essential.

Clinical Practice in Military Settings: A Unique and Demanding Environment

The military clearly represents a unique clinical setting, one that requires flexible and effective interventions. This is particularly true during deployments and combat. The unique environment presented by a military setting is further compounded by potentially serious psychopathology. As the above review suggests, much of what we know about the treatment of suicidality can be applied to the military setting. Whether or not that includes deployments and combat zones is another issue entirely. The empirical evidence suggests practical and concrete solutions for the treatment and ongoing management of suicidality. If I had to sum up the empirical literature in this area it would be the following: Simple things work. Simple interventions tend to be portable and flexible, across both clinical contexts and providers. Available data indicate this to be the case for proven CBT approaches for suicidality. They are understandable and specifically target suicidality, cutting across a range of Axis I and II diagnostic entities, with particularly salience for PTSD, depression, and substance abuse.

Continuing with the theme of simple things work, let me summarize what we know to date. First, simple and straightforward theoretical models are needed, ones that facilitate patient understanding and active engagement in the treatment process. An easy test is to try to diagram the model on a chalkboard or piece of paper. Simple and straightforward models are also likely to impact treatment fidelity. In the majority of cases, the manualized nature of the CBT approaches improves training of nondoctoral providers and eventual service delivery. For those struggling with PTSD and depression, the primary themes of hopelessness (i.e., the suicidal belief system) that must be targeted are somewhat unique but include the following: (1) guilt secondary to combat action ("I've taken a life," "I didn't do enough," or "Why did I survive?"), (2) shame (i.e., a sense of worthlessness secondary to the themes noted in the above item), (3) perceived burdensomeness (e.g., "I'm supposed to be strong and I can't even get through the day now without falling apart"), (4) helplessness (e.g., "I can't fix these problems"), and (5) poor distress tolerance (e.g., "I can't stand the way I feel").

Simply helping soldiers recognize and understand the emergence of these hopeless themes can be extremely powerful. Translating these themes into a simple and understandable model of the suicidal crisis (as mentioned previously) helps create a sense of control and mastery that can go a long way toward diffusing the crisis nature of the situation. Similarly, with the large numbers of soldiers, the ability to normalize the experience within the context of PTSD or depression is not difficult and entirely accurate; for example, it is not unusual for someone with repeated combat exposures to experience PTSD symptoms and, if untreated for this long, it is not unusual to start to believe some of things we have talked about. Providing an understandable model of the suicidal crisis lays the foundation for articulating and understanding treatment targets, including the critical issue of symptom management (including the role of medications) during the early phase of care. The better a soldier understands why something is happening the more likely he or she is to cooperate with treatment.

It is also important to address issues of motivation and commitment to treatment from the outset of care. Talk with patients directly and consistently about their motivation for care. Although somewhat obvious, targeting compliance (motivation for care) improves the eventual outcome. With suicidality this is essential because a bad treatment outcome can involve death. Define the construct of a crisis, identify steps to self-management, and practice them to make sure patients possess the basic skills set necessary to implement them. Identify what skills must be targeted for recovery. Recovery stretches well beyond symptom improvement. What is essential is response to future crises. Finally, always address personal responsibility as critical to recovery. In some ways this overlaps with treatment compliance, but it is a distinct and clearly important aspect of individual motivation.

Of concern in combat and deployment environments is the risk of retaining a solider who has experienced a serious depression or suicidal crisis. The issue of risk extends well beyond the individual to the broader unit, potentially jeopardizing the safety of others in unforeseen ways markedly dissimilar from civilian settings. Take for example the issue of access to firearms, something critical in wartime and combat settings. No one would argue that restriction of access to firearms in combat zones is at best difficult, if not practically impossible. It may well be that soldiers experiencing severe depressive spectrum problems and suicidal crises in a combat zone simply are not capable and do not have adequate functional capacity to be retained. Although there is no empirical evidence to support treatment efficacy in such an environment, we do have quite a few anecdotal cases of quick recovery and reintegration into a combat unit during wartime. What we do know on the empirical end, though, is that simple things work, sometimes more quickly than anticipated for a smaller percentage. For those with chronic histories, it is highly unlikely there would be any meaningful recovery in such a context. Ultimately, only science will reveal the answers, but it would be expected that a very small percentage of those experiencing suicide attempts could be effectively treated and safely reintegrated in such a context. This is one problem with implications that stretch well beyond the individual in the military setting, hence the importance of careful and thorough screening upfront.

References

Abramson, L. Y., Metalsky, G. I., & Alloy, L. B. (1989). Hopelessness depression: A theory-based subtype of depression. *Psychological Review*, 96, 358–372.

Beck, A. T. (1967). *Depression: Causes and treatment*. Philadelphia: University of Pennsylvania Press.

Beck, A. T. (1976). *Cognitive therapy and the emotional disorders*. New York: International Universities Press.

Beck, A. T. (1996). Beyond belief: A theory of modes, personality and psychopathology. In P. Salkovkis (Ed.), *Frontiers of cognitive therapy* (pp. 1–25). New York: Guilford Press.

Beck, A. T., Brown, G., Berchick, R. J., & Stewart, B. L. (1990). Relationship between hopelessness and ultimate suicide: A replication with psychiatric outpatients. *American Journal of Psychiatry*, 147, 190–195.

Blair-West, G. W., Mellsop, G. W., & Eyeson-Annan, M. L. (1997). Down-rating lifetime suicide risk in major depression. *Acta Psychiatrica Scandinavica*, 95, 259–263.

Blair-West, G. W., Cantor, C. H., Mellsop, G. W., & Eyeson-Annan, M. L. (1999). Lifetime suicide risk in major depression: Sex and age determinants. *Journal of Affective Disorders*, 55, 171–178.

Brown, G. K., Have, T. T., Henriques, G. R., Xie, S. X., Hollander, J., & Beck, A. T. (2006). Cognitive therapy for the prevention of suicide attempts: A randomized controlled trial. *Journal of the American Medical Association*, 294(5), 563–570.

Clark, D. A., & Beck, A. T. (1999). *Scientific foundations of cognitive theory and therapy of Depression*. New York: Wiley.

Crawford, M. J., Thomas, O., Khan, N., & Kulinskaya, E. (2007). Psychosocial interventions following self-harm: A systematic review of their efficacy in preventing suicide. *British Journal of Psychiatry*, 190, 11–17.

DeLao, D., Bertolote, J., & Lester, D. (2002). Self-directed violence. In E. G. Krugg, L. L. Dahlberg, J. A. Mercy, A. B. Zwi, & R. Lozano (Eds.), *World report on violence and health*, pp. 183–240. Geneva, Switzerland: World Health Organization.

DHHS. (2001). *The national strategy for suicide prevention*. Washington, D.C.: Department of Health and Human Services.

Dixon, W. A., Heppner, P. P., & Rudd, M. D. (1994). Problem-solving appraisal, hopelessness, and suicidal ideation: Evidence for a meditational model. *Journal of Counseling Psychology*, 41, 91–98.

Fazza, A. R. (1987). *Bodies under siege: Self-mutilation in culture and psychiatry.* Baltimore, MD: The Johns Hopkins University Press.

Gunnell, D., & Frankel, S. (1994). Prevention of suicide: Aspirations and evidence. *British Medical Journal*, 308, 1227–1233.

Guthrie, E., Navneet, K., Moorey, J., Mackway-Jones, K., Chew-Graham, C., & Moorey, J. (2001). Randomized controlled trial of brief psychological intervention after deliberate self poisoning. *British Medical Journal*, 323, 135–138.

Guze, S. B., & Robins, E. (1970). Generation of stress in the course of unipolar depression. *Journal of Abnormal Psychology*, 100, 555–561.

Hawton, K., Arensman, E., Townsend, E., Bremner, S., Feldman, E., & Goldney, R. (1998). Deliberate self-harm: Systematic review of efficacy of psychosocial and pharmacological treatment in preventing repetition. *British Medical Journal*, 317, 441–447.

Hawton, K., Townsend, E., Arensman, E., Gunnell, D., Hazell, P., & Housed, A. (2005). Psychosocial and pharmacological treatments for deliberate self-harm. In *The Cochrane Library*, Issue 3. Oxford: Update Software.

Hepp, U., Wittmann, L., Schnyder, U., & Michel, K. (2004). Psychological and psychosocial interventions after attempted suicide. *Crisis*, 25(3), 108–117.

Institute of Medicine. (2002). *Reducing suicide: A national imperative.* Washington, D.C.: National Academies Press.

Joiner, T. E. (2005). *Why people die by suicide.* Cambridge, MA: Harvard University Press.

Joiner, T. E., Brown, J. S., & Wingate, L. R. (2005). The psychology and neurobiology of suicidal behavior. *Annual Review of Psychology*, 56, 287–314.

Joiner, T. E., Van Orden, K., & Rudd, M. D. (2008). *Clinical work with suicidal patients: The interpersonal–psychological theory of suicidality as guide.* Washington, D.C.: APA Press.

Kessler, R. C., Borges, G., & Walters, E. E. (1999). Prevalence and risk factors for lifetime suicide attempts in the National Comorbidity Study. *Archives of General Psychiatry*, 56(7), 617–626.

Koons, C. R., Robins, C. J., Tweed, J. L., Lynch, T. R., Gonzalez, A. M., Morse, J. Q., Bishop, G. K., Butterfield, M. I., & Bastian, L. A. (2001). Efficacy of dialectical behavior therapy in women veterans with borderline personality disorder. *Behavior Therapy*, 32, 371–390.

Linehan, M. M. (1982). Suicidal behaviors among clients in an outpatient clinic versus the general population. *Suicide and Life-Threatening Behavior*, 12, 234–239.

Linehan, M. M. (1993). *Cognitive–behavioral treatment of borderline personality disorder.* New York: Guilford Press.

Linehan, M. M. (1997). Behavioral treatments of suicidal behaviors: Definitional obfuscation and treatment outcomes. *Annals of the New York Academy of Sciences*, 836, 302–328.

Linehan, M. M., Armstrong, H. E., Suarez, A., Allmon, D., & Heard, H. L. (1991). Cognitive–behavioral treatment of chronically parasuicidal borderline patients. *Archives of General Psychiatry*, 48, 1060–1064.

Linehan, M. M., Comotois, K. A., & Korslund, K. E. (2004). Dialectical behavior therapy versus non-behavioral treatment by experts in the community: Clinical outcomes. Paper presented at the 112th Convention of the American Psychological Association, Honolulu, HI.

Linehan, M. M., Comtois, K. A., Murray, A. M., Brown, M. Z., Gallop, R. J., Heard, H. L., Korslund, K. E., Tutek, D. A., Reynolds, S. K., & Lindenboim, N. (2006). Two-year randomized controlled trial and follow-up of dialectical behavior therapy vs. therapy by experts for suicidal behaviors and borderline personality disorder. *Archives of General Psychiatry*, 63, 757–766.

Lonnqvist, J. K. (2000). Psychiatric aspects of suicidal behavior: Depression. In K. Hawton & K. van Heeringen (Eds.), *Handbook of suicide and attempted suicide* (pp. 107–120). Chichester: John Wiley & Sons.

McLeavey, B. C., Daly, R. J., Ludgate, J. W., & Murray, C. M. (1994). Interpersonal problem-solving skills training in the treatment of self-poisoning patients. *Suicide and Life-Threatening Behavior*, 24, 382–394.

Nordentoft, M., Jeppesen, P., Abel, M., Kassow, P., Petersen, L., Thorup, A., Krarup, G., Hemmingsen, R., & Jorgensen, P. (2002). OPUS study: Suicidal behavior, suicidal ideation and hopelessness among patients with first-episode psychosis: One year follow-up of a randomized controlled trial. *British Journal of Psychiatry*, 181(Suppl. 43), s98–s106.

Priest, D. (2008). Soldier suicides at record levels. *Washington Post*, January 31, p. A01.

Rudd, M. D. (2006a). Fluid vulnerability theory: A cognitive approach to understanding the process of acute and chronic suicide risk. In P. T. Ellis (Ed.), *Cognition and suicide: Theory, research, and therapy* (pp. 355–368). Washington, D.C.: American Psychological Association.

Rudd, M. D. (2006b). An update on the psychotherapeutic treatment of suicidal behavior. In J. Trafton & W. Gordon (Eds.). *Best practices in the behavioral management of chronic disease*, Los Altos, CA: Institute for Brain Potential.

Rudd, M. D. (2006c). Treating suicidal behavior: A review. In J. A. Trafton & W. P. Gordon (Eds.), *Best practices in the behavioral management of chronic disease* (pp. 171–190). Los Altos: Institute for Brain Potential.

Rudd, M. D. (2007). Inaccurate conclusions based on limited data. *The British Journal of Psychiatry Online*, 190 (http://bjp.rcpsych.org/cgi/eletters?lookup=by_dateand days=30).

Rudd, M. D., Rajab, M. H., & Dahm, P. F. (1994). Problem-solving appraisal in suicide ideators and attempters. *American Journal of Orthopsychiatry*, 64(1), 136–149.

Rudd, M. D., Joiner, T. E., & Rajab, M. H. (1995). Help negation after acute suicidal crisis. *Journal of Consulting and Clinical Psychology*, 63(3), 499–503.

Rudd, M. D., Joiner, T. E., & Rajab, M. H. (1996a). Relationships among suicide ideators, attempters, and multiple attempters in a young-adult sample. *Journal of Abnormal Psychology*, 105, 541–550.

Rudd, M. D., Rajab, M. H., Orman, D. T., Stulman, D., Joiner, T., & Dixon, W. (1996b). Effectiveness of an outpatient intervention targeting suicidal young adults: Preliminary results. *Journal of Consulting and Clinical Psychology*, 64, 179–190.

Rudd, M. D., Joiner, T. E., & Rajab, M. H. (2004). *Treating suicidal behavior* (2nd ed.). New York: Guilford Press.

Rudd, M. D., Mandrusiak, M., & Joiner, T. E. (2006). The case against no-suicide contracts: The commitment to treatment statement as an alternative for clinical practice. *Journal of Clinical Psychology: In Session*, 62(2), 243–251.

Rudd, M. D., Joiner, T. E., Trotter, D., Williams, B., & Cordero, L. (2008). The psychological and behavioral treatment of suicidal behavior: A critique of what we know (and don't know). In P. Kleespies (Ed.), *Evaluating and managing behavioral emergencies: An evidence-based resource for the mental health practitioner*, Washington, D.C.: American Psychological Association.

Salkovkis, P. M., Atha, C., & Storer, D. (1990). Cognitive–behavioral problem solving in the treatment of patients who repeated attempt suicide: A controlled trial. *British Journal of Psychiatry*, 157, 871–876.

Schotte, D. E., & Clum, G. A. (1982). Suicide ideation in a college population: A test of a model. *Journal of Consulting and Clinical Psychology*, 50, 690–696.

Suicide Prevention Resource Center. (2006). *Assessing and managing suicide risk: Core competencies for mental health professionals*. Newton, MA: Author.

Turner, R. M. (2000). Naturalistic evaluation of dialectical behavior therapy-oriented treatment for borderline personality disorder. *Cognitive and Behavioral Practice*, 7, 413–419.

van den Bosch, L. M., Koeter, M. W., Stijnen, T., Verheul, R., & van den Brink, W. (2005). Sustained efficacy of dialectical behavior therapy for borderline personality disorder. *Behavior Research and Therapy*, 43, 1231–1241.

van der Sande, R., van Rooijen, L., Buskens, E., Allart, E., Hawton, K., & van der Graaf, Y. (1997). Intensive inpatient and community intervention versus routine care after attempted suicide: A randomized controlled intervention study. *British Journal of Psychiatry*, 171, 35–41.

Walsh, B. W., & Rosen, P. M. (1988). *Self-mutilation: Theory, research, and treatment*. New York: Guilford Press.

Wenzel, A., Brown, G., & Beck, A. T. (2008). *Cognitive therapy for suicidal patients: Scientific and clinical applications*. Washington, D.C.: American Psychological Association.

Williams, J. M. G. (1996). Depression and the specificity of autobiographical memory. In D. C. Rubin (Ed.), *Remembering our past: Studies in autobiographical memory*, (pp. 244–267). New York: Cambridge University Press.

Williams, J. M. G. (2001). *Suicide and attempted suicide*. London: Penguin.

Williams, J. M. G., Barnhoffer, T., Crane, C., & Duggan, D. S. (2006). The role of over-general memory in suicidality. In T. E. Ellis (Ed.), *Cognition and suicide: Theory, research, and therapy* (pp. 173–192). Washington, D.C.: American Psychological Association.

Wood, A., Trainor, G., Rothwell, J., Moore, J., & Harrington, R. (2001). Randomized trial of group therapy for repeated deliberate self-harm in adolescents. *Child and Adolescent Psychiatry*, 40, 1246–1253.

14

Substance Use, Misuse, and Abuse: Impaired Problem Solving and Coping

SHARON MORGILLO FREEMAN and MICHAEL R. HURST

Reviewed by HUGH RUESSER

Contents

> Drinking is the soldier's pleasure:
> Rich the treasure,
> Sweet the pleasure,
> Sweet is pleasure after pain.

John Dryden (1631–1700)

The U.S. military is made up of healthy, mostly young, strong, ambitious, intelligent, motivated, and energetic men and women in the prime of life. These very normal, healthy men and women experiment and use the same substances that other normal, healthy men and women might use in the same age group. One can find references to alcohol- and drug-related incidents recorded since the American Revolution. The impact of substances on our military members during active duty, both deployed and nondeployed, as well as on our military members who have redeployed out of theater, retired, completed their terms of service, or been discharged requires careful consideration by the treatment community.

The U.S. military has been an all-volunteer force since the 1970s and is steeped in traditions of selfless service, duty, sacrifice, honor, integrity, and personal courage and initiative, all characteristics that we recognize through our national experience. The mission of one particularly elite subset, the Special Forces of the military, is to free the oppressed. The Special Forces volunteer for this level of service, and their missions consist of building schools, communities, and governments. Ultimately, these most elite of our service members defend and protect anyone deemed worthy of being defended against an oppressor (*de oppresso liber*, Latin for "to liberate the oppressed"). Special Forces units are tasked with seven primary missions: (1) unconventional warfare, (2) foreign internal defense, (3) special reconnaissance, (4) direct action, (5) counterterrorism and counterproliferation, (6) psychological operations, and (7) information operations. The first two emphasize language, cultural, and training skills in working with foreign troops. Other duties include coalition warfare and support, combat search and rescue (CSAR), security assistance, peacekeeping, humanitarian assistance, humanitarian demining, and counter-drug operations.

Since the American Revolution, citizen soldiers have raised their right hands and taken the oath of allegiance to "support and defend the Constitution of the United States" (Center of Military History, 2004). Because of this long tradition and pride, the U.S. military takes a hard line when it comes to the use and abuse of chemicals among its members. This chapter provides a brief historical view of substance use disorders in the military, as well as a discussion about the appropriate assessment, discipline, and treatment. The clinician will find information geared toward developing a better understanding of these disorders and preparing appropriate treatment plans for service members. Several aspects of therapy are discussed, as are various issues that may arise when working with these patients.

Background and History

The consumption of alcohol, most commonly rum, was a well-known fact during the American Civil War. Alcohol was often used medically as well as to provide the necessary coping skills in camp and before engaging in battle. Daily consumption of rum was not uncommon at that time, and reports of behaviors of individuals who drank heavily were documented along with the consequences of those behaviors. In 1862, the U.S. Navy stopped allowing rum rations in part due to the influence of the temperance movement in the civilian population and laws prohibiting alcohol consumption in 18 states (Chambers, 1999). The Union army imposed similar restrictions shortly thereafter that limited rations for soldiers who previously had tended to drink their alcohol as fast as it was rationed out to them. The Army at that time found that excessive alcohol consumption was correlated with an increased use of local prostitution, resulting in thousands of cases of venereal disease outbreaks. Their physicians began using whiskey mixed with elderberries, mercury, pokeweed, silver nitrate, and zinc sulfate to try to control the outbreak among the troops, but the Army finally decided it was easier to simply ration the whiskey and then eliminate it altogether (Robertson, 1984).

In the Civil War, 282,000 Union soldiers were wounded. Medications such as opium, morphine sulfate, laudanum (whisky and opium mixture), and paregoric were used to treat the majority of casualties; unfortunately, the records for the Confederate forces are incomplete due to loss and destruction of records. Generally, soldiers were treated with opiates or alcohol only when a limb was amputated. These amputations saved lives but left thousands of individuals in pain and crippled both physically and mentally. According to the *Medical and Surgical History of the War of the Rebellion* (MSHWR), soldiers in both armies were given generous amounts of opiate analgesics during the active and recovery phases of treatment. Opiate analgesics used during the Civil War by the Union army included 10,000,000 opium pills and 2,840,000 ounces of opium compounds. It was estimated that 1.3 million soldiers in both armies were given opiates as a part of their medical therapy. Not surprisingly, an estimated 120,000 individuals went on to develop significant opiate dependency. Soldiers who were identified as such were termed "opiate eaters" and were denied their Civil War pensions (Barnes, 1870).

Vietnam Era Substance Abuse History

Other than alcohol, the primary substance used by U.S. military personnel in Vietnam was marijuana, until the latter part of 1970, when heroin became the primary drug used. Estimates in 1971 were that heroin was used by 25,000 to 37,000 (10 to 15%) of the lower ranking enlisted service members in Vietnam (Jonnes, 1996). Heroin use by service members in Vietnam was identified through routine urine testing. In 1971, President Nixon requested that Dr. Jerome Jaffe develop a program for all active service members returning from Vietnam to the United States. Service members who tested positive for substances were required to attend a 30-day treatment program. In the first 6 months after initiation of this program, positive urine drug screens dropped from 10% to 2%, theoretically due to service members choosing to discontinue use to avoid disciplinary action and mandatory treatment (Jonnes, 1996). The next step was creation of the position of the Deputy Assistant Secretary of Defense (Drug and Alcohol Abuse). This position was given to Major General Singlaub, whose responsibility it was to plan and coordinate major programs to address the heroin problem in Vietnam. These continuing efforts were extremely successful in reducing the overall use among service members (Wilbur, 1973).

Substance Use by Vietnam Era Soldiers After Vietnam

Contrary to common belief, only 5% of military members were found to be dependent on heroin after serving in Vietnam, according to research reports of Vietnam veterans; 12% of these relapsed within 3 years of their return. Those who were dependent prior to deployment to Vietnam relapsed approximately 10 months after returning to the United States. Some felt that the relapse rates were due to the affordability of heroin and the fact that users found this substance to be enjoyable and that it made the time pass quickly (Burkett & Whitley, 1998).

Many treatment professionals today forget that the Vietnam conflict coincided with increased consumption of mood-altering substances all over the world. It was not surprising, then, that the two populations, military and civilian, would intersect. The majority of service members classified as heavy users of marijuana, for example, had been such prior to arriving in Vietnam (Roffman & Sapol, 1970). Most of the redeployed service members reported having used marijuana (61.0%) at least once in Vietnam, and at least 31.7% during their lifetimes (Roffman & Sapol, 1970). A study comparing substance use among deployed and nondeployed soldiers in 1971 at Fort Riley in Kansas demonstrated little difference in the reported use of illicit drugs between Vietnam returnees and nondeployed service members (Robins, Helzer, & Davis, 1975).

During the Vietnam era, the Department of Defense (DoD) documented the success of formal intervention efforts regarding alcohol abuse among service members, as shown in Table 14.1. Of the 8000 service members who were discharged from the military, 7000 continued treatment with the Department of Veterans Affairs (Singlaub, 1991). The DoD addressed the issue of alcohol abuse during the same time period by providing treatment for 27,000 military personnel (Singlaub, 1991). This record of achievement was the foundation for subsequent highly successful DoD substance abuse programs.

In the post-Vietnam all-volunteer military, the problem of drug abuse has continued to be an issue for unit leaders at all levels of command. Some reports have noted that the overall success rates for substance use treatment of military service members is consistently better than among the civilian population (Chambers, 1999). In 1980, the DoD began conducting a survey of health-related behaviors among active-duty military personnel, and the trends in military alcohol and drug abuse are shown in Table 14.2.

For comparison, alcohol and drug use among civilian populations as reported by the Substance Abuse and Mental Health Services Administration (SAMHSA) National Survey on Drug Use and Health (NSDUH) examined the efficacy of their "drug-free workplace" efforts in 2005 and 2006 (see Table 14.3).

Table 14.1 Department of Defense Vietnam Era Intervention Program

Screened for Drug Abuse	(+) Drug Screen	Entry into Treatment	Return to Duty after Treatment
5,300,000	70,000	70,000	62,000

Table 14.2 Substance Use Trends in the Military

	Military Illicit Drug Use		Military Alcohol Abuse	
Year	Past 30 Days	Past 12 Months	Moderate/ Heavy	Heavy
1980	27.6%	36.7%	32.4%	20.8%
1982	19.0%	26.6%	29.6%	24.1%
1985	8.9%	13.4%	28.5%	23.0%
1988	4.8%	8.9%	28.8%	17.2%
1992	3.4%	6.2%	26.3%	15.5%
1995	3.0%	6.5%	24.5%	17.4%
1998	2.7%	6.0%	23.2%	15.4%
2002	3.4%	6.9%	22.7%	18.1%
2005	5.0%	10.9%	23.5%	18.5%

Source: Research Triangle Institute, *Department of Defense Survey of Health Related Behaviors Among Active Duty Military Personnel*, RTI International, Research Triangle Park, NC, 2006. With permission.

Table 14.3 Illegal Drug Use

	2005	2006
Civilians		
Ages 18 to 20		
Past 30 days	22.3%	22.2%
Past 12 months	37.9%	37.6%
Ages 21 to 25		
Past 30 days	18.7%	18.3%
Past 12 months	31.8%	32.3%
Military		
Past 30 days	5.0%	—
Past 12 months	10.9%	—

Source: Data courtesy of the Substance Abuse and Mental Health Services Administration (SAMHSA). Note that statistical analysis has not been conducted on these samples.

Table 14.4 Reasons for Starting to Use Substances

Recreation	Rite of passage
Socialize	New experience
Conformation	Prove sexuality
Reduce stress	Relieve anxiety
Pleasure	Rebellion
Response to an impulse	Self-exploration
Relieve depression	Relieve fatigue
Relieve boredom	Solve personnel problems

Source: Milhorn, H.T., *Drug and Alcohol Abuse: The Authoritative Guide for Parents, Teachers, and Counselors,* Perseus Books, Reading, MA, 1994. With permission.

Prerecruitment Issues

Evaluation of individuals for military service must be undertaken with great care. Most military members entering the service are between the ages of 18 and 25, which coincide with the greatest use of alcohol and other mood-altering substances in the general population. Individuals in this age group cite numerous reasons for beginning substance use, and these reasons for doing so would also apply to individuals this age who may be interested in a military career. Table 14.4 lists some common examples.

Understanding Substance Use Evaluation in Modern Times

Our understanding of substance use disorders (e.g., substance dependence and substance misuse) has grown a great deal since the 1970s. Education and training in brain diseases and behavioral disorders have taken quantum leaps due to the work of researchers and clinicians in the field. The Army maintains a training course at Fort Sam Houston at their Army Medical Education Department Division (AMEDD) under the Army Substance Abuse Program (ASAP). The website for nomination of military members eligible to be trained through these programs is https://www.cs.amedd.army.mil/sfsb/AD_Courses. asp. The largest amount of material available to civilian practitioners to assist them in understanding the nuances of evaluating military service members is the 2006 book *Assessing Fitness for Military Enlistment: Physical, Medical and Mental Health Standards,* by the Physical, Medical, and Mental Health Standards Committee on Youth Population and Military Recruitment and the National Research Council. It is available in print or online at http://books. nap.edu/catalog.php?record_id=11511.

Screening for Recruitment

Drug trafficking, the most severe abuse, disqualifies applicants for all services, as does a history of alcohol dependence (except for the Navy, which requires a waiver if the applicant is no longer dependent on alcohol). All services agree that limited or recreational use of marijuana does not require a waiver; in other words, an individual who admits to the use of marijuana may still enter the service. The following paragraph is taken directly from the text *Assessing Fitness for Military Enlistment: Physical, Medical and Mental Health Standards* (p. 153):

> Limited pre-service use of drugs other than marijuana and alcohol has the most diverse standards. The Army does not require a waiver, while the Marine Corps will issue a waiver after the applicant fills out a drug abuse screening form. The Air Force will issue a waiver for non-narcotics, such as amphetamines and barbiturates, but narcotics are disqualifying. Finally, the Navy also distinguishes between non-narcotic and narcotic drugs. In the case of non-narcotics, the Navy does not require a waiver if the use was more than one year prior to screening, but narcotics use requires a waiver if use was over one year prior. Use within the past six months is disqualifying.

Another good resource for the civilian clinician is the 2004 *Iraq War Clinician Guide*, 2nd ed., published by the National Center for Post-Traumatic Stress Disorder, Department of Veterans Affairs (http://www.ncptsd.va.gov/ncmain/ncdocs/manuals/iraq_clinician_guide_ch_12.pdf). The chapter by Lande et al., "Substance Abuse in the Deployment Environment" (pp. 79–82), contains excellent information for the clinician who needs a succinct guide to military recommendations regarding substance use and its treatment. This publication recommends the use of the standard CAGE questionnaire as a screening tool:

C—Have you tried to *cut* down your use?
A—Have people *annoyed* you by asking you to stop using?
G—Have you ever felt *guilty* about your use?
E—Have you needed an *eye-opener* to get going (or have you used to stop withdrawal symptoms)?

The CAGE questionnaire is one of the easiest screening tools to use because it is portable and reliable, has a strong research proven base, and does not require pen or paper. The CAGE can reliably predict 70 to 80% of individuals with a substance dependency issue (Friedman, Saitz, Gogieni, Zhang, & Stein, 2001).

A standing military order prohibits the use of alcohol or illegal substances in a deployed environment. Rules and guidelines are also in place for the post-deployment and nondeployment environments, as well. Service members who are identified as having used either alcohol or illegal substances in a deployed environment are subject to military discipline. Civilians need to be aware that there are consequences to reporting certain activities and what those consequences entail.

Military Code of Conduct Regarding Substance Abuse

Alcohol is a legal substance, and its use and abuse impact social, family, financial, and legal situations (Freeman & Storie, 2007). Given that alcohol is readily available, inexpensive, and, in most cases, socially acceptable, many service members do not view the use or abuse of alcohol as harmful; however, factors that discourage the abuse of alcohol include the threat of military discipline, standards of conduct, and pride in the military.

Military regulations are in place to address service members who exhibit an inability to control their behavior to the point that their behavior discredits the uniform of the United States. Several steps required before serious consequences ensue are elucidated in Chapter Nine of the Uniform Code of Military Justice, which includes the following:

Alcohol or Other Drug Abuse Rehabilitation Failure:

This chapter provides the authority and outlines the procedures for discharging soldiers for alcohol and other drug abuse rehabilitation failure. Soldier is entitled to request a hearing before an administrative separation board if he/she has six or more years of total active and reserve military service per paragraph 2-2c(5). A soldier who has less than six years is not entitled to a board. Discharge is based upon alcohol or other drug abuse such as illegal, wrongful, or improper use of any controlled substance, alcohol, or other drug when:

- The soldier is enrolled in the Army Substance Abuse Program (ASAP).
- The commander determines that further rehabilitation efforts are not practical, rendering the soldier a rehabilitation failure. This determination will be made in consultation with the rehabilitation team. (See AR 600-85.)

Effect of Structure on Substance Use Choices

The U.S. military service member lives a very structured and disciplined life while on duty. The off-duty service member is allowed to relax and pursue various hobbies and interests, just as a civilian would do; however, service members remain responsible to the military for their behavior regardless of whether or not they are on duty, a situation that is very different from that of civilian workers. Service members are representing the United States of America at all times and must demonstrate appropriate behavior at all times. If their behavior is not in the best interest of the uniform of our nation or if the behavior involves certain risky behaviors, disciplinary consequences are possible.

These disciplinary consequences become more complex if the service member is deployed on active duty (in theater). The stress that service members experience at any given time is magnified because they may fear not only the consequences of immediate disciplinary actions should they find themselves in a difficult situation but also the loss of their deployment status or career.

Particularly since September 11, 2001, multiple deployments are not uncommon. Many service members who have deployed at least once with a unit experience an extremely strong bond with their units. It is often nearly incomprehensible to consider not being able to deploy with their units due to an error in personal judgment, especially one generally considered by their peers as stupid, such as drinking to excess and committing a behavioral error.

A complicating factor of postdeployment stress is the use of alcohol or other substances to cope with stress, sleep disruption, intrusive recollections, and other problems. Sometimes the increased use of alcohol and other substances may result in disciplinary action and referral for treatment. Service members are often reluctant to seek assistance on their own for stress-related issues, mood disorders, or substance use problems. Their need for help is often not discovered until a crisis occurs.

Prior to August 2007, regulations were in place stating that: "Second-time offenders, and first-time offenders in pay grades E5–E9, *must* be processed for separation. For first-time offenders in pay grades E1–E4, the separation authority will decide whether to separate, based on recommendations from the immediate and intermediate commanders." These regulations were amended by the Senate Committee on Veterans Affairs to address posttraumatic stress disorder (PTSD) and the problematic use of alcohol and other substances to cope with PTSD. The Committee is a statutorily created panel of clinicians that reports annually to Veterans Affairs and to Congress. The postdeployment issues that the Committee has been charged to deal with include major depression, alcohol abuse, narcotic addiction, generalized anxiety disorder, job loss, family dissolution, homelessness, violence toward self and others, and incarceration. The Committee has advised that

> Rather than set up an endless maze of specialty programs, each geared to a separate diagnosis and facility, the VA needs to create a progressive system of engagement and care that meets veterans and their families where they live. The emphasis should be on wellness rather than pathology; on training rather than treatment. The bottom line is prevention and, when necessary, recovery.

An Unsuccessful Outcome: Veteran Case Study

One of the best documented cases of a veteran with substance abuse and PTSD is the story of America's longest-held prisoner of war (POW) from the Vietnam conflict, Colonel Floyd "Jim" Thompson (Philpott, 2002). Then-Captain Thompson was captured on March 26, 1964, and held captive until his release on March 16, 1973. He survived an aircraft crash, broken back, bullet wound across the cheek, and burns before his capture. He was held in solitary confinement for 5 of the 9 years of his capture. Upon debriefing it was learned he was tortured and starved and suffered unspeakable cruelties.

Born July 8, 1933, in Bergenfield, New Jersey, he was the younger of two brothers. He grew up in a physically abusive home, although neither parent used alcohol. In high school, he worked at the local grocery store and spent weekends drinking in New York City. He married Alyce in July 1953. According to his wife, the first few years of marriage were happy. Alyce said that Jim drank during the week and would return home late at night and intoxicated. She thought he was being rebellious toward his parents.

Thompson was drafted into the U.S. Army in June of 1956. He adjusted well to military life; he was in excellent health and was a committed soldier, but he spent his free time drinking at least twice a week. He volunteered for Special Forces (Green Berets) in June 1963 and began preparing for deployment to Vietnam in December of that year. Alyce was pregnant at that time, and they already had three young daughters. Alyce gave birth to a son on March 27, 1964, the day after Thompson's capture. Nine years later, upon his release, Thompson was in very poor physical condition with many health concerns, and he had great difficulty adjusting to the changes in his environment that had occurred since his incarceration. He began to use alcohol more heavily, which adversely affected his physical condition even further. His stress level escalated upon learning of his wife's infidelity during his imprisonment. (She had taken their children and moved in with another man.) Thompson reported that he was having nightmares every night about his captivity. Doctors prescribed Valium® for depression and hypnotics for sleep even though they were aware that he was using significant amounts of alcohol. In October 1974, he returned to South Vietnam to receive South Vietnam's Medal of Honor, but this public spotlight was a constant reminder of his captivity.

Thompson continued his career in the Army. Despite his efforts to adjust, he and his wife divorced in 1974. His use of alcohol escalated to the point that in 1976 the Army ordered him into an inpatient treatment program. He attended the program with minimal commitment and motivation and subsequently relapsed shortly after discharge from the hospital. Following the relapse, he was given an ultimatum by his commanding officer to attend a tougher alcohol and drug program or resign from the U.S. Army. He made several attempts at treatment but was not committed to remaining sober so he had numerous relapses. His supervisor threatened to declare him a rehabilitation failure. Shortly after returning to Fort Benning, Georgia, he called his psychologist and informed him that he had attempted suicide by overdosing on a combination of alcohol, benzodiazepines, and barbiturates. Thompson was rushed to Martin Army Hospital and given emergency lavage treatments. Thompson retired from the U.S. Army in 1981 after 25 years of service, and he died on July 16, 2002, in Key West, Florida. He was recommended for the Congressional Medal of Honor but it was never approved (Special Forces Association, 2004).

Motivation to Change

Therapists who listen carefully to their patients to help them find their own motivations to make their lives better are more likely to assist these individuals to take action. Prochaska and DiClemente (1992) developed an elegant and simple model for understanding the stages through which an individual theoretically passes when undergoing a change process. Although no individual follows these stages in a stepwise fashion, the model provides a "talking point" for the therapist and patient to discuss motivation behavior and thought processes during their treatment. In the real world, patients will regularly move back and forth, in and out of these stages. An adept and well-trained therapist has learned to accept this as the natural progression of change, as long as the overall progress is in the forward direction. Prochaska and DiClemente's five-stage model has been widely accepted and utilized in the addiction profession. It is easy to remember and easy to put into practice. The basic model includes the stages of *precontemplation, contemplation, preparation, action,* and *maintenance.* The revised model (Freeman & Dolan, 2002) expands upon the original by adding the components of *noncontemplation, anticontemplation, precontemplation, contemplation, action planning, action, lapse activation and redirection, relapse and redirection, termination,* and *maintenance.* By increasing specificity, the expanded model helps the patient as well as the therapist identify certain subcomponents that are critical in the change process, something that is the hallmark of cognitive–behavioral therapy (CBT). The additional stages reflect the experiences of clients and therapists both in and out of the therapeutic process (see Table 14.5).

Therapists and clients may find that the stages of change (SOC) of precontemplation, contemplation, and maintainence can serve as "talking points" in their sessions, but it is important to appreciate the considerable controversy found in the literature regarding the strength of validation for the SOC model (D. Meichenbaum, 2009, pers. comm.). As Littell and Girvin (2002), Sutton (2001), and Josephs, Breslin, and Skinner (1999) observed, supportive data for distinct stages remain weak and inconsistent. Moreover, few data exist for the hypothesized relationship between the stages of change and specific processes of change; therefore, clinicians must be circumspect in their discussions when advocating the SOC model with their clients.

Impediments to Change

In addition to understanding that there are specific stages that an individual experiences when undergoing a change process, the therapist must also understand that there are often roadblocks, or impediments, that may stand in the way of change (Freeman & Storie, 2007). Unfortunately, most substance abuse treatment literature contains only references to problem behavior that place the major emphasis on treatment failure on the patient while ignoring the

Table 14.5 Revised Stages of Change

Prochaska and DiClemente (1992)	Freeman and Dolan (2002)
Precontemplation	Noncontemplation: "I didn't realize …"
	Anticontemplation (willful or nonwillful): "Leave me alone!"
	Precontemplation: "I am willing to think …"
Contemplation	Contemplation: "I need to do something about …"
Preparation	Action planning: "What can I do?"
Action	Action: "I need a plan."
	Prelapse (redirection, cognitive and metacognitive): "I keep thinking about using …"
	Lapse (redirection, behavioral): "I slipped."
	Relapse (redirection, cognitive and behavioral): "I slipped again …"
Maintenance	Maintenance: "I have been sober for … months now."

impediments to change that derive from the therapist, the environment, and even the neurobiology of the patient. As an example, derogatory terms such as *noncompliance, denial,* and *manipulative behaviors* are labels that place the responsibility for a lack of progress in treatment squarely on the shoulders of the patient. Applying such labels not only is overly simplistic but may also point to something more serious—therapist incompetence. The term *incompetence* may sound harsh, but a basic tenet of professional practice is to only practice those therapeutic techniques for which competence has been attained, maintained, and retained through ongoing training (Freeman, 2005).

The Freeman Impediments to Change Scale—Substance Use (FITCS-SU) outlines various factors that may slow or even stop the therapeutic process (Freeman, 2005) (see Table 14.6). The scale allows for evaluation of these factors in an objective, nonjudgmental, and targeted fashion. Once the impediments have been identified and discussed, a treatment plan can be developed that addresses these impediments to change. If biological or family factors are in play, they must be incorporated in the plan to prevent or limit their negative pressure on the progress of the therapy (Freeman & Storie, 2007).

Parallel Presentations With Personality Disorders

When the brain is compromised by an injury or chemical use, it is imperative that this be taken into account when conducting an evaluation of that individual. This is especially true with individuals who present to therapy currently using or abusing substances. The presence of noxious mood-altering and psychoactive substances creates an artificial cognitive atmosphere that does not give an accurate evaluative picture. These service members may display behaviors that mimic certain personality disorders. If clinicians are not careful, they may be too quick to diagnose these individuals with Axis II disorders. A good rule of thumb is to wait at least 6 months to a year for the brain to heal and clear itself of substance-related damage to determine what baseline brain functioning is really like before evaluating for an Axis II disorder. Table 14.7 outlines those parallel presentations that often confuse the diagnostic picture for substance abusing persons.

To differentiate, it is important to conduct an evaluation that is significantly different from evaluations for other disorders, such as affective disorders or psychotic disorders. The following guidelines are recommended in addition to those above for evaluating co-occurring Axis II disorders (Freeman, 2005; Freeman & Storie, 2007):

1. Do not rely on patient self-report for information, which is likely to be skewed, protected, and limited despite a presentation of cooperation.
2. Obtain collateral information from all sources connected with the patient. Freeman recommends obtaining information from the family, friends, medical practitioners, and previous therapists, as well as from the patient's pharmacy. There is much information to be learned from a history of prescription medications.
3. Use *timeline followback* (Sobell & Sobell, 1998) to allow patients to identify their own use progression as well as behavioral escalation in a nonthreatening, collaborative manner.
4. Evaluate physical symptoms with a thorough history and physical evaluation.
5. Always rule out neurological problems. Many substance abusing persons have suffered head trauma, may have mild mental retardation problems, or may have a history of learning disabilities that has never been diagnosed.

Do's, Don'ts, and Other Challenges

It is important to incorporate the following list of recommendations from the chapter on substance misuse in *Cognitive Behavior Therapy in Nursing Practice* (Freeman, 2005) when working with these patients in general:

1. Limit expectations for patient-generated problem solving until problem-solving techniques are in place.
2. Use a concrete approach and avoid any abstract conceptual expectations.
3. The primary substance will likely point to the schematic structure underlying the substance misuse personality developments. Test this theory out in the assessment.
4. Integrate the information provided by the schema uncovered in therapy when planning outcomes related to the use of substances, as the schema will likely activate craving to use the drug.

Table 14.6 Freeman Impediments to Change Scale—Substance Use

Instructions: For each of the following impediments, identify the contribution of that issue to the problems being encountered in therapy. It is essential for the therapist to review all of the areas with the patient.

0, no importance; 1, some importance; 2, moderate importance; 3, great importance; 4, major importance.

Patient Factors

1.	Skill deficit regarding techniques to control substance use and/or comply with therapeutic regimen/expectations	0	1	2	3	4
2.	Negative cognitions regarding previous treatment experience or abstinence failure(s)	0	1	2	3	4
3.	Negative cognitions regarding the consequences to others about altering substance use	0	1	2	3	4
4.	Patient experiencing secondary gain from disease symptoms	0	1	2	3	4
5.	Patient experiencing significant primary gain from substance use	0	1	2	3	4
6.	Fear of changing one's actions, thoughts, feelings	0	1	2	3	4
7.	Motivation to discontinue substance use not at contemplative stage	0	1	2	3	4
8.	General negative set regarding ability to control use or stop use	0	1	2	3	4
9.	Limited or restricted self-monitoring/monitoring of others	0	1	2	3	4
10.	Patient frustration with lack of treatment progress over time or perceived stigma of being in therapy	0	1	2	3	4
11.	Insufficient personal resources (physical, cognitive, or intellectual) to control substance use	0	1	2	3	4

Practitioner/Therapist Factors

1.	Insufficient therapist skill/experience with substance use	0	1	2	3	4
2.	Congruence of patient and practitioner distortions	0	1	2	3	4
3.	Limited or insufficient socialization of patient to treatment generally and to specific treatment model	0	1	2	3	4
4.	Incomplete or absent collaboration and working alliance	0	1	2	3	4
5.	Insufficient or inadequate data regarding individual's history	0	1	2	3	4
6.	Therapeutic narcissism	0	1	2	3	4
7.	Timing of intervention not aligned with patient or does not match motivation level	0	1	2	3	4
8.	Therapy goals unstated, unrealistic, or vague; misalignment of patient goals with therapy goals	0	1	2	3	4

	0	1	2	3	4
9. Evaluation of developmental process not taking into account temporal factor of substance abuse or overestimated	0	1	2	3	4
10. Generalized negative beliefs (discriminatory) about substance use or unrealistic expectations of patient	0	1	2	3	4
11. Insufficient flexibility and creativity in treatment planning	0	1	2	3	4

Environmental Factors

	0	1	2	3	4
1. Environmental stressors precluding change	0	1	2	3	4
2. Significant others actively or passively sabotaging therapy	0	1	2	3	4
3. Agency reinforcement of pathology and illness via compensation or benefits	0	1	2	3	4
4. Cultural or family issues regarding help seeking	0	1	2	3	4
5. Significant family pathology or active substance use in the home	0	1	2	3	4
6. Demands made by family members or significant others directly conflicting with the therapeutic plans or activities	0	1	2	3	4
7. Unrealistic or conflicting demands on patient by institutions or other external source	0	1	2	3	4
8. Financial factors limit change (inability to access care)	0	1	2	3	4
9. System homeostasis	0	1	2	3	4
10. Inadequate or limited support network	0	1	2	3	4

Pathology Factors

	0	1	2	3	4
1. Patient flexibility severely restricted resulting in movement constraints in treatment compliance	0	1	2	3	4
2. Significant medical and/or physiological problems	0	1	2	3	4
3. Difficulty in establishing trust	0	1	2	3	4
4. Autonomy press	0	1	2	3	4
5. Severe impulsive response pattern independent of substance use	0	1	2	3	4
6. Confusion, dementia, or limited cognitive ability	0	1	2	3	4
7. Symptom profusion	0	1	2	3	4
8. Dependence on external controls	0	1	2	3	4
9. Severe self-devaluation	0	1	2	3	4
10. Severely compromised energy	0	1	2	3	4

TABLE 14.7 Parallel Diagnostic Presentation:
Personality Disorder and Substance Dependence Disorder

1. There is often support system exhaustion (family, employment, financial, social).
2. Among the primary reactions and defenses are "other-blaming."
3. The changes in behavior can usually be traced to early to mid-teens.
4. Both disorders tend to progress over time as behaviors and drug use escalate.
5. There are often exacerbation and remissions with crises.
6. Interactions tend to be manipulative in nature.
7. Both disorders are at risk for affective disorders.
8. There are limited or inadequate problem-solving skills.
9. There tends to be a history of multiple failure experiences across multiple aspects of life.

Source: Freeman, S.M., in *Cognitive Behavior Therapy in Nursing Practice*, Freeman, S.M. and Freeman, A., Eds., Springer, New York, 2005, pp. 113–144. With permission.

5. Identify external sources that may reinforce the person's use of substances or increase the likelihood of abstinence.
6. Refuse to participate in storytelling that includes euphoric recall of substance ingestion.
7. Be sure that therapeutic goals are proximal, realistic, directive, and patient driven.
8. Although therapy must include an education component, include it without lecturing.
9. Assist the patient in identifying automatic thoughts that trigger, activate, or fuel craving responses.
10. Overemphasize the use of self-instruction techniques (especially 12-step slogans such as "This too shall pass" and "Let it go").
11. Identify areas of vulnerability and develop specific plans of deflection (people, places, things).
12. Remain cognizant of both positive and negative transference.

Challenges

1. Remember that the patient will not be cured. The patient will need external support, possibly for the rest of his or her life. Therapists must be very active in helping the patient to establish supports and then work at maintaining them.
2. Inasmuch as these patients often "other-blame," care must be taken to walk the fine line between confrontation of denial and taking responsibility for their actions.
3. The substance misuse culture may have become a central part of the patient's social and family life. Leaving or minimizing contact with this social system may be experienced as a great loss.

4. Problem-solving training and the use of concrete, focused rules (e.g., 12 steps) are useful in obviating the need to engage in complex activity.

5. Discomfort anxiety is a major contributor to maintenance of drug abuse. Use anxiety management and reduction techniques liberally.

6. Minimization of the pain or embarrassment of demotion or being passed over or even discharged from the military may impede therapy. Other pain-related issues in the family or in finances may also hinder treatment. Carefully, concretely, and convincingly, the therapist must address the patient's avoidance techniques.

7. It is not unusual to hear, "I can stop or control my use whenever I want to." The patient's unrealistic view of being able to maintain control must be addressed through many personal experiments.

8. Patients may have an unrealistic view of their ability to change or affect their life circumstances. Often so much interpersonal damage has been done that jobs are no longer secure and relationships no longer viable.

9. Treatment is a process. This process can be long, difficult, and energy consuming. Ideally, it would benefit all if there was a simple and easily effected cure. Until a magic bullet is discovered, therapy will require hard work by both therapist and patient.

10. Some patients expend vast amounts of time and energy to maintain their addiction without getting into legal or health-related difficulties. Therapy, detoxification, or treatment may be sought as a quick way out of discomfort. Also, the withdrawal process may become intolerable because of the symptoms.

11. The comorbidity of other health-related problems makes treatment more complicated. Several professionals may be involved, requiring a coordinator for all of the medications, time, therapies, and support.

12. Patient recollections may be absent, skewed, or revised by the fact that they may have been intoxicated at the time an event occurred or they may have developed cognitive dysfunction (e.g., memory impairment) related to long-term use.

13. Some medications, even those prescribed, may potentiate or simulate the effect of the substance of choice. Further, there is the potential for self-harm in the mixing of prescription, nonprescription (over-the-counter), and street drugs.

14. The substance of choice is self-reinforcing in that it can eliminate anxiety, trigger the brain reward system, or distance the patient from unpleasant situations.

15. Moving patients away from self-medication to a more organized, controlled regimen is problematic, frightening, complicated, and difficult.

16. Patients with long treatment histories may be confused by the various treatment models to which they have been subjected. Is it their family? Their biology? Their thinking? Their brain chemistry? It is important to

integrate and synthesize the previous treatments and assist the patient in moving on in the present treatment setting and CBT model.

17. The complications of diagnoses on Axes I, II, and III, with severe psychosocial stressors on Axis IV, must be addressed in a comprehensive treatment protocol.

18. The patient may seem challenging, adversarial, or even threatening. Experienced therapists have learned when to push and when to back off.

19. Through experience, the patient has learned the language, techniques, and courses of detoxification, recovery, and relapse. The patient may appear to know more than the therapist. Here, a collaborative approach works best. By inviting knowledgeable patients to use their knowledge in support of change rather than using their knowledge as a defense mechanism can make effective use of previous learning.

20. In an outpatient setting or during stressful events, the patient may come to the treatment session high, resulting in poor focus and state-dependent learning. If the patient is high, the session should be canceled until a time when the patient is able to use the time effectively.

21. In the military, all of the notes are subject to possible review. There is no confidentiality unless the service member is being seen by a non-service-related provider. When the patient has legal problems, the therapist must be very clear as to the lines of control and report. Who has access to the session notes? What can and cannot be reported? To whom are reports made? These lines must be clarified and discussed with the patient at the onset of treatment.

22. Finally, the therapist should be educated, trained, and certified in substance misuse work. It is not for all therapists, and many can do more harm than due to their lack of understanding of the disease.

Specific Discussion Regarding Service Members

Most individuals who develop problems with substance use disorders (SUDs) move through phases of argument with themselves, others, and the disease itself. The argument usually involves a process of trying to control their drinking/using behavior, explaining their drinking/using behavior, using rational or logical explanations regarding the consequences of their behavior related to their drinking/using, and finally accepting that the drinking/using behaviors are self-destructive and out of their control, requiring assistance from the outside. This is no different for service members; however, the situation has the potential to become more complicated because these are individuals who have learned that they must maintain control and honor at all times. This factor must be kept in mind by therapists who work with service members. Another common trait of individuals with substance use disorders is the mindset that "I am different than others with this problem." This is especially true with service members who may attend 12-step meetings that do not include other service members. Even among service members, some will distinguish, for example, between those who have experienced action and those who have not. It is important to assist these individuals in focusing on current consequences as opposed to circumstances that may have led to stress, depression, anxiety, or other problems involved in their use behaviors.

The U.S. Army has a program in place to help soldiers remember the impact of alcohol and drug use. The WARRIOR Pride system emphasizes that the warrior ethos and Army values are incompatible with alcohol abuse and drug use. Components of this system include

- Have the personal courage not to use or abuse drugs and alcohol and to notify the chain of command of soldiers who do.
- Respect your Army, unit, fellow soldiers, and yourself by staying drug free and drinking responsibly.
- Have the integrity to stay true to the Army values and warrior ethos by supporting the Army's drug and alcohol policies.
- Do your duty as a soldier and stay mentally and physically tough by not abusing drugs or alcohol.
- Maintain excellence by exhibiting honorable behavior on and off duty. Don't be a substance abuser!

Service members can be taught specific skills to develop resilience to substance use. Masten, Best, and Garmezy (1990) defined resilience as "a process, capacity, or outcome of successful adaptation despite challenges or threatening circumstances" (p. 426). Resilience emerges and increases or decreases over time as the person interacts with social and cultural environments (Glantz & Johnston, 1999); therefore, peers, training, command factors, and therapy can play important roles in assisting service members to avoid or change the course of substance misuse behaviors.

CBT Techniques for Substance Misusing Service Members

1. Be aware that structure and boundaries are comfortable and expected. Each session should address overall goals and specific session goals.
2. Use a concrete rather than abstract approach (aphorisms are useful).
3. Limit demands for problem solving until problem-solving skills are in place. Remember that substance misusing service members are used to command giving them duties, responsibilities, and direction.
4. Because the substance of choice often points to the relevant schematic structure, use this information to help identify the schema.
5. Focus on the purpose and meaning of the substance misuse. How and why did this behavior begin? How does this behavior affect the patient's current mission? Use terms that are familiar.
6. Identify schema that fuel substance misuse.
7. Identify enablers (personal, institutional, or group).

8. Work to disempower enablers (e.g., WARRIOR concept).
9. Identify supporters.
10. Let the patient know that he or she must agree to change.
11. Be sure that the therapeutic goals are realistic and proximal. Maintain focus on goals that are concrete and achievable between sessions. Use the buddy system, as this is again a familiar concept.
12. Make the therapy directive.
13. Include psychoeducational concepts without being boring or preachy.
14. Encourage the patient to learn how to identify automatic thoughts (ATs) related to craving.
15. Encourage the patient to learn how to identify ATs related to seeking (drive) behavior.
16. Encourage the patient to learn how to identify ATs related to using substances.
17. Make extensive use of self-instructional techniques.
18. Emphasize how important it is to develop and maintain motivation.
19. Identify and develop prescriptions for vulnerability factors.
20. Identify thresholds for substance use.
21. Focus on the therapeutic relationship.
22. Be aware of both negative and positive countertransference.

Pharmacology

Many useful pharmacologic options are available today for the treatment of substance dependence that were not available even a few years ago, such as naltrexone for the treatment of alcohol cravings (in conjunction with psychosocial intervention), acamprosate calcium for the treatment of post-acute withdrawal (Kranzler & Van Kirk, 2001), and buprenorphine for the treatment of opiate dependence (Johnson, Jaffe, & Fudala, 1992). A study by Oslin et al. (2008) examined the impact of three types of psychosocial treatments combined with either naltrexone or placebo treatment on alcohol dependency over 24 weeks of treatment: (1) CBT + medication clinic, (2) BRENDA (an intervention promoting pharmacotherapy) + medication clinic, and (3) a medication clinic model. The study demonstrated that medication combined with behavioral interventions resulted in significantly reduced potential for relapse.

Summary and Future Directions

The U.S. military, all branches, have a no-tolerance policy regarding the use of illegal substances. This means that individuals who are found to use illegal substances will be helped out of the armed services. Active-duty service members will be assisted with treatment for dependence-related problems with legal substances such as alcohol. Reserve members and National Guard members do not necessarily have this same consideration and often have to use their personal civilian insurance benefits for treatment. The therapist must keep these issues in mind when treating members of the military. Substance use and misuse are not unusual and are much more prevalent in individuals who have experienced or are currently experiencing trauma. Practitioners who will be treating service members must be adept at identifying substance use disorders and the complications that occur when combat stress accompanies these disorders. The practitioner must include an assessment of the individual's motivation for change using a model such as that developed by Prochaska and DiClemente (1992) and later expanded upon by Freeman and Dolan (2002).

References

APA. (1994). *Diagnostic and statistical manual of mental disorders* (4th ed.). Washington, D.C.: American Psychiatric Association.

Barnes, J. K. (1870; reprinted 1991). *Medical and surgical history of the war of the rebellion 1861–1865.* Washington, D.C.: U.S. Army Surgeon General's Office.

Bolton, C. K. (1902). *The private soldiers under washington.* New York: Charles Scribner's Sons.

Burkett, B. G., & Whitley, G. (1998). *Stolen valor: How the Vietnam generation was robbed of its heroes and its history.* Dallas, TX: Verity Press.

Chambers II, J. W. (1999). *The Oxford companion to american military history.* Oxford, U.K.: Oxford University Press.

Davis, B. (1982). *The Civil War: Strange and fascinating facts.* New York: Fairfax Press.

Freeman, A., & Dolan, M. (2002). Revisiting Prochaska and DiClemente's stages of change: An expansion and specification to aid in treatment planning and outcome evaluation. *Cognitive and Behavioral Practice, 8*(3), 224–234.

Freeman, S. M. (2005). CBT with substance misusing patients. In S. M. Freeman & A. Freeman (Eds.), *Cognitive behavior therapy in nursing practice* (pp. 113–144). New York: Springer.

Freeman, S. M., & Storie, M. (2007). Substance misuse and dependency: Crisis as process or outcome. In F. M. Dattilio & A. Freeman (Eds.), *Cognitive–behavioral strategies in crisis intervention* (3rd ed., pp. 175–198). New York: Guilford Press.

Friedman, P. D., Saitz, R., Gogieni, A., Zhang, J. X., & Stein, M. D. (2001). Validation of the screening strategy in the NIAAA physicians' guide to helping patients with alcohol problems. *Journal of Studies on Alcohol, 62*(2), 234–238.

GAO. (1971). *Alcoholism among military personnel,* B-164031(2). Washington, D.C.: General Accounting Office.

GAO. (1972). *Drug abuse control program activities in Vietnam,* B-164031(2). Washington, D.C.: General Accounting Office.

Glantz, M., & Johnston, J. L. (Eds.) (1999). *Resilience and development: Positive life adaptation.* New York: Kluwer Academic/Plenum Press.

Johnson, R. E., Jaffe, J. H., & Fudala, P. J. (1992). A controlled trial of buprenorphine treatment for opioid dependence. *Journal of the American Medical Association, 267*(20), 2750–2755.

Jonnes, J. (1996). *Hep-cats, narcs, and pipe dreams.* Baltimore, MD: The Johns Hopkins University Press.

Joseph, J., Breslin, C., & Skinner, H. (1999). Critical perspectives on the transtheoretical model and stages of change. In J. A. Tucker, D. M. Donovan, & G. A. Marlatt (Eds.), *Changing addictive behavior* (pp. 160–190). New York: Guilford Press.

Kopperman, P. E. (1996). The cheapest pay: Alcohol abuse in the eighteenth century British army. *The Journal of Military History*, 60, 445–470 (http://www.jstor.org/stable/pdfplus/2944520.pdf).

Kranzler, H. R., & Van Kirk, J. (2001). Efficacy of naltrexone and acamprosate for alcoholism treatment: A meta-analysis. *Alcohol Clinical and Experimental Research*. 25(9), 1335–1341.

Littell, J. H., & Girvin, H. (2002). Stages of change: A critique. *Behavior Modification*, 26(2), 223–273.

Masten, A. S., Best, K. M., & Garmezy, N. (1990). Resilience and development: Contributions from the study of children who overcome adversity. *Development and Psychopathology*. 2(4), 425–444.

McCullough, D. (2005). *1776*. New York: Simon & Schuster.

Miles, B. (2004). *HIPPIE*. New York: Sterling Publishing.

Milhorn, Jr., H. T. (1994). *Drug and alcohol abuse: The authoritative guide for parents, teachers, and counselors*. Reading, MA: Perseus Books.

Morgillo Freeman, S. (2008). *Cognitive behavior therapy*. In K. Wheeler (Ed.), *Psychotherapy for the advanced practice psychiatric nurse* (pp. 177–202). St. Louis, MO: Mosby.

Oslin, D. W. et al. (2008). A placebo-controlled randomized clinical trial of naltrexone in the context of different levels of psychosocial intervention. *Alcoholism: Clinical and Experimental Research*, 32(7), 1299–1308

Philpott, T. (2002). *Glory denied: The saga of Vietnam veteran Jim Thompson, America's longest-held prisoner of war*. New York: Penguin Putnam.

Prochaska, J. O., & DiClemente, C. C. (1992). Stages of change in the modification of problem behaviors. *Psychology of Addictive Behaviors*, 10, 81–89.

Research Triangle Institute. (2006). *Department of Defense survey of health-related behaviors among active duty military personnel*. Research Triangle Park, NC: RTI International.

Robertson, J. R. (1984). *Tenting tonight: The soldier's life*. Alexandria, VA: Time-Life Books.

Robins, L. N., Helzer, J. E., & Davis, D. H. (1975). Narcotic use in Southeast Asia and afterward: An interview study of 898 Vietnam returnees. *Archives of General Psychiatry*, 32(8), 955–961.

Roffman, R. A., & Sapol, E. (1970). Marijuana in Vietnam: A survey of use among Army enlisted men in the two Southern corps. *International Journal of the Addictions*, 5(1), 1–42.

SAMHSA. (2007). *2006 national survey on drug use and health*. Rockville, MD: Substance Abuse and Mental Health Services Administration (http://www.oas.samhsa.gov/nsduh/2k6nsduh/2k6Results.cfm#TOC).

Singlaub, J. K. (1991). *Hazardous duty*. New York: Summit Books.

Sobell, M. B., & Sobell, L. C. (1998). Guiding self change. In W. R. Miller & N. Heather (Eds.), *Treating addictive behaviors* (pp. 189–202). New York: Plenum Press.

Special Forces Association. (2004). Col. Floyd Thompson dedication ceremony. *The Drop*, Winter 2004.

Sutton, S. (2001). Back to the drawing board? A review of applications of the transtheoretical model to substance abuse. *Addiction*, 96, 175–186.

U.S. Army Center for Military History. (2004). *Oaths of enlistment and oaths of office*. Washington, D.C.: U.S. Army. (http://www.history.army.mil/html/faq/oaths.html).

Wilbur, R. S. (1973). Distinguished service medal citation and letter of commendation on duty performance of John K. Singlaub, Major General, U.S. Army.

15

Characteristics, Effects, and Treatment of Sleep Disorders in Service Members

BRET A. MOORE and BARRY KRAKOW

Contents

> O sleep, O gentle sleep,
> Nature's soft nurse, how have I frighted thee,
> That thou no more wilt weigh mine eyelids down
> And steep my senses in forgetfulness?

William Shakespeare (*King Henry IV*, Part II)

The impact of sleep disorders on the individual and society can be enormous. Whether it be insomnia, nightmares, daytime sleepiness, or even snoring, virtually every human has been affected by sleep disturbances to some degree. Consider that just insomnia (the single most common sleep complaint) affects one third of Americans in the general population on any given night. In clinical settings, estimates indicate that one half of patients are affected (NIH, 2005). High prevalence rates have also been found globally (Soldatos, Allaert, Ohta, & Dikeos, 2005). With annual health costs estimated to be tens of billions of dollars and associated quality-of-life issues, insomnia has awakened the general public and the healthcare community to a new "epidemic."

The main goals of this chapter are to give the practitioner who treats service members a better understanding of sleep disorders, information on how sleep disorders impact the men and women of our Armed Forces, and guidance on effective and practical methods of treatment. Although no sleep disorders exclusively arise in service members, this population manifests unique assessment and treatment issues. A brief review of sleep architecture precedes a focused discussion on the characteristics of sleep disorders and the evidence-based strategies for their effective management.

The Basics of Sleep Architecture

Sleep is a natural, recurring process that temporarily reduces the environmental and situational awareness of the individual. Although originally described as a passive activity, researchers and clinicians now characterize sleep as a very active and dynamic physiological and neurophysiological process. Sleep architecture is divided into non-rapid eye movement (NREM) and rapid eye movement (REM) stages. NREM sleep is divided into three separate stages (I, II, and III/IV). Stages I and II are characterized by drowsiness and the onset of light sleep, with slow but steady decreases in heart rate and body temperature. In contrast, stage III/IV consists of delta waves and is typically referred to as deep sleep. It is during this stage when it may be difficult to arouse the individual. If the person is awakened, he or she may be disoriented and confused for several minutes. Delta sleep is associated with the most frequent occurrences of bedwetting, sleepwalking, and night terrors.

The REM phase typically accounts for one fourth (90 to 120 minutes) of total time spent in sleep, usually emerging four to six times during the night at progressively longer durations for each subsequent period. It is characterized by increased heart rate and respiration, paralysis in voluntary muscles, and rapid rhythmical shifting of the eyes. Dream awareness seems to be greatest in REM sleep, although research has shown conclusively that we dream in all stages of sleep (Cavallero, Foulkes, Hollifield, & Terry, 1990). Approximately 85% of individuals who are awakened during REM sleep will remember their dreams; however, few will recall any significant dream content once the REM phase is completed (Silber, Krahn, & Morgenthaler, 2004).

The purpose of sleep has been debated by researchers and philosophers for centuries, but it still remains a relative mystery. A number of propositions have been put forth to include the need for the body and brain to restore itself (cell growth and protein synthesis), the conservation of energy to aid survival, and protecting the psyche from emotional overload. Sleep also assists with the process of immunocompetence, or the body's ability to produce an immune response after exposure to some antigen (Toth & Opp, 2002). The theory that has attracted the most investigation of late is the role of sleep in memory, particularly with regard to memory consolidation (Stickgold, 2005; Stickgold & Walker, 2005; Walker, Brakefield, Hobson, & Stickgold, 2003). Regardless of the actual purpose and function of sleep (it is likely a combination of several factors), it is without question that the body, brain, and mind function decays when adequate slumber is not received.

Sleep and the Military Culture

For the service member, sleep is friend or foe. From the initial days of basic training, newly minted recruits are conditioned to view sleep as a privilege meted out by those responsible for their training. Although it may be seen as cruel and unnecessary by those outside of the military culture, this approach to training prepares service members for the realities of sleep deprivation that invariably occur during sustained military operations (for a detailed review of challenges during continuous operations, see Belenky et al., 1987).

Although it depends on the location, military occupational specialty (e.g., infantryman, mechanic, medic), and atmosphere of the service member's unit, the standard 8 hours of sleep is not a common reality. Ideally, a service member is provided a minimum of 4 hours of sleep for every 24-hour period; however, troops often function and survive on much less for extended periods of time. They learn to accept the agitation, mood swings, decreases in memory and concentration, and changes in overall efficiency and proficiency caused by lack of sleep and engage in compensatory behavior such as using excessive amounts of caffeine and nicotine.

In many cases, the residual effects of disturbed sleep can have significant consequences for the service member in both operational environments (e.g., Iraq, Afghanistan, Bosnia) and garrison environments (e.g., stateside, foreign coalition bases). In the combat zone, service members may be at greater risk of injury and death due to decreased reaction time (both physically and cognitively), decreased awareness and attention, and overall decrements in decision-making and critical-thinking skills. In the garrison environment, the service member is at increased risk for accidents during routine training missions and human errors during the operation of military and civilian vehicles. The service member may also experience a worsening of psychiatric and substance use disorders and increased marital and family discord.

To understand this population it is imperative that the nonmilitary practitioner appreciate that many service members take pride in their ability to function without adequate sleep and associate it with strength and perseverance (two qualities highly valued within the military culture). Within combat units, requiring or succumbing to sleep can be seen as weak or selfish. Obviously, this mentality has significant implications for the identification and treatment of sleep disorders, as well as overall prognosis; therefore, the clinician working with service members must be aware of these perceptions and beliefs.

Sleep Disorders and Effective Treatments

The sleep disorders nosology from the *Diagnostic and Statistical Manual of Mental Disorders*, 4th edition, text revision (DSM-IV-TR) (APA, 2000) may be one that many clinicians are familiar with, but the newest version of the International Classification of Sleep Disorders (ICSD-2) (AASM, 2005) is a much more practically suited format for aiding the clinician in developing a clear schematic on how to approach the sleep disordered patient. In general, the nosology from the American Academy of Sleep Medicine divides sleep disorders into six main categories:

- Insomnia (unwanted sleeplessness)
- Parasomnias (unusual behaviors in or around the sleep period)
- Sleep breathing disorders
- Circadian disorders (problems due to an undesirable timing of the sleep period)
- Sleep movement disorders
- Hypersomnias (excessive sleep or sleepiness)

These constructs are extremely important and pragmatic, because the clinician will soon discover that the majority of patients seeking help for sleep problems rarely fit into just one category. The average patient suffers problems in at least two or three dimensions; for example, the nightmare patient (a parasomnia) will often present with the complaint of insomnia. As many as 20% of patients with insomnia will report restless legs or leg jerks (sleep movement disorders). Approximately 15% of patients with sleep breathing disorders will demonstrate leg jerks on their sleep tests, and up to 50% of patients with sleep breathing disorders will report insomnia. Thus, by keeping these six categories in mind, the clinician has an easy tool to rapidly clarify which areas require attention, and they are less likely to miss a disorder by recognizing that the average patient rarely has only one condition. Furthermore, holding this paradigm in mind promotes a more judicious approach to prescribing sleep aids.

Contributing Factors to Sleep Problems in Service Members

Sleep disorders are some of the most common ailments present in our society today. They cut across race, sex, ethnic, and socioeconomic lines and can lead to decrements in physical, emotional, and occupational functioning. This is certainly true for the service member. In addition to the six main diagnostic categories described above, treating clinicians should be aware of various factors that cause or contribute to any of the specific sleep disorders. For the nonmilitary clinician, providing care in an operational environment is not likely to happen; however, with the current operational and deployment tempo of our military, it is extremely likely that clinicians will provide treatment to service members who have recently returned from a deployment. Being aware of the most common contributing factors to sleep disorders will allow for greater understanding, identification, and amelioration of problems.

Psychiatric Factors

Although sleep disturbances secondary to a psychiatric disorder are not unique to service members, troops returning from combat are susceptible to various psychiatric illness. This problem is particularly important for military subgroups at even greater risk such as women, enlisted and Army members, and those who are single and less educated (Riddle et al., 2007).

The largest study to date on the prevalence of psychiatric disorders in returning service members from Iraq revealed positive screens for posttraumatic stress disorder (PTSD), generalized anxiety, or major depression in 15 to 17% of soldiers and Marines, with the highest proportion reporting symptoms consistent with PTSD (Hoge et al., 2004). In a much smaller sample, relatively high rates of PTSD and substance use disorders were found (Erbes, Westermeyer, & Engdahl, 2007). These findings are noteworthy because sleep disturbances are one of the DSM-IV-TR diagnostic criteria for PTSD (distressing dreams and/or difficulty falling or staying asleep), major depression (insomnia or hypersomnia), and generalized anxiety disorder (difficulty falling or staying asleep, or restless unsatisfying sleep) (APA, 2000). Thus, as is often the case in the setting of what might be called posttraumatic sleep disturbance, the patient frequently suffers from two classes of problems: nightmares (a parasomnia) and insomnia. Moreover, some proportion of PTSD patients suffer from sleep breathing disorders as well as sleep movement conditions. These outcomes are prime examples of how the clinician really must stay attuned to the multiple contributing diagnoses or influences on the patient's sleep in order to devise a complete treatment plan.

Although not a specific criterion for any of the substance use disorders, the literature is replete with evidence that the use of alcohol and illicit substances can wreak havoc on sleep architecture (Landolt & Gillin, 2001; Morgan et al., 2006; Yules, Lippman, & Freedman, 1967). Moreover, many people use alcohol or various drugs for insomnia as well as other agents to counter daytime fatigue and sleepiness. Other disorders to be aware of that may have a direct impact on sleep include relationship-based disorders (e.g., partner relational problem) and adjustment disorders, which are extremely common in operational environments and in returning service members attempting to reintegrate into their stateside life.

Environmental Factors

Environmental factors cause havoc with sleep for the service member in both combat and garrison settings. During deployment, work schedules can be unpredictable from week to week. It is not uncommon to alternate between day and night shifts while working 12-hour shifts 7 days a week. Circadian rhythm disorders (sleep cycle) may develop, with sleep deprivation, fatigue, agitation, and depression being the more significant consequences. Moreover, the service member may be required to maintain this schedule for up to 15 consecutive months. Although to a lesser extent, these same problems can arise in the garrison environment due to alternating work schedules and field exercises that may last for several weeks.

Service members must deal with other unique environmental factors in a combat environment. The constant noise and commotion are prime examples. Whether it be helicopters flying overhead, heavy machinery roaring past the service members' sleeping quarters, or explosions from mortars, rockets, or improvised explosive devices, the continuous disruption can challenge the most sound of sleepers. On a related note, the fear of being hit by a rocket or mortar during the night while sleeping can create insomnia and sleep fragmentation in its own right. Other factors to contend with include harsh weather, noxious fumes and smells, and cramped living spaces (three or four service members living together in a 10 by 20-foot area) (Nash, 2006).

Pharmacotherapy

Pharmacotherapy is a reality in the military. Due to intense training schedules, long and grueling work days, and intense emotional and physical demands, service members often benefit from pharmacological assistance. In the garrison environment, service members suffer from the same psychiatric disturbances that their civilian counterparts face, and pharmacotherapy provides clear benefits. In a deployed setting, medication can help alleviate the debilitating effects of combat and operational stress such as hyperarousal, anxiety, and nightmares (Clayton & Nash, 2006). Common examples include hypnotics such as zolpidem (Ambien®, Sanofi Aventis), antidepressants such as sertraline (Zoloft®, Pfizer), low-dose antipsychotics such as quetiapine (Seroquel®, AstraZeneca), and antiadrenergic antihypertensives such as prazosin (Minipress®, Pfizer). Many of these medications, however, have significant effects on sleep architecture (Gimenez et al., 2007; Winokur et al., 2001) and may hinder natural healing processes induced by sleep (Clayton & Nash, 2006). The medications themselves are likely to cause significant sleep-related side effects, including insomnia, nightmares, and other parasomnias, such as REM behavior disorder, as well as worsening some sleep breathing or sleep movement conditions. For a service member already taking medication, the first step in the management of sleep disturbances may have to be a referral for a medication evaluation.

Insomnia

Characteristics and Effects

Before we discuss insomnia in any detail, it is important to have a working definition of this term. Insomnia is a disorder or a symptom of a disorder that is characterized by difficulty initiating or maintaining sleep and/or experiencing nonrestorative sleep resulting in daytime fatigue or sleepiness and/or a general sense of feeling unrested during the day, which may be related to an underlying medical or psychiatric condition and highly influenced by environmental and behavioral factors. Insomnia is typically not conceptualized as a unitary medical or psychiatric diagnosis. In most cases, it is a multilayered dynamic disturbance that can potentially cause significant distress and frustration for the individual. Current thinking in the field of sleep medicine casts aside terms such as primary or secondary insomnia in favor of the conceptualization of comorbid insomnia, which then creates a clearer perspective on the needs and the means for direct treatment of the condition in tandem with other ongoing treatments.

Insomnia typically has ties to hyperarousal and cognitive ruminations seen in various anxiety and depressive disorders as well as to the behavioral and environmental factors mentioned above. For most insomniacs, the common culprit is the inability to bring about closure to the mind at the end of the day. Unless someone is able to calm the mind both emotionally and intellectually, initiating and maintaining sleep will be extremely difficult, and insomnia may become a chronic and debilitating learned behavior.

Understanding insomnia from a duration perspective can also be useful. Chronicity can be conceptualized as a transient, short-term, or chronic problem, with each type having its own treatment implications. Transient insomnia is typically viewed as only lasting a few days and less than a week. Short-term insomnia lasts up to approximately 3 weeks, and chronic insomnia is anything longer than 1 month but which often lasts for years or decades (Silber et al., 2004).

If left untreated, insomnia can cause myriad problems. It is not uncommon for chronic insomniacs to exhibit decreases in memory and concentration, deficits in cognitive and motor efficiency, and disturbed mood (Bonnet, 1985; Durmer & Dinges, 2005). Some individuals also experience difficulties fulfilling occupational and social roles, and they report an overall decline in pleasure received from interpersonal contact (Roth & Ancoli-Israel, 1999).

For the service member, particularly in the combat environment, insomnia can lead to catastrophic consequences. Sleep deprivation, which can occur with extended periods of insomnia, can easily turn a physically and mentally strong service member into a stress casualty on the battlefield. Not only does this increase the risk for the individual service member, but it also increases the risk of the entire unit. In the highly acclaimed book *On Combat* (Grossman

& Christensen, 2004), Lt. Col. Dave Grossman writes about an experiment the U.S. Army conducted on an artillery battalion. The battalion was divided into four groups, which were asked to perform artillery fire missions every hour for 20 days straight. The control group received 7 hours of sleep per night, and each group thereafter received progressively larger curtailments (the fourth and most restricted group received only 4 hours per night). Results revealed that group one fired at a 98% accuracy/hit rate, group two fired at 50%, group three at 28%, and group four performed at only 15% of optimal efficiency. This experiment demonstrates the harsh reality that sleep disturbances lead to decreased effectiveness on the battlefield and promote inefficiency that may result in injury or loss of life; therefore, prompt and effective treatment of sleep disorders in service members is crucial.

Treatment

Ideally, effective treatment of insomnia would result in improvements in overall quality and quantity of sleep, increased daytime functioning, and an overall sense of well-being (Ramakrishnan & Scheid, 2007). To accomplish these goals, a number of methods must be employed in the treatment of insomnia in service members. Depending on factors such as the clinician's theoretical orientation, degree, and training, psychological, pharmacological, or a combination of both methods may be most appropriate. It also depends on the service member's symptom presentation (e.g., acute vs. chronic, sleep onset vs. sleep maintenance) and often the service member's job and mission requirements. Regardless of these factors, for the nonmilitary clinician to provide the most appropriate intervention for insomnia, he or she must be aware of the many effective options that are available from both a psychological and pharmacological standpoint.

Psychological Therapy   In our opinion, unless the service member is for some reason at risk of being injured or killed if immediate sleep is not initiated, a psychological approach to treating insomnia should almost always be the first line of treatment. Nonpharmacological treatments as first-line interventions have also been cited by others in the fields of medicine and psychology (Morin, 2003, 2006; Murtagh & Greenwood, 1995); therefore, a cognitive–behavioral approach will likely be the best option when treating service members.

Education about Sleep Practices   Inadequate sleep hygiene (poor sleep habits) is a contributing factor to creating and maintaining insomnia (Bootzin & Rider, 1997). For the service member, varied and unpredictable schedules, crowded living quarters, and maximizing limited downtime by playing video games, watching movies, and talking with friends can create poor sleep practices and lead to insomnia. For this reason, education about behavioral practices that lead to insomnia and best practices to reverse any bad sleep habits

Table 15.1 Sleep Behaviors

Maladaptive	Adaptive
Use caffeine excessively.	Do not use caffeine at least 4 hours before bedtime.
Exercise too close to bedtime.	Exercise in the morning, afternoon, or early evening.
Smoke before going to bed.	Don't smoke at least an hour before going to sleep.
Watch horror or action movies before bed.	Watch a comedy or something "light."
Eat foods that cause heartburn.	Eat earlier in the evening and avoid rich and spicy foods.
Watch the clock.	Turn clock away to lose awareness of time.
Take over-the-counter sleep aides.	Practice relaxation and imagery exercises.
Watch television or playing video games in bed.	Only use bed for sex and sleep.
Stay in bed awake for more than 15 minutes.	Get into bed only when your eyes feel heavy.
Take naps during the day.	Stay awake until it is time to go to bed.

that have developed are crucial. In some cases, it is the only intervention needed. Table 15.1 provides a list of common maladaptive and adaptive sleep behaviors.

In this author's (BAM) clinical experience in working with service members, the first maladaptive behavior listed in Table 15.1 (using caffeine excessively) can be a significant health issue that may have to be addressed with the individual in greater depth. An alarmingly high number of service members use energy drinks (often in high quantities) to combat fatigue and drowsiness. Alone, this behavior is concerning, but, considering that they may also be drinking coffee, sodas, and taking over-the-counter muscle-building supplements throughout the day for extended periods of time, worse outcomes may follow such as dehydration, kidney problems, and cardiac arrhythmias. It can also complicate clinical presentations when anxiety symptoms are a primary complaint, so specific education about caffeine and other supplements may be warranted.

In general, although sleep hygiene tools are very pragmatic, they often fail to achieve satisfactory results because the maladaptive behaviors are invariably fueled by maladaptive perspectives about sleep in general, poor emotional processing skills or other inadequate coping techniques, or ingrained psychophysiological conditioning. Thus, many insomnia patients, especially those with more moderate to severe conditions, require advanced techniques best delivered in the context of a motivational interviewing style in which the patient discovers a cause for the maladaptive behavior and seeks to address the cause in the process of changing to a more adaptive behavior.

Time monitoring is probably the single best example of this phenomenon (Krakow, 2007). In mild cases, removal of the clock may cause the patient to worry less about time, and insomnia lessens rapidly. More commonly, patients turn the clock around but then continue to worry about lost sleep time and start looking for environmental cues to guess the time, which then leads to a stream of consciousness about "how much time is left in the night; if I fall asleep in 30 minutes, how much sleep can I still gain or how will I feel tomorrow if I sleep 5, 4, or 3 hours" and so on. In most of these situations, the insomniac needs to complete an in-depth discussion of either how time monitoring of any sort is ingraining psychophysiological conditioning or how the patient is using time monitoring to ward off deeper, more vexing emotional duress.

Sleep Restriction Therapy   Sleep restriction therapy (SRT) is a highly effective behavioral technique used for the treatment of insomnia and is based on the assumption that excessive time spent in bed is the primary factor that maintains insomnia (Spielman, Saskin, & Thorpy, 1987). In individuals suffering from insomnia, the faulty logic of "if I spend more time in bed, I will sleep more" is often present. The goal of SRT is to decrease the gap between time in bed and time asleep so the patient's sleep period is much more consolidated even though total sleep time is temporarily reduced. In the beginning, the patient maintains a sleep log to track his or her time spent in bed. After a few days to a week, an estimate of actual sleep time is calculated based on the sleep diary, and the patient is then instructed to reduce his or her time in bed to the estimated time spent sleeping (not less than 4 hours). Once the patient successfully sleeps for this period of time for a few nights, the time spent in bed is gradually increased in 15- to 20-minute increments each night over the next couple of weeks or until the individual reaches an acceptable or optimal sleep time.

Relaxation Training   Relaxation training is an easy and effective method for teaching service members how to manage insomnia, particularly in the combat environment. Relaxation training is portable and able to be applied when privacy or space is an issue. Numerous effective methods can be employed, such as progressive muscle relaxation (PMR), imagery, biofeedback, diaphragmatic breathing, and meditation (Lundh, 1998). The one to use depends on the specific needs of the patient (i.e., progressive muscle relaxation for muscle tension or imagery for cognitive ruminations) and the training or skills of the therapist. Due to limited time, space, and privacy, however, PMR, diaphragmatic breathing, and imagery are likely to be the most practical and beneficial for the service member.

Progressive muscle relaxation is a simple technique used to reduce muscle tension and has been found to be very effective with chronic and severe insomnia (Lick & Heffler, 1977). Because muscle tension is associated with increased stress and anxiety, relaxing of the muscles can decrease both stress and anxiety levels. The process starts with having the patient find a comfortable sitting position and taking a few deep breaths. After a general sense of relaxation has been reached, the patient is then instructed to tense a particular muscle, hold that tension for at least 10 seconds, relax for approximately 10 seconds, and repeat the tension–release process two or three more times with the same muscle. The tension–release process is then used sequentially with various muscles until the major muscles of the body are relaxed. The sequence is left up to the individual's preference. Patients who specifically believe, feel, or notice that their muscles are not tense to begin with are less likely to respond to the technique.

Diaphragmatic breathing or deep breathing is the process of breathing fully into the lungs by extending the diaphragm as opposed to breathing shallowly by extending the rib cage. Generally considered a more efficient way of bringing oxygen into the body, deep breathing quickly influences the autonomic nervous system, essential when dealing with anxiety, panic symptoms, or hyperarousal symptoms (Somer, Tamir, Maguen, & Litz, 2005), so common in service members suffering from insomnia. It can be easily demonstrated by asking the patient to place one hand on his or her chest and the other on the stomach. The goal is to ensure that the hand on the chest does not move while the hand on the stomach does. If practiced for 10 to 15 minutes a day for a couple of weeks, diaphragmatic breathing can become second nature.

Imagery is a cognitive process in which the individual creates relatively detailed images of a pleasant experience to focus attention away from cognitive ruminations, negative self-talk, or racing thoughts that may interfere with sleep. It is usually taught as a self-guided tool. The patient is asked to close his or her eyes and imagine a time when he or she felt at ease or at peace or a great sense of calm and relaxation; however, the instructions are not geared toward creating photographic replications of past scenes. Rather, the goal is to tap into the natural flow of images that traverse the mind's eye, which may mimic the hypnagogic imagery that precedes sleep onset. The rationale behind how imagery works (i.e., the brain does not know that it is not actually at that past pleasurable event or place) can be provided and suggestions offered if the service member is having a difficult time conjuring up an image. Imagery is highly effective with regard to the concept mentioned above of bringing closure to the mind. It can assist the individual in letting go of the emotionally and cognitively charged events of the day and ward off the apprehension and anxiety of the upcoming events of tomorrow. Furthermore, it promotes a receptive mode of consciousness in which the individual becomes more inclined toward the natural signals and cues that lead to sleep initiation; however, patients with severe PTSD and concomitant unstable imagery systems may see a worsening of intrusive symptoms and are more likely to respond best to imagery techniques in the setting of exposure therapies, not as an isolated treatment for insomnia.

Cognitive Therapy   Cognitive therapy (CT) is a common and effective method used to treat individuals suffering from insomnia and has been shown to be superior to pharmacotherapy in numerous clinical trials (Harvey, Sharpley, Ree, Stinson, & Clark, 2007; Jacobs, Pace-Schott, Stickgold, & Otto, 2004; Sivertsen et al., 2006). The rationale for the use of CT for insomnia is not unlike that for other clinical disorders in which distorted and dysfunctional thoughts create and maintain affective and behavioral disturbances. As a result of ruminations, worry, exaggeration of daytime and sleep dysfunction, and irrational beliefs about sleep in general, individuals can create significant sleep disturbances. This schema is certainly true for service members struggling with intense anxiety about deploying to a combat environment, trying to assuage guilt about leaving behind their families, feeling ambivalence about relationship difficulties exacerbated by deployment, or coping with a clear set of symptoms of a more serious psychiatric disorder (e.g., PTSD, major depression).

Although slight variations can be found among different cognitive model interpretations for the management of insomnia, most include key cognitive components such as thought stopping or blocking and cognitive restructuring. Cognitive restructuring is especially helpful in that it challenges false assumptions and beliefs about sleep (e.g., quantity of sleep is more important than quality or that you need 8 hours each night), confronts irrational beliefs that may be contributing to depressive or anxiety symptoms that may be fueling the insomnia, and replaces maladaptive appraisals of their ability to handle their current difficulties and regain control over their sleep routine.

Sleep Dynamic Therapy   Sleep dynamic therapy (SDT) is an integrated mind–body program of primarily evidenced-based sleep therapy that minimizes the use of medications in the amelioration of insomnia and assumes that both mental and physical factors cause insomnia (Krakow et al., 2002a). As developed by Barry Krakow and described in his book, *Sound Sleep, Sound Mind* (2007), SDT consists of seven key components for sleeping well: (1) focus on quality over quantity of sleep, (2) gain awareness of self-talk and how it fuels insomnia, (3) utilize your natural imagery system to encourage a receptive mode of consciousness at night, (4) address unresolved emotions during the day so they do not perplex you at night, (5) learn to sleep without medication, (6) assess physical causes of symptoms such as sleep breathing and sleep movement conditions, and (7) develop a plan with a sleep medical center to use evidence-based treatments for physical sleep conditions.

Pharmacotherapy   In a combat environment, pharmacological intervention is typically the first-line treatment for insomnia. Considering the clinical and cost-effectiveness of cognitive and behavioral interventions, this practice is unfortunate; however, it is understandable due to the fact that face-to-face time with the individual is limited and the "quick fix" (although not typically the long-term fix) may be the only practical choice. For the nonmilitary practitioner, it is important to understand that the returning service member may possess that "quick fix" mentality and expect complaints of sleep to be addressed by dispensing medication. When psychological interventions are unsuccessful, several classes of medications are available that can be effective in combating insomnia.

Antihistamines   Although relatively little evidence supports their efficacy, antihistamines are a popular choice for the treatment of insomnia due to their sedating effects and their lack of pharmacologic dependency. Two common antihistamines used for insomnia include diphenhydramine (Benadryl®, McNeil-PPC) and hydroxyzine (Vistaril®, Pfizer). Diphenhydramine has been shown to be effective in treating insomnia (Kudo & Kurihara, 1990) and is especially popular with service members, as it can be purchased over the counter and bought at most base exchanges (BXs) and post exchanges (PXs) in combat zones. Antihistamines are most likely to be effective over shorter periods of time, as tolerance to the sedating effects can develop rapidly (Richardson, Roehrs, Rosenthal, Koshorek, & Roth, 2002). These medications may also cause unwanted side effects such as dry mouth, tremor, and respiratory distress (Stahl, 2006).

Antidepressants   Sedation is a side effect of a variety of antidepressant medications; however, two of the more common ones used specifically for insomnia and not necessarily for depression are trazodone (Desyrel®, Apothecon) and amitriptyline (Elavil®, AstraZeneca). Both trazodone and amitriptyline are effective in treating insomnia (Jacobsen, 1990; Stahl, 2006; Wheatley, 1984) and are typically available in a deployed environment. If a trial of an antihistamine does not produce any significant relief, then one of these medications may very well be an appropriate choice. It should be noted that, due to a fairly extensive list of potential negative side effects, medication compliance may be an issue. Examples include gastrointestinal distress, fatigue, hypotension, and cardiac arrhythmias. Also, amitriptyline should be used with caution in those service members with suicidal ideation, as it can be lethal in overdose (Stahl, 2006).

Nonbenzodiazepine Hypnotics   The most likely prescribed medications for the treatment of insomnia in service members are nonbenzodiazepine hypnotics such as zolpidem (Ambien®) and zaleplon (Sonata®, Jones). Ask any soldier, sailor, airmen, or Marine if they have heard of Ambien and most will answer affirmatively. These medications are fast acting, extremely sedating, and relatively effective. Many see these medications as ideal for the airplane ride to and from the combat environment, while serving in the combat environment, and after the return home. These medications are classified as controlled substances and their use should be monitored closely. Although there is a relatively low potential for abuse with these medications, physical and psychological dependence can develop, particularly in those with histories of substance abuse and dependence (Hajak, Muller, Wittchen, Pittrow, & Kirch, 2003). It should also be noted that when taking these medications the patient should be able to dedicate a sufficient amount of time (6 to 8 hours) to sleep, as a hangover effect may be present the next morning. For service members that are often on call and likely to be roused for a mission, these medications may not be the most prudent choice. These medications are also not without side effects. Zolpidem has been shown to cause dizziness, amnesia, excitability, and hallucinations (Stahl, 2006), and recently parasomnia behavior involving eating or driving while sleeping has been reported.

Other Medications Depending on the prescribing clinician and drug formulary, other medications may be used in the treatment of insomnia. Two examples are benzodiazepines and antipsychotics. Benzodiazepines such as triazolam (Halcion®, Pharmacia and Upjohn) and temazepam (Restoril®, Mallinckrodt Pharmaceuticals) are very sedating and have been shown to have both strengths and limitations with regard to treating insomnia (Stahl, 2006). As with all benzodiazepines, tolerance and dependence can be an issue. Although little research supports their use for insomnia, newer or atypical antipsychotics such as quetiapine (Seroquel®, AstraZeneca) and olanzapine (Zyprexa®, Eli Lilly) in low dosages are used in clinical practice. Specific to service members, quetiapine and olanzapine may be useful in alleviating nightmares and insomnia in patients suffering from combat-related PTSD (Jakovljevic, Sagud, & Mihaljevic-Peles, 2003; Robert et al., 2005). Unwanted side effects may include dizziness, constipation, dyspepsia, tachycardia, and orthostatic hypotension (Stahl, 2006).

Parasomnias

Parasomnias are sleep disorders in which odd and sometimes dangerous sensory and motor behavior occur during sleep. Often they are frightening to the individual as well as to those who share the bed or bedroom. The more common and relevant parasomnias experienced by service members are nightmares, REM behavior disorder, sleepwalking, and sleep terrors.

Nightmares

Nightmares are characterized by repeated awakenings during sleep (typically REM sleep) due to unusually vivid and disturbing dreams that consist of terrifying content, although some nightmare patients seem to develop a tendency toward less awakenings over time while still suffering from frequent disturbing dreams. Nightmares not only cause anxiety and fear but can also rouse various dysphoric emotions and may disrupt social, occupational, and other important areas of functioning (AASM, 2005; APA, 2000).

Nightmares are a common occurrence, and the lifetime incidence rate is most likely near 100% (Silber et al., 2004). Studies have shown that between 8 and 25% of adults report at least one nightmare per month (Belicki & Belicki, 1982, 1986; Feldman & Hersen, 1967; Levin, 1994; Wood & Bootzin, 1990), while 2 to 6% report at least one nightmare each week (Belicki & Cuddy, 1991; Feldman & Hersen, 1967; Levin, 1994).

Increased incidences of nightmares have also been found in those exposed to a wide range of traumatic experiences (Barrett, 1996; Lifton & Olsen, 1976; Low et al., 2003) and in clinical populations, most notably PTSD patients (Kilpatrick et al, 1998; Krakow et al., 2002b; Ross, Ball, Sullivan, & Caroff, 1989). For any practitioner involved in the rehabilitative care of service members, it is reasonable to assume that a significant portion of patients will present with complaints of nightmares, although the overwhelming majority of nightmare sufferers do not usually seek treatment for this condition nor do most such individuals imagine that treatment exists for the condition. Nonetheless, an understanding of effective and efficient treatment methods for nightmares is crucial for those willing to pursue therapy.

Psychological Therapy

Imagery Rehearsal Therapy   Imagery rehearsal therapy (IRT) is a two-component cognitive–behavioral individual or group treatment that views nightmares as a learned sleep disorder—similar to a learned behavior—and due in part to a damaged imagery system (Krakow & Zadra, 2006). Numerous controlled studies have shown IRT to be effective in reducing nightmare frequency and associated distress while maintaining long-term gains (Kellner, Neidhardt, Krakow, & Pathak, 1992; Krakow, Kellner, Neidhardt, Pathak, & Lambert, 1993; Krakow, Kellner, Pathak, & Lambert, 1996; Neidhardt, Krakow, Kellner, & Pathak, 1992). It has been shown to be effective with nightmares specific to PTSD (Kellner et al., 1992; Krakow, Kellner, Pathak, & Lambert, 1995; Krakow et al., 2001a–d; Neidhardt et al., 1992) and with veterans (Forbes, Phelps, & McHugh, 2001; Forbes et al., 2003). With regard to the latter, a recent case series revealed that IRT was even effective for acute nightmares in combat soldiers while serving in Iraq (Moore & Krakow, 2007). Without question, IRT is the most researched treatment for the management of nightmares.

Although there is some variation in how IRT is applied, particularly with regard to number of sessions, Krakow and Zadra (2006) provide a comprehensive description of IRT that includes four group treatment sessions of roughly 2 hours each. The first two sessions focus on how nightmares promote learned insomnia and take on a life of their own. The final two sessions focus on the human imagery system and the primary intervention of IRT, which is selecting a nightmare, altering the content of the nightmare to create a new dream, and rehearsing the new dream via imagery (or through writing if the patient is unable to fully access their visual imagery system). Table 15.2 provides an overview of the treatment content as originally described by Krakow and Zadra (2006) but adapted to an individual format for soldiers serving in a combat environment.

Other Psychological Therapies   Other treatments purported to be effective for nightmares include various adaptations of cognitive–behavioral therapy (Davis & Wright, 2007), psychodynamic psychotherapy (Gorton, 1988), hypnosis (Kennedy, 2002; Kingsbury, 1993), and systematic desensitization (Eccles, Wilde, & Marshall, 1988).

Psychopharmacological Therapy   The evidence for the use of medications to specifically target nightmares is growing at a significant rate. Much of this research is fueled by the need to address posttraumatic stress symptoms (of which nightmares are one) in service members returning from Iraq and Afghanistan. To date, the most promising of these medications is prazosin (Minipress®, Pfizer). Prazosin is an α1-adrenergic antagonist that is typically used for the treatment of hypertension. Studies have shown that prazosin can be effective for the treatment of nightmares and can positively impact other important sleep processes such as total and REM sleep time without the sedative effects. In a recent study by Taylor and colleagues (2007), prazosin was shown to effectively reduce nighttime PTSD symptoms and improve the overall quality of sleep in a civilian sample with PTSD. Specific to nightmares as a result of combat exposure, a recent series of studies has shown that prazosin can effectively reduce nightmares, sleep disturbances, and PTSD symptoms (Daly, Doyle, Radkind, Raskind, & Daniels, 2005; Raskind et al., 2003, 2007). As a result of these studies and outcomes in clinical practice, prazosin has become the first-line pharmacological treatment for nightmares for many military clinicians and is used in both garrison and deployed environments. Within the context of PTSD, other pharmacological agents have been found to be effective in reducing nightmares such as nefazodone (Gillin et al., 2001) and trazodone (Warner, Dorn, & Peabody, 2001). These authors are not aware of any evidence indicating that these medications totally eliminate the nightmares; in fact, the nightmares often return after cessation of medication treatment.

REM Behavior Disorder

The REM behavior disorder (RBD) is characterized by the loss of normal voluntary muscle paralysis during REM sleep and is associated with complex behavior in tandem with dreaming. An easy way to conceptualize RBD is by viewing it as a severe type of nightmare in which the individual acts out the distressing dream content. It is often violent and typically very frightening to the individual's bedmate.

Table 15.2 Imagery Rehearsal Therapy (IRT)

Session 1

Emphasize that IRT does not address past traumatic events or traumatic content of nightmares.

Provide education about nightmares, insomnia, and sleep hygiene.

Discuss treatment expectations and higher levels of care in a combat environment.

Discuss risks unique for soldiers with nightmares (e.g., safety, mission focus, PTSD).

Discuss differences among combat stress, acute stress reaction, and PTSD.

Session 2

Discuss why nightmares persist after combat stressor.

Discuss nightmares as a learned behavior and as a normal response.

Discuss the basic principles of imagery and how to apply imagery in a war zone.

Teach how to access personal imagery skills.

Practice personal imagery.

Help the patient learn about the potential for change from a nightmare-sufferer identity to a good-dreamer identity.

Session 3

Develop plan for regular use of IRT for nightmares.

Select a nightmare.

Change the nightmare to a new dream.

Rehearse the new dream.

Explain paths for follow-up care in the combat environment and at home.

The prevalence rate of RBD has been estimated to be approximately 0.5% in individuals between the ages of 15 and 100 (Ohayon et al., 1997) with a mean age of onset somewhere in the 50s or 60s (Olson, Boeve, & Silber, 2000; Schenck, Hurwitz, & Mahowald, 1993). It has also been estimated that around 50 to 60% of sufferers have some form of subsequent neurological deficits years after the initial RBD diagnosis (Schenck & Mahowald, 1996).

Although the average age of onset is considerably higher than the average age of the service member, RBD is relevant for this population due to the fact that service members are susceptible to brain injury on the battlefield. A recent study showed that parasomnias were present in about 25% of those suffering from chronic brain injury. Approximately one fourth of those parasomnias were RBD (Verma, Anand, & Verma, 2007); therefore, it is logical to posit that with the increased risk of brain injury comes an increased risk of RBD.

In addition, psychotropic medications such as SSRIs, tricyclics, and some tetracyclic antidepressants can carry the side effect of inducing RBD. This side effect may prove very difficult to distinguish in some PTSD patients, who may already be prone to violent nightmares with some excess motor activity in either REM or NREM sleep. Thus, a careful history evaluating the time frame prior to prescribing psychotropics in a PTSD patient with nightmares is crucial to being able to monitor whether the medication might be worsening parasomnia behavior.

With regard to management of RBD, restructuring the person's sleeping environment and educating his or her bedmates or roommates about the disorder are initially required to avoid harm to self or others. For the service member suffering from RBD, ensuring that access to any weapons is restricted is of utmost importance. Clonazepam (Klonopin®, Roche Laboratories) is the treatment of choice, usually in doses of 0.25 to 2.0 mg administered prior to bedtime. Obviously, for the service member dealing with RBD, any future deployment to a combat environment should be appropriately questioned.

Sleepwalking

Sleepwalking is an unconscious and involuntary process that typically involves the person leaving the bed and moving about his or her environment and performing tasks as if he or she were awake. It is believed that approximately 2% of adults sleepwalk (Ohayon, Guilleminault, & Priest, 1999). For the clinician working with the service member, it is imperative that the individual prone to sleepwalking be referred to a sleep specialist prior to deployment, particularly because recent evidence suggests higher than expected rates of sleep disordered breathing in adults with sleepwalking (Guilleminault et al., 2005). Most service members in combat zones have a weapon and ammunition within arm's reach 24 hours a day. This combination, plus the potential of wandering into a restricted area, can have dire consequences. With regard to management, as with RBD the first step is to create a safe environment for the service member (clear objects from pathways, secure weapons and ammunition in a locked cabinet, lock doors and windows). From a pharmacological standpoint, benzodiazepines or other sedatives may be warranted. Depending on severity, removal from the combat environment should be considered.

Sleep Disordered Breathing

Sleep disordered breathing (SDB) is a general term that refers to a group of disorders that cause disruptions in respiratory patterns and overall quantity of oxygen levels in the body. Studies supporting the ill effects of sleep breathing difficulties on health, particularly cardiac health, are overwhelming. Two common disorders of this group are snoring and sleep apnea.

Snoring is an extremely common event that can cause significant distress for the individual and those who share the same sleeping quarters. Relevant to service members, snoring may develop or be exacerbated by harsh environmental conditions during deployments and training exercises that can increase nasal congestion. If this is the case, treatment for the congestion via medications or the use of nasal dilators may be all that is needed.

Sleep apnea is simply defined as obstructive breathing during sleep—in its most severe form, as apnea (complete cessation) or in its most subtle form as upper airway resistance. This continuum of breathing disruption is now designated by the umbrella term *obstructive sleep apnea* (OSA), although *sleep disordered breathing* appears to be the more commonly used term in clinical practice. Studies suggest that approximately 2% of women and 4% of men over the age of 50 have some level of OSA (Strollo & Rogers, 1996), but these data only looked at patients with the classic apneas and hypopneas (roughly 50% of collapsibility) instead of the full spectrum of breathing disruption. Other studies suggest much higher rates.

In the past, service members with sleep apnea were often medically retired from service. This practice has changed considerably over the past few years as increased infrastructure in combat environments such as Iraq and Afghanistan have made it possible to use continuous positive airway pressure (CPAP) machines, the standard treatment. Due to harsh environmental conditions (sand, dirt, mud), however, maintaining optimal functioning of the machines can be a challenge.

Regardless of whether the service member is suffering from snoring, sleep apnea, or some other biological or physically based sleep disorder, referral for a sleep study should be considered. Most military medical centers will have a full-time board-certified sleep physician on staff or will maintain a contract with a provider within the community.

Circadian Rhythm Disorders

Circadian rhythm disorders (CRDs) occur when the body's natural sleep–wake cycle is disrupted and the individual is not able to sleep at the appropriate time; consequently, as a result of insomnia, excessive daytime sleepiness may develop. The most common CRDs are jet lag, delayed sleep phase disorder, and shift work sleep disorder. The latter is the most relevant to service members.

Shift Work Sleep Disorder   Shift work sleep disorder (SWSD) is a sleep disorder that affects people who work erratic work schedules, which wreaks havoc on the body's natural circadian rhythm (sleep–wake cycle). Extended periods of circadian sleep cycle disruption often results in insomnia or excessive sleepiness, which can result in decrements in awake time functioning. Depending on the military occupational specialty of the service member, varying degrees of risks to themselves or others may be an issue that must be addressed. Obviously, special attention should be given to occupational specialties such as aviator, combat infantryman, truck driver, or anyone who provides checkpoint security. Management of SWSD is primarily behavioral and may include decreasing the number of night shifts worked in a row, avoiding reliance on stimulants or learning to use them more judiciously, practicing proper sleep hygiene, getting adequate sleep on days off or when the shift is over, and creating a nonstimulating and sleep-conducive environment. In some cases, sedative medications may be necessary to regulate sleep patterns, and this approach is most likely to succeed when the medication is only used two or three times in a week or less.

Sleep Movement Disorders

Sleep movement disorders are characterized by excessive and uncomfortable movement during sleep or during wakefulness in ways that interfere with sleep. The two most common are restless legs syndrome (RLS), a waking condition, and periodic limb movement disorder (PLMD), a sleep disorder. The essential feature of RLS is an intense desire to move the limbs, typically the lower extremities. Patients often describe uncomfortable sensations such as a creeping feeling or a restless agitation that is only relieved by moving the affected limbs. The symptoms are usually most troublesome in bed as the person is trying to fall asleep or after waking during the night; consequently, insomnia due to sleep onset and maintenance problems is common (Silber et al., 2004). PLMD is characterized by behavior during sleep ranging from shallow, continual movement of the ankle or toes to wild and frantic kicking and flailing of the legs and arms. In addition, nasal, oral, and abdominal movement sometimes accompanies the disorder. As with RLS, movement of the legs is more typical than movement of the arms in PLMD. Insomnia can also be a compounding factor in PLMD. The most pertinent issue related to these disorders in service members is that certain antidepressant medications, particularly selective serotonin reuptake inhibitors (SSRIs), can cause or worsen PLMD and RLS symptoms (Agargun, 2002; Bakshi, 1996; Hargrave & Beckley, 1998; Perroud et al., 2007). Considering that SSRIs are a common treatment for PTSD, a thorough evaluation of when the PLMS and RLS symptoms started is crucial.

Hypersomnia

Hypersomnia, it might be argued, is something that every combat solider or active-duty personnel can readily attest to, except they only know it as severe sleepiness; they do not get to act on the sleepiness to experience more sleep. As previously described, necessary sleep deprivation is the norm in the military so someone with a condition such as narcolepsy would presumably never make it past basic training. Two conditions are of most pressing concern: (1) insufficient sleep syndrome (ISS) significant enough to impair performance, at which point the patient must be given the opportunity for more sleep or the excessive daytime sleepiness must be treated with stimulants (e.g., modafinil, caffeine) if deemed essential to survival, and (2) posttraumatic narcolepsy, which may arise following head trauma and is treated similarly to the standards for narcolepsy.

Conclusions

Knowledge regarding the causes, characteristics, effects, and management of sleep disorders in service members is crucial for the practitioner who works with service members. It is also imperative to be cognizant of the fact that many of these disorders coexist. As discussed earlier, service members may be at greater risk for developing particular sleep disturbances due to a variety of unique factors. Other disorders may not be common in service members, but they may have more dire consequences if left untreated. Whether the service member is preparing for a training exercise at home or overseas, readying for a deployment to a hostile environment, or returning home after a 12-month combat tour, effective treatments for sleep disorders using a medical, psychological, and pharmacological approach are available and likely to improve quality of life, enhance productivity and performance, and decrease accidents and errors.

References

AASS. (2005). *International classification of sleep disorders: Diagnostic and coding manual* (2nd ed.). Westchester, IL: American Academy of Sleep Medicine.

Agargun, M. Y., Kara, H., Ozbek, H., Tombul, T., & Ozer, O. A. (2002). Restless legs syndrome induced by mirtazapine. *Journal of Clinical Psychiatry, 63,* 1179.

APA. (2000). *Diagnostic and statistical manual of mental disorders* (4th ed., text revision). Washington, D.C.: American Psychiatric Association.

Bakshi, R. (1996). Fluoxetine and restless legs syndrome. *Journal of the Neurological Sciences, 142,* 151–152.

Barrett, D. (Ed.). (1996). *Trauma and dreams.* Cambridge, MA: Harvard University Press.

Belenky, G. L., Kreuger, G. P., Balking, T. J., Headley, D. B., & Solick, R. E. (1987). *Effect of continuous operations (CONOPS) on soldier and unit performance: Review of the literature and strategies for sustaining the soldier in CONOPS,* WRAIR Report BB-87-1. Washington, D.C.: Walter Reed Army Institute of Research.

Belicki, D., & Belicki, K. (1982). Nightmares in a university population. *Sleep Research, 11,* 116.

Belicki, K., & Belicki, D. (1986). Predisposition for nightmares: A study of hypnotic ability, vividness of imagery, and absorption. *Journal of Clinical Psychology, 42,* 714–718.

Belicki, K., & Cuddy, M. A. (1991). Nightmares: Facts, fictions and future directions. In J. Gackenbach & A. A. Sheikh (Eds.), *Dream images: A call to mental arms* (pp. 99–115). Amityville, NY: Baywood.

Bonnet, M. H. (1985). Recovery of performance during sleep following sleep deprivation in older normal and insomniac adult males. *Perceptual and Motor Skills, 60,* 323–334.

Bootzin, R. R., & Rider, S. P. (1997). Behavioral techniques and biofeedback for insomnia. In M. R. Pressman & W. C. Orr (Eds.), *Understanding sleep: The evaluation and treatment of sleep disorders* (pp. 315–338). Washington, D.C.: American Psychological Association.

Cavallero, C., Foulkes, D., Hollifield, M., & Terry, R. (1990). Memory sources of REM and NREM dreams. *Sleep, 13,* 449–455.

Clayton, N. M., & Nash, W. P. (2006). Medication management of combat and operational stress injuries in active duty service members. In C. R. Figley & W. P. Nash (Eds.), *Combat stress injury: Theory, research, and management* (pp. 219–238). New York: Routledge.

Daly, C. M., Doyle, M. E., Radkind, M., Raskind, E., and Daniels, C. (2005). Clinical case series: The use of prazosin for combat-related recurrent nightmares among Operation Iraqi Freedom combat veterans. *Military Medicine*, 170, 513–515.

Davis, J. L., & Wright, D. C. (2007). Randomized clinical trial for treatment of chronic nightmares in trauma-exposed adults. *Journal of Traumatic Stress*, 20, 123–133.

Durmer, J. S., & Dinges, D. F. (2005). Neurocognitive consequences of sleep deprivation. *Seminars in Neurology*, 25, 117–129.

Eccles, A., Wilde, A., & Marshall, W. L. (1988). *In vivo* desensitization in the treatment of recurrent nightmares. *Journal of Behavior Therapy and Experimental Psychiatry*, 19, 285–288.

Erbes, C., Westermeyer, J., & Engdahl, B. (2007). Posttraumatic stress disorder and service utilization in a sample of service members from Iraq and Afghanistan. *Military Medicine*, 172, 359–363.

Feldman, M. J., & Hersen, M. (1967). Attitudes toward death in nightmare subjects. *Journal of Abnormal Psychology*, 72, 421–425.

Forbes, D., Phelps, A. J., & McHugh, A. F. (2001). Treatment of combat-related nightmares using imagery rehearsal: A pilot study. *Journal of Traumatic Stress*, 14, 433–442.

Forbes, D., Phelps, A. J., McHugh, A. F., Debenham, P., Hopwood, M., & Creamer, M. (2003). Imagery rehearsal in the treatment of posttraumatic nightmares in Australian veterans with chronic combat-related PTSD: 12-month follow-up data. *Journal of Traumatic Stress*, 16, 509–513.

Gillin, J. C., Smith-Vaniz, A., Schnierow, B., Rapaport, M. H., Kelsoe, J., Raimo, E., Marler, M. R., Goyette, L. M., Stein, M. B., & Zisook, S. (2001). An open-label, 12-week clinical and sleep EEG study of nefazodone in chronic combat-related posttraumatic stress disorder. *Journal of Clinical Psychiatry*, 62, 789–796.

Gimenez, S., Clos, S., Romero, S., Grasa, E., Morte, A., & Barbanoj, M. J. (2007). Effects of olanzapine, risperidone and haloperidol on sleep after a single oral morning dose in health volunteers. *Psychopharmacology*, 190, 507–516.

Gorton, G. E. (1988). Life-long nightmares: An eclectic treatment approach. *American Journal of Psychotherapy*, 42, 610–618.

Grossman, D., & Christensen, L. W. (2004). *On combat: The psychology and physiology of deadly conflict in war and in peace*. St. Louis, MO: PPCT Research Publications.

Guilleminault, C., Kirisoglu, C., Bao, G., Arias, V., Chan, A., & Li, K. K. (2005). Adult chronic sleepwalking and its treatment based on polysomnography. *Brain*, 128, 1062–1069.

Hajak, G., Muller, W. E., Wittchen, H. U., Pittrow, D., & Kirch, W. (2003). Abuse and dependence potential for the non-benzodiazepine hypnotics zolpidem and zopiclone: A review of case reports and epidemiological data. *Addiction*, 98, 1371–1378.

Hargrave, R., & Beckley, D. J. (1998). Restless legs syndrome exacerbated by sertraline. *Psychosomatics*, 39, 177–178.

Harvey, A. G., Sharpley, A. L., Ree, M. J., Stinson, K., & Clark, D. M. (2007). An open trial of cognitive therapy for chronic insomnia. *Behaviour Research and Therapy*, 45, 2491–2501.

Hoge, C. W., Castro, C. A., Messer, S. C., McGurk, D., Cotting, D. I., & Koffman, R. L. (2004). Combat duty in Iraq and Afghanistan, mental health problems, and barriers to care. *New England Journal of Medicine, 351*, 13–22.

Jacobs, G. D., Pace-Schott, E. F., Stickgold, R., & Otto, M. W. (2004). Cognitive behavior therapy and pharmacotherapy for insomnia: A randomized controlled trial and direct comparison. *Archives of Internal Medicine, 164*, 1888–1896.

Jacobsen, F. M. (1990). Low-dose trazodone as a hypnotic in patients treated with MAOIs and other psychotropics: A pilot study. *Journal of Clinical Psychiatry, 51*, 298–302.

Jakovljevic, M., Sagud, M., & Mihaljevic-Peles, A. (2003). Olanzapine in the treatment-resistant, combat-related PTSD—a series of case reports. *Acta Psychiatrica Scandinavica, 107*, 394–396.

Kellner, R., Neidhardt, J., Krakow, B., & Pathak, D. (1992). Changes in chronic nightmares after one session of desensitization or rehearsal instructions. *American Journal of Psychiatry, 149*, 659–663.

Kennedy, G. A. (2002). A review of hypnosis in the treatment of parasomnias: Nightmare, sleepwalking, and sleep terror disorders. *Australian Journal of Clinical and Experimental Hypnosis, 30*, 99–155.

Kilpatrick, D. G., Resnick, H. S., Freedy, J. R., Pelcovitz, D., Resick, P., Roth, S. et al. (1998). Post-traumatic stress disorder field trial: Evaluation of the PTSD construct— criteria A though E. In T. A. Widiger, A. J. Frances, H. A. Pincus, R. Ross, M. B. First, W. Davis, & M. Kline (Eds.), *DSM-IV sourcebook* (4th ed., Vol. 4, pp. 803–844). Washington, D.C.: APA Press.

Kingsbury, S. J. (1993). Brief hypnotic treatment of repetitive nightmares. *American Journal of Clinical Hypnosis, 35*, 161–169.

Krakow, B. (2007). *Sound sleep, sound mind: 7 keys to sleeping through the night.* Hoboken, NJ: John Wiley & Sons.

Krakow, B., & Zadra, A. (2006). Clinical management of chronic nightmares: Imagery rehearsal therapy. *Behavioral Sleep Medicine, 4*, 45–70.

Krakow, B., Kellner, R., Neidhardt, J., Pathak, D., & Lambert, L. (1993). Imagery rehearsal treatment of chronic nightmares with a thirty month follow-up. *Journal of Behavior Therapy and Experimental Psychiatry, 24*, 325–330.

Krakow, B., Kellner, R., Pathak, D., & Lambert, L. (1995). Imagery rehearsal treatment for chronic nightmares. *Behaviour Research and Therapy, 33*, 837–843.

Krakow, B., Kellner, R., Pathak, D., & Lambert, L. (1996). Long term reduction of nightmares with imagery rehearsal treatment. *Behavioural and Cognitive Psychotherapy, 24*, 135–148.

Krakow, B., Germain, A.,Warner, T. D., Schrader, R., Koss, M., Hollifield, M. et al. (2001a). The relationship of sleep quality and posttraumatic stress to potential sleep disorders in sexual assault survivors with nightmares, insomnia, and PTSD. *Journal of Traumatic Stress, 14*, 647–665.

Krakow, B., Hollifield, M., Johnston, L., Koss, M., Schrader, R., Warner, T. D. et al. (2001b). Imagery rehearsal therapy for chronic nightmares in sexual assault survivors with posttraumatic stress disorder: A randomized controlled trial. *Journal of the American Medical Association, 286*, 537–545.

Krakow, B., Johnston, L., Melendrez, D., Hollifield, M., Warner, T. D., Chavez-Kennedy, D. et al. (2001c). An open-label trial of evidence-based cognitive behavior therapy for nightmares and insomnia in crime victims with PTSD. *American Journal of Psychiatry, 158*, 2043–2047.

Krakow, B., Sandoval, D., Schrader, R., Kuehne, B., McBride, L., Yau, C. L. et al. (2001d). Treatment of chronic nightmares in adjudicated adolescent girls in a residential facility. *Journal of Adolescent Health*, 29, 94–100.

Krakow, B., Melendrez, D. C., Johnston, L. G., Clark, J. O., Santana, E. M., Warner, T. D., Hollifield, M. A., Schrader, R., Sisley, B. N., & Lee, S. A. (2002a). Sleep dynamic therapy for Cerro Grande Fire evacuees with posttraumatic stress symptoms: A preliminary report. *Journal of Clinical Psychiatry*, 63, 673–684.

Krakow, B., Schrader, R., Tandberg, D., Hollifield, M., Koss, M. P., Yau, C. L. et al. (2002b). Nightmare frequency in sexual assault survivors with PTSD. *Journal of Anxiety Disorders*, 16, 175–190.

Kudo, Y. and Kurihara, M. (1990). Clinical evaluation of diphenhydramine hydrochloride for the treatment of insomnia in psychiatric patients: A double-blind study. *Journal of Clinical Pharmacology*, 30, 1041–1048.

Landolt, H. P., & Gillin, J. C. (2001). Sleep abnormalities during abstinence in alcohol-dependent patients: Aetiology and management. *CNS Drugs*, 15, 413–425.

Levin, R. (1994). Sleep and dreaming characteristics of frequent nightmare subjects in a university population. *Dreaming*, 4, 127–137.

Lick, J. R., & Heffler, D. (1977). Relaxation training and attention placebo in the treatment of severe insomnia. *Journal of Consulting and Clinical Psychology*, 45, 153–161.

Lifton, J. R., & Olsen, E. (1976). The human meaning of total disaster: The Buffalo Creek experience. *Psychiatry*, 39, 1–18.

Low, J. F., Dyster-Aas, J.,Willebrand, M., Kildal, M., Gerdin, B., & Ekselius, L. (2003). Chronic nightmares after severe burns: Risk factors and implications for treatment. *Journal of Burn Care and Rehabilitation*, 24, 260–267.

Lundh, L.-G. (1998). Cognitive–behavioural analysis and treatment of insomnia. *Scandinavian Journal of Behaviour Therapy*, 27, 10–29.

Moore, B. A., & Krakow, B. (2007). Imagery rehearsal therapy for acute posttraumatic nightmares among combat soldiers in Iraq. *American Journal of Psychiatry*, 164, 683–684.

Morgan, P. T., Pace-Schott, E. F., Sahul, Z. H., Coric, V., Stickgold, R., & Malison, R. T. (2006). Sleep, sleep-dependent procedural learning and vigilance in chronic cocaine users: Evidence for occult insomnia. *Drug and Alcohol Dependence*, 82, 238–249.

Morin, C. M. (2003). Treating insomnia with behavioral approaches: Evidence for efficacy, effectiveness, and practicality. In M. P. Szuba, J. D. Kloss, & D. F. Dingess (Eds.), *Insomnia: Principles and management* (pp. 83–95). New York: Cambridge University Press.

Morin, C. M. (2006). Cognitive–behavioral therapy of insomnia. *Sleep Medicine Clinics*, 1, 375–386.

Murtagh, D. R., & Greenwood, K. M. (1995). Identifying effective psychological treatments for insomnia: A meta-analysis. *Journal of Consulting and Clinical Psychology*, 63, 79–89.

Nash, W. P. (2006). The stressors of war. In C. R. Figley & W. P. Nash (Eds.), *Combat stress injury: Theory, research, and management* (pp. 11–32). New York: Routledge.

Neidhardt, E. J., Krakow, B., Kellner, R., & Pathak, D. (1992). The beneficial effects of one treatment session and recording of nightmares on chronic nightmare sufferers. *Sleep*, 15, 470–473.

NIH. (2005). *NIH State-of-the-science conference statement on manifestations and management of chronic insomnia in adults*. Bethesda, MD: National Institutes of Health.

Ohayon, M. M., Guilleminault, C., Paiva, T., Priest, R. G., Rapoport, D. M., Sagales, T., Smirne, S., & Zulley, J. (1997). An international study on sleep disorders in the general population: Methodological aspects of the use of the Sleep-EVAL system. *Sleep*, 20, 1086–1092.

Ohayon, M. M., Guilleminault, C., & Priest, R. G. (1999). Night terrors, sleepwalking, and confusional arousals in the general population: Their frequency and relationship to other sleep and mental disorders. *Journal of Clinical Psychiatry*, 60, 268–276.

Olson, E. J., Boeve, B. F., & Silber, M. H. (2000). Rapid eye movement sleep behavior disorder: Demographic, clinical and laboratory findings in 93 cases. *Brain*, 123, 331–339.

Perroud, N., Coralie, L., Baleydier, B., Andrei, C., Maris, S., & Damsa, C. (2007). Restless legs syndrome induced by citalopram: A psychiatric emergency? *General Hospital Psychiatry*, 29, 72–74.

Ramakrishnan, K., & Scheid, D. C. (2007). Treatment options for insomnia. *American Family Physician*, 76, 517–526.

Raskind, M. A., Peskind, E. R., Kanter, E. D., Petrie, E. C., Radant, A., Thompson, C. E., Dobie, D. J., Hoff, D., Rein, R. J., Straits-Troster, K., Thomas, R. G., & McFall, M. M. (2003). Reduction of nightmares and other PTSD symptoms in combat veterans by prazosin: A placebo-controlled study. *American Journal of Psychiatry*, 160, 371–373.

Raskind, M. A., Peskind, E. R., Hoff, D. J., Hart, K. L., Holmes, H. A., Warren, D., Shofer, J., O'Connell, J., Taylor, F., Gross, C., Rohde, K., & McFall, M. E. (2007). A parallel group placebo controlled study of prazosin for trauma nightmares and sleep disturbance in combat veterans with post-traumatic stress disorder. *Biological Psychiatry*, 61, 928–934.

Richardson, G. S., Roehrs, T. A., Rosenthal, L., Koshorek, G., & Roth, T. (2002). Tolerance to daytime sedative effects of H1 antihistamines. *Journal of Clinical Psychopharmacology*, 22, 511–515.

Riddle, J. R., Smith, T. C., Smith, B., Corbeil, T. E., Engel, C. C., Wells, T. S., Hoge, C. W., Adkins, J., Zamorski, M., & Blazer, D. (2007). Millennium cohort: The 2001–2003 baseline prevalence of mental disorders in the U.S. military. *Journal of Clinical Epidemiology*, 60, 192–201.

Robert, S., Hammer, M. B., Kose, S., Ulmer, H. G., Deitsch, S. E., & Lorberbaum, J. P. (2005). Quetiapine improves sleep disturbances in combat veterans with PTSD: Sleep data from a prospective, open-label study. *Journal of Clinical Psychopharmacology*, 25, 387–388.

Ross, R. J., Ball, W. A., Sullivan, K. A., & Caroff, S. N. (1989). Sleep disturbance as the hallmark of posttraumatic stress disorder. *American Journal of Psychiatry*, 146, 697–707.

Roth, T., & Ancoli-Israel, S. (1999). Daytime consequences and correlates of insomnia in the United States: Results of the 1991 National Sleep Foundation Survey. II. *Sleep*, 22, 354–358.

Schenck, C. H., & Mahowald, M. W. (1996). REM sleep parasomnias. *Neurology Clinics*, 14, 697–720.

Schenck, C. H., Hurwitz, T. D., & Mahowald, M. W. (1993). REM sleep behaviour disorder: An update on a series of 96 patients and a review of the world literature. *Journal of Sleep Research*, 2, 224–231.

Silber, M. H., Krahn, L. E., & Morgenthaler, T. I. (2004). *Sleep medicine in clinical practice*. New York: Taylor & Francis.

Sivertsen, B., Omvik, S., Pallesen, S., Bjorvatn, B., Havik, O. E., Kvale, G., Nielsen, G. H., & Nordhus, I. H. (2006). Cognitive behavioral therapy vs. zopiclone for treatment of chronic insomnia in older adults: A randomized controlled trial. *Journal of the American Medical Association*, 295, 2851–2858.

Soldatos, C. R., Allaert, F. A., Ohta, T., & Dikeos, D. G. (2005). How do individuals sleep around the world? Results from a single-day survey in ten countries. *Sleep Medicine*, 6, 5–13.

Somer, E., Tamir, E., Maguen, S., & Litz, B. T. (2005). Brief cognitive–behavioral phone-based intervention targeting anxiety about the threat of attack: A pilot study. *Behaviour Research and Therapy*, 43, 669–679.

Spielman, A. J., Saskin, P., and Thorpy, M. J. (1987). Treatment of chronic insomnia by restriction of time in bed. *Sleep*, 10, 45–56.

Stahl, S. M. (2006). *Essential psychopharmacology: The prescriber's guide*. New York: Cambridge University Press.

Stickgold, R. (2005). Sleep-dependent memory consolidation. *Nature*, 437, 1272–1278.

Stickgold, R., & Walker, M. P. (2005). Sleep and memory: The ongoing debate. *Sleep*, 28, 1225–1227.

Strollo, P. J., & Rogers, R. M. (1996). Obstructive sleep apnea. *New England Journal of Medicine*, 334, 99–104.

Taylor, F. B., Martin, P., Thompson, C., Williams, J., Mellman, T. A., Gross, C., Peskind, E. R., & Raskind, M. A. (2007). Prazosin effects on objective sleep measures and clinical symptoms in civilian trauma posttraumatic stress disorder: A placebo-controlled study. *Biological Psychiatry*, 63(6), 629–632.

Toth, L. A. and Opp, M. R. (2002). Infection and sleep. In T. Lee-Chiong, M. A., Carskadon, & M. Sateia (Eds.), *Sleep medicine* (pp. 77–84). Philadelphia, PA: Hanley & Belfus.

Verma, A., Anand, V., & Verma, N. P. (2007). Sleep disorders in chronic traumatic brain injury. *Journal of Clinical Sleep Medicine*, 15, 357–362.

Walker, M. P., Brakefield, T., Hobson, J. A., & Stickgold, R. (2003). Dissociable stages of human memory consolidation and reconsolidation. *Nature*, 425, 616–620.

Warner, M. D., Dorn, M. R., & Peabody, C. A. (2001). Survey on the usefulness of trazodone in patients with PTSD with insomnia or nightmares. *Pharmacopsychiatry*, 34, 128–131.

Wheatley, D. (1984). Trazodone: Alternative dose regimens and sleep. *Pharmatherapeutica*, 3, 607–612.

Winokur, A., Gary, K. A., Rodner, S., Rae-Red, C., Fernando, A. T., & Szuba, M. P. (2001). Depression, sleep physiology, and antidepressant drugs. *Depression and anxiety*, 14, 19–28.

Wood, J. M., & Bootzin, R. R. (1990). The prevalence of nightmares and their independence from anxiety. *Journal of Abnormal Psychology*, 99, 64–68.

Yules, R. B., Lippman, M. E., & Freedman, D. X. (1967). Alcohol administration prior to sleep: The effect on EEG sleep stages. *Archives of General Psychiatry*, 16, 94–97.

16
After the Battle:
Violence and the Warrior*

BRET A. MOORE, C. ALAN HOPEWELL,
and DAVE GROSSMAN

Contents

Combat changes a person. This seems like an obvious proposition, but it is a proposition that is not fully appreciated by the average citizen. The only profession that explicitly trains its employees to harm, disable, and destroy another human being is the professional warrior—the soldier, sailor, airman, and Marine. More directly put, the primary objective of the combat troop is to kill. The primary objective of the combat support troop is to make sure the combat troop can complete its objective. For those who have never worked with a professional warrior, this is a concept that can be somewhat unsettling. Appreciating the cognitive and emotional sequelae that occur from struggling with the social, moral, and religious implications of taking another's life based on a job description is hardly an easy accomplishment. But, for the

---

* Much of the content of this chapter is based on the highly acclaimed book *On Combat: The Psychology and Physiology of Deadly Conflict in War and in Peace*, by Lt. Col. (Ret.) Dave Grossman and Loren W. Christensen.

nonmilitary practitioner, it is important at least to appreciate the warrior's beginnings to understand where he* is headed.

Somewhat limited in depth, the psychological literature does provide examples of increased risk of violence perpetration among some, but certainly not all, combat veterans. The risk of domestic violence is particularly high in this population (Taft et al., 2005, 2007); however, most studies have looked at this issue within the context of posttraumatic stress disorder (PTSD) and Vietnam era veterans. Although certainly not as common as partner abuse, murder committed by veterans is not unheard of, particularly in the Special Forces (SF) community. Several cases have gained considerable attention in the media over the past few decades. One of the most extreme examples is that of Captain Jeff MacDonald, a U.S. Army physician and Green Beret who was convicted of killing his family in Ft. Bragg, North Carolina, in the early 1970s. It should be remembered that MacDonald, as a physician, had none of the military training of the Green Beret unit to which he was assigned, and he had never been to Vietnam. The most plausible explanation for his killings was dissatisfaction with his family and concomitant substance use; no one ever theorized that the killings had anything to do with MacDonald's military experience, which was negligible.

Even more disturbing was the series of killings in 2002 at Ft. Bragg when the wives of four Army soldiers were killed within a matter of 6 weeks. Three of these soldiers had recently returned from tours in Afghanistan. These events brought headlines pointing to such causal factors as antimalarial drugs, combat stress, and partner infidelity. The apparent rise in suicide rates, spiking in 2007, also raises the question of increased self-homicide by means of suicide (from the Latin *sui* for "self" and *cidium* for "a killing").

Although the adaptability and necessity of aggression in the combat environment are discussed in greater detail later, the point that aggression is an integral part of the warrior armamentarium as much as the rifle, ammunition, and the armored vest cannot be stated enough. Clinicians who treat service members usually ask: What happens when the warrior returns home from the battlefield? How does he compartmentalize this fundamental aspect of the warrior mentality? How does one go from a trained killer on the battlefield to an ordinary Joe in the civilian world? How can I prevent another headline?

In the following pages, it is our goal to illuminate the difficulties that may arise once the warrior returns from the battlefield. Specifically, we address how a soldier learns to manage the aggression, hostility, and, yes, even the desire to kill, which is considered to be adaptive in the combat environment. We discuss what it means to be a warrior and highlight some of the characteristics unique

---

* For consistency and due to the fact that combat infantrymen are males, the authors use the masculine form when describing the service member. It is not meant to take away from the courageous sacrifice made by women in combat support roles often found on the frontlines.

to these individuals. We delve into the mentality of the warrior and how hostility and aggression are key components of his psyche. We discuss the myths and misperceptions related to veterans and violence, and, finally, we review factors that likely contribute to violence and aggression in the warrior after his return and ways to protect the service member and those around him.

Who Are These Men?

The third author (Grossman) introduced the concept of the *universal human phobia* in a series of papers he presented at the annual conventions of the American Psychological Association, American Psychiatric Association, and the International Congress of Critical Incident Stress Management. The concept is by no means a novel one, but it does attach a new name to something that is generally well known, nor is it truly universal, as it probably affects around 98% of the population. This universal human phobia is *interpersonal human aggression*.

In their book *On Combat*, Grossman and Christensen (2007) provide the example of a stranger walking into a crowded room and emptying a pistol into a random person. Up to 98% of the average audience would experience an extreme phobic reaction. Due to the fight instinct hardwired into our evolutionary make-up and the desperate nature of the situation, some may actually defend themselves against the attacker and become the aggressor. A few brave people may even risk their lives to care for the wounded, but most of the room would flee from the gunman and seek safety. In spite of the fear and the instinctual disposition to move away from the threat, and when every other sane, rational creature does what every ounce of their physical and psychological being tells them to, the warrior almost always moves toward the universal phobia. More accurately, he rushes toward it.

Today's soldiers, sailors, airmen, and Marines are the knights of old. Everyday they are asked to don their armor, secure their weapons, and purposefully confront danger to protect those they have sworn to protect. They are paladins in the most literal and contemporary sense. These men are typically not products of royalty or privilege, but represent the heartland of a country and the traditionalism of a nation. They are patriotic and loyal. They are honest and hardworking. More importantly, they are proud, self-sacrificing, and voluntary.

In the same work, the authors wrote about an anonymous military leader and his leader's thoughts about his men after watching them engage in uncommon acts of valor (Grossman & Christensen, 2007, p. xxii):

Dear God, where do we get such men? What loving God has provided, that each generation, afresh, there should arise new giants in the land. Were we to go but a single generation without such men, we should surely be both damned and doomed.

It is difficult to argue his point. If we were somehow forced to endure a single generation without the men who have so honorably agreed to be the point of the spear in defending a nation, we would certainly be both damned and doomed. Although difficult and costly, we could go for a generation without doctors, teachers, ministers, and, yes, even the guy that makes that perfect nonfat, sugar-free vanilla latte at our favorite coffee house, but we would still survive. One generation without warriors to defend our collective welfare against aggression would be catastrophic.

Although it is debatable, warriors are not born but rather are developed. They are built, trained, and nurtured. On the first day of basic training (or boot camp) these new professionals immerse themselves in a culture with a history and a tradition like no other. They learn to live by creeds and are taught about concepts such as honor, duty, courage, and sacrifice. These are not just words but real concepts that have genuine meanings and implications far beyond what the average citizen will ever know. They are trained in warrior tasks such as hand-to-hand combat, weapons proficiency, and battlefield medical care. They are also taught that aggression and the ability to kill without hesitation are part and parcel of being a warrior. The soldier who embraces these concepts and learns to control his impulses and behavior and to direct them against the enemy when called upon will be the ultimate warrior; in other words, controlled and deliberate aggression is key to being successful in this profession.

The Warrior Mentality

In *On Combat*, Grossman and Christensen (2007) discuss two types of people found on the battlefield: warriors and sheep. Sheep avoid the battle or refuse to participate, but warriors seem to have two basic attitudes as they go into combat. One group appears to look forward to it. The other group does not really want to do it, but since it has to be done their attitude is one of biting the bullet and getting the job done. Both are healthy responses. The following is an e-mail exchange with a reporter embedded with U.S. forces preparing to invade Iraq in 2003 (Grossman & Christensen, 2007, p. 139):

> I know that a lot of what I'm hearing is bravado. You often hear things, like "I just want to get in there, get it over with and get the job done" or "It's just part of the job," both of which indicate a more detached view. How does one explain these two attitudes from a psychological point of view? Do you really buy it when you read about soldiers who say they want to go to war? What is driving these men? And, also, how does one account for the more detached attitude?

The reply was that a sizable number of warriors really do want to see combat. Some of this might be mindless bravado, but some of it is not. These warriors have trained for the "big fight." For many, anything less than a full-out brawl will be a letdown. What fuels this desire? Training, training, and more training.

As mentioned earlier, warriors are groomed, nurtured, and developed. They are taught that aggression and killing are acceptable, but contextual. They are taught that their job is to protect those who are innocent and to annihilate those who pose a threat. First, one deters and then one stops the threat. The most effective way to stop someone is to kill them. But, as mentioned earlier, this is contextual. There are rules of engagement in which deadly force can only be used under specific circumstances. When it is done right, as he has been trained, the threat may die, a possibility the warrior must accept and embrace.

For the warrior, accepting the need to kill is protective. In the chaos and physiological intensity that unfolds in any firefight, the warrior will not respond with panic. He will slow his breathing, scan his environment, and engage his target. What gets people killed is the individual who cannot control his emotions and who debates the moral and religious implications of taking another's life while bullets fly over his head. The correct response is this: "I think I'm going to have to kill this guy. I knew it might come to this some day. This is what I trained for."

In his Pulitzer Prize-nominated book, *On Killing: The Psychological Cost of Learning to Kill in War and Society*, Grossman (1995) discusses a set of stages that one goes through when taking a life. First is the exhilaration stage in which there might be joy. The common psychological term for this is "survivor euphoria" and combat survivors know that there can be an intense relief and satisfaction that comes from killing your opponent and knowing that you will live another day. Then there is an overwhelming rush of remorse and guilt. Sometimes they say, "I just killed a man, and I enjoyed it. What's wrong with me? Is this normal?" Finally, there is a lifelong process of rationalization. If this process fails, it can be one of the paths to posttraumatic stress disorder (PTSD) or, at a minimum, a path to lifelong doubt, guilt, and ruminations about the event. Unfortunately, many of these men will never come to grips with taking another's life.

There is nothing wrong with those who are not disturbed by the killing or who may even derive joy from the act during combat. Let me say it again: There is nothing wrong with the men who find peace and satisfaction in killing on the battlefield. *On Combat* advances the idea that believing that a person will be irreparably damaged from a mental and psychological standpoint by the act of killing during combat is primarily a modern-day cognitive and emotional concept. Based on personal interviews with hundreds that have killed, Grossman (1995, p. 170) posits

If you tell yourself that killing will be an earth-shattering, traumatic event, then it probably will be. But if you prepare yourself mentally and can rationalize and accept that killing is lawful and justified during combat then using deadly force does not have to be a psychologically damaging event.

Again, there is nothing wrong with those who find killing distasteful and unsettling, and such individuals deserve compassion and support. But, there is absolutely nothing wrong with those who are not troubled or disturbed by it. Combat kills enough. It is senseless to let combat experience ruin a warrior's life, especially if he could have prevented it through prior preparation. The key is mental preparation, to have the warrior spirit and the Kevlar mind before stepping on to the battlefield. This is the warrior.

Violence and the Warrior: Myths and Misperceptions

Is a new generation of crazed, suicidal, and otherwise dysfunctional veterans about to be unleashed on an unsuspecting homefront population? The answer is yes—but only if you believe a recent front-page *The New York Times* story. According to the paper, tens of thousands of vets are returning from Iraq "with serious mental-health problems brought on by the stress and carnage of war." The number of soldiers eventually requiring treatment for posttraumatic stress disorder or the like, says *The Times*, could top 100,000. If that conjures up the image of the Vietnam vet—unable to cope with life and threatening either to kill himself or to go postal on innocent folks—well, it's probably meant to.

Such is the beginning of a *New York Post* editorial that ran in 2004 (Anon., 2004). Only a little more than a year into Operation Iraqi Freedom (OIF), a plethora of articles and publications began to emphasize the numbers of soldiers who supposedly would return from Iraq and Afghanistan psychologically scarred from their service in combat. A January 13, 2008, article in *The New York Times* was subsequently widely circulated after it claimed that 121 OIF and Operation Enduring Freedom (OEF) veterans had been charged with murder after returning from combat; however, the article was widely criticized as it failed to give any professional analysis of such claims. This included the fact that adjusted for murders per 100,000, killings in the District of Columbia itself accounted for at least twice the number of deaths as the veteran figures, and this was after a decline of murders in the District which had spanned at least 15 years. Indeed, cities in which the yearly murder rate probably exceeds that of OIF veterans include Detroit, Baltimore, New Orleans, Newark, St. Louis, and Oakland, along with the District of Columbia (Wikipedia, 2008).

Soon added to accounts of those needing treatment were the numbers of veterans who experienced concussions, often dramatically being described as "brain damaged," and those with amputations and other disabling injuries that undoubtedly take a psychological toll. Many of these claims were unfounded, however, and exaggerated the real problems that did exist. Articles in *USA Today*, for example, quoted data from Walter Reed Army Medical Center, which seemed to imply that over 60% of returning veterans would be brain damaged, misrepresenting data that were from a highly specialized medical unit that accepted only those patients most at risk for such injuries (Zoroya, 2005). *USA Today* also claimed at one point that more "brain injured" soldiers were to be found at Ft. Hood, Texas, than had been medically evacuated out of the entire OIF theater of operations (Zoroya, 2007). This "fact" was quite surprising to the second author of this chapter (Hopewell), who, as Co-Director of the Ft. Hood Brain Injury team, was in charge of monitoring such injuries at Ft. Hood and who found the claims of *USA Today* to be quite exaggerated. The initial findings of Hopewell and Christopher (2007) that fewer concussions were occurring among the armed forces than were often reported in the media and that the majority of those with concussions showed cognitive improvement and returned to duty have now been confirmed in a survey of 2525 infantry soldiers after their return from a year-long deployment to Iraq (Hoge et al., 2008).

Such embellishments by the media present two generally clear conclusions: Hundreds of thousands of veterans (extrapolating from *The Times* story) would return mentally unstable from the Global War on Terror and would show the potential for violence once back in society, and these veterans would require substantial mental health treatment. The overall theme is an extension of that repeated during the Vietnam conflict—that otherwise normal and well-adjusted American youth are trained by the military to kill and are then sent into the trauma of combat which leaves them scarred for life. Mentally crippled, they are then unable to adjust upon their return home and are unusually prone to violence, substance abuse, and mental instability.

The *New York Post* editorial continued, however: "Don't get us wrong here: Wars—all wars—take a psychological toll on those who fight them. That's been true throughout history. ...We don't mean to belittle the psychological trauma that war can—and, sadly, does—produce. But the myth of the dysfunctional vet that began with Vietnam has been created and spread, in large measure, by groups bitterly opposed to all U.S. military action. ...This latest attempt at myth making needs to be challenged and discredited before it becomes, once again, received wisdom."

Such Vietnam myths were finally and thoroughly discredited by the exhaustive work done by Burkett and Whitley (1998), who showed that as a group Vietnam veterans did not show nearly the problems that one would believe by reading newspapers or watching television or film. Indeed, people began to grow suspicious of the Vietnam PTSD figures as the numbers of veterans diagnosed and compensated for PTSD grew to more than double the number of total service members known to be involved in actual combat operations in Vietnam. The numbers of patients with PTSD from Vietnam are now acknowledged to have been inflated about threefold, with the number of severe cases at 18.7% after American withdrawal and 9.1% 12 years after the communist North invaded and ended the war. This was in sharp contrast to the estimate of 30.9% which had been reported for years (Dohrenwend et al., 2006).

Indeed, Burkett and Whitley (1998) showed that the vast majority of veterans were as well adjusted or even more successful than their nonserving civilian peers, and estimates of violence and incarceration of these veterans did not seem to differ from their nonserving peers. Yet, no one can dispute that some veterans show signs and symptoms of mental scarring as casualties of war, and for some these scars eventually erupt into their lives and the lives of their family members as violent action, violence that appears to have followed the veteran home from the field of combat and explodes as an unwelcome guest when the veteran should, in theory, be at rest from violence.

The problem here is delineation of a truthful estimate of the amount, severity, and nature of violence that may occur among combat veterans and devising an accurate way to identify, treat, and predict such trends without embellishment for political gain, unwarranted mitigation to avoid personal responsibility, or the disparagement of true heroes. This chapter, therefore, reviews the factors that contribute to violence, how this may occur in a veteran population, how to avoid the excesses of the "Whacko Vet" myths, and how to identify and intervene effectively in such disorders.

Risk Factors

As has been noted, many factors contribute to the fact that the majority of warriors return from the battlefield, resume normal lives, adjust, and often succeed better than their nonserving peers and are never unusually violent. Despite the most rigorous training and monitoring in the world, things still can and do go wrong, and some violent acts will be committed by a minority of veterans after leaving the battlefield and returning to civilian life. In addition to the "habit of violence," the acute and sustained stress of combat and the potential neuropsychological changes resulting from brain injury can change the nervous system and possibly the mind and behavior of the warrior.

Impact of Sustained Stress

Early in the 20th century, one of the most influential of all psychologists, Clark Hull, believed that human behavior is a result of the constant interaction between the organism and its environment; therefore, even with substantial training and overlearning, stimulation occurring in the environment can trigger individuals to react in ways they normally would not, prodding them at times to violence. Hull noted "when survival is in jeopardy, the organism is in a state of need (when the biological requirements for survival are not being met) so the organism behaves in a fashion to reduce that need" (Schultz & Schultz, 1987, p. 238). Simply, the organism behaves in such a way that reinforces the optimal biological conditions that are required for survival. A number of classic Hull experiments were even designed mathematically to predict the very point at which an organism (in Hull's case, rats) would inhibit a response or would become overwhelmed with anxiety to the point that it would attack or engender risk of self-harm to achieve a goal. To extrapolate to humans, if threat or anxiety becomes too much for the veteran to bear, he or she may therefore revert to a protective mode, which may include violence as a way to cope with stress. Of course, a number of other factors may also lower this threshold, such as physical or mental illness, fatigue, and alcohol or substance abuse. It now appears that it is this sustained or even multidetermined stress that may be the cause of at least some, if not much, of the long-term residuals from concussion rather than the cognitive impact of the concussion itself, such as those experienced in blast injuries (Hoge et al., 2008).

Hull also recognized that organisms were motivated by other forces, those known as *secondary reinforcements*. This means that previously neutral stimuli (such as fireworks) may assume drive characteristics because they are capable of eliciting responses that are similar to those aroused by the original need state or primary drive (such as incoming mortar and rocket explosions) (Schultz & Schultz, 1987, p. 240). The veteran therefore begins to react to the fireworks on the Fourth of July as if they were a mortar barrage, showing increased anxiety, fear, startle reactions, and perhaps even fleeing, fighting, or hiding.

The phrase "fight or flight" is often used to describe the body's appropriate response to a stressful stimuli. When an individual is exposed to real or perceived danger, a series of complex, interactive neurophysiological reactions occur in the brain, the autonomic nervous system, the hypothalamic–pituitary–adrenocortical (HPA) axis, and the immune system. These responses are thought to have initially developed during evolution to provide the vital total body mobilization required for the individual to survive a life-threatening danger. During the alarm reaction and the stage of resistance to acute stress, the parts of the brain involved in arousal, attention, and concentration functions become activated. This results in hypervigilance to the threat and a decrease in attention to less pressing environmental stimuli—the warrior in the midst of a firefight, for example, may not know he has been wounded until the end of the fight. This also contributes to the telescopic vision and perceptual distortions well documented by Grossman and Christensen (2007).

The degree of anxiety varies with the degree of threat, ranging from jitters to outright panic and terror, and stress reactions have often been considered a *spectrum reaction*, merely annoying to some but creating substantial disability in those either more vulnerable or exposed to more severe and cumulative stress. In addition to changes in brain functioning, other organ systems are involved in this same systemic reaction. The sympathetic division of the autonomic nervous system (SNS), which originates in the brain and distributes throughout the rest of the body, implements the brain's mobilization of the rest of the body. The activation of the SNS increases blood pressure and pulse, dilates the pupils, increases respiratory rate, increases the blood supply to the muscles, and inhibits digestion. The HPA axis is activated, thereby releasing a variety of stress-related hormones. Neural and hormonal signals activate the adrenal glands, which release important stress-related hormones, including epinephrine (or adrenaline) and cortisol. These hormones enter the bloodstream rapidly, acting in all organ systems to prepare the body to fight or flee. The cost of such an adaptive hyperarousal mechanism can be substantial; the alarm reaction consumes energy and depletes stores of available neurotransmitters and hormones.

With sufficient time between threatening events, the body usually makes substantial progress in returning to a previous homeostasis or equilibrium by replenishing the stores of neurotransmitter, hormone, glucose, and other important chemicals. When the stressful event is of a sufficient duration, intensity, or frequency, however, the body does not have the capability to sustain this high state of arousal—the stress-responding apparatus becomes fatigued. One theory is also that under extreme stress of repeated stress (such as repeated attacks or deployments), neurotransmission may become overly sensitized. A type of kindling, or cascade, effect is also thought to occur when anxiety spikes and sets off the entire surge of HPA activation in an abnormal and maladaptive manner; consequently, changes in mood states and behavior may occur. It is now also known that more severe combat exposure increases later risk of risk taking, in that warriors who have been exposed to more violent levels of combat, those who have killed others (depending in part on the physical proximity in which the killing occurs), and those in contact with a high level of human trauma are those who show the highest levels of risk-taking, impulsivity, feelings of invincibility, etc. (Kilgore et al., 2008). Such factors, of course, may or may not translate into violence, but it is presumed that when combined with ingredients such as anger, substance abuse, irritability, etc., violence may well result.

The Contribution of Brain Injury

During the course of the current conflicts in Afghanistan and Iraq, traumatic brain injury (TBI) has emerged as a significant cause of injury to our warriors and has at times been designated the "signature injury" of the Global War on Terror. Although penetrating and severe closed head injuries are typically identified and cared for immediately, mild TBI (mTBI or concussion) may be missed, particularly in the presence of other more obvious injuries. Due to numerous deployments and the nature of enemy tactics, warriors are at risk for sustaining more than one mild brain injury or concussion in a short time-frame (DVBIC, 2006). This is particularly true for those serving in high-risk jobs such as route clearance and bomb disposal units. Mild TBI, or as it will be termed here *concussion*, is not felt by itself to be a significant risk factor for violence. Yet, the question arises, how much, if at all, will these combat concussions contribute to aggression among returning veterans? We predict, that like PTSD and herbicide orange in the Vietnam era, OIF concussions will soon be blamed for everything from homelessness to murder.

It is true that violence can occur as a result of brain damage, but such damage usually must be rather severe and normally involves damage to areas that either control emotional responses or serve as inhibitory centers (braking systems), or both. Uncharacteristically violent behavior has been documented in patients after the onset of metabolic disease, dementia, and tumors and following head trauma or stroke, for example (Paradis, Horn, Lazar, & Schwartz, 1994). Such damage, though, normally must be significant, and one study done at the University of Southern California indicated that as many as six areas of the brain had to demonstrate significant abnormalities in order for a level of violence to occur that led to murder, conditions that rarely occur in concussion and do not occur in most veterans returning with concussion from Iraq or Afghanistan (Hopewell & Christopher, 2007). As with any other problem, the risk of aggression might be elevated by comorbid problems such as alcohol abuse, PTSD, or multiple concussions that are severe enough to cause more substantial damage. Further research into the correlation of concussion and PTSD confirm the 2007 findings of Hopewell and Christopher and suggest that many of the persistent difficulties shown by such patients are much more related to PTSD issues than the original concussion, the latter usually resolving (Hoge et al., 2008). Violent behavior after simple, uncomplicated concussion, therefore, should not be expected.

Among actually violent offenders, frontal lobe damage may result in a loss of inhibitory control over other brain centers, to include those areas that modulate fear, arousal, and emotion (Donavon-Westby & Ferraro, 1999). In addition to prefrontal lesions, areas of damage in the hypothalamus of the brain have been shown to be related to problems such as intermittent explosive disorder (Tonkonogy & Geller, 1992). In contrast, most concussions contribute to what is felt to be minute damage at the cellular level, such as axonal shearing, or, with blast injuries, cell death deep within areas such as memory centers (Taber, Warden, & Hurley, 2006). Blast injuries can occur as a direct result of blast-wave-induced changes in atmospheric pressure (primary blast injury), from people being struck by primary or secondary fragments (secondary blast injury), or by people being forcefully put in motion by the blast (tertiary blast injury) (Taber et al., 2006). Such injuries may result in diffuse axonal injury, contusion, or even subdural hemorrhage. Diffuse axonal injuries are very common following closed head injuries and can result when shearing, stretching, or angular forces pull on axons and small vessels. Impaired axonal transport leads to focal axonal swelling, which, after several hours, may result in axonal disconnection. The most common locations are the corticomedullary (gray matter–white matter) junction (particularly in the frontal and temporal areas), internal capsule, deep gray matter, upper brainstem, and corpus callosum. Concussion also leads to a mismatch between cerebral demand for glucose as opposed to a drop in cerebral blood flow and decreased oxygen metabolic rate, the occurrence of which may result in metabolic and neurotransmitter dysfunction. This leads to a cascade of

- Nonspecific depolarization
- Release of excitatory neurotransmitters
- Massive efflux of potassium
- Increased activity of membrane ionic pumps to restore homeostasis
- Hyperglycolysis to generate adenosine triphosphate (ATP)
- Lactate accumulation
- Calcium influx and sequestration in mitochondria leading to decreased oxidative metabolism
- Decreased ATP production
- Calpain activation and initiation of apoptosis
- Axonal swelling and eventual axotomy

Although irritability can occur, the most frequent symptoms seen after concussion are those of headache, impaired information processing (which people interpret as memory dysfunction), and photo/audio phobia. Most people with concussion want to withdraw and reduce stimulation, and few are violently aggressive. The vast majority also recovers well and returns to normal function. If concussion is related to aggression and violence, it is most likely due to more severe damage from cumulative concussions being present, along with probably comorbid disorders such as noted above.

The Nature of the Job

Creature of Habit

On December 22, 2007, the U.S. Central Command Rest and Recuperation Pass Program announced its 150,000th participant since its inception in 2004. Often coming straight from combat duties in Iraq, warriors on pass are able to have four days of respite from the field. "The best thing is not having to worry about anything," said Paul Harris, an infantryman from Valdosta, Georgia. "I was riding in a vehicle through Qatar with no equipment, no rifle—it felt great. …But I caught myself still looking along the roadsides, scanning rooftops … it becomes a habit.

Stars and Stripes (Anon., 2007)

Much of a warrior's life becomes and remains habit. Habit is as necessary to a combat troop as are food, water, fuel, and ammunition. Without habit, one is untrained, one is vulnerable, and one probably dies. From the first day of boot camp or officer's training, the recruit is drilled and drilled and drilled again in habit. It was only with the drilling of von Steuben at Valley Forge that the Continental Army began to become effective, with the soldiers marching, forming, firing, and reloading all in accordance with habit. One hundred and twenty-five years later, at the beginning of the 20th century, William James, one of the founders of psychology, defined as formal scientific principles that which von Steuben knew as common sense in training men (James, 2003, p. 48)

- Habit is second nature, or rather, ten times nature.
- Ninety-nine hundredths or, possibly, nine hundred and ninety-nine thousandths of our activity is purely automatic and habitual, from our rising in the morning to our lying down each night.
- We are stereotyped creatures, imitators, and copiers of our past selves.
- The teacher's prime concern should be to ingrain into the pupil that assortment of habits that shall be most useful to him throughout life. Education is for behavior, and habits are the stuff of which behavior consists.
- We are mere bundles of habits.

Habits, however, are merely tools and means to an end. A hockey stick is essential to the game and provides hours of active sport for the entire team. Holding and manipulating the stick are so much a habit that a player may even perform the actions unconsciously; however, when swung in anger at a player, the very same hockey stick becomes a weapon. Warrior habits are much the same. They are critical and vital to survival and, just as in hockey, the game of survival in combat could not be won without extensive training that produces the instant and flawless habits needed when the warrior's life is in danger. On the other hand, hypervigilance while on pass in Qatar, much less at home in River City, America, is of less use and may be counterproductive.

In the Zone

Mihaly Csikszentmihalyi (1990) coined the term *flow* as the mental state in which a person is fully immersed in what he or she is doing. Flow is characterized by a feeling of energized focus, full involvement, and success in the process of the activity. Other terms for this or similar mental states include being *on the ball, in the zone,* or *in the groove.* Athletes generally talk about being in the zone or on an aerobic high when in this state. Some of the psychological aspects created in this state include

- A distorted sense of time, as one's subjective experience of time is altered
- Concentrating and focusing, with a high degree of concentration on a limited field of attention
- A loss of the feeling of self-consciousness, the merging of action and awareness
- People becoming absorbed in their activity and the focus of awareness being narrowed down to the activity itself, with action awareness merging

Such a state is a clear example of what psychologists have long termed *over-learning* and has profound implications for survival in a dangerous situation such as combat.

Habits and skills learned for combat survival are considered by psychologists to be functions of motor or procedural memory and are quite different from other types of memory such as recalling a poem or when one was married (declarative memory). Procedural memory is also known to be encoded in the brain in an entirely different manner than are other memories. Whereas declarative memory uses primarily areas of the brain that process verbal and visual memories such as the hippocampus, procedural memories are largely processed through the brain centers used for muscle movement and coordination such as the basal ganglia, cerebellum, and motor cortex. Memories processed in this manner become so ingrained and permanent that they are virtually resistant to decay; for example, no one ever forgets how to ride a bicycle, drive a car, or play the piano, unless a progressive neurological illness literally destroys these areas of the brain.

It should now be obvious that critical warrior training, so absolutely essential to surviving combat, requires extensive overlearning and makes use of procedural/motor learning. The end result places the warrior in the zone, which is needed for survival. Such learning, however, is extremely difficult to suppress or unlearn, and the application of these skills in a civilian or home environment may place the veteran at risk. Fortunately, however, the human brain possesses the capability to take old habits and procedural learning and to make modifications to them to develop new ways of responding and adapting. Bike riders learn to ride motorcycles, high school quarterbacks may be switched to college receivers, and piano players often learn to play other instruments.

But, does such overlearning, this being in the zone, produce men who are ultimately killers when returned to society? Does such training increase the probability of violence? It is unlikely. Experts point out that the returning veteran is often a more integrated member of society and is less likely to use his skills inappropriately than someone who has never served in the military. What the veteran learns, as does the police officer, are restraint and discipline rather than impulsivity and carelessness. Research has continually shown that from the time of World War I, veterans as a group are less likely to be incarcerated, have higher educations, and generally achieve more success upon return to the civilian world than their nonserving peers (Burkett & Whitley, 1998; Grossman & Christensen, 2007). This is because, beginning with World War I, not only are recruits screened rigorously for mental attitudes, but their further training in discipline and restraint to use violence only in specific circumstances is also extensive and ongoing. As Grossman and Christensen (2007, p. 250) point out, veterans "were less likely to use those skills [of violence] than a nonveteran. The reason is clear: combined with learning to kill, they acquired a steely, warrior discipline—and that is the safeguard."

Management of Aggressive Risk

Now that the primary factors related to aggression and violence in a veteran population have been identified and reviewed, what can be done to manage these risk factors? Current thinking formulates this risk management in terms of the three main interventions of:

- Prediction and communication of risk
- Post-deployment screening and service utilization
- Evidence-based therapeutic interventions

The Prediction and Communication of Risk

The first of these, prediction of risk, has already been reviewed to some extent. The importance of this issue is that risk management begins with prediction and identification. If some veterans are felt to be at risk for violence, what factors contribute to this and what issues would be predictive of violence? Secondarily, how might this risk be managed therapeutically? To investigate the nature and causes of violent risk among veterans, it is hoped that one would be able to identify risk factors as well as be able to make at least a modicum of accurate prediction of risk.

In the general population, some risk factors have been posited. Although these factors are not specific to service members, they may be useful when assessing and communicating risk. They include juvenile delinquency and family problems (Bonta, Law, & Hanson, 1998); a history of harm and injury to others (Wolfgang, Figlio, Tracy, & Singer, 1985); factors that suggest exploitation of others and a chronically unstable lifestyle (Hare, 1991); deviant sexual arousal and a history of maladaptive personality factors (Hanson and Bussier, 1998); substance use (Epperson, Kaul, & Huot, 1995); severe brain and frontal lobe damage (Donavon-Westby & Ferraro, 1999); spousal assault, criminal history, and poor previous psychosocial adjustment (Kropp, Hart, Webster, & Eaves, 1999); and a remote and recent history of violence, opportunity, and triggers (HOT) (Hall & Ebert, 2002).

Some argue, however, that the prediction of violent behavior in any circumstance is unreliable. Indeed, mental-health professionals have been cast from the one extreme of being essentially omniscient and being able to predict far-flung behavioral events with great accuracy to being completely incompetent to predict anything and no better than the average person in terms of predicting violence. The truth, as usual, seems to lie somewhere in between, and prediction has been enhanced by recent progress in both our technical tools as well as how we think about the problem.

Advances in predictive procedures have been developed that allow for relatively accurate predictions of illegal behavior, at least within reasonable time frames. (Harris & Rice, 1997). These measures are specific to different types of persons such as criminal offenders, antisocial personality disorders, the mentally ill, etc. The predictive accuracy of these procedures depends on the collection of high-quality objective information, both historical and current. In addition, risk appraisal research indicates that violence is well predictable in some populations, such as those noted previously (Monahan, 1996). Interestingly, military service attempts to screen for all of the most predictive variables with the exception of sex. Ongoing programs of risk assessment research in psychology seek to improve the precision with which psychologists can estimate the risk of harmful behavior under specified conditions (Monahan & Steadman, 1994).

In addition to predicting risk, perilous factors must be communicated. Risk assessment by itself is useless if the assessment is not communicated and subsequently acted upon. An ideal system for communicating assessments of risk would provide clear, precise, and complete information regarding those assessments in a form that would be fully accessible to the parties who must make decisions and take action on the basis of those assessments. The system would also communicate this information in a manner that would reflect and facilitate the appropriate allocation and discharge of responsibilities among the participants in light of their competence and authority (Schopp, 1996). For example, a clinician might discern evidence of fear and agitation associated with delusions of being persecuted in a veteran who has previously engaged in assaultive behavior in response to similar states of emotional stress. The veteran has also been drinking heavily and his spouse has threatened to leave. Such observations must then be communicated to family, healthcare givers, and pertinent authorities, who could arrange for a restraining order, shelter and a security plan for the wife and children, and treatment for the veteran.

Thus, clinicians might be able to provide reasonable estimates of risk by virtue of their ability to describe and explain: (1) psychological processes, (2) impairment of those processes, (3) the relationship between the current impairment and the person's impairment at the time of past dangerous conduct, and (4) the relationship between that past dangerous conduct and the impairment at that time. In this pattern of clinical risk assessment, professionals provide information relevant to an estimate of the risk represented by the individual in the current circumstances by describing and explaining that person's history of risky or harmful conduct in similar circumstances, to either mental health or civil authorities who are empowered to undertake appropriate intervention, or both. The person's psychopathology constitutes one type of relevant circumstance in which risk may be assessed and subsequently communicated (Schopp & Quattrocchi, 1995).

Finally, the communication should ideally structure a prescriptive narrative that (1) includes a categorization of level of risk, (2) provides an estimate of time structure, and (3) communicates a prescriptive narrative. In simple terms, the communication should inform those who are responsible for completing the action plan how significant the risk is, when it might happen, and what to do about it. A recommendation for a judge for a restraining order, shelter and a security plan for the wife and children, and treatment for the veteran as outlined in the scenario above would be an example of a narrative prescription.

Postdeployment Screening and Redeployment Services

The movie *We Were Soldiers* (Paramount Pictures, 2002) has several moving scenes in which both families and the soldiers themselves experience the pain and frustration of being unprepared for many of the emotional issues that accompany combat and the return of the warrior from the combat environment. Undoubtedly the most moving of these scenes is when Mrs. Hal Moore, the Commander's wife, has death notices delivered to her by taxicab because the military was completely unprepared to handle this tragedy in a better manner. The Commander's wife and other officer's wives then had to organize their own support system and procedures themselves.

Partly as a result of such experiences early during the Vietnam conflict, considerably more preparation has occurred in the ensuing years to cope with such emotional and family stresses. Even before leaving the battlefield, each warrior is now processed through what is known as a Post-Deployment Health Assessment (PDHA) and then reassessment (PDHRA). This is conceptualized as part of a complete "deployment cycle" that prepares both service member and family to anticipate the challenges of a deployment assignment, supports them through the training and predeployment phase, provides family support through family readiness groups and rear detachment operations throughout the deployment itself, and then provides for postdeployment healthcare assessment, to include mental health care upon return, such as transfer to the Veterans Affairs healthcare system if the service member moves on to civilian life.

The PDHA consists of a thorough health screening with a credentialed provider conducted for all personnel from 90 to 180 days after their return to their home station from deployment. The assessment is designed to be completed before the end of 180 days to afford Reserve Component members the option of receiving further treatment using their TRICARE health benefit. Reserve Component members also have the option to seek treatment in a military treatment facility, to use their TRICARE benefit, or to seek care through the Veterans' Administration. In addition to the mandatory screenings for PTSD already done for all redeploying personnel, all service members are now screened thoroughly for possible mild TBI per Department of Defense mandate. All service members who are separated administratively are also mandated to be screened for both PTSD and mild TBI. And, finally, all service members and their dependents are authorized to complete up to six counseling sessions on any issue and with minimal red tape with Military OneSource, a benefit usually managed by the TRICARE contractor.

BATTLEMIND training, developed by the Walter Reed Army Institute of Research (Castro, 2006), was created specifically for the "acquisition of a new habit, or the leaving off of an old one" (James, 2003) as mentioned above (e.g., the return of OIF and OEF veterans to civilian life upon redeployment). BATTLEMIND training was designed to be administered immediately at redeployment as part of the Deployment Cycle Support Program with a follow-up training module at 3 to 6 months after deployment. The warrior is taught that BATTLEMIND is his inner strength to face fear and adversity in combat, with courage, and that he has well acquired and demonstrated these strengths during his combat tour. Psychological experiences, including combat stress and in-the-zone issues are normalized as a normal reaction to an abnormal environment and the warrior is taught to rechannel to new, more adaptive habits. This reorientation is begun as the warrior winds down his combat tour and begins the medical review process to return to his home garrison station. Emphasis is upon relearning adaptive civilian habits as an ongoing process upon return to the United States while at the same time retaining the discipline, safety habits, and mental focus that were the determinants of a true warrior in the first place.

In addition to the BATTLEMIND program, a number of classes and briefings are offered throughout the postdeployment adjustment cycle. These include such handouts as "ACS One Source Brief," "Educators Guide," "Family Reunion Handbook," "Family Support Group Leader Basic Handbook," "Homecoming Card," "I Can do That," and "Personal and Family Handbook," among many others. Classes and briefings include, again among many additional ones, "Home for the Holidays," "Homecoming and Going," "Normalization of Experiences," "Reunion—Soldiers' Brief," "Signs and Symptoms of Distress Briefing," and "Family Reintegration Briefing."

Evidence-Based Therapeutic Interventions

The subject of therapeutic intervention, of course, is vast, and the majority of such material is beyond the scope of this chapter but is covered in other chapters of this volume. Although this section has shown that aggression and violence are multidetermined behaviors, a logical place to begin a review of intervention strategies is with those geared toward the warrior exhibiting PTSD. This is because research among military veterans has shown that those with PTSD are higher in anger, hostility, aggression, general violence, and relationship violence and abuse than those without the disorder (Taft & Niles, 2004), and PTSD will exacerbate the effects of other disorders such as concussion (Hoge et al., 2008, Hopewell & Christopher, 2007). Also, irritability and outbursts of anger represent one of the diagnostic criteria for PTSD. When PTSD is comorbid with other factors known to contribute to aggression as previously outlined, such as substance abuse, mental health interventions are virtually mandated if aggression is to be ameliorated or avoided altogether. Best-practices recommendations by the Department of Veterans Affairs' National Center for PTSD (NCPTSD, 2004) specific for the treatment of anger associated with PTSD now include anger management, psycho-education about PTSD, self-monitoring techniques, assertiveness training, stress management, and communication skills training. Best practices for the treatment of PTSD in general include exposure therapy, cognitive therapy, eye movement desensitization and reprocessing (EMDR), and family counseling. Of course, veterans with comorbid disorders, such as substance abuse, violent histories, personality disorders, and other mental illnesses or behavioral problems, should be treated with specific interventions for those issues as well. For those suspected of experiencing concussion, proper neuropsychological assessment and treatment should be sought.

In conclusion, our modern soldiers, sailors, airmen, and Marines really are our new knights in every sense of the word. As we have noted, every day they are asked to confront danger to protect those whom they have sworn to defend. In doing so, they represent the heartland of our country and the traditionalism of our nation. Truly we could not survive a generation without them. They undergo strenuous and constant training as well as discipline in the art of arms and warfare. They then subject themselves selflessly to levels of danger and stress unimaginable to those who never have served, and never will, in the Armed Forces. The great majority does well upon return to civilian life, and their overall adjustment and civilian achievements equal or exceed those of their civilian counterparts; however, combat and stress take a toll upon all, and some warriors from every conflict will be expected to show adjustment problems and to be at some risk for violence. Such aggression is usually directed at families or significant others, less frequently in alternative ways, and is rendered more probable by the presence of comorbid disorders such as substance abuse or preexisting personality problems. Many of these risk factors are known, and an estimate of risk, access to available resources postdeployment, and therapeutic mental health strategies all help to manage such risk and to treat those warriors who are most vulnerable.

References

Anon. (2004). Return of the "wacko vet" myth [editorial], *New York Post*, December 19, p. 28.

Anon. (2007). R&R program hits No. 150,000. *Stars and Stripes*, Mideast Edition, December 22, p. 2.

Bonta, J., Law, M., & Hanson, K. (1998). The prediction of criminal and violent recidivism among mentally disordered offenders: A meta-analysis. *Psychological Bulletin*, 123, 123–142.

Burkett, B. G., & Whitley, G. (1998). *Stolen valor: How the Vietnam generation was robbed of its heroes and its history*. Dallas, TX: Verity Press.

Castro, C. (2006). *BATTLEMIND*. Washington, D.C.: WRAIR Land Combat Study Team, Walter Reed Army Medical Center.

Csikszentmihalyi, M. (1990). *Flow: The psychology of optimal experience.* New York: Harper & Row.

DVBIC. (2006). *Defense and veterans brain injury center working group on the acute management of mild traumatic brain injury in military operational settings: Clinical practice guideline and recommendations.* Washington, D.C.: Defense and Veterans Brain Injury Center

Dohrenwend, B. P., Turner, J. B., Turse, N. A., Adams, B. G., Koenen, K. C., & Marshall, R. (2006). The psychological risks of Vietnam for U.S. veterans: A revisit with new data and methods. *Science, 313,* 979–982.

Donavon-Westby, M. D., & Ferraro, R. F. (1999). Frontal lobe deficits in domestic violence offenders. *Genetic, Social, and General Psychology Monographs, 125,* 75–102.

Epperson, D. L., Kaul, J. D., & Huot, S. J. (1995). Predicting risk for recidivism for incarcerated sex offenders: Updated development on the sex offender screening tool (SOST). Poster session presented at the Annual Conference of the Association for the Treatment of Sexual Abusers, New Orleans, LA.

Grossman, D. (1995). *On killing: The psychological cost of learning to kill in war and society.* Boston, MA: Back Bay.

Grossman, D., & Christensen, L. W. (2007). *On combat: The psychology and physiology of deadly conflict in war and in peace* (2nd ed.). Bellville, IL: PPCT Research Publications.

Hall, H. V., & Ebert, R. (2002). *Violence prediction: Guidelines for the forensic practitioner.* Springfield, IL: Charles C Thomas.

Hanson, R. K., & Bussiere, M. (1998). Predicting relapse: A meta-analysis of sexual offender recidivism studies. *Journal of Consulting and Clinical Psychology, 66,* 348–362.

Hare, R. (1991). *Manual for the Hare psychopathy checklist (revised).* Toronto: Multi-Health System.

Harris, G. T., & Rice, M. E. (1997). An overview of research on the prediction of dangerousness. *Psychiatric Services, 48,* 1168–1176.

Hodierne, R. (2007). War story told by former sailor disputed. *Navy Times,* March 25 (http://www.navytimes.com/news/2007/03/navy_timesmagazine_veteranrape_070322w/).

Hoge, C. W., McGurk, D., Thomas, J. L., Cox, A. L., Engle, C. C., & Castro, C. A. (2008). Mild traumatic brain injury in U.S. soldiers returning from Iraq. *New England Journal of Medicine, 358*(5), 453–463.

Hopewell, C. A., & Christopher, R. (2007). *Military personnel and combat trauma: Operation Iraqi Freedom; Operation Enduring Freedom.* Sparks, NV: Professional, Clinical, and Forensic Assessments.

James, W. (2003/1890). *Habit.* Whitefish, MT: Kessinger Publishing.

Kilgore, W. D., Cotting, D. I., Thomas, J. L., Cox, A. L., McGurk, D., Vo, A. H., Castro, C. A., & Hoge, C.W. (2008). Post-combat invincibility: Violent combat experiences are associated with increased risk-taking propensity following deployment. *Journal of Psychiatric Research, 42*(13), 1112–1121.

Kropp. P., Hart, S., Webster, C., & Eaves D. (1999). *Spousal assault risk assessment guide user's manual.* Toronto: Multi-Health Systems and British Columbia Institute Against Family Violence.

Monahan, J. (1996). Risk appraisal and management of violent behavior. *Criminal Justice and Behavior, 23,* 107–120.

Monahan, J., & Steadman, H. (1994). *Violence and mental disorder: Developments in risk assessment.* Chicago, IL: University of Chicago Press.

NCPTSD. (2004). *Iraq war clinician guide*. Washington, D.C.: Department of Veterans Affairs National Center for PTSD.

Paradis, C. M., Horn, L., Lazar, R. M., & Schwartz, D.W. (1994). Brain dysfunction and violent behavior in a man with a congenital subarachnoid cyst. *Hospital and Community Psychiatry*, 45, 714–716.

Schopp, R. F. (1996). Communicating risk assessments. *American Psychologist*, 51, 939–944.

Schopp, R. F., & Quattrocchi, M. (1995). Predicting the present: expert testimony and civil commitment. *Behavioral Sciences and the Law*, 13, 159–181.

Schultz, D. P., & Schultz, S. E. (1987). *A history of modern psychology*. Orlando, FL: Harcourt Brace.

Taber, K. H., Warden, D. L., & Hurley, R. A. (2006). Blast-related traumatic brain injury: What is known? *Journal of Neuropsychiatry and Clinical Neuroscience*, 18, 141–145.

Taft, C. T., & Niles, B. L. (2004). Assessment and treatment of anger in combat-related PTSD. In *Iraq war clinician guide* (2nd ed., pp. 70–74). Washington, D.C.: National Center for PTSD and the Department of Defense.

Taft, C. T., Pless, A., Stalans, L., Koenen, K., King, L., & King, D. (2005). Risk factors for partner violence among a national sample of combat veterans. *Journal of Consulting and Clinical Psychology*, 73, 151–159.

Taft, C. T., Street, A. E., Marshall, A. D., Dowdall, D. J., & Riggs, D. S. (2007). Posttraumatic stress disorder, anger, and partner abuse among Vietnam combat veterans. *Journal of Family Psychology*, 21, 270–277.

Tonkonogy, J. M., & Geller, J. L. (1992). Hypothalamic lesions and intermittent explosive disorder. *Journal of Neuropsychiatry and Clinical Neurosciences*, 4, 45–50.

Wikipedia. (2008). *Crime in Washington, D.C.*, http://en.wikipedia.org/wiki/Crime_in_ Washington,_D.C.

Wolfgang, J., Figlio, R., Tracy, P., & Singer, S. (1985). *The national survey of crime severity*, NCJ-96017. Washington, D.C.: U.S. Government Printing Office.

Zoroya, G. (2005). Key Iraq wound: Brain trauma. *USA Today*, March 3, p. 1.

Zoroya, G. (2007). Combat injuries multiply: 20,000 vets' brain injuries not listed in Pentagon tally. *USA Today*, November 22, p. 3.

17

Myths and Realities of Pharmacotherapy in the Military

SHARON MORGILLO FREEMAN, LESLIE LUNDT,
EDWARD J. SWANTON, and BRET A. MOORE

Contents

Introduction

The goal of this chapter is to assist the civilian practitioner to understand the importance of military culture and its impact on the service member before considering medication as a treatment option. Each situation requires an understanding of that service member's duty status, job description, and potential for activation into military duty both at home and abroad. As the choice of pharmacologic agent may have a significant impact on the service member's career, the mental health practitioner must choose wisely between psychological interventions only and categories of medication that may be considered nonpsychiatric, or psychiatric depending on the situation at hand. The overriding goal is always to keep our service members safe and, as a result, our civilians, safe. The mental health practitioner must walk a fine line when treating one of these heroes, but they must keep in mind that while treating these brave men and women it is important to help them maintain their courage, self-esteem, and, whenever possible, careers.

Psychiatric Assessment

The purpose of any psychiatric assessment is to explain the service member's behavior within his or her unique context. Differential diagnosis requires both art and science. The military recommends a thorough medical examination, with special emphasis on medical disorders that can manifest as psychiatric symptoms (e.g., subdural hematoma, hyperthyroidism) in addition to the usual possible psychiatric disorders (e.g., acute stress disorder, depression, psychotic disorders, panic disorder) and the use of alcohol and substances of abuse, use of prescribed and over-the-counter medication, and possible drug allergies.

Differential Diagnosis of Behavioral Disorders: Evaluation of Medication Use

It is important to evaluate for the presence of current medications that may be confounding the psychiatric presentation. Often the individual may have tried to self-treat sleep disorders, the common cold, an annoying cough, or a host of other ailments. It is also possible that the individual may have developed an addiction to over-the-counter medications or is actively using or abusing or is dependent on common substances of abuse. A relatively recent issue for service members is the use of muscle-building and fat-burning supplements. In the deployed environment, it is not uncommon for service members to present to a mental health provider complaining of a recent development or exacerbation of anxiety, insomnia, agitation, or even paranoia after beginning a self-prescribed supplement regimen. Table 17.1 lists common over-the-counter medications and their effects when used in excess, as well as behavioral manifestations of chronic use or overuse.

Differential Diagnosis: Physical Injuries

Many symptomatic service members will be at an age when first episodes of schizophrenia, mania, depression, or panic disorder are often seen. They may also have suffered a traumatic brain injury that they have chosen not to disclose. It is therefore important to inquire about *any* history of head injury— before military service and while in training or during combat operations. These individuals may be very reluctant to reveal histories of such injuries for fear of losing their military career or being deemed nondeployable, which can lead to social stigma and interfere with their career progression. It takes a skilled and compassionate clinician to assist the individual in uncovering a physical cause for symptoms. It is also very important to avoid premature conclusions in this population. In other words, just because a young man was in a car accident or in a fight at age 15 or 16 that rendered him temporarily unconscious does not mean that he suffered a mild traumatic brain injury that is causing a behavioral disorder today. Any information uncovered must be thoroughly verified before any diagnosis is determined. An inaccurate diagnosis made too quickly may ruin a very promising career and a bell once rung in the military cannot be unrung. Other physical comorbid conditions might include idiopathic tremors as opposed to withdrawal tremors; mitral valve prolapse or hyperthyroidism as opposed to anxiety or panic disorder, etc. Each potential psychiatric disorder has many potential medical causes and it is incumbent upon the clinician to rule these out regardless of the odds ratios.

Medically Disqualifying Conditions

Many psychiatric disorders can potentially disqualify an individual from enlisting in the Armed Forces. Although some are listed as disqualifiers, most allow individuals to enlist with an appropriate waiver. Each service has its own regulations governing the waiver process; however, the Army Regulation (AR) is typical of the other services. AR 40-501 (Standards of Medical Fitness) describes the disorders that are disqualifiers and the time course of other disorders for which a waiver is not required. The full Army Regulation is available at http://www.army.mil/usapa/epubs/40_Series_Collection_1.html.

For obvious reasons, psychotic disorders are particularly troublesome. Current or a history of such disorders with psychotic features such as schizophrenia, paranoid disorder, or other unspecified psychosis are disqualifying. Other disqualifying current or a history of any such disorders (regardless of the time course) include, but are not limited to, the following: conduct or behavior disorders, personality disorders, suicidal behaviors, self-mutilations, anxiety disorders (including posttraumatic stress disorder, panic, social phobia, and obsessive–compulsive disorder), dissociative disorders, somatoform disorders, psychosexual conditions, and alcohol or drug dependence. With respect to mood disorders, current mood disorders including, but not limited to, major depression, bipolar disorder, affective psychoses, and depressive not otherwise specified are disqualifying; however, a history of these disorders is only disqualifying if the individual required outpatient care for longer than 6 months or ever required inpatient treatment in a hospital or residential facility. Although this list may seem overly restrictive, it is important to realize that a waiver can be requested for any disorder. The typical factors that the waiver authority will consider include, but again are not limited to, the time course of the disorder, severity of the disorder, and whether or not the disorder or history of the disorder will impair the individual's capacity to adapt to military service.

TABLE 17.1 Over the Counter Medications and Effects

Drug Name	CNS Effect	Excessive Doses	Common Effects	Common Uses
Dextromethorphan—found in some formulations of cough remedies sold under such trade names as Robitussin, Delsym, Pertussin, Drixoral, Vicks Formula 44, Triaminic, Coricidin, Sudafed, and Contac; also sold as generic brands	Not associated with CNS depression, analgesia, or respiratory depression	Dissociative anesthetic, similar to phencyclidine (PCP) and ketamine	Hyperexcitability, lethargy, ataxia, slurred speech, diaphoresis, hypertension, nystagmus, mydriasis, nausea, eye pain, puritis, vomiting	Antitussive, cough medication
Antihistamines (e.g., diphenhydramine)	CNS depressant	Blocks alpha-adrenergic, muscarinic, and serotonin receptors; overdose can result in psychosis, hallucinations, agitation, insomnia, lethargy, tremors, and seizures; severe effects may include cardiac arrhythmias due to QTc prolongation	Drowsiness so severe that, in some states, the penalty for driving while taking diphenhydramine is the same as the penalty for driving under the influence of illicit drugs, prescription drugs of abuse, and alcohol; slurred speech, nystagmus, somnolence, decreased reflexes, unsteady gait, impaired attention, impaired memory, impaired judgment, inappropriate behavior, mood lability, impaired functioning	Anxiolytic, sleep aid, antipruritic, antitussive, hypnotic, antiemetic, anti-Parkinsonian agent

Anticholinergics	—	Tachycardia, hyperexia, mydriasis, vasodilation, decreased exocrine secretions, urinary retention, decreased gastrointestinal motility	Clumsiness or unsteadiness; severe dryness of mouth, nose, or throat; flushing or redness of face; shortness of breath or troubled breathing	
Ephedrine, ephedra, ma huang, desoxyephedrine, pseudoephedrine, phenylephrine	Sympathomimetic CNS stimulants (see Table 17.2 for list of ingredients for synthesizing methamphetamine from ephedrine)	Palpitations common, sometimes progressing to chest pain, submyocardial infarction, arrhythmia, paranoia, and seizures	Increased heart rate, elevated mood, decreased appetite; vasoconstriction due to alpha agonist activity which helps relieve nasal congestion; bronchodilation caused by beta stimulation; agitation and weakness; irritability or anger, sensitivity; paranoia; acute withdrawal symptoms, including increased appetite, psychomotor retardation, vivid dreams, hypersomnia, and dysphoria	Appetite suppressant

Table 17.2 Products Used to Synthesize Methamphetamine

Common cold pill containing ephedrine or pseudoephedrine
Acetone
Alcohol (gasoline additives or rubbing alcohol)
Toluene (in the form of brake cleaner)
Drain cleaner or automobile battery acid (sulfuric acid)
Engine starter (contains ether)
Coffee filters
Iodine
Salt (table/rock)
Batteries (lithium)
Propane tank (anhydrous ammonia)
Lye (sodium hydroxide)
Stick matches (red phosphorous)
Dishes (Pyrex®)
Muriatic acid pool chemicals (hydrochloric acid)

Attitudes Among Military Members About Psychiatric Medications

Service members are no different than civilians when it comes to attitudes about psychiatric evaluations, diagnoses, and treatments. The mantra of military mental health providers has always been: "It is not the mental health visit nor treatment that will limit one's ability to hold certain positions, but the behaviors associated with untreated disorders." So, the service member suffering from depression but treated with cognitive–behavioral therapy (CBT) and a selective serotonin reuptake inhibitor (SSRI) will do a lot better professionally than the service member who misses morning formation due to his insomnia or hypersomnia associated with his depressive disorder. Nonetheless, service members seeking behavioral health care are typically perceived as weak. It is not just the old-school soldiers, sailors, airmen, and Marines who maintain and propagate this stigma; it is also rampant among the newer recruits, as well. It is also much more prevalent among the more testosterone-driven combat specialties.

Many service members are concerned about the effect involvement with mental health will have on their security clearances. Again, very few disorders will absolutely disqualify a service member from obtaining or maintaining a specific security clearance, either secret or top secret. It is important that service members and potential service members be honest with their recruiters and security clearance investigators, as dishonesty is an absolute disqualifier. Applications for initial or recertification of security clearances that contain a current or history of mental health treatment will typically go to adjudication. The adjudication service will ask the applicant to undergo a mental health evaluation. The adjudication service provides the identified clinician, typically a psychiatrist, with all of the available supporting data and asks the clinician to comment on the applicant's judgment and reliability.

Neurobiology, Stress, and Warrior Training

The limbic system first appeared in mammals 150 million years ago. This part of the brain is responsible for recording memories of events and behaviors that create both pleasant and disagreeable experiences, so it is responsible for what we call *emotions* (INMHA, 2008a). From a structural standpoint, the primary components of the limbic system include the amygdala, hippocampus, and the hypothalamus. The neurochemical basis of the stress response reaction is noradrenalin, otherwise known as norepinephrine (NE), which is the prominent neurotransmitter for this process. Almost half of the brain's NE is generated in an area of the brainstem known as the locus ceruleus (LC). The brain structure that starts the ball rolling is the almond-shaped amygdala, the part of the brain designed to process emotions, particularly those emotions connected with potential threats or dangers. The amygdala is activated when we are startled or frightened. No surprise that the amygdala and the LC are closely connected. Brain-imaging studies of combat veterans with posttraumatic stress disorder (PTSD) indicate that these veterans' amygdalas become overactive when they listen to recordings that bring back their worst memories (INMHA, 2008b). The amygdala has connections to parts of the brainstem that control breathing and heart rate. As a result, when we are stressed our respiratory rate and pulse increase.

The hippocampus, near the amygdala, has a primary memory function. This area draws upon previous experience to tell us if the threat is potentially dangerous and, if it is a new experience, places a "memory stamp" on the experience to file it away as a permanent "life threat" memory. In other words, the hippocampus and the amygdala cooperate to create a permanent "danger image" of the situation for future reference. A representative evolutionary example is how humans learned to avoid noxious and dangerous stimuli within the environment (e.g., avoiding predatory animals and hostile human adversaries). Interestingly, patients with PTSD have smaller hippocampal volumes (Figure 17.1). Perhaps this is why we so commonly see memory problems in veterans who have been exposed to extreme trauma. We call this *stress-induced dysfunction of the hippocampus*.

The part of our brain that can place some control over the amygdala is the prefrontal cortex. This structure, along with the anterior cingulate cortex, imparts reason by drawing on rational thought, mediating cortical neurons, and allowing us to stand firm, thereby behaving rationally even when we might be afraid, startled, or uncertain. Psychotherapies such as CBT attempt to get to the amygdala via the prefrontal cortex by using strategies such as

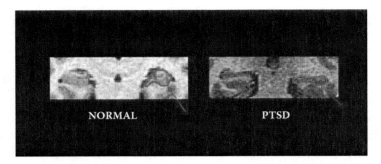

Figure 17.1 Patients with PTSD have smaller hippocampal volumes. (From Bremner, J.D. et al., *American Journal of Psychiatry*, 152, 973–981, 1995. With permission.)

drilling or practice. This is a top–down approach (from cortex to deeper brain structures). Fear-driven circuits are not necessarily removed by successful CBT; rather, alternative pathways originating in the cortex are strengthened. This is why a trauma can be reexperienced; given a strong enough trigger or cue, the conditioned fear in the amygdala can overpower the calming influence of the cortex.

When the ignition switch is hit at the amygdala, the resultant NE release from the LC can produce either an adaptive response (e.g., alertness, fight) or a maladaptive one (e.g., hypervigilance, autonomic arousal, exaggerated startle response, flashbacks). The maladaptive response is considered to be an acute stress response, but if it lasts longer than 4 weeks it is PTSD. It should be noted, however, that typical maladaptive behaviors such as hypervigilance, heightened arousal, and exaggerated startle response may be considered adaptive in the combat environment. The important message for a civilian practitioner to remember is that symptoms should be understood from a contextual viewpoint. The service member's body has been trained *purposely* to respond in a Pavlovian manner with this type of a supercharged NE response; therefore, it may be necessary to retrain behaviorally rather than pharmacologically and to normalize the service member's response patterns rather than pathologize them.

Service members who are experiencing NE surges may need to physically drain the adrenaline in what Grossman and Christensen (2007, p. 16) call "burning off an adrenaline dump." Learning how to manage these changes in sympathetic and parasympathetic nervous systems can help the veteran return to homeostasis. Furthermore, Grossman and Christensen advise that transitioning to noncombat work environments can be mistaken for "detachment, isolation, and pessimism." They attribute this to the greater potential for increased sympathetic nervous system mobilization and therefore a strong likelihood that a parasympathetic backlash will occur as soon as there is downtime. This downtime may result in family discord, marital dissatisfaction, general unhappiness, and the potential for many other disorders secondary to these resultant problems if the mental health practitioner is not aware of such problems.

A Brief Review of Other Relevant Neurotransmitters

Serotonin   Serotonin is important as a mediator during stress or possible threat. When there is a deficiency of serotonin, individuals may feel suicidal, impulsive, aggressive, and even violent. Some may experience depression, anxiety disorders, and even panic. As a result of the blocking of serotonin reuptake, SSRIs increase the amount of serotonin available in the synapses. More importantly, SSRIs promote neurogenesis, literally the formation of new neurons, which in time can result in improvement in mood, decreased anxiety, increased appetite, normalization of sleep–wake cycles, and reversal of suicidal ideation.

Dopamine   Dopamine (DA) functions as both a neurotransmitter and a neurohormone. As a neurotransmitter, it is a mediator of cognition, motivation, pleasure, movement, mood, sleep, and attention. As a neurohormone released from the hypothalamus, it is involved in inhibiting the release of prolactin from the pituitary gland. Relevant to the treatment of service members, particularly in a combat environment, DA reuptake inhibition may be instrumental in alleviating symptoms of attention-deficit hyperactivity disorder (ADHD) (Wilens et al., 2005) and depression (Nieuwstraten & Dolovich, 2001) and assist with smoking cessation (Ferry & Johnston, 2003). Specific to ADHD, the clinician is able to initiate a trial of bupropion, which has been shown to have fewer side effects than psychostimulants and is not subject to abuse. This is relevant considering that the use of psychostimulants may be contraindicated in a deployed environment.

Gamma-Aminobutyric Acid   Gamma-aminobutyric acid (GABA) is a fast-acting amino acid neurotransmitter and the major inhibitory transmitter in the brain; in other words, GABA is the main brake of the brain. Medications that act on GABA are categorized as antianxiety, anticonvulsants, or mood stabilizers because they either potentiate GABA or diminish glutamate (discussed below). The brains of people with PTSD have too much corticotropin-releasing factor (CRF), and GABA can inhibit the release of CRF. Too much CRF will activate the stress system, but too little GABA also allows this activation. In acute stress the inhibitory action of GABA is sometimes compromised so CRF and adrenergic activity are unopposed. GABAergic medications include the mood stabilizer divalproex and benzodiazepines, which are anxiolytics. Alprazolam (Xanax®), a commonly used short-acting benzodiazepine, reduces CRF; however, patients taking benzodiazepines may experience tolerance, dependence, and even withdrawal symptoms when they stop, especially for short-acting benzodiazepines. As a result, these prescriptions should be limited to short-term use.

Glutamate   Glutamate is a fast-acting amino acid neurotransmitter and the major excitatory transmitter in the brain. If GABA is the brake, glutamate is the gas. An important model of PTSD is that particularly the limbic nuclei, the amygdala, and associated areas become sensitized to traumatic reminders, and activating these pathways by these traumatic reminders can set off a PTSD episode or even a relapse. The hypothesis is that glutamate is implicated in the cascade effect which is mediated by N-methyl-D-aspartic acid (NMDA) receptors. The use of anticonvulsant or mood stabilizer medications inhibits these mechanisms which can significantly reduce PTSD symptoms in many individuals (Mula, 2007).

Neuropeptides   Neurotransmitters such as norepinephrine, serotonin, GABA, and glutamate are not the only brain chemicals that have an effect on anxiety. Neuropeptides such as cholecystokinin (CCK) and corticotropin-releasing hormone (CRH) can also produce anxiety. Neuropeptide Y, on the other hand, can relieve anxiety. These compounds are under active investigation as new targets for future medications.

The Truth About Psychotropic Medications and the Military

Prescribing a specific medication rarely, if ever, affects a service member's ability to enlist, obtain, or recertify a security clearance or remain in the military. The important factor is the actual diagnosis and even more importantly the manifestations of the disorder. This is true for pilots as well. Military aviation has been very slow to jump on the mental health bandwagon; however, they recently made significant progress with regard to not penalizing aviators for seeking mental health care. The following is a link to the policy letter that lists the diagnoses for which a waiver may be requested and the requirements for obtaining a waiver to remain on flight status: https://aamaweb.usaama.rucker.amedd.army.mil/AAMAWeb/policyltrs/Army_APLs_Mar06_v3.pdf.

Similar procedures are followed by the Federal Aviation Administration (FAA). Pilots will potentially lose their license to fly aircraft for life if they are ever diagnosed with a psychotic episode; however, a waiver may be granted depending on the etiology of the psychotic symptoms. If the symptoms were due to a primary thought disorder, it is very unlikely that a waiver will ever be granted, but if the symptoms were due to a distant major depressive episode or associated with a legitimate severe combat-related disorder then a waiver may be obtained. The factors to be considered by the waiver authority include acuity, disease-free (remission) time, response to treatment, and typical prognosis. A good example would be a mid-level helicopter pilot who developed panic disorder in his 30s. Prior to the waiver policy, this type of disorder would disqualify (or ground) the pilot; however, he was treated very successfully with an SSRI and can now pursue a waiver, which only requires follow up every 6 months or sooner, if needed.

Medical Certificates

§67.411 Medical certificates by flight surgeons of Armed Forces:

(a) The FAA has designated flight surgeons of the Armed Forces on specified military posts, stations, and facilities, as aviation medical examiners.
(b) An aviation medical examiner described in paragraph (a) of this section may give physical examinations for the FAA medical certificates to persons who are on active duty or who are, under Department of Defense medical programs, eligible for FAA medical certification as civil airmen. In addition, such an examiner may issue or deny an appropriate FAA medical certificate in accordance with the regulations of this chapter and the policies of the FAA.

Any interested person may obtain a list of the military posts, stations, and facilities at which a flight surgeon has been designated as an aviation medical examiner from the Surgeon General of the Armed Force concerned or from the Manager, Aeromedical Education Division, AAM-400, Federal Aviation Administration, P.O. Box 26082, Oklahoma City, OK 73125.

Pharmacologic Treatment of Posttrauma Syndromes

Treatment with a β-adrenergic blocker within 6 hours of an acute psychologically traumatic event and continued treatment for 10 days has been shown to reduce subsequent development of PTSD symptoms (Pitman et al., 2002). Benzodiazepines are not recommended for PTSD by the military based on two studies, one conducted in Israel by Gelpin et al. (1996) and another in the United States by Mellman, Clark, and Peacock (1998). In both of these studies, a benzodiazepine was used in the acute aftermath of trauma as both an anxiolytic and a hypnotic. The hypothesis was that reduction of immediate anxiety and induction of sleep after the trauma might reduce the development of PTSD. Unfortunately, no difference was found in the development of PTSD in the individuals who received this treatment compared to baseline.

Treatment for PTSD involves a thorough understanding of the brain system: The negative-feedback-looking cortisol activates the dorsal rafae nucleus, which is where the serotonergic neurons are. There are three avenues for effective medication if one considers only the amygdala. The first is to block the glutamate receptors so threatening information is not transmitted to the central nucleus of the amygdala. The second is to increase GABA activity, as

GABA is inhibitory and higher levels of GABA are going to inhibit the action at the glutamate receptor. Third is to increase serotonin, which will activate GABA and produce additional inhibitory information. Any of these choices will reduce activity at the level of the amygdala, which is the primary target for the warrior suffering from chronic trauma syndromes.

In an open trial of 10 participants, carbamazepine (Tegretol®) demonstrated excellent efficacy for symptoms of PTSD as well as decreased impulsivity and some efficacy regarding violent angry outbursts (Ford, 1996). Topiramate (Topamax®) blocks glutamate and increases GABA activity. It is currently being studied as a medication to decrease alcohol craving in alcoholics at doses of 150 mg twice daily (Johnson et al., 2003). Research on the use of topiramate for PTSD is cautiously optimistic in small trials (Berlant, 2002). It is unclear why larger double-blind trials have not been done.

Clonidine (Catapress®) and guanfacine (Tenex®) are alpha-2 agonists that reduce presynaptic release of norepinephrine. These medications show promise for reducing dissociative symptoms (Kinzie & Leung, 1989). Atypical antipsychotics, including risperidone (Risperdal®), olanzapine (Zyprexa®), and quetiapine (Seroquel®), have shown promise when used in conjunction with selective serotonin reuptake inhibitors (SSRIs) for treatment failures or partial responders to SSRIs (Petty et al., 2001; Prior, 2001; Roefaro & Mukherjee, 2001; Stein et al., 2002). The approved military pharmacotherapy treatment table for recommendations to reduce the development of PTSD can be found at: http://www.oqp.med.va.gov/cpg/PTSD/PTSD_cpg/content/interventions/evidence.htm.

Pharmacologic Treatment of Anxiety Disorders Other Than PTSD

The rule of thumb in the military is to treat anxiety disorders that coexist with PTSD with SSRIs. SSRIs have U.S. Food and Drug Administration (FDA) approval for anxiety disorders, but these medications may require a month or more before significant symptomatic improvement is apparent. Another option to consider that may treat both PTSD and other anxiety disorders in a matter of hours without concern for dependence or tolerance is valproic acid (VPA). VPA requires monitoring of liver enzymes as well as blood levels but may get the service member back in the field quickly without the sexual side effects of the SSRIs. As clinicians we often fail to consider this very important side effect, especially when treating young people with an active and healthy sex life. Very often an SSRI will put an end to the ability to obtain or maintain an erection and in females the ability to experience arousal or achieve orgasm. Education for the use of VPA includes

- Minimize alcohol use.
- Do not discontinue medication abruptly.
- Do not take more than prescribed.

- Do not operate heavy machinery until you know how this medication will affect you when doses are changed.
- Do not become pregnant while taking VPA, as there is a risk of neural tube defects in the fetus.

Pharmacologic Treatment of Depression

Treatment of depression should vary depending on presenting symptoms. Individuals with primarily serotonergic deficiencies may respond best to an SSRI; however, if the individual complains of difficulty concentrating then an SSRI may not be adequate. Serotonin–norepinephrine reuptake inhibitors (SNRIs), which work on both serotonin and norepinephrine, may be a better choice in these patients. If individuals complain of weight gain as a symptom of their depression, the serotonergic actions of most antidepressants may become problematic. The prescriber may choose bupropion (Wellbutrin®), which does not act upon serotonin. Bupropion has both norepinephrine and dopaminergic action which can cause anxiety in some patients. If this is the case, the selegeline patch (Emsam®), which is a monoamine oxidase inhibitor acting on serotonin, norepinephrine, and dopamine, may be the best answer. As you can see, it is important to thoroughly assess and address the type of depression the individual is experiencing. Adjunctive medications are also chosen to help the individual through the initial phase of treatment; for example, individuals with severe anhedonia, anergy, and somnolence may have a circadian rhythm disturbance as a factor of their depression. A short-term course of modafinil (Provigil®), 100 to 200 mg, may be useful to minimize excessive daytime sleepiness.

With regard to stimulants such as modafinil and methylphenidate, caution is advised for their use in active-duty populations due to their high potential for abuse. Medications that require special storage or monitoring of laboratory values should also be avoided, especially in the soon to be deployed population. Accordingly, providers must use great caution and coordinate with the active-duty commanders and physicians when considering the use of lithium, tricyclic antidepressants, antipsychotics, and valproic acid. An Assistant Secretary of Defense for Health Affairs policy dated April 24, 2007, delineates "deployment-limiting psychiatric conditions." The policy affirms the above precautions about certain medications and directs that, "Disorders not meeting the threshold for a MEB [Medical Evaluation Board, the process by which a service member is medically separated from the military] should demonstrate a pattern of stability without significant symptoms for at least 3 months prior to deployment." The complete policy can be found at http://www.oqp.med.va.gov/cpg/PTSD/PTSD_cpg/content/interventions/evidence.htm. Military guidelines for treatment of major depressive disorder can be found at: http://www.oqp.med.va.gov/cpg/MDD/MDD_Base.htm. *The Pharmacologic Management of Major Depression in the Primary Care Setting*

(VA Medical Advisory Panel for the Pharmacy Benefits Management Strategic Health Group, 2000) can be downloaded at http://odphp.osophs.dhhs.gov/pubs/guidecps/text/CH49.txt.

Pharmacologic Treatment of Bipolar Disorder

The military does not provide guidelines for the treatment of individuals with a diagnosis in the bipolar spectrum; however, research is being conducted at the Michael E. DeBakey VA Medical Center on two of the most common medications used for bipolar I, or bipolar, manic phase according to the DSM-IV-TR. The first of these two medications is lithium, which was the first mood-stabilizing medication approved by the FDA for the treatment of mania. Lithium is often very effective in controlling mania and preventing the recurrence of both manic and depressive episodes. Additionally, valproic acid, an anticonvulsant medication approved by the FDA in 1995 for the treatment of mania, has mood-stabilizing effects and may be especially useful for bipolar episodes that are difficult to treat. Lithium can have significant side effects, including ataxia, dysarthria, delirium, tremor, memory problems, diarrhea, weight gain, nausea, acne, rashes, alopecia, leukocytosis, euthyroid goiter, increased thyroid-stimulating hormone (TSH) and reduced thyroxine levels, hypothyroid goiter, polyuria, and polydypsia. All of these problems would be devastating to an active-duty soldier. Valproic acid has its own set of side effects, including sedation, dizziness, ataxia, headache, asthenia, nausea, constipation, weight gain, and polycystic ovaries. Neither lithium nor VPA should be used in women wishing to become pregnant, and we must be vigilant about ensuring that women are on effective birth control.

Another medication that is particularly helpful for the depressed phase of bipolar illness is lamotrigine (Lamictal®), also an anticonvulsant. Lamotrigine can be used in conjunction with VPA but in much lower doses because VPA increases the plasma concentration of lamotrigine which can lead to serious side effects. This medication also has the potential for causing Stevens–Johnson syndrome (SJS), a rare but life-threatening rash that can lead to multiple organ failure. We advise all patients beginning on lamotrigine to not change their personal hygiene products while we are titrating this medicine. It can be difficult to determine the etiology of rashes, so we want to minimize confounding effects such as a recent change in soap or laundry detergent. Any rash that develops while a patient is on lamotrigine must be evaluated for the possibility of SJS. Until the cause of a rash is clear, discontinue the medication.

A very promising combination medication, olanzapine and fluoxetine (Symbyax®), has demonstrated extremely robust clinical efficacy in individuals with irritable, depressed-phase, cyclical disorders such as bipolar II, or bipolar depression. In acute and a long-term follow-up studies, there has been a very low rate of induction of mania or mixed states which was the one major concern of using a combination medication that included an SSRI (Shelton, 2006).

Over-the-Counter Medications That May Cause Problems

According to various estimates, 15 to 40% of the general population suffers from transient, short-term, or chronic insomnia associated with subjective complaints of prolonged sleep latency, frequent nighttime awakenings, long periods of nighttime wakefulness, or early-morning awakening. Because altered nighttime sleep leads to daytime fatigue and sleepiness, any type of insomnia negatively affects performance and thus increases the risk of accidents as well as health and mood problems, most commonly depression and anxiety. Many individuals, military or not, choose over-the-counter (OTC) sleep aids in an attempt to shorten sleep latency or maintain sleep. The most widely used OTCs are the first-generation antihistamines (i.e., diphenhydramine, brompheniramine, and chlorpheniramine). These medications induce sedation via central antihistaminergic and anticholinergic mechanisms. Unfortunately, as discussed above, this leaves the individual at risk for possible unwanted side effects such as impaired cognition and daytime hangover.

The most commonly used OTC is caffeine. Always inquire about the consumption of coffee, tea, energy drinks such as Red Bull®, and other caffeine-containing drinks. Caffeine can trigger anxiety reactions and cause disturbed sleep, so its consumption should be minimized. Energy drinks and dietary supplements (especially those used by weightlifters and other personnel involved in physical fitness training) contain a significant amount of caffeine and other stimulants, as well as dehydrating agents. These types of products are very popular among service members and are widely available, including at deployed locations. In fact, energy drinks are available at no cost to deployed service members at the dining facilities.

Pharmacologic Treatment of Sleep Problems

Insomnia is a complex and common problem (see Table 17.3). Traditional hypnotic agents such as barbiturates and benzodiazepines are to be avoided because of their addiction potential. Sleep hygiene education is an important part of the treatment of any patient with insomnia. Fortunately, we now have a nonaddicting alternative. Ramelteon (Rozerem®) is a melatonin receptor agonist that effectively turns off the alerting signal generated in the suprachiasmatic nucleus (SCN). Decreasing this alerting signal should then allow sleep to occur more quickly without sedation. Studies by Griffith and Seuss (2005) have shown that substance-abusing patients do not find ramelteon attractive, as it creates no pleasurable effects; however, the proper use of ramelteon includes patient education strategies, as many people expect to be knocked out by a sleeping medication. The potential for prazosin (Minipress®), an alpha-1 antagonist, to reduce nightmares has been investigated by Raskind et al. (2003). They have found that it may reduce not only nightmares but also global PTSD symptoms.

Table 17.3 Sleep Behaviors

Maladaptive	Adaptive
Use caffeine excessively.	Do not use caffeine at least 4 hours before bedtime.
Exercise too close to bedtime.	Exercise in the morning, afternoon, or early evening.
Smoke before going to bed.	Don't smoke at least an hour before going to sleep.
Watch horror or action movies before bed.	Watch a comedy or something "light."
Eat foods that cause heartburn.	Eat earlier in the evening and avoid rich and spicy foods.
Watch the clock.	Turn clock away to lose awareness of time.
Take over-the-counter sleep aides.	Practice relaxation and imagery exercises.
Watch television or playing video games in bed.	Only use bed for sex and sleep.
Stay in bed awake more than 15 minutes.	Get into bed only when your eyes feel heavy.
Take naps during the day.	Stay awake until it is time to go to bed.

Pharmacologic Treatment of Substance Use Disorders

Unfortunately, only two medications are approved by the military for the treatment of active-duty military members for alcohol dependence: naltrexone (Trexan®) and disulfiram (Antabuse®). The newest treatment on the market in the United States, acamprosate calcium (Campral®), is not yet approved by the military, although it has shown excellent efficacy for post-acute withdrawal and alcohol cravings. It does not yet come in a generic form, which may play a part in the decision not to place it on the Department of Defense Formulary. Naltrexone may be effective in the reduction of alcohol cravings and should be dosed based on both liver function and gender. Starting dose is generally 25 mg for the first 2 to 3 days to limit nausea and vomiting followed by 50 mg daily. Some individuals require higher doses for efficacy. Females in general require higher doses (McClellan, 2002, pers. comm.). Disulfiram does not have any anticraving properties or any other assistive factors for the individual who is attempting to quit. Many individuals drink on top of disulfiram which is not only dangerous but also defeats the purpose of the medication, which is to provide negative reinforcement (it makes the person ill if alcohol is added while taking the medication). Both medications should be used adjunctively for the greatest likelihood of success. For more information on treatment of substance use disorders, see Chapter 12 in this volume. For active-duty military members, addiction-focused treatment follow-up may be mandated for a period of 6 to 12 months from the time of initial referral to a healthcare practitioner (this may be referred to as "aftercare"). It is unfortunate that opioid agonist therapy (OAT) is not a treatment option for active-duty personnel who have an opioid dependence. These individuals can be treated *only* with naltrexone and clonidine. The approved military guidelines for the treatment of substance use disorders can be found at http://www.oqp.med. va.gov/cpg/SUD/SUD_Base.htm.

Emergency Psychiatric Medicine

The acute use of medications may be necessary when the service member is dangerous, extremely agitated, or even psychotic. In such circumstances, as noted earlier, the service member's long-term career implications should be kept in mind, and every opportunity should be taken to use the least potentially career-ending choice of medication possible. The individual should be taken to an emergency room where short-acting benzodiazepines (e.g., lorazepam) or, if absolutely necessary, haloperidol or atypical neuroleptics may be administered to treat aggression. Early and prolonged use of benzodiazepines is contraindicated, because benzodiazepines have been associated with a higher rate of subsequent PTSD. Although antiadrenergic agents, including clonidine, guanfacine, prazosin, and propranolol, have been recommended (primarily through open, non-placebo-controlled treatment trials) for the treatment of hyperarousal, irritable aggression, intrusive memories, nightmares, and insomnia in survivors with chronic PTSD, there is only suggestive preliminary evidence of their efficacy as an acute treatment. Of importance, antiadrenergic agents should be prescribed judiciously for trauma survivors with cardiovascular disease due to potential hypotensive effects, and these agents should also be tapered, rather than discontinued abruptly, to avoid rebound hypertension.

References

Ancoli-Israel, S., & Roth, T. (2000). Characteristics of insomnia in the United States: Results of the 1991 National Sleep Foundation Survey. *Sleep*, 22, S347–S353.

Berlant, J., & van Kammen, D. P. (2002). Open-label topiramate as primary or adjunctive therapy in chronic civilian posttraumatic stress disorder: A preliminary report. *Journal of Clinical Psychiatry*. 63(1), 15–20.

Bremner, J. D., Randall, P. R., Scott, T. M., Bronen, R. A., Delaney, R. C., Seibyl, J. P. et al. (1995). MRI-based measurement of hippocampal volume in patients with combat-related posttraumatic stress disorder. *American Journal of Psychiatry*, 152, 973–981.

Ferry, L., & Johnston, J. A. (2003). Efficacy and safety of buproprion SR for smoking cessation: Data from clinical trials and five years of postmarketing experience. *International Journal of Clinical Practice*, 57, 224–230.

Ford, J. D., Ruzek, J. I., & Niles, B. L. (1996). Identifying and treating VA medical care patients with undetected sequelae of psychological trauma and post-traumatic stress disorder. *NCP Clinical Quarterly*, 6(4), 77–82.

Gelpin, E., Bonne, O., and Peri, T. et al. (1996). Treatment of recent trauma survivors with benzodiazepines: A prospective study. *Journal of Clinical Psychiatry*, 57, 390–394.

Gengo, F., Gabos, C., & Miller, J. K. (1989). The pharmacodynamics of diphenhydramine-induced drowsiness and changes in mental performance. *Clinical Pharmacology and Therapeutics*, 45, 15–21.

Griffiths, R., & Seuss, P. (2005). Ramelteon and triazolam in humans: Behavioral effects and abuse potential (poster). Atlanta, GA: APA Annual Meeting.

INMHA. (2008a). *The evolutionary layers of the human brain.* Quebec: Institute of Neurosciences, Mental Health, and Addiction (http://thebrain.mcgill.ca/flash/d/d_05/d_05_cr/d_05_cr_her/d_05_cr_her.html).

INMHA. (2008b). *Brain abnormalities associated with anxiety disorders.* Quebec: Institute of Neurosciences, Mental Health, and Addiction (http://thebrain.mcgill.ca/flash/i/i_08/i_08_cr/i_08_cr_anx/i_08_cr_anx.html).

Johnson, B. A., Ait-Daoud, N., Bowden, C. L. et al. (2003). Oral topiramate for treatment of alcohol dependence: A randomised controlled trial. *Lancet*, 361(9370), 1677–1685.

Kinzie, J. D., & Leung, P. (1989). Clonidine in Cambodian patients with posttraumatic stress disorder. *Journal of Nervous and Mental Disorders*, 177, 546–550.

Mellman, T. A., Clark, R. E., & Peacock, W. J. (1998). Prescribing patterns for patients with posttraumatic stress disorder. *Psychiatric Services*, 54, 1618–1621.

Mula, M. (2007). The role of anticonvulsant drugs in anxiety disorders. *Journal of Clinical Psychopharmacology*, 27(3), 262–272.

National Sleep Foundation. (1991). *Sleep in America.* Washington, D.C.: Gallup.

Nieuwstraten, C. E., & Dolovich, L. R. (2001). Bupropion versus selective serotonin-reuptake inhibitors for treatment of depression. *Annals of Pharmacotherapy*, 35, 1608–1613.

Petty, F., Brannan, S., Casada, J., Davis, L. L., Gajewski, V., Kramer, G. L. et al. (2001). Olanzapine treatment for post-traumatic stress disorder: An open-label study. *International Clinical Psychopharmacology*, 16, 331–337.

Pitman, R. K., Sandersa, K. M., Zusmana, R. M., Healya, A. R., Cheemaa, F., Laskoa, N. B. et al. (2002). Pilot study of secondary prevention of posttraumatic stress disorder with propranolol. *Biological Psychiatry*, 51(2), 189–192.

Prior, T. I. (2001). Treatment of posttraumatic stress disorder with olanzapine. *Canadian Journal of Psychiatry*, 46(2), 182.

Raskind, M. A., Peskind, E. R., Kanter, E. D. et al. (2003). Reduction of nightmares and other PTSD symptoms in combat veterans by prazosin: A placebo-controlled study. *American Journal of Psychiatry*, 160, 371–373.

Roefaro, J., & Mukherjee, S. M. (2001). Olanzapine-induced hyperglycemic nonketonic coma. *Annals of Pharmacotherapy*, 35(3), 300–302.

Shelton, R. C. (2006). Olanzapine/fluoxetine combination for bipolar depression. *Expert Review of Neurotherapeutics*, 6(1), 33–39.

Stein, D. J., Zungu-Dirwayi, N., van Der Linden, G. J. et al. (2000). Pharmaco-therapy for posttraumatic stress disorder. *Cochrane Database Systematic Reviews* (CD002795).

VA Medical Advisory Panel for the Pharmacy Benefits Management Strategic Health Group. (2000). *The pharmacologic management of major depression in the primary care setting.* Washington, D.C.: Veterans Affairs.

Wilens, T. E., Haight, B. R., Horrigan, J. P., Hudziak, J. J., Rosenthal, N. E., Connor, D. F., Hampton, K. D., Richard, N. E., & Modell, J. G. (2005). Bupropion XL in adults with attention-deficit/hyperactivity disorder: A randomized, placebo-controlled study. *Biological Psychiatry*, 57, 793–801.

Witek, T. J., Jr., Canestrari, D. A., Miller, R. D., Yang, J. Y., & Riker, D. K. (1995). Characterization of daytime sleepiness and psychomotor performance following H1 receptor antagonists. *Annals of Allergy, Asthma, and Immunology*, 74, 419–426.

IV
The Service Member's Family and Community—Intervention

18

War and Children Coping With Parental Deployment

P. ALEX MABE

Contents

Introduction

As the military campaigns in Iraq and Afghanistan have continued for the past 5 years, the American society has endured enormous challenges entailing considerable demands for resources and personnel to carry out a mission that continues to be under intense political debate. The spearhead for these campaigns has been American military service personnel, who have received considerable attention with regard to the costs of waging war. Perhaps in response to societal shortcomings with regard to support of American troops in the Korean and Vietnam campaigns, it appears that in contrast to previous wars much more attention has been directed toward the health and well-being of our service members as they fight in this global conflict. Indeed, this attention is much needed and deserved as the risks of death and injury have been very real for the approximately 1.5 million American troops thus far deployed in support of the war effort. One third of them have served at least two tours in a combat zone, 70,000 having been deployed three times, and 20,000 having been deployed at least 5 times (Presidential Task Force, 2007). As of December 2006, the Iraq war alone accounted for over 3000 fatalities and over 18,000 American casualties (Hoshmand & Hoshmand, 2007). The psychological impact of combat exposure has long been a hidden morbidity of war, but awareness and concern about this form of debilitation are growing.

The risks for psychological morbidity in the current campaigns are very real, and many service members return from battle with problems of major depression, generalized anxiety, or posttraumatic stress disorder (PTSD). On a positive note, however, the increased attention being paid to combat veterans and their stresses has brought about significant progress in identifying ways to mitigate and treat combat-related problems. Unprecedented has been the new light cast on the welfare of military families as they endure the strain of war. Since Operation Desert Storm (ODS) in the early 1990s, there has been a growing awareness that the children and families of military service members are also worthy of our attention because of their unique stresses and challenges during times of war. This increased attention to the needs of military children and families has derived, in part, from the changing demographics of the military as well as what appears to be the strategic impact of family stress on an all voluntary military.

Since the 1990s, military demographics have reflected the military's increasing responsibility for military dependents; for example, in 2005 there were 1,373,534 activity-duty members compared to 1,865,058 family members (Caliber Associates, 2007a). Nearly 38% of women and 44% of men in the active-duty force have children (U.S. Congress Joint Economic Committee, 2007). Approximately 11% of women in the military are single mothers compared to 4% of single fathers, and dual military marriages are growing among military service personnel (10% of women and 2% of men) (U.S. Congress Joint Economic Committee, 2007). Concerns about childcare issues have become particularly significant as a result of the increased percentage of women in the military who are struggling to balance their careers and motherhood (Kelley et al., 2001). In 2005, women represented 14.6% of active-duty personnel, up from the 11.5% of officers and 10.9% of enlisted in 1990 (Caliber Associates, 2007a). With the increased participation of women in combat deployments many more children will be separated from their mothers, a phenomenon that has received little study. The demographics of the military indicate that a significant proportion of military personnel are parents. Approximately 700,000 children in America today have at least one parent deployed away from their family in military service (Presidential Task Force, 2007), so it behooves us as a society to learn how best to assist military children as they face the stresses of war.

In an era of an all-voluntary military, retention of personnel is paramount, particularly in a time of war when repeated and extended tours of combat duty are required. Thus, it is noteworthy that in a metaanalysis of the relationships between family factors and military retention Ethridge (1989) found that military members' satisfaction with raising their families in the military was a significant predictor of retention. Increasingly, it is has become clear that it is strategically necessary to support military children and families to maintain the military's ability to sustain its operations around the world. It should not be surprising that the U.S. military provides health care for nearly 2 million children, operates the largest employee-provided childcare system in the country, and sponsors a number of programs to support families and their children (Caliber Associates, 2007b). Beyond these military program efforts, however, relatively little has been accomplished with regard to understanding and responding to the unique needs of military children and their families during times of war.

This chapter provides an overview of what is currently known about the impact of deployments on military children and their families. Moreover, a model of preventive and therapeutic efforts is proposed as a preliminary step toward the ultimate goal of fostering behavioral health in military children enduring the stresses of war deployments. To date, prevention and treatment efforts for military children and their families have been quite limited, adapted from general psychoeducational/resilience-based interventions unrelated to the military experience and untested with regard to their efficacy. The proposed prevention/intervention model attempts to more carefully address key aspects of the deployment experience for military children and their families.

The Impact of Deployment on Military Children and Their Families

When addressing the impact of deployment on military children and their families, the complexity of this issue cannot be overstated. Inherent in all definitions or models of stress is the notion that the demands of any circumstance must be weighed against the resources and vulnerabilities of the individual. In the context of war deployment, the demands, resources, and vulnerabilities that come into play are enormously diverse and similar to an individual's fingerprints in that no two deployment experiences are alike. The nature of the deployment can be quite varied, including such potentially important variables as length of deployment, frequency of deployment, time for preparation, exposure to combat and risk of injury or death, form and level of the communication available during deployment, perceived purpose of the deployment, and societal support for the mission. At the individual child level, differences that can influence coping include age, gender, temperament, coping style and skills, intelligence, premorbid psychological disorder, and prior deployment experience. At the family level, relevant differences include whether one or two parents are deployed, gender of the deployed parent, marital status of the deployed parent, number and age of siblings, quality of parent–child relationship, premorbid psychological disorder in the parent, military rank of the parent, socioeconomic status, and prior deployment experience.

A child's community may vary in its availability of social supports (extended family and friends), access to military-related information and programs, degree to which local schools understand the military family experience, and deployment-specific assistance. Needless to say, empirical studies have been unable to absorb all of the complex variables that could and perhaps should be considered in an attempt to understand the impact of deployment on children and their families. Furthermore, of the limited studies available regarding the impact of deployment on children, many of the findings are based on subjective descriptions rather than quantitative measures of child outcomes; therefore, the findings presented provide only a rudimentary picture of the impact of deployment on military children and their families.

Impact on Children

Studies have linked parental deployment to a variety of adverse child outcomes, including increased levels of anxiety and depression (Huebner & Mancini, 2005; Jensen, Martin, & Watanabe, 1996; Kelley et al., 2001; Levai, Kaplan, Ackerman, & Hammock, 1995; Medway, Davis, Cafferty, Chappell, & O'Hearn, 1995; Pincus, House, Christensen, & Adler, 2001; Rosen, Teitelbaum, & Westhuis, 1993), sleep disturbance (Caliber Associates, 1992; Kelley et al., 2001), discipline problems or externalizing behavior problems (Kelley et al., 2001; Levai et al., 1995; Rosen et al., 1993), declines in school performance (Hiew, 1992), and increased irritability, anger, lashing out, and demands for attention (Huebner & Mancini, 2005; Rosen et al., 1993).

Although many studies report the pervasiveness of stress reactions in children during parental deployment, on average the symptom severity has been reported to be relatively mild. For example, Rosen et al. (1993) studied 934 Army spouses with 1798 children whose partners belonged to units that deployed during Operation Desert Storm. According to parental reports, more than half of the children (ages 3 to 12 years) experienced sadness or home disciplinary problems during deployment. Nevertheless, few (6%) of the children had symptoms severe enough to warrant treatment. Similarly, Jensen et al. (1996) studied 383 children and the remaining parent whose partners were deployed in ODS and found that, although internalizing symptoms were more commonly reported by children with deployed parents, their levels of internalizing symptoms on average did not reach suggested clinical cutoffs. Finally, Pisano (1992) examined the California Achievement scores of 157 sixth-grade children during ODS and noted that, although daughters of deployed personnel manifested a significant decrease in reading comprehension scores, all other California Achievement scores showed no statistically significant differences, either between the sexes or between the subgroups of children with parents deployed or nondeployed.

It should be cautioned that all three of these studies reporting relatively mild responses to deployment were conducted during ODS, a conflict that was relatively short lived and resulted in fewer casualties and deaths in comparison to the current campaigns in Afghanistan and Iraq. General findings in stress research point to the fundamental importance of recognizing "dose effects" as well as the specific form of stress in predicting the impact of any given stressor on children. So, even in combat deployment, not all deployments are likely to have the same impact on children. Research has demonstrated, for example, that the length of deployment is an important factor in predicting stress responses, with longer deployments having more adverse effects on children (Cozza, Chun, & Polo, 2005). Of great concern to children is the possibility of injury or death of the deployed parent, and it seems likely that should injury or death actually occur the psychological impact would be quite negative. Surprisingly, the impact of parental death or injury on military children has not been broadly examined. Nevertheless, studies of parental death have consistently identified the surviving children as being at a higher risk for poor behavioral health (Dowdney, 2000). Given the high psychological cost of combat injuries on military personnel (Presidential Task Force, 2007), it would seem that deployments resulting in combat injuries are also likely to have more adverse effects on children than deployments in which injuries do not occur.

Impact on Families

The extent to which deployment affects the stay-behind parent and overall family functioning, of course, will have profound effects on the well-being of children. Studies of deployment in OSD consistently have noted the close association of parental distress and child distress (Jensen et al., 1996; Medway et al., 1995; Rosen et al., 1993). Nearly universal in studies examining the impact of deployment on stay-behind spouses is the emotion of loneliness which few describe as a stress reaction that is handled well (Caliber Associates, 2007b). Deployment of one's spouse has been associated with significant increases in depression, anxiety, anger and irritability, sleep disturbance, and physical symptoms or health complaints (Caliber Associates, 1992; Jensen et al., 1996; Pincus et al., 2001; Wright, Burrell, Schroeder, & Thomas, 2005). Stay-behind parents have reported problems with emotional withdrawal, significant increases in parenting stress, and disruptions in parenting rules and expectations for children (Kelley, 2002; Kelley, Herzog-Simmer, & Harris, 1994). Stay-behind women have reported increased anxiety, emotional withdrawal, and disruptions in parenting rules and expectations for children during the deployment period (Amen, Jellen, Merves, & Lee, 1988; Kelley et al., 1994).

Two family outcomes of deployment of particular concern involve increased rates of child abuse and intimate partner violence. Gibbs, Martin, Kupper, and Johnson (2007) examined a sample of 1771 families of enlisted U.S. Army soldiers who were combat deployed between 2001 and 2004 and reported that the rate of child maltreatment was 42% higher during deployment than during non-deployment. This increased rate of maltreatment pertained primarily to neglect by civilian mothers which increased threefold during deployment of the father.

The rate of physical abuse was actually lower during deployment but incidents of physical abuse were depicted as being more severe. Rates of sexual abuse were unaffected by deployment. Rentz et al. (2007) reported in a time-series analysis of Texas maltreatment rates that increases in maltreatment among military families (although lower than civilian rates of maltreatment) were associated with both departures to and returns from operational deployment during ODS.

With regard to intimate partner violence, the data have been inconsistent regarding the impact of deployment. Although spousal violence among military families has been reported to be comparable to that of the civilian population, findings regarding such violence after combat deployment have varied, with some studies reporting modest increases in spousal violence (Gimbel & Booth, 1994; McCarroll et al., 2000; Orcutt, King, & King, 2003) and others not supporting this conclusion (McCarroll et al., 2003; Newby et al., 2005). Angrist and Johnson (2000) reported that deployment of female soldiers was associated with an increased likelihood of divorce, but the deployment of men did not have a similar effect. In contrast, in a 2004 survey sponsored by the Kaiser Family Foundation, 58% of Army spouses believed deployment had strengthened their marriages, while 31% believed it had no effect, and only 10% felt it had weakened their marriages (Caliber Associates, 2007b).

At-Risk Children and Families

Research on moderators of the impact of deployment in children and their families is quite limited. Nevertheless, a few child and family characteristics have been linked to greater risk for adverse deployment outcomes. Researchers have noted that boys tend to experience more adverse reactions to deployment than girls (Jensen et al., 1996; Levai et al., 1995). This relative vulnerability of boys may be related to the fact that fathers are disproportionately more likely to be the deployed parent, and as has been observed in divorce research the absence of fathers seems to have more detrimental impacts on boys. Younger children (especially under 6 years of age) and younger families also seem to be more adversely affected by deployment (Jensen et al., 1996; Orthner & Rose, 2005). The child coping literature suggests that young children have fewer available coping strategies, are less flexible and adaptive in dealing with stress, and tend to use more avoidant rather than active approaches to coping (Amirkhan & Auyeung, 2007)—all factors that could contribute to their experiencing more stress symptoms during parental deployment. Younger military families have fewer resources (e.g., economic, psychological, social) than older military families, have less knowledge and experience, have less affiliation with their military units, and are more likely to move away from military installations (and away from military resources) during deployment (Caliber Associates, 2007b).

A fundamental principle in behavioral health is that premorbid functioning is a good predictor of responses to stresses; therefore, it is not surprising that children and their families who are experiencing difficulties prior to deployment (e.g., mood or behavior disturbance, developmental or learning problems, marital or family conflicts, parental mood disturbance) are more vulnerable to the adverse effects of deployment (Caliber Associates, 2007b; Presidential Task Force, 2007). Parents with dual military careers and single military parents often experience more family stress than military personnel with one civilian parent due to complications associated with arranging and providing child care when deployments occur. Consequently, parents and children in single-parent or dual-parent military homes tend to experience more adverse deployment outcomes (Kelley, 2006). Finally, women in the military appear to be at greater risk for having adverse reactions to combat deployment than men (Kelley et al., 2002); thus, it seems likely that that the families of military women are also at greater risk for adverse deployment outcomes. This concern is heightened by the reality that women in the military are also more likely to be a single parent or married to another member of the military and thus face the possibility of dual deployment. In fact, women service members with children more so than single women and their male counterparts report a substantial decline in health and well-being after deployment, and their children report more emotional distress during their mother's absence (U.S. Congress Joint Economic Committee, 2007).

A Prevention/Treatment Model

Proposed interventions to reduce the adverse effects of parental deployment on children and their families are built on key conceptualizations of the deployment experience from the child and parent perspective. Drawing from the descriptive and survey data of the past 15 to 20 years for families enduring military deployments, this section addresses key conceptualizations of what wartime deployment entails and interventions linked to relevant theory and data regarding child and family coping.

Loss

A central feature of deployment is separation of the military service member from his or her family. This separation results to varying degrees in losses in companionship (e.g., spending time together in work or play), affection and intimacy, and emotional and instrumental support (e.g., opportunities to exchange empathic understanding and advice, help with homework, transportation, recreational pursuits). For the stay-behind parent, separations are associated with the temporary loss of a spouse who assisted in maintaining the household, caring for children, and solving family problems (Caliber Associates, 1992). For the child left behind, the unique parenting contributions of the deployed parent are to some extent lost during the deployment. Although research suggest that fathers and mothers are comparably capable of being competent and nurturing caregivers, they differ in the amount of time spent and the kind of interactions they have with their children; for example, fathers tend to focus more on problem-solving and instrumental communications, whereas mothers tend to focus more on relationship and emotional support communications (Rohner & Veneziano, 2001). Thus, mothers and fathers may provide unique contributions to parenting that benefit their children, and the loss of either's contribution could be a hindrance to the psychological and social development of their children. The sense of loss that the children experience can be compounded by the anxiety and depression experienced by the stay-behind parents that now renders them less emotionally present for their children. Likewise, some children of deployed parents have been described as engaging in social and emotional withdrawal (Presidential Task Force, 2007), so they, too, may become less available to the stay-behind parent. Some stay-behind families opt for changing residence during deployment to be closer to extended family support. Although this move may mitigate the loss experience in some respects, it may compound the loss in other ways such as removing access to valuable military resources and the familiarity of one's home, neighbors, and schools. For some families, deployment can result in a loss in income and other economic supports that might be available for the family. Particularly in lower ranked service members it is not unusual for the parents to hold second jobs to supplement the family's income or to share childcare responsibilities so both parents can be employed. Deployment, however, may remove these financial opportunities/support.

Therapeutic Response to Loss To guide the construction of therapeutic responses to parental deployment, the literature pertaining to loss of a parent as a result of death is considered relevant and will be used to develop treatment strategies (although losses resulting from deployment tend to be templossorary and thus differ in important ways from the bereavement experience). From bereavement studies and guidelines, the following therapeutic strategies would seem relevant for the loss associated with deployment:

- Encouragement of open and honest communication is essential in responding to loss experiences (Black & Urbanowicz, 1987; Greeff & Human, 2004; Sandler et al., 2003). Children are likely to benefit from discussions about the loss experience that set into motion the coping resources of the child and the family to deal with the losses. Likewise, open and honest discussion of the losses associated with deployment gives children the opportunity to express feelings of apprehension or fear and sadness about the separation, help increase children's beliefs that their feelings are understood by caregivers, and reduce their need to inhibit the negative expressions of grief-related feelings. Moreover, typical grief reactions can be normalized, and problematic thoughts and feelings about the separation can be identified and addressed; for example, due to exposure to the media coverage of war it is not unusual for children and their families to develop distorted beliefs regarding the risk of parental death in combat deployment. Yet, in the current Afghanistan and Iraq campaigns the casualty rate for American service members has been less than 1% (Cozza et al., 2005); therefore, fears of the death of deployed parents can be ameliorated by discussions about and reassurances of the actual risks involved.

- Efforts to minimize the loss can be valuable when dealing with child bereavement (Black & Urbanowicz, 1987; Lohnes & Kalter, 1994). For the bereaved child, efforts to minimize the loss can entail encouragement for the child to maintain a relationship with the deceased parent through pictures, stories, and aspirations to carry on favorable traits of the deceased. Also, surrogates of parental care can be provided to shore up the losses of instrumental and emotional support that result when a parent dies. In the context of deployment, the loss can be reduced by maintaining effective communication with the deployed parent and coaching the deployed parents in how to maintain effective instrumental and emotional support of their children during their physical absence (e.g., inquiring about the ups and downs of life and assisting with advice on life problems, helping with long-term school assignments, reading short books or telling stories to young children via phone or video communications). In modern warfare, families now have virtually immediate communication with loved ones who are deployed, even in combat areas, and this increased access can greatly facilitate efforts to maintain the child and deployed parent relationships. Children of deployed parents can also benefit from surrogates for parental care such as extended family members, community friends, church members, and teachers or coaches who can support children by addressing their important emotional and instrumental needs.
- Facilitation of a nurturing and effective parent–child relation with the surviving parent and child has been shown empirically to improve child bereavement coping (Sandler et al., 2003). Time and time again, research has punctuated the finding that when a child has at least one warm, loving parent or surrogate caregiver (grandparent, foster parent) who also provides firm limits and boundaries then that child is able to endure incredible hardships (Masten & Coatsworth, 1998). For families dealing with parental deployment, it is important for the stay-behind parents to be supported while dealing with their own separation stresses and feelings of loneliness and depression so they can function better and be more positive and accessible for their children in their coping efforts. Direct attention to providing

instrumental support in maintaining the household would also reduce the overall sense of loss during deployment. As observed in bereaved families, psychoeducational materials and sessions that provide guidelines and instructions on effective parenting (e.g., setting aside quality time, using empathic communications, using positive attention and praise, authoritative discipline skills) can foster a strong parent–child relationship that can help the family deal with the temporary loss of the deployed parent.

• Instruction in positive coping skills for both the parent and the child has shown promise in helping families cope with parental death (Sandler et al., 2003). For the stay-behind parents, the ability to express positive emotions (e.g., love and appreciation) and negative submissive emotions (e.g., worry and sorrow) while limiting negative dominant emotions should be emphasized when helping their children cope with such feelings as anger and hostility (Valiente, Fabes, Eisenberg, & Spinrad, 2004). Also, assistance might be needed to help the stay-behind parent identify, initiate, and maintain social supports that could aid the family in the emotional and instrumental tasks of dealing with deployment. For children dealing with loss associated with a deployed parent, it is important to teach coping skills that can assist in dealing with troubling negative feelings (e.g., sadness, fear, anger)—support seeking, self-soothing, cognitive restructuring (i.e., constructing more positive thoughts about a potentially negative event), and distraction (i.e., activities that would draw attention away from the negativity of the moment). As with bereaved children, it would be helpful to instruct children on how to maintain a relationship with the deployed parent (e.g., contributing to a family photo album or scrapbook, contributing to e-mails sent by the stay-behind parent, sending letters and packages, preparing for phone or video communications with the deployed parent).

• Development of meaning can play an important role for children trying to understand the loss of a parent (Antonovsky & Sourani, 1988). How families make sense of any crisis, such as the loss of a family member, and endow it with meaning is an important component for family resilience. For children dealing with the loss of a parent deployed in combat, their sense of what the deployment is all about is important for their coping efforts. If a child believes that the deployment is somehow his or her fault, is related to problems in the parents' marital relationship, or is a mission of little value, then the separation is likely to have a greater negative impact on the child. Conversely, if the child and the family understand that the deployment is a comprehensible and manageable part of military life and the mission is an important one, then the separation stresses are significantly reduced.

In fact, research has indicated that families cope better with deployment when they understand the mission and believe that the sacrifices that they are making are worthwhile (Caliber Associates, 2007b; therefore, parents should be encouraged to convey a sense of the value of the mission as well as general beliefs that the stresses the family is facing are comprehensible and manageable. Also, a general understanding should be provided of where the deployed parent is going to be, what he or she will be doing, and why it is important.

Uncertainty

A sense of uncertainty is present in all phases of deployment to combat areas. When deployment is announced, there is uncertainty as to the location, nature, and projected start date for the mission. A consistent finding in surveys is that family members desire a fixed departure date so they can plan their lives (Schumm, Bell, & Knott, 2001), yet false alarms on start dates are quite common. Deployments may begin with relatively little advance notice and can extend well beyond the original planned duration if the military situation requires. During deployment, the location of the operation and level of danger present may have to be kept secret; thus, in combat zone operations the family knows that there are inherent risks but does not know when and how significant the dangers are. The family also has financial uncertainties with regard to taxes, travel advances, loss of income, additional childcare costs, and the high expenses of staying in touch with the deployed service member. Huebner, Mancini, Wilcox, Grass, and Grass (2007) noted that in war ambiguous loss occurs because the deployed parent is alive but not physically present in the home and alive for the moment but facing the possibility of imminent death. During deployment the availability of communication with family members can be quite variable as plans for phone or video contacts have to be changed as the conditions of combat shift. The stay-behind family has to make many adjustments that may represent very new roles and responsibilities in the absence of the military service member, and there is uncertainty as to what extent roles and responsibilities adopted will be changed when the deployed parent returns. Military families want predictability in return dates, but again the rapidly changing conditions of war make certainty about such dates unlikely; changes in return dates wreak havoc on family reunion plans and cause great frustration and emotional upheaval (Hosek, Kavanaugh, & Miller, 2006). In light of the potential physical and emotional costs of battle, who exactly will be returning home can be of concern (e.g., "Will my father be the same person when he comes home?"). The stresses of the uncertainties about deployment are often fueled by rumors before and during deployment that tend to add to the confusion experienced rather than bringing clarity to what can to be expected. Overall, the conditions of uncertainty create conditions ripe for fear and anxiety in the military family dealing with deployment in a combat zone, and their efforts to plan for and cope with the stresses of deployment become much more difficult. This sense of uncertainty appears to be heightened for young and inexperienced families who do not know what to expect and do not have experience coping with the multiple stresses entailed in deployment (Blount, Curry, & Lubin, 1992).

Therapeutic Response to Uncertainty   Uncertainty is a fundamental theme in the experience of childhood cancer because of the uncertainty of cure, the ongoing risk for late relapse and additional cancers, and the unpredictability of other late effects of curative treatment regimens (Stewart, 2003). And, like deployment, the uncertainties of childhood cancer engage the entire family; therefore, from the childhood cancer literature, the following therapeutic strategies would seem relevant for addressing the uncertainties associated with deployment.

Information   The provision of information can be invaluable in reducing a child's sense of uncertainty in stressful situations. In childhood cancer studies, the children who received open information from their parents at the initial stage of diagnosis were significantly less anxious and depressed and had higher self-esteem than children who were provided with information at a later stage (Ishibashi, 2001). Likewise, military families would be expected to better handle the stresses of deployment when accurate and timely information is provided. With regard to the type of information that is needed, McCormick (2002) described three characteristics that are central to the concept of uncertainty: (1) *probability*, not knowing the likelihood that something is going to happen; (2) *temporality*, not knowing when something is going to take place; and (3) *perception*, not having a frame of reference from past experiences that help to guide the individual in coping with the current stressor.

In the context of military deployments, children would likely benefit from information pertinent for each of these characteristics of uncertainty. Probabilities of adverse events such as injury or death of the deployed parent are often significantly lower than what is feared, and sharing of this information could soothe children's fears. Definite timing of events in war, such as returning home, can be difficult, but tolerance for this uncertainty can be reduced when children have approximate time expectations for events. Finally, although circumstances during deployment may appear quite novel to children, parents can help their children recognize that in fact there have been past experiences that the family has endured that could help them discover ways of coping with the present. From the childhood cancer literature, effective information is provided sooner rather than later, is geared to the development level of understanding of the child, and is repeated. On this latter point, parents will need to remember that updates and repetition of information will be necessary to soothe children's fears and anxieties across the various phases of deployment. To address possible distortions in the available information, parents should periodically ask about what the child feels, thinks, and believes regarding the family experience of deployment. Finally, the manner of providing information should convey a sense of calm and confidence that the child and family can handle the challenges before them.

Present Certainties   Focusing on present certainties appears to be a common and successful coping response for children dealing with childhood cancer (Parry, 2003). Inevitability additional information will not be able to resolve or even mitigate many of the uncertainties in wartime deployment. Similar to children coping with cancer, however, children and their families may better endure these uncertainties by focusing on what is certain. When responding to the incertitude of reunion following deployment, for example, the stay-behind parent could communicate, "We don't know when Dad will be coming home exactly, but we do know that tomorrow we will make plans for a wonderful party when he returns."

Embracing Uncertainty   Embracing uncertainty may seem a contradictory response to reducing one's fears and anxieties, but as Parry (2003) pointed out some families who face and survive the adversities of childhood cancer experience psychological and spiritual growth. Likewise, within the stresses of deployment children and their families no doubt will experience keen distress at not knowing, yet the very same uncertainties of deployment can also become opportunities for children and their families to assess their values and priorities, to learn to take risks for personal growth, and to become more appreciative of what they do have because they are more fully aware of the uncertain and fragile nature of life. Frank and open discussions about the uncertain outcomes of deployment can thus become opportunities for inspiring in children a sense of what is truly important and worth embracing in such defining moments of life.

Uncontrollability

The literature on coping and resilience emphasizes that an important dimension of any stressful situation is the degree to which individuals believe that they can by their own efforts positively impact change in the situation. This belief in controllability has been associated with more active coping responses and more positive outcomes in children dealing with stress (Clarke, 2006; Compas, Connor-Smith, Saltzman, Thomsen, & Wadsworth, 2002). Many of the challenges faced by children and their families during wartime deployments, however, are beyond their direct control. Obviously, children have no control over the nature and timing of deployment, cannot protect their deployed parent from injury, have little control over the frequency of communications with the deployed parent, and do not determine the duration of the deployment. The list of uncontrollable features of deployment is quite extensive and predictability could exacerbate the distress that children experience.

Therapeutic Response to Uncontrollability Child coping and resilience research provides general guidelines in dealing with stresses that are perceived to be uncontrollable and two of these strategies appear to be particularly relevant for military families:

- Coping responses that reflect engagement rather than disengagement appear to be generally more effective for children in dealing with all stresses including those perceived to be uncontrollable (Clarke, 2006; Compas et al., 2002). Specifically coping strategies that involve problem solving, cognitive restructuring, and positive reappraisal of the stressor have been reported to be generally effective. These coping strategies entail a careful analysis of the stressful situation, selective attention to positive aspects of the situation, and generation of thoughts that are positive and hopeful. Compas et al. (2002) contrast these copings responses to more passive and disengaged coping strategies that include avoidance, social withdrawal, resigned acceptance, emotional ventilation, wishful thinking, and self-blame. Even for stresses perceived to be uncontrollable, engaged coping responses can help children address their negative thoughts and feelings about the situation or the self more adaptively; therefore, in military families dealing with the many uncontrollable events of deployment, parents should model and encourage their children to use coping strategies that actively engage them with stresses and their reactions to the stresses.
- Focusing on what is controllable appears to be an important qualification for the benefits of engaged coping strategies. In a metaanalysis of children coping with interpersonal stress, Clarke (2006) observed that young people who attempt to actively resolve uncontrollable interpersonal stressors, such as parental conflict or illness, are more likely to demonstrate poorer social competence and greater behavioral problems. This finding suggests that children need to discriminate between stressful situations that are amenable to direct influence and those that are not, and accordingly they need to adjust the focus of their coping efforts based on an accurate interpretation of controllability. This qualifying principle does not dictate that the child must become passive or disengaged when a situation is uncontrollable, but rather the focus of coping must shift from changing the situation to coping with a situation that cannot be changed. In the context of deployment stresses, parents should teach children how to assess the situation with regard to its controllability in addition to teaching them specific engaged coping responses.

Strengths and Resilience

This review would be remiss without acknowledging the strengths of military families as they face wartime deployments. In contrast to popular films and books that collectively portray the experiences of children in military families negatively (Caliber Associates, 2007b), research has failed to support such conclusions (Cozza et al., 2005; Ryan-Wenger, 2001). In fact, important strengths within military families must be acknowledged and fostered in families as they face deployment stresses. Military families often have a strong sense of community, and most enjoy valuable community support and services as they live among others who share a commitment to the mission, face common life challenges, and have access to many formal and informal family support services associated with the U.S. Armed Forces (Hoshmand & Hoshmand, 2007). Behavioral expectations within the military culture include requirements for order and civility among service members and their families (Caliber Associates, 2007b). Although these expectations may be perceived as being intrusive to family life independence at times, they also encourage a certain degree of order and discipline that can be healthy for families. Also, since ODS, innovations and improvements in community partnerships have provided various additional means of social support for military families (e.g., Boys and Girls Clubs of America, YMCA). Overall, this sense of community, commitment to the mission, culture of order and discipline, and availability of social support systems have been quite beneficial in helping military families cope with deployment (Hoshmand & Hosmand, 2007).

The U.S. Department of Defense (DoD) manages one of the largest integrated healthcare systems in the world, and the comprehensive healthcare services available for military dependents include mental health services (Caliber Associates, 2007b). By definition, at least one parent is fully employed in military families, and the U.S. military service offers job security along with healthcare benefits and community resources. For these reasons, the military is considered a family-friendly environment by many because it provides financial stability, health benefits, and social supports that are conducive to relationship stability (Lundquist & Smith, 2005). With major technological advances, communication capabilities during deployment have dramatically improved in recent years, and surveys have suggested that most military spouses manage to obtain the communication they need, regardless of the deployment status of the military service member (Caliber Associates, 2007b). Finally, it should be emphasized that, although deployments can be quite stressful, many families report that outcomes of these deployments have included the development of new skills and competencies as well as a sense of independence and self-reliance (Caliber Associates, 1992, 2007b).

Therapeutic Response to Strengths and Resilience   It behooves policymakers and clinicians to recognize that when confronted with the challenges of wartime deployment military families can draw upon important strengths that should be identified and fostered as they face the hardships of deployment. Child resilience research offers the following additional points of emphasis with regard to fostering children's strengths during deployment of a parent:

- A focus on strengths enables children and their families to recognize their strengths as they can be brought to bear on the challenges at hand (Podorefsky, McDonald-Dowdell, & Beardslee, 2001). Children and their families who are able to identify their strengths are encouraged to believe that they are more capable of addressing the challenges of deployment and maintain a sense of hope for positive outcomes in the stressful experience.
- Proactive orientation involves taking initiative in one's own life and believing in one's own efficacy, and it has been identified in the literature as being a primary characteristic defining resilience (Alvord & Grados, 2005). An extension of engagement coping responses described above, proactive orientation encourages individuals to anticipate problems and to positively engage in problem solving or other active coping strategies before the challenges occur. Many of the stresses of wartime deployment can be anticipated, and children and families should be encouraged to take the initiative to prepare for them, thus fostering a sense of self-efficacy and increasing the chances of successful coping when stresses do occur.
- Teaching self-regulation, defined as learning to gain control over attention, emotions, and behavior, can be valuable in helping children learn to cope with stress (Alvord & Grados, 2005). Under the broader umbrella of positive coping skills described above, self-regulation emphasizes the perspective that children can and should be taught to have a sense of ownership about the regulation of their attention, emotions, and behavior. The relevant self-regulation skills to be taught and methods to teach them are the focus of many developing therapies designed for children and adults and thus are beyond the scope of this review. The emphasis here is that parents and care providers need to recognize the value of encouraging children to be aware of their responses to stress and to come to believe that they can and should be responsible for learning ways to cope with stress.

Case Vignette

Jeremy is an 8-year-old boy whose father is a noncommissioned officer in the U.S. Army. The father's military occupational specialty is intelligence, and he has had multiple deployments away from his family but none to a combat zone.

Now preparing for the father's deployment to a combat zone in Afghanistan, Jeremy is brought to the outpatient psychiatric clinic with severe symptoms of fearfulness and anxiety. Specifically, the family reported that Jeremy has had difficulty going to bed on his own, complains of nightmares, and has frequently refused to go to school. Jeremy's parents also reported that his teachers have indicated that in the last few weeks he has been inattentive, often does not complete his schoolwork, and has exhibited tearfulness. Jeremy lives with his father, mother, and a 5-year-old sister. On inquiry, the parents reported that Jeremy has had a history of symptoms of attention deficit hyperactivity disorder that have been reasonably well controlled with stimulant medication. In the past, Jeremy has exhibited some mild symptoms of anxiety usually related to separation from his parents or shyness and worries about other children not liking him, but these anxieties had been short lived. The parents reported that Jeremy had been told that the father was going to be deployed but they had been reluctant to tell him much about the deployment details because he had not handled previous deployments very well and they feared that his symptoms would become even more unmanageable if he knew that the father was going into a combat area. In past deployments, Jeremy became very lonely without his father, had to sleep with his mother due to nighttime fearfulness, and became oppositional with the mother and combative with his sister. Within 5 weeks of the anticipated departure date the parents decided that they had to tell Jeremy that his father was going to Afghanistan because he obviously was making preparations for his deployment. In the interview with Jeremy, the child stated that his father was going to Afghanistan and admitted to fears that his father would be killed because he had seen on the television news bombs that blew up and killed lots of people. He also related that he did not like it when his father left the home because it made him sad and there was no one to play with after school. He also complained that his sister was a brat and, unlike the father, the mother always took his sister's side.

Therapeutic Response to Key Deployment Stresses

Jeremy's case vignette illustrates many of the key issues that trouble children during parental deployment and below are listed these issues along with an illustrative treatment plan:

- *Loss*—Jeremy was anticipating the loss of his father as it related to his relationship with his father (e.g., someone to play with after school) and his vague understanding that his mother did not handle the children as well without his father around. The family was encouraged to be much more up front about the imminent departure of Jeremy's father so they could begin to share their feelings about the stresses ahead of them. By sharing the family's feelings about the departure it was hoped that Jeremy would be able to express his own feelings better and receive the reassurance and support that he needed. Jeremy and his family were to discuss a specific communication plan before the father's departure so Jeremy would understand fully that he would still be connected to his father. The mother was encouraged to sign Jeremy up for soccer when the season began, and in the meantime she should identify some peers with whom he could begin to establish some play dates. The mother also agreed to participate in some parenting consultations so she could be better prepared to handle the parenting problems that may result with the father gone. Finally, the mother agreed to work on establishing some one-on-one time with Jeremy after his sister went to bed so she could improve the quality of her relationship with Jeremy and elicit better cooperation from him.

- *Uncertainty*—Jeremy had been provided with little information about the deployment, and his anxieties and fears were being driven by his own fantasies about what might happen. His parents were encouraged to be more upfront about what the deployment was going to involve. In particular, his father was encouraged to get a map of Afghanistan to show Jeremy the region where he would be working, to provide a show-and-tell related to some of the equipment he would be using to keep himself safe, and to remind Jeremy of past deployments and the similarities of these deployment experiences with the one upcoming. Jeremy's father was to communicate the importance of his mission and how Jeremy could do his part by helping his father get ready for the departure.

- *Uncontrollability*—Jeremy had little control over the timing, nature, and duration of his father's deployment. The family was encouraged to focus instead on what each of the children could do to help the family manage the challenges of the father's deployment. Tasks that were initially identified for Jeremy included efforts to stay on task and do his work in school, to be more cooperative and helpful with his mother around the home, and to help his sister cope with their father's departure by playing nicely with her. Also, Jeremy agreed to come up with some ideas of things that he could do when he felt afraid or sad such as playing with his action figures, playing fetch with his dog, and looking at the family photo album.

- *Strengths and resilience*—Specific strengths and supports within the family were identified to help the family cope with the deployment. Jeremy had an enviable collection of action figures that was quite enticing for peers who might come to his house and play. Jeremy was learning to be skilled on the computer, and it was decided that he could help his mother and sister use e-mail to correspond with and send pictures to his father during his deployment. The neighbors were very supportive and had a swimming pool, so additional emotional and social support and recreation opportunities could be found right next door. Recently, the family had joined a church that had several military families, and the mother appreciated the opportunity to worship with and get support from these families.

Conclusion

As is evident in this review, military families face tremendous challenges in times of war. Moreover, there is some evidence that for children and their families adverse reactions and outcomes are common when dealing with parental deployment into combat areas. It has been proposed that paramount when dealing with the deployment of a parent or spouse is the sense of loss, uncertainty, and uncontrollability regarding the experience. Yet, we are also learning that military families have many strengths and often demonstrate remarkable resilience in facing the challenges of war. Scientifically, our knowledge of military families' experiences and outcomes of wartime deployments has been limited by the lack of rigorous studies in the field. In particular, very few studies have been done on children facing parental deployment that use quantitative outcome measures and adequate comparison groups. Likewise, interventions for children and their families facing the challenges of wartime parental deployment have lacked clear theory, wartime relevance, and efficacy. As noted, however, due to the broad and unpredictable nature of war stresses the scientific enterprise of understanding and intervening with children facing parental deployment is staggering in its methodological complexities and logistical difficulties; therefore, from this review few solid conclusions can be formed. Nevertheless, two basic starting points are suggested. First, the phenomenology and epidemiology of wartime stress must include the study of military children and their families. For the most part, only the mediating effects of age, gender, and length of deployment have received any significant attention. Studies of such psychological constructs as loss, uncertainty, and uncontrollability as mediators of deployment stress outcomes would be useful. Certainly, federal support for broad research initiatives will be needed to address the enormous complexities of the issues at hand. Second, efficacy studies are necessary to determine the benefits of proposed prevention and intervention strategies emphasizing conceptually driven models such as the one proposed in this review. Given the complexity and unique nature of the deployment experience for military families, treatment efficacy studies should be based on interventions built on clear conceptualizations of the deployment experience rather than just broad coping strategies.

References

Alvord, M. K., & Grados, J. J. (2005). Enhancing resilience in children: A proactive approach. *Professional Psychology: Research and Practice, 36,* 238–245.

Amen, D., Jellen, L., Merves, E., & Lee, R. (1988). Minimizing the impact of deployment separation on military children: Stages, current preventive efforts, and system recommendations. *Military Medicine, 153,* 441–446.

Amirkhan, J., & Auyeung B. (2007). Coping with stress across the lifespan: Absolute vs. relative changes in strategy. *Journal of Applied Developmental Psychology, 28,* 298–317

Angrist, J. D., & Johnson, J. H. (2000). Effects of work-related absences on families: Evidence from the Gulf War. *Industrial and Labor Relations Review, 54,* 41–58.

Antonovsky, A., & Sourani, T. (1988). Family sense of coherence and family adaptation. *Journal of Marriage and the Family, 50,* 79–92.

Black, D., & Urbanowicz, M. A. (1987). Family intervention with bereaved children. *Journal of Child Psychology and Psychiatry, 28,* 467–476.

Blount, B. W., Curry, A., & Lubin, G. (1992). Family separations in the military. *Military Medicine, 157*(2), 76–80.

Caliber Associates. (1992). *A study of the effectiveness of family assistance programs in the Air Force during Operation Desert Shield/Storm,* Final Report, Eric #ED373293. Prepared for the Air Force and Family Matters Office. Fairfax, VA: Author.

Caliber Associates. (2007a). *2005 demographics report.* Prepared for the Office of the Deputy Under Secretary of Defense (Military Community and Family Policy). Fairfax, VA: Author (http://www.cfs.purdue.edu/mfri/pages/military/2005_Demographics_Report.pdf).

Caliber Associates. (2007b). *What we know about Army families: 2007 update.* Prepared for the Family and Morale, Welfare, and Recreation Command. Fairfax, VA: Author (http://www.army.mil/fmwrc/documents/research/whatweknow2007.pdf).

Clarke, A. T. (2006). Coping with interpersonal stress and psychosocial health among children and adolescents: A meta-analysis. *Journal of Youth and Adolescence, 35,* 11–24.

Compas, B. E., Connor-Smith, J. K., Saltzman, H., Thomsen, A. H., & Wadsworth, M. E. (2002). Coping with stress during childhood and adolescence: Problems, progress, and potential in theory and research. *Psychological Bulletin, 127,* 87–127.

Cozza, S. J., Chun, R. S., & Polo, J. A. (2005). Military families and children during Operation Iraqi Freedom. *Psychiatric Quarterly, 76,* 371–378.

Dowdney, L. (2000). Childhood bereavement following parental death. *Journal of Child Psychology and Psychiatry, 41,* 819–830.

Ethridge, R. M. (1989). *Family factors affecting retention: A review of the literature.* Research Triangle Park, NC: Research Triangle Institute.

Gibbs, D. A., Martin, S. L., Kupper, L. L., & Johnson, R. E. (2007). Child maltreatment in enlisted soldiers' families during combat-related deployment. *Journal of the American Medical Association, 298,* 528–535.

Gimbel, C., & Booth, A. (1994). Why does the military combat experience adversely affect marital relations? *Journal of Marriage and the Family, 56,* 691–703.

Greeff, A. P., & Human, B. (2004). Resilience in families in which a parent has died. *American Journal of Family Therapy, 32,* 27–42.

Hiew, C. C. (1992). Separated by their work: Families with fathers living apart. *Environment and Behavior, 24,* 206–225.

Hosek, J., Kavanagh, J., & Miller, L. (2006). *How deployments affect service members.* Santa Monica, CA: The Rand Corporation.

Hoshmand, L. T., & Hoshmand, A. L. (2007). Support for military families and communities. *Journal of Community Psychology, 35,* 171–180.

Huebner, A. J., & Mancini, J. A. (2005). *Adjustments among adolescents in military families when a parent is deployed.* Prepared for Military Family Research Institute and Department of Defense Quality of Life Office (http://www.cfs.purdue.edu/mfri/pages/research/Adjustments_in_adolescents.pdf).

Huebner, A. J., Mancini, J. A., Wilcox, R. M., Grass, S. R., & Grass, G. A. (2007). Parental deployment and youth in military families: Exploring uncertainty and ambiguous loss. *Family Relations*, 56, 112–122.

Ishibashi, A. (2001). The needs of children and adolescents with cancer for information and social support. *Cancer Nursing*, 24, 61–67.

Jensen, P. S., Martin, D., & Watanabe, H. (1996). Children's response to separation during Operation Desert Storm. *Journal of the American Academy of Child and Adolescent Psychiatry*, 35, 433–441.

Kelley, M. L. (2002). The effects of deployment on traditional and nontraditional military families: Navy mothers and their children. In M. G. Ender (Ed.), *Military brats and other global nomads: Growing up in organization families* (pp. 229–253). Westport, CT: Praeger.

Kelley, M. L. (2006). Single military parents in the new millennium. In C. A. Castro, A. B. Adler, & T. W. Britt (Eds.), *Military life: The psychology of serving in peace and combat* (Vol. 3, pp. 193–219). Bridgeport, CT: Praeger.

Kelley, M. L., Herzog-Simmer, P. A., & Harris, M. A. (1994). Effects of military-induced separation on the parenting stress and family functioning of deploying mothers. *Military Psychology*, 6, 125–138.

Kelley, M. L., Hock, E., Smith, K. M., Jarvis, M. S., Bonney, J. F., & Gaffney, M. A. (2001). Internalizing and externalizing behavior of children with enlisted navy mothers experiencing military-induced separation. *Journal of American Academy of Child Adolescent Psychiatry*, 40, 464–471.

Kelley, M. L., Hock, E., Jarvis, M. S. et al. (2002). Psychological adjustment of Navy mothers experiencing deployment. *Military Psychology*, 14, 199–216.

Levai, M., Kaplan, S., Ackerman, R., & Hammock, M. (1995). The effect of father absence on the psychiatric hospitalization of Navy children. *Military Medicine*, 160, 103–106.

Lohnes, K. L., & Kalter, N. (1994). Preventive intervention groups for parentally bereaved children. *American Journal of Orthopsychiatry*, 64, 594–603.

Lundquist, J. H., & Smith, H. L. (2005). Family formation among women in the U.S. military: Evidence from the NLSY. *Journal of Marriage and the Family*, 67, 1–37.

Masten, A. S., & Coatsworth, J. D. (1998). The development of competence in favorable and unfavorable environments: Lessons on successful children. *American Psychologist*, 53, 205–220.

McCarroll, J. E., Ursano, R. J., Liu, X., Thayer, L. E., Newby, J. H., Norwood, A. E., & Fullerton, C. S. (2000). Deployment and the probability of spousal violence by U.S. Army soldiers. *Military Medicine*, 165, 41–44.

McCarroll, J. E., Ursano, R. J., Newby, J. H. et al. (2003). Domestic violence and deployment of U.S. Army soldiers. *Journal of Nervous and Mental Disease*, 191, 3–9.

McCormick, K. M. (2002). A concept analysis of uncertainty in illness. *Journal of Nursing Scholarship*, 34, 127–131.

Medway, F. J., Davis, K. E., Cafferty, T. P., Chappell, K. D., & O'Hearn, R. E. (1995). Family disruption and adult attachment correlates of spouse and child reactions to separation and reunion due to Operation Desert Storm. *Journal of Social and Clinical Psychology*, 14, 97–118.

Newby, J. H., Ursano, R. J., McCarroll, J. E. et al. (2005). Postdeployment domestic violence by U.S. Army soldiers. *Military Medicine*, 170, 643–647.

Orcutt, H. K., King, L. A., & King, D. W. (2003). Male-perpetrated violence among Vietnam veteran couples: Relationships with veteran's early life characteristics, trauma history, and PTSD symptomatology. *Journal of Traumatic Stress*, 16, 381–390.

Orthner, D. K., & Rose, R. (2005). *SAF V survey report: Adjustment of Army children to deployment separations*. Prepared for the U.S. Army Community and Family Support Center. Chapel Hill, NC: University of North Carolina.

Parry, C. (2003). Embracing uncertainty: An exploration of the experiences of childhood cancer survivors. *Qualitative Health Research*, 13, 227–246.

Pincus, S. H., House, R., Christensen, J., & Adler, L. E. (2001). The emotional cycle of deployment: A military family perspective. *Journal of the Army Medical Department*, April–June, 615–623.

Pisano, M. C. (1992). The children of Operation Desert Storm: An analysis of California achievement test scores in sixth graders of deployed and nondeployed parents. Dissertation Abstracts International 54/06-A, p. 2126.

Podorefsky, D. L., McDonald-Dowdell, M., & Beardslee, W. R. (2001). Adaptation of preventive interventions for a low-income, culturally diverse community. *Journal of the American Academy of Child and Adolescent Psychiatry*, 40,879–886.

Presidential Task Force on Military Deployment Services for Youth, Families, and Service Members. (2007). *The psychological needs of U.S. military service members and their families: A preliminary report*. Washington, D.C.: American Psychological Association. (http://www.apa.org/releases/MilitaryDeploymentTaskForceReport.pdf).

Rentz, E. D., Marshall, S. W., Loomis, D. et al. (2007). Effect of deployment on the occurrence of child maltreatment in military and nonmilitary families. *American Journal of Epidemiology*, 165, 1199–1206.

Rohner, R. P., & Veneziano R. A. (2001). The importance of father love: History and contemporary evidence. *Review of General Psychology*, 5, 382–405.

Rosen, L. N., Teitelbaum, J. M., & Westhuis, D. J. (1993). Children's reactions to the Desert Storm deployment: Initial findings from a survey of Army families. *Military Medicine*, 158465–158469.

Ryan-Wenger, N. A. (2001). Impact of the threat of war on children in military families. *American Journal of Orthopsychiatry*, 71, 236–244.

Sandler, I. N., Ayers, T. S., Wolchik, S. A., Tein, J. Y., Kwok, O. M. et al. (2003). The Family Bereavement Program: Efficacy evaluation of a theory-based prevention program for parentally bereaved children and adolescents. *Journal of Consulting and Clinical Psychology*, 71, 587–600.

Schumm, W. R., Bell, D. B., & Knott, B. (2001). Predicting the extent and stressfulness of problem rumors at home among Army wives of soldiers deployed overseas on a humanitarian mission. *Psychological Reports*, 89, 123–134.

Stewart, J. L. (2003). "Getting used to it": Children finding the ordinary and routine in the uncertain context of cancer. *Qualitative Health Research*, 13, 394–407.

U.S. Congress Joint Economic Committee. (2007). *This Mother's Day: Helping military moms balance family and longer deployments*. Washington, D.C.: Author (http://www.jec.senate.gov/Documents/Reports/MilitaryMoms05.11.07Final.pdf).

U.S. Department of Defense. (2006). *Men and women in active duty by rank and race*. Washington, D.C.: Defense Manpower Data Center.

Valiente, C., Fabes, R. A., Eisenberg, N., & Spinrad, T. L. (2004). The relations of parental expressivity and support to children's coping with daily stress. *Journal of Family Psychology*, 18, 97–106.

Wright, K. M., Burrell, L. M., Schroeder, E. D., & Thomas, J. L. (2005). Military spouses: Coping with the fear and reality of service member injury and death. In C. A. Castro, A. B. Adler, & T. W. Britt (Eds.), *Military life: The psychology of serving in peace and combat* (Vol. 1, pp. 185–201). Bridgeport, CT: Praeger.

19

Intimate Relationships and the Military[*]

JUDITH A. LYONS

Contents

Intimate Relationships and the Military

In a previous review, Galovski and Lyons (2004) concluded that the impact of combat exposure on the family is mediated by posttraumatic stress disorder (PTSD). Studies from the 1980s to mid-1990s comprised the bulk of the evidence from which that conclusion was derived. At that time, nearly all veterans were men, a literature on deployment stress was just beginning to develop, and there were few studies of family interventions for PTSD. Research from that period found that psychopathology (PTSD more so than other diagnoses) consistently predicted the problems observed in or reported by partners (wives) and children. Of the various symptom clusters that comprise PTSD, numbing/avoidance symptoms were found to be most disruptive of relationship functioning. This chapter builds on our previous examination of marital/partner issues associated with military service and reexamines our prior conclusions in light of recent studies. Increased attention is given to deployment stress and to sexual trauma, although combat trauma remains a major

---

[*] This chapter is the result of work supported with resources and the use of facilities at the G.V. (Sonny) Montgomery VA Medical Center, Jackson, MS. The views expressed here represent those of the author and do not necessarily represent the views of the Department of Veterans Affairs or the University of Mississippi Medical Center.

emphasis. Interventions to strengthen and repair relationships are discussed. Recommendations for future directions are offered.

Identifying Risk: Stressors Associated With Relationship Functioning

Military couples face numerous challenges, particularly in times of war; however, they often embody characteristics and have access to resources that promote resilience (Jensen, Martin, & Watanabe, 1996; Rentz et al., 2007). Thus, the admonition of Cozza, Chun, & Polo (2005) is echoed: It is important to sift through the data with openness to both strengths and liabilities in military couples and to eschew preconceptions or common stereotypes. Beginning with a broad view, the following sections look at deployment as a risk factor, narrow to trauma exposure and PTSD, and, finally, zoom in on individual risk factors among families where PTSD is present.

Stressors Shared With Other Occupations

Particularly during times of war, the challenges facing military families garner public attention. Some of these challenges—such as shift work, relocation, periods of separation, dangerous assignments—are shared with various civilian occupations. Although often assumed to be hardships, such challenges are not uniform in their effect; for example, research on the impact of family relocation has yielded mixed results, with some studies showing negative effects (Wood, Halfon, Scarlata, Newacheck, & Nessim, 1993) and others reporting increasing resilience across moves (Weber & Weber, 2005). Research with both civilian and military samples has shown that shift work often places a strain on relationships via fatigue, interference with social opportunities, unavailability for children's school activities or church and club activities, etc., as perceived by both workers and their partners (Demerouti, Geurts, Bakker, & Euwema, 2004; Smith & Folkard, 1993). Readers, however, can probably think of families in their own experience where shift work was used to an advantage—for example, enabling a couple to provide round-the-clock child care via complementary shifts.

Difficulties can increase when an occupation requires extended absences. Such occupations often involve a concomitant degree of physical risk. Researchers have studied the adjustments in role functioning and coping required by periods of marital separation in the partners of offshore workers, merchant marines, commercial fishermen, and long-haul truckers (Zvonkovic, Solomon, Humble, & Manoogian, 2005; Zvonkovic, Moon, & Manoogian-O'Dell, 2007). The strain of dangerous occupations is further augmented when an acute crisis occurs, as evidenced by long-term stress responses among partners of civilian firefighters involved in rescue and recovery efforts after terrorist attacks (Menendez, Molloy, & Magaldi, 2006; Pfefferbaum et al., 2006). When acute circumstances result in retirement of the worker, the need to adjust coping strategies is further compounded (Menendez et al., 2006). In communities where families associated with a particular occupation socialize together, peer support can be a crucial source of strength to deal with these strains. That community may become particularly vulnerable to vicarious traumatization and impairment of the entire social network when disaster strikes the community as a whole, as when a mine or burning building collapses or a vessel is lost at sea. It is important to keep in perspective the fact that military families are not the only occupational group that shares many of these challenges and that both dysfunction and resilience can result.

Deployment as a Stressor

Whereas some civilian occupations share characteristics as previously noted, it is also true that military deployment during wartime or for dangerous peacekeeping missions creates a convergence of the aforementioned stressors. The literature on deployment stress was sparse at the time of our earlier review but has mushroomed since the Global War on Terror began. Reports of high symptom rates among returning troops are causing alarm. It is important to maintain our perspective, as many reports are based on brief screening tools that were designed to be risk sensitive, erring on the side of over-identification with the understanding that identified cases will subsequently be evaluated more thoroughly to rule in or out clinically significant problems. Additionally, the huge scale of many recent studies makes them statistically sensitive to even small numerical differences.

McCarroll et al. (2000) examined surveys collected from 1990 to 1994 from 26,835 married active-duty soldiers, 43% of whom had been on deployment within the previous year, and found that longer deployments were associated with small but incremental increases in spousal aggression. Newby et al. (2005) analyzed 951 soldiers' reports of positive and negative consequences of a peacekeeping deployment. Married soldiers were more likely to report negative consequences (70% vs. 55% of single soldiers), and 15.4% of married soldiers (vs. 6.3% of single soldiers) reported that they missed important family events while away. Across topics, 63% of all soldiers surveyed reported some positive consequence of deployment (including income, self-improvement, time to think), but married soldiers reported fewer positive consequences (72%) than single soldiers (82%). Interestingly, 9.4% of married soldiers reported that deployment led to improvement in their relationships with their significant others (1.2% of single soldiers reported such improvement).

Hoge, Auchterlonie, and Milliken (2006) analyzed routine postdeployment health assessment data for 303,905 U.S. soldiers and Marines. The data were collected within a year of return from deployment. Rates of mental health problems varied by deployment site: Iraq, 19.1%; Afghanistan, 11.3%; other locales, 8.5%. Deployment to Iraq and screening positive for mental health problems were both associated with attrition from military service. One implication of this finding is that veterans who are having problems may also be leaving the structure and social support network that might otherwise help sustain them, adding other major life events (job change, relocation, change of peer group, change of access to military resources including health care) to the burden they and their families carry.

The U.S. military's fourth Mental Health Advisory Team report (MHAT-IV) (Mental Health Advisory Team, 2006) studied the impact of war on American combat troops. They analyzed survey data from 1320 soldiers, 447 Marines, and 460 members of health/ministry teams. Various focus groups, interviews, and behavioral observations were also incorporated in this wide-ranging report. The team found that service personnel in Iraq longer than 6 months are 1.5 to 1.6 times more likely to screen positive for mental health problems. They are more likely to have marital concerns (31% vs. 19%), report problems with infidelity (17% vs. 10%), and be planning marital separation or divorce (22% vs. 14%). Multiple deployments are associated with a 50% greater prevalence of mental health problems (9% among repeaters vs. 6% among first-time deployers). High (vs. low) combat exposure is associated with 2.4 times more cases screening positive for anxiety, 2.6 times more depression, and 3.5 times more acute stress or PTSD. Readers interested in broader commentary on the MHAT-IV report are referred to the December 2007 special issue of *Traumatology*, 13(4).

Sareen et al. (2007) analyzed interview data from 8441 active Canadian military personnel (29% of the sample were reservists who had been on active duty within the previous 6 months). The sample included troops who had deployed for a variety of combat or peacekeeping operations. Both combat and peacekeeping deployments were associated with increased mental health problems; however, when combat exposure and witnessing atrocities were added to the regression analysis, peacekeeping deployments were no longer associated with increased mental health problems. Thus, similar to the MHAT-IV results, trauma exposure (not merely deployment) appeared to be the key risk factor.

To study the time course of postdeployment problems, Milliken, Auchterlonie, and Hoge (2007) reviewed the postdeployment health assessment and reassessment forms of 88,235 U.S. soldiers who served in Iraq. They found that overall mental health risk indices were higher at 4 to 10 months postdeployment than they had been at homecoming, irrespective of whether the soldier had redeployed in the interim. The item "having thoughts or concerns that you may have serious conflicts with your spouse, family members, or close friends" was broadened at follow-up to also include conflicts at work. Endorsement of this item increased from 3.5% to 14.0% at follow-up within the active-duty subgroup and increased within the National Guard and reserve subgroup from 4.2% to 21.1%. The highest rate of positive screens at follow-up was for PTSD (16.7% of active-duty soldiers and 24.5% of reserve group); 10.3% and 13.0%, respectively, were identified as at risk for depression and 11.8% and 15.0%, respectively, for alcohol abuse. Overall mental health risk increased from 17.0% to 27.1% in the active group and from 17.5% to 35.5% in the reserve group. The authors speculate that the higher symptom rates among Guard and reserve troops may be due transitioning back to civilian employment, reduced access to services, and diminished peer support. As found in their 2006 study (previously described), attrition from the military was more common among soldiers who screened positive for mental health risk (7.4% left military service vs. 5.7% of soldiers who did not screen positive).

Toomey et al. (2007) examined a longer time span. Their follow-up of veterans from the 1991 Gulf War found 18% of veterans who had deployed ($n = 1061$) retrospectively reported war-era onset of mental disorders, in contrast to less than 9% of those who did not deploy ($n = 1128$). Surprisingly, deployment but not combat/war-zone stressor exposure predicted PTSD, whereas war-zone stressor exposure was significantly associated with other anxiety disorders and depression. Some reassurance can be drawn from the finding that prevalence decreased in the decade following deployment. Among the deployed group, the prevalence of depression reduced from 7.1% during the Gulf War period to 3.2% a decade later, PTSD shrank from 6.2% to 1.8%, and other anxiety disorders decreased from 4.3% to 2.8%. The significance of risk factors of age, gender, and other demographic variables varied by diagnosis.

Analysis of a national Veteran Affairs (VA) database of diagnoses listed from October 2001 to December 2005 ($n = 103,788$) (Seal, Bertenthal, Miner, Sen, & Marmar, 2007) found 25% of veterans who served in Iraq or Afghanistan received at least one mental health diagnosis. The percentage identified as having mental health problems increased to 31% if V-codes for psychosocial problems were included along with diagnoses of psychopathology. Veterans ages 24 or younger were found to be at significantly increased risk. Seal and colleagues acknowledge that, because their database includes only those who sought VA care, the percentage may overestimate problems in active-duty troops and veterans who have not felt a need for VA care.

A study of 10,272 British veterans (Browne et al., 2007) revealed that reservists who deployed to Iraq ($n = 786$) were at greater risk for mental health problems than regular military personnel who deployed ($n = 3936$). The study speculated this may be because most reservists did not deploy with their parent unit and lacked the cohesive support network that might have helped sustain troops in regular units. Marital satisfaction was high among both reservists and regulars overall; however, satisfaction was lower among reservists who had deployed to Iraq, in contrast to regulars who deployed and also in contrast to reservists and regulars who did not deploy.

Smith et al. (2008) recently published a prospective study of 50,128 U.S. military personnel. They found that Reserve and National Guard members (as opposed to active-duty personnel), divorced individuals, healthcare specialists, and smokers or problem drinkers were at higher risk for both new-onset PTSD and for persistent PTSD (i.e., already had PTSD prior to deployment that had not remitted by follow-up). Several other variables—including being female, less educated, younger, single, black non-Hispanic, enlisted—also predicted new onset of PTSD following deployment.

The impact of deployment also affects the spouse. In a study of 45 Army couples following the husband's deployment to Iraq or Afghanistan (Goff et al., 2007), 24% of wives reported that their husbands' deployment was the "most traumatic" event they had ever experienced. Other studies, however, report nonsignificant findings, findings that do not rise to the level of clinical significance or are significant only for select subgroups where other risk factors are present or where the effect is temporary (Jensen et al., 1996; Ruger, Wilson, & Waddoups, 2002; Schumm, Bell, Knott, & Rice, 1996). Thus, it would be inaccurate to conclude that the picture is uniformly dire. Identifying which individuals and relationships are at risk is the current challenge.

The adjustment of partners of deployed personnel has been found to correlate with that of their children. In a study of 383 children during deployment of a parent to Operation Desert Storm, Jensen and colleagues (1996) found children to be quite resilient, consistent with data regarding World War II separations (Pesonen et al., 2007). Although Jensen et al. found some increase in symptoms associated with parental deployment, scores tended to remain within the normal range; however, the finding that caretaking parents who are distressed tend to have children who are distressed supports the value of identifying and assisting the families who are in need of services. This point is further highlighted by analyses of a child maltreatment database (n = 1399 military children and 146,583 nonmilitary children) in Texas (Rentz et al., 2007). Overall, rates of substantiated child maltreatment were lower in military families than civilian families; however, rates spiked among military families both when statewide deployment rates were high and again when troops were returning from deployment. This echoes the finding of Milliken and colleagues (2007) that postdeployment, not merely deployment itself, is a time of risk. Rentz et al. (2007) found that the largest proportion of substantiated mistreatment claims among military families was perpetrated by the nonmilitary caretaker and occurred while most troops were still deployed. The research team interprets this finding as emblematic of the stress experienced by the caregiver or parent left at home.

Gibbs, Martin, Kupper, & Johnson (2007) identified 1771 U.S. Army married families in which there was an enlisted (noncommissioned) soldier–parent who had been deployed for some portion of the 40-month study period, a civilian parent, and substantiated child maltreatment by a parent had been recorded during the study interval. For the overall sample, physical and emotional abuse rates were actually lower while the soldier was deployed (the soldier was the perpetrator during 59% of incidents that occurred during nondeployment); however, neglect constituted the majority of all maltreatment incidents, and rates of neglect during deployment nearly doubled. The increase in maltreatment rates associated with deployment was highest when the civilian parent was female. Among civilian mothers, neglect rates were nearly four times greater and physical abuse rates nearly doubled while their husband was deployed.

Trauma and PTSD as Family Risk Factors

Studies from Britain (Iverson et al., 2005), Canada (Sareen et al., 2007), and the United States (Hoge et al., 2002) report that PTSD is less common among military personnel than several other mental health problems. Most studies, however, continue to support the conclusion expressed by Galovski and Lyons (2004) that exposure to combat and atrocities is associated with increased mental health risk and PTSD is one of the best predictors of relationship problems.

Ruger et al. (2002) analyzed archival data from the National Survey of Families and Households. They examined the responses of 3800 men, including veterans who served in World War II, Korea, or Vietnam. Combat exposure was associated with a 62% increase in rate of marital dissolution. Dekel and Solomon (2006) studied Israeli veterans of the 1973 Yom Kippur War. They compared cohabitating female partners ("wives") of men who had PTSD attributed to their being POWs in 1973 ($n = 18$), wives of POWs without PTSD ($n = 64$), and wives of combat veterans who were neither captured nor had PTSD ($n = 72$). The highest distress was found among the female partners of men with PTSD. Captivity without PTSD was associated with an intermediate level of overall distress in the wives but was not associated with marital distress. Similarly, Dekel's (2007) analyses of data from 79 ex-POWs and 74 non-POW combat veterans found that higher PTSD symptom scores for the veteran were associated with greater distress reported by the wife.

A study by Riggs, Byrne, Weathers, & Litz (1998) found that the Vietnam veterans who had PTSD symptoms ($n = 26$) reported more marital distress than the non-PTSD group ($n = 24$). Across three indices of relationship distress, 71 to 75% of the PTSD group reported marital distress on each measure. This was 35 to 48 points higher than the percentage of the non-PTSD group who reported distress on each corresponding measure. A recent study of World War II ex-POWs by Cook, Riggs, Thompson, Coyne, & Sheikh (2004) compared 125 men with PTSD to 206 who did not have PTSD. Similar to Riggs et al., Cook et al. found that the subset with PTSD reported much greater marital problems; however, the percentage of each group that reported problems was lower than in the study by Riggs et al. Using a cut-off score of 98 on the Dyadic Adjustment Scale (DAS) (Spanier, 1976) to define distress, the following percentages in each study were categorized as distressed: Vietnam veterans with PTSD, 75%; Vietnam veterans without PTSD, 32%; World War II ex-POWs with PTSD, 31%; World War II ex-POWs without PTSD, 11%. Both studies found that the PTSD symptom of emotional numbing was particularly associated with relationship problems.

Dirkzwager, Bramsen, Adèr, and van der Ploeg (2005) analyzed survey responses from 708 partners (99% of whom were women) and 332 parents (approximately 60% of whom were mothers) of male Dutch peacekeepers. Most partners (61%) knew the peacekeeper prior to deployment, some (32%) had cohabited prior to deployment, and 91% were living with the peacekeeper at the time of the study. In cases in which a parent was the respondent, the peacekeeper had almost always been residing with the parent prior to deployment (88%) but only 54% were currently living with the parent. Partners of those with PTSD reported less marital satisfaction, having more symptoms themselves, and receiving negative social support. Significant effects were limited to partners, with parents not appearing to be as severely impacted by the peacekeeper's PTSD symptoms. It should be noted, however, that there is a reported confound in location/recency of deployment in that 67% of parents were responding regarding a peacekeeper who served in the most recent theater, some within the past year, whereas only 10% of parents were responding about a peacekeeper who served in the 1970s to 1980s. Thus, the partners and parents were referring to peacekeepers whose PTSD may not be of comparable chronicity, and the respondent's exposure to the symptoms may also be of different durations. The finding of more distress among partners is consistent with our findings at two Mississippi VA trauma recovery programs (Lyons & Root, 2001), discussed in greater detail later in this chapter.

Beckham, Lytle, & Feldman (1996) may have been the first to apply a caregiver burden measure to the study of families coping with PTSD. They found that 50% of partners felt on the verge of a nervous breakdown. Burden reported by the partner, measured by the Burden Interview (a 22-item questionnaire measure) (Zarit, Reever, & Bach-Peterson, 1980), correlated with the severity of PTSD of the veteran. Similarly, Calhoun, Beckham, and Bosworth (2002) found that partners of Vietnam veterans with PTSD (n = 51) reported more burden and distress than partners of veterans without PTSD (n = 20). PTSD was the best predictor of partners' burden scores.

To further assess partners' burden, we contacted 89 female cohabiting partners of veterans who were in ongoing treatment for service-connected PTSD. Partners scored above the 90th percentile on anxiety, depression, and somatization subscales of the Brief Symptom Inventory-18 (BSI) (Derogatis, 2000), with BSI global mean score of 26.3 (Manguno-Mire et al., 2007); 15% of the women reported recent suicidal ideation.

A maladaptive cognitive appraisal style can further increase the partner's risk of distress. In a study of wives of veterans with PTSD, Dollar, Lyons, Kibler, and Ma (2007) found that nearly half (15 of 31) perceived that the strain the veteran's PTSD imposed exceeded their ability to cope (threat appraisal group). Compared to the 16 women who perceived their coping abilities as being equal to or greater than the strain the PTSD imposed (challenge appraisal group), the threat appraisal group reported greater depression and burden.

Much as our research group found partners' appraisal style to be associated with increased distress, Israeli research found differences associated with partners' attachment style. Dekel (2007) found that more severe PTSD in POWs was associated with greater distress in their wives, as previously discussed; however, the relationship was stronger among women with certain attachment styles than among others. An even more striking differentiation was found in levels of posttraumatic growth among the wives. For wives who demonstrated a secure attachment style, greater PTSD symptoms in the veteran predicted more growth in the wife, whereas the direction of the correlation was reversed (more PTSD in husband was associated with less growth in the wife) among wives with an insecure/avoidant attachment style.

The effects of PTSD and family distress can spiral. In the large British study by Browne et al. (2007), described earlier, it was found that most postdeployment mental health problems were predicted by variables associated with experiences during deployment (such as unit cohesion and combat events). Counterintuitively, PTSD did not conform to this pattern. PTSD symptoms were more strongly associated with problems at home during and since deployment. Authors of that study speculate that PTSD symptoms may have contributed to problems at home. Alternatively, they also suggest that symptoms that might otherwise have remitted spontaneously instead become entrenched if there are adverse domestic experiences. This latter interpretation is consistent with the findings of Tarrier, Sommerfield, & Pilgrim (1999) that negative family relationships accounted for nearly 20% of the variance in PTSD treatment outcome in a civilian sample.

Numbing/Withdrawal and Anger   Galovski and Lyons (2004) concluded that numbing/withdrawal accounts for much of the difficulty traumatized veterans have with relationships. A recent study by Lunney and Schnurr (2007) further implicates emotional numbing. They examined the impact of PTSD symptom clusters on various aspects of quality of life in 319 male Vietnam veterans with PTSD before and after treatment. Only changes in numbing were significantly related to changes in relationship satisfaction across analyses.

Even though evidence pointing to numbing as the primary impediment to relationship functioning continues to grow, there is reason to continue to evaluate anger and the toll it takes on families after deployment. As previously noted, Milliken et al. (2007) identified conflict-management concerns in 14% of their postdeployment sample (21% of the reservist subsample). Jakupcak et al. (2007) report that hostility and anger are particularly evident among returning veterans who have PTSD. In our phone survey of partners of Vietnam veterans diagnosed with PTSD, 60% reported feeling physically threatened by the veteran (Manguno-Mire et al., 2007).

Among 179 couples (veterans' ages ranging from 23 to 83) seeking relationship therapy (Sherman, Sautter, Jackson, Lyons, & Han, 2006), 81% of veterans with PTSD or depression had perpetrated domestic violence in the past year (significantly more than the 46% of the comparison group diagnosed with adjustment disorder or partner relational problem). Furthermore, 42% and

45% of veterans with depression or PTSD, respectively, reportedly committed at least one act of domestic violence in the past year. In spite of how alarmingly high these rates are, they are likely to underestimate the true populations rates of violence because most referral sources knew that the particular clinic in which the data were collected would not enroll couples who were actively abusing substances or engaging in domestic violence.

Dekel and Solomon's (2006) study of Israeli veterans found that the man's aggression and female partner's level of self-disclosure were both associated with the woman's distress level. The veterans who had been POWs and had PTSD had the poorest marital adjustment and were more physically aggressive than ex-POWs without PTSD and combat veterans who had neither been prisoners nor had PTSD.

It is possible that which variable (numbing or anger) seems more important may be a result of which person in the relationship you ask and how precisely you define the variables; for example, Riggs et al. (1998) emphasized numbing and Evans, McHugh, Hopwood, & Watt (2003) emphasized anger in their publications. The findings of both studies, however, are much more similar than they appear at first glance. Riggs et al. recruited U.S. Vietnam veterans and their female partners through newspaper ads and flyers in a VA medical center. Of the 50 men, 26 met diagnostic criteria for PTSD and 24 did not. Calculated from the published group means, the overall average score on the PTSD Checklist–Military Version (PCL-M) (Weathers, Litz, Herman, Huska, & Keane, 1993) was 36. Evans et al. studied Australian Vietnam veterans at intake to a PTSD treatment program. Their treatment-seeking sample had a higher average PCL-M score than that of Riggs et al. (calculated from the published cluster scores, the average total was 66). Riggs et al. compared the PCL-M to measures of dyadic relationship quality, whereas Evans et al. compared the PCL-M to a measure of overall family functioning.

Both studies found that the avoidance cluster of symptoms was significantly associated with the veteran's rating of relationship/family variables but did not significantly predict the partner's ratings. Riggs et al. refined this analysis further and found that the emotional numbing subset of items within the avoidance symptom cluster (constricted affect, detachment, loss of interest in pleasurable activities) significantly predicted the relationship ratings of both the veteran and his partner. Evans et al. did not test whether partner ratings could be predicted by the numbing subset of symptoms; however, Evans et al. included additional anger measures that were not studied by Riggs et al. and found that the veteran's anger did predict both veteran and partner ratings of family functioning. Thus, the two studies are not as contradictory as they first appear, but instead yield surprisingly similar results on the analyses that were parallel, in spite of differences in symptom severity and different instruments used to assess interpersonal relationships.

Sexual Trauma   The Mental Health Advisory Team (2006) found that 12 to 13% of female troops in Iraq reported acute stress/PTSD symptoms, regardless of level of combat exposure. Sexual trauma was not reported but may account for this otherwise anomalous finding (Lyons, 2007a). Exact rates of sexual assault in the military are unclear, as 69 to 90% of the incidents may go unreported (Hunter, 2008). Physical sexual assault is reported by 23 to 41% of female troops and veterans (Hunter, 2008; Suris, Lind, Kashner, Borman, & Petty, 2004; Yaeger, Himmelfarb, Cammack,, & Mintz, 2006). Of reported military sexual assault victims, 9% are men, and 85% of assaults against men involve multiple assailants (Hunt, 2008). In a study of 196 women veterans, Yaeger et al. (2006) found that military sexual trauma led to PTSD in more cases (60%) than did other types of trauma (43%). Similarly, Suris et al. (2004) found that women who experienced military sexual assault were three times more likely to have PTSD than women who were sexually assaulted in other contexts (childhood, civilian) and nine times more likely than women who had no sexual assault history. Sexual trauma can elicit feelings of anger, helplessness, and guilt in male partners of women who were assaulted by others (Smith, 2005). Both men and women report that a trauma history interferes with sexual intimacy, but this finding is not limited to sexual trauma (Goff, Crow, Reisbig, & Hamilton, 2006).

Striking a Balance: Issues in Screening and Data Interpretation

Research is striving to refine screening strategies; for example, in a study of PTSD in Dutch infantry troops, Engelhard et al. (2007) found that questionnaires significantly overestimated rates of pathology relative to clinical interviews. The rates determined by interview were 41% lower than those indicated by symptom questionnaires. Wright et al. (2007) contrast the merits of various measures and offer recommendations on how to balance the competing goals of sensitivity and specificity. Smith et al. (2008) compare different scoring strategies when using the PCL as a screening tool. Smith et al. (2008) also point out that the rate of PTSD and other disorders attributed to deployment and combat will be over-inflated if predeployment/noncombat rates are not subtracted. In their study of 50,128 U.S. troops, they concluded that up to 76% of new-onset cases of PTSD during deployment can be attributed to combat exposure.

Milliken et al. (2007) found that the incidence of positive PTSD screens increased across several months of postdeployment, as previously described, but the same individuals were not necessarily identified as at risk at the two points in time. Of soldiers who screened positive at the initial postdeployment assessment, most (59.2% of the active group and 49.4% of the reserve group) had improved before the second assessment without treatment. Even higher percentages of depressive symptoms had resolved from time 1 to time 2 (62.2% active and 56.0% reserve). Such improvement is fully consistent with the symptom course DSM-IV outlines for PTSD: "complete recovery occurring within

3 months in approximately half of cases" (APA, 1994, p. 426). It is also similar to the finding by Smith et al. (2008) that more than half of military personnel who screened positive for PTSD prior to deployment no longer screened positive at postdeployment follow-up. The increase in absolute numbers screening positive in the Milliken et al. (2007) study was due to the fact that the number improved was more than doubly offset by new cases who screened negative at time 1 but positive at time 2.

Goff et al. (2007) demonstrate several of the ambiguities present in the literature, so their study is presented here in some detail. This is not meant as a criticism of the study, as its authors fairly acknowledge most of these points in their article. The study assessed, via questionnaire and interview, couples' relationship satisfaction after the husbands' recent deployments to Iraq or Afghanistan. The research team recruited 45 couples from the communities near two Army posts. They examined relationship satisfaction relative to levels of trauma and trauma-related symptom endorsement. Specific hypotheses were tested regarding which symptom variables would predict marital satisfaction, but these predictions were not supported. Other statistical associations contrary to prediction were found to be significant. There appear to be two very different ways of describing the study's findings, thus it provides an excellent example of the role of cognitive appraisal in helping couples frame their perspective of stressors. The study's abstract summarizes the findings as follows (p. 344): "The results indicated that increased trauma symptoms … in the soldiers significantly predicted lower marital/relationship satisfaction for both soldiers and their female partners. The results suggest that individual trauma symptoms negatively impact relationship satisfaction in military couples in which the husband has been exposed to war trauma." This is an accurate depiction of the correlations and regression analyses conducted. Given that the war experiences cannot be erased, however, it presents a disheartening view that may not be the only way of looking at the data. Another slant on the findings, based on a subset of results reported in the text as a "limitation" of the study, offers a much more positive perspective. In spite of the fact that they had just undergone the strain of deployment, combat trauma exposure, and reunion, the couples participating in the study showed high average DAS scores (greater than 113 for both wives and soldiers), had scores on PTSD symptom checklists that were well below those of clinical samples, and were "overall, a highly satisfied sample of couples" (p. 352). Thus, rather than evidence that trauma creates marital problems (even though there were significant correlations, as previously reported), this sample can actually be reinterpreted as demonstrating the resilience of military couples in overcoming the challenges of deployment separation and war trauma. This study is also an example of why clinicians must be careful not to assume that all trauma among service personnel is combat trauma. Even though 82% of the men listed war trauma as their most traumatic experience, other types of trauma were also reported (including 4 of the 45 men reporting childhood sexual abuse).

Interventions

Deployment/Postdeployment

One of the primary findings from the MHAT-IV report (Mental Health Advisory Team, 2006, p. 79) was that positive leadership may be the "panacea, a silver bullet for sustaining the mental health and well-being of the deployed force." Rates of mental health problems among high combat troops who had a good noncommissioned officer (NCO) were barely higher than those of low combat troops with a poor NCO and were approximately half the rate seen when high combat troops have a poor NCO. Ambiguous or false information can contribute to family and community stress during difficult times (Walsh, 2007). Good leadership at the unit level can help address this problem and is associated with reduced stress reported by wives during their husbands' deployment on a humanitarian mission (Schumm et al., 2001).

Good leaders can also set the tone with which sexual assault issues are dealt. The reactions that sexual assault survivors receive from others are associated with both posttraumatic symptoms and positive adjustment and growth, although exactly what is most helpful from whom is less clear (Borja, Callahan, & Long, 2006; Campbell, Ahrens, Sefl, Wasco, & Barnes, 2001; Campbell & Raja, 2005). Regarding military sexual assaults, it is of concern that Campbell and Raja found that efforts to report such assaults were often rebuffed by military officials. Of the 39 female veterans in their sample who reported assaults during the military, 83% said their experiences with military legal personnel made them reluctant to seek additional help, and negative reactions on the part of legal authorities were significantly correlated with increased symptoms. The assaults in question took place an average of 11.6 years prior to data collection. Sensitivity to this type of trauma has likely improved in the intervening years, but this remains a topic about which leadership should remain vigilant.

The military acknowledges that family access to mental health care has been less than optimal in the past and efforts to improve access are ongoing (Pueschel, 2007). Much of the emphasis to date has been on psychoeducational/growth-oriented support groups during deployment (Jensen et al., 1996). Most such groups are conducted and other informational materials disseminated without collection of outcome data.

Contemporary interventions for helping families cope with deployment stress remain similar to the guidelines presented by Black (1993) and Peebles-Kleiger and Kleiger (1994). Most are organized in stages, akin to the phases of deployment stress identified by Peebles-Kleiger and Kleiger (i.e., initial notification, departure, emotional disorganization of the family during the initial weeks of the loved one's deployment, family stabilization, anticipation of homecoming, reunion, and reintegration/stabilization). Recent data have supported Black's warning that the reunion can be more stressful than the separation.

Many of the issues and strategies that Walsh (2007) advocates to promote resilience after disasters can be applied to deal with deployment stress, the impact of physical and emotional disabilities, combat deaths, and other trauma. Walsh emphasizes the importance of developing individual and community meaning, "weaving the experience of loss and recovery into the fabric of individual and collective identity and life passage" (p. 210). The importance of restoring routines and bolstering existing networks is emphasized. Walsh highlights the following: "(1) to normalize and contextualize distress, (2) to draw out strengths and active coping strategies for empowerment, (3) to offer follow-up sessions and mental health services for those in severe or persistent distress, and (4) to mobilize family and social support for ongoing recovery" (p. 217). Guidelines are provided for group facilitators that are specific enough to be helpful without being rigidly manualized (Walsh, 2007, p. 218). Similar to issues identified in the MHAT-IV report, Walsh highlights the importance of effective leadership (e.g., provision of accurate and unambiguous information) and the need for creativity to deliver services in nonclinical settings where individuals feel more at ease (e.g., local community center).

Recognizing that reserve troops often have limited access to mental health services, efforts such as the Strategic Outreach to Families of all Reservists (SOFAR) project (Darwin & Reich, 2006) have been launched. The SOFAR project has recruited psychoanalysts and psychodynamic therapists in private practice to volunteer at least one hour weekly to offer *pro bono* services. Darwin and Reich's article offers tips on setting up similar programs, and additional training materials are offered by contacting help@sofausa.org. Give an Hour (giveanhour.org), another rapidly growing *pro bono* mental health program, is available to troops and families affected by either reserve or regular military duty in Iraq and Afghanistan.

Numerous resources are now available online for families and clinicians. BATTLEMIND (www.battlemind.org) offers a series of pamphlets and training programs to help troops and families. The Deployment Health Clinical Center (www.pdhealth.mil) and National Center for Posttraumatic Stress Disorder (www.ncptsd.org) are also good starting points. Figley (2005) calls attention to a helpful online handout, "Becoming a Couple Again: How to Create a Shared Sense of Purpose After Deployment" (www.usuhs.mil/psy/RFSMC.pdf). Organizations such as Hearts Toward Home International market packaged workshops for returning troops who have experienced trauma and their families, including participant workbooks and facilitators manuals (www.heartstowardhome.com).

Interventions for Families Where PTSD Is Present

After interviewing couples engaged in therapy, Goff et al. (2006, p. 458) concluded: "Previous trauma appears to act as a 'phantom' in the couple relationship, always present but not always seen or understood by either partner." Therapy cannot undo the trauma exposure, but it can effectively treat the PTSD. The *Journal of Clinical Psychiatry* published treatment recommendations (Foa & Davidson, 1999) that include a synopsis written specifically for clients and families. For clinicians, more technical information is available in a textbook by Foa, Keane, & Friedman (2000). Many of the therapies earned an "A" rating on the A to F scale employed in the text by Foa et al., indicating that the evidence supporting their use is from well-controlled, randomized clinical trials. Links to additional practice guidelines for PTSD are available at http://www. ncptsd.va.gov/ncmain/resources/treatment.jsp. Specialized PTSD services are widely available within military and VA healthcare systems. Readers should be aware that care is even more accessible for sexual trauma than for other types of trauma; any veteran who experienced military sexual trauma is eligible for free counseling from Veterans Affairs Medical Centers/Vet Centers regardless of gender, era of service, income, or disability rating.

Riggs (2000) reviewed marital and family therapies for PTSD and identified two major treatment approaches: systemic and support. Systemic approaches treat the relationship to decrease conflict or strengthen bonds; the outcome of interest is relationship quality. In contrast, the aim of support treatments is to enhance the trauma survivor's PTSD treatment by increasing social support and family treatment engagement, with reduced PTSD being the target outcome. Support treatments often educate family about PTSD and help them find ways to cope with their loved one's PTSD symptoms. Using the A to F rating system to evaluate the strength of the evidence for their impact, Riggs rated marital and family therapy for PTSD as an "E." This reflects Riggs' judgment that the interventions were based on long-standing clinical practice by select groups of clinicians but were not in widespread use, nor had they been empirically tested for use with PTSD.

Glynn et al. (1999) published a study of behavioral family therapy as a supplement to exposure therapy for veterans with PTSD. The emphasis was on coping skills training with the primary target outcome being a reduction in the veteran's PTSD, thus the treatment would be categorized by Riggs as a support intervention. The family therapy component consisted of 3 sessions of orientation and evaluation, 2 informational sessions about PTSD/mental health treatment, and 11 to 13 skills training sessions (communication, anger control, problem solving). The family/exposure therapy combination ($n = 11$) resulted in approximately twice as large a reduction in PTSD symptoms as was achieved by exposure therapy alone ($n = 12$), but the difference was not statistically significant. Glynn et al. (1995) detail the interventions evaluated later in the 1999 empirical report and present case examples. The 1995 article also offers recommendations for dealing with avoidance, aggression, alexithymia, substance use, and trauma disclosure during behavioral family therapy.

Monson, Schnurr, Stevens, and Guthrie (2004) reported the impact of cognitive–behavioral couple's therapy on the PTSD symptoms and relationship quality of seven Vietnam veterans. The 15-session manualized intervention, described in detail by Monson, Guthrie, and Stevens (2003), included a strong systemic emphasis. Two psychoeducational sessions provided information about PTSD and associated relationship problems. The remaining sessions addressed communication skills training and cognitive interventions. Some veterans reported improved PTSD and relationships but others reported deterioration. Ratings by the clinicians and wives were more positive, as were veterans' reports of reduced anxiety and depression. Interpretation is limited by the small sample size and lack of a comparison group or multiple baseline. Larger, controlled studies are needed.

Neither systemic or support approaches focus on the needs of the significant other, but recent studies indicate that spouses and other partners are seeking that treatment emphasis (Lyons & Root, 2001). We conducted a survey at workshops held for families of veterans who were in treatment for PTSD. Nonspouses (children, parents, siblings, friends) told us that their role in helping the veteran deal with the PTSD symptoms was limited and they asked for the types of supportive and systemic services that had traditionally been offered; however, the 30 spouses we surveyed reported a different need. They described their role in managing their husband's PTSD symptoms as large or "very large … more than the treatment team." They help the veterans get to their appointments and adhere to their medication regimens, step in to fill roles vacated by the veterans, and structure the family's lifestyle around sheltering these veterans from situations that might spark a relapse. Many reported exhaustion from being the breadwinner plus primary caretaker for children, aged parents, and the veteran. Most were already well versed about PTSD and had no desire for informational sessions. One third of the spouses expressed interest in systemic therapy to enhance their relationship or reduce shared stress. The most frequent request, however, was for therapy to address their individual needs. Many spouses also wanted social activities to offset the isolation imposed by their husband's social avoidance.

When we conducted a multisite phone survey to further explore partners' needs (Sherman et al., 2005), 89 women partners of Vietnam veterans disabled by PTSD were asked to describe the services that could help them "better support their loved ones that have PTSD." Of these women, 54% asked for a partners-only group, 19% requested individual therapy for themselves, and 13% asked for couples therapy. Somewhat contrary to our earlier results (perhaps due to the veteran-focused wording of the question), 20% were seeking more information about trauma and PTSD.

The high level of caregiver burden partners reported in our phone survey was correlated with feeling incapable of controlling the veteran's emotional difficulties or their own coping with his symptoms (Sautter et al., 2006). Their ratings of burden also correlated with their reports of barriers in access to clinical care (e.g., distance, cost, bureaucratic hurdles). Such barriers vary across sites (Lyons & Root, 2001). In an effort to reduce barriers (Lyons, 2003), we developed a workbook-based treatment that partners can complete at home. The intervention is specifically designed to help the subset of partners identified as having a dysfunctional cognitive appraisal style. Pilot data to date are encouraging (Dollar et al., 2007). Feedback from participants indicates that the workbook (supplemented by weekly phone calls) helps women to view stressors more adaptively and helps them feel empowered to consider coping options.

Several good informational resources are available. A 29-minute DVD or videotape, *PTSD: Families Matter* (Abrams & Freeman, 2000), was developed by the VA's South Central Mental Illness Research, Education, and Clinical Center and is available to licensed clinicians by contacting Michael. Kauth@va.gov. It depicts common issues couples face when one partner has PTSD. Another video, *Living with PTSD: Lessons for Partners, Friends, and Supporters*, is available online at http://www.giftfromwithin.org. The National Center for PTSD offers information for families at http://www.ncptsd.va.gov/ncmain/veterans/. A free book, *Veterans and Families' Guide to Recovering from PTSD* (Lanham, 2005), provides helpful information, as well as essays by veterans and family members and a resource directory. Lyons (2007b), Mason (1990, 1998), and Matsakis (1996, 1998) offer additional advice for families.

Discussion

Military couples face many challenges but also demonstrate many strengths. Of the studies reviewed, many found military couples—on average—to be as well adjusted or functioning better than comparison groups in spite of stressors they encountered. Nonetheless, on the individual family level, a significant subset of couples clearly are suffering. Many more intervention options are now available than were offered during previous war eras. We have empirical confirmation that intervention begins in the field, as good military leadership has been found to be a powerful buffer against the effects of stressors and trauma. The needs of the partner and family (rather than merely the service member) are gaining emphasis.

We know some of the characteristics and situations that place individuals and relationships at risk. We see that reservists and their families are at increased risk. We know that some indices of deployment stress peak rather than resolve at homecoming and after deployment. Recent studies reconfirm that combat and atrocity exposure and PTSD are associated with significant relationship problems. Anger and numbing are both implicated as threats to relationship satisfaction and family functioning.

Several topic areas call for more research. Specific treatment matching is encouraged, wherein targeted interventions are offered to couples and individuals based on prescreening. We need more controlled studies to evaluate intervention effectiveness. We are missing an important opportunity to learn about the prevalence and impact of military sexual trauma if we fail to assess for that in the many epidemiological databases currently being developed. Many aspects of posttraumatic growth remain a mystery, such as how seemingly positive and negative outcomes can both occur in a relationship (Goff et al., 2006; Nelson, Wangsgaard, Yorgason, Kessler, & Carter-Vassol, 2002) and what reactions from loved ones are helpful vs. hurtful (Borja et al., 2006; Campbell et al., 2001; Dorval et al., 2005). Ongoing advocacy is also needed to ensure that policymakers continue to recognize that healthy families and support networks promote resilience and facilitate recovery from the stress of deployment and from traumatic experiences.

References

Abrams, P., & Freeman, T. (2000). *PTSD: Families matter* (DVD/videotape). North Little Rock, AR: U.S. Department of Veterans Affairs, VA South Central MIRECC (Michael.Kauth@med.va.gov).

APA. (1994). *Diagnostic and statistical manual of mental disorders* (4th ed.). Washington, D.C.: American Psychiatric Association.

Beckham, J. C., Lytle, B. L., & Feldman, M. E. (1996). Caregiver burden in partners of Vietnam war veterans with posttraumatic stress. *Journal of Consulting and Clinical Psychology, 64*, 1068–1072.

Black, W. G., Jr. (1993). Military-induced family separation: A stress reduction intervention. *Social Work, 38*, 273–280.

Borja, S. E., Callahan, J. L., & Long, P. J. (2006). Positive and negative adjustment and social support of sexual assault survivors. *Journal of Traumatic Stress, 19*, 905–914.

Browne, T., Hull, L., Horn, O. Jones, M., Murphy, D., Fear, N. T., Greenberg, N., French, C., Rona, R. J., Wessely, S., & Hotopf, M. (2007). Explanations for the increase in mental health problems in UK reserve forces who have served in Iraq. *British Journal of Psychiatry, 190*, 484–489.

Calhoun, P. S., Beckham, J. C., & Bosworth, H. B. (2002). Caregiver burden and psychological distress in partners of veterans with chronic posttraumatic stress disorder. *Journal of Traumatic Stress, 15*, 205–212.

Campbell, R., & Raja, S. (2005). The sexual assault and secondary victimization of female veterans: Help-seeking experiences with military and civilian social systems. *Psychology of Women Quarterly, 29*, 97–106.

Campbell, R., Ahrens, C. E., Sefl, T., Wasco, S. M., & Barnes, H. E. (2001). Social reactions to rape victims: Healing and hurtful effects on psychological and physical health outcomes. *Violence and Victims, 16*, 287–302.

Cook, J. M., Riggs, D. S., Thompson, R., Coyne, J. C., & Sheikh, J. I. (2004). Posttraumatic stress disorder and current relationship functioning among World War II ex-prisoners of war. *Journal of Family Psychology, 18*, 36–45.

Cozza, S. J., Chun, R. S., & Polo, J. A. (2005). Military families and children during Operation Iraqi Freedom. *Psychiatric Quarterly, 76*, 371–378.

Darwin, J. L., & Reich, K. I. (2006). Reaching out to the families of those who serve: The SOFAR project. *Professional Psychology: Research and Practice, 37*, 481–484.

Dekel, R. (2007). Posttraumatic distress and growth among wives of prisoners of war: The contribution of husbands' posttraumatic stress disorder and wives' own attachments. *American Journal of Orthopsychiatry*, 77, 419–426.

Dekel, R., & Solomon, Z. (2006). Secondary traumatization among wives of Israeli POWs: The role of POWs' distress. *Social Psychiatry and Psychiatric Epidemiology*, 41, 27–33.

Demerouti, E., Geurts, S. A., Bakker, A. B., & Euwema, M. (2004). The impact of shiftwork on work–home conflict, job attitudes and health. *Ergonomics*, 47, 987–1002.

Derogatis LR (2000). *Brief Symptom Inventory-18* (BSI-18). Minneapolis, MN: Pearson Assessments.

Dirkzwager, A. J. E., Bramsen, I., Adèr, H., & van der Ploeg, H. M. (2005). Secondary traumatization in partners and parents of Dutch peacekeeping soldiers. *Journal of Family Psychology*, 19, 217–226.

Dollar, K. M., Lyons, J., Kibler, J. L., & Ma, M. (2007, March). Feasibility and efficacy of a home study workbook for caregivers of veterans with PTSD: A pilot intervention. Poster presented at the annual meeting of the Society of Behavioral Medicine, Washington, D.C.

Dorval, M., Guay, S. Mondor, M., Mâsse, B., Falardeau, M., Robidoux, A., Deschênes, L., & Maunsell, E. (2005). Couples who get closer after breast cancer: Frequency and predictors in a prospective investigation. *Journal of Clinical Oncology*, 23, 3588–3596.

Engelhard, I. M., van den Hout, M. A. M, Weerts, J., Arntz, A., Hox, J. J. C. M., & McNally, R. J. (2007). Deployment-related stress and trauma in Dutch soldiers returning from Iraq. *British Journal of Psychiatry*, 191, 140–145.

Evans, L., McHugh, T., Hopwood, M., & Watt, C. (2003). Chronic posttraumatic stress disorder and family functioning of Vietnam veterans and their partners. *Australian and New Zealand Journal of Psychiatry*, 37, 765–772.

Figley, C. R. (2005). Strangers at home: Comment on Dirkzwager, Bramsen, Adèr, and van der Ploeg. *Journal of Family Psychology*, 19, 227–229.

Foa, E. B., & Davidson, J. R. T. (1999). Treatment of posttraumatic stress disorder (expert consensus guideline series). *Journal of Clinical Psychiatry*, 60(Suppl. 16), 1–76 (full guidelines, http://www.psychguides.com/ecgs10.php; family guide only, http://www.psychguides.com/pfg12.php).

Foa, E. B., Keane, T. M., & Friedman, M. J. (2000). *Effective treatments for PTSD: Practice guidelines from the International Society of Traumatic Stress Studies*. New York: Guilford Press.

Galovski, T., & Lyons, J. A. (2004). Psychological sequelae of combat violence: A review of the impact of PTSD on the veteran's family and possible interventions. *Aggression and Violent Behavior*, 9, 477–501.

Gibbs, D. A., Martin, S. L., Kupper, L. L., & Johnson, R. E. (2007). Child maltreatment in enlisted soldiers' families during combat-related deployments. *Journal of the American Medical Association*, 298, 528–535.

Glynn, S. M., Eth, S., Randolph, E. T., Foy, D. W., Leong, G. B., Paz, G. G. et al. (1995). Behavioral family therapy for Vietnam veterans with posttraumatic stress disorder. *Journal of Psychotherapy Practice*, 4, 214–223.

Glynn, S. M., Eth, S., Randolph, E. T., Foy, D. W., Urbaitis, M., Boxer, L. et al. (1999). A test of behavioral family therapy to augment exposure for combat-related posttraumatic stress disorder. *Journal of Consulting and Clinical Psychology*, 67, 243–251.

Goff, B. S. N., Crow, J. R., Reisbig, A. M. J., & Hamilton, S. (2007). The impact of individual trauma symptoms of deployed soldiers on relationship satisfaction. *Journal of Family Psychology*, 21, 344–353.

Goff, B. S. N., Reisberg, A. M. J., Bole, A., Scheer, T., Hayes, E., Archuleta, K. L., Henry, S. B., Hoheisel, C. B., Nye, B., Osby, J., Sanders-Hahs, E., Schwerdtfeger, K. L., & Smith, D. B. (2006). The effects of trauma on intimate relationships: A qualitative study with clinical couples. *American Journal of Orthopsychiatry, 76*, 451–460.

Hoge, C. W., Lesikar, S. E., Guevara, R., Lange, J., Brundage, J. F., Engel, C. C., Messer, S. C., & Orman, D. T. (2002). Mental disorders among U.S. military personnel in the 1990s: Association with high levels of health care utilization and early military attrition. *American Journal of Psychiatry, 159*, 1576–1583.

Hoge, C. W., Auchterlonie, J. L., & Milliken, C. S. (2006). Mental health problems, use of mental health services, and attrition from military service after returning from deployment to Iraq or Afghanistan. *Journal of the American Medical Association, 295*, 1023–1032.

Hunter, M. (2008). Sexual abuse: Another source of military trauma. *The National Psychologist, 17*(1), 11.

Iversen, A., Dyson, C., Smith, N., Greenberg, N., Walwyn, R., Unwin, C., Hull, L., Hotopf, M., Dandeker, C., Ross, J., & Wessely, S. (2005). "Goodbye and good luck": The mental health needs and treatment experiences of British ex-service personnel. *British Journal of Psychiatry, 186*, 480–486.

Jakupcak, M., Conybeare, D., Phelps, L., Hunt, S., Holmes, H. A., Felker, B., Klevens, M., & McFall, M. E. (2007). Anger, hostility, and aggression among Iraq and Afghanistan war veterans reporting PTSD and subthreshold PTSD. *Journal of Traumatic Stress, 20*, 945–954.

Jensen, P. S., Martin, D., & Watanabe, H. (1996). Children's response to parental separation during Operation Desert Storm. *Journal of the American Academy of Child and Adolescent Psychiatry, 35*, 433–441.

Lanham, S. L. (2005). *Veterans and families' guide to recovering from PTSD* (3rd ed.). Annandale, VA: Purple Heart Service Foundation (while supplies last, a free copy should be available through local veteran centers; visit http://www.vetcenter.va.gov/ to find the nearest one).

Lunney, C. A., & Schnurr, P. P. (2007). Domains of quality of life and symptoms in male veterans treated for posttraumatic stress disorder. *Journal of Traumatic Stress, 20*, 955–964.

Lyons, J. A. (2003). Veterans Health Administration: Reducing barriers to access. In B. H. Stamm (Ed.), *Rural behavioral health care: An interdisciplinary guide* (pp. 217–229). Washington, D.C.: American Psychological Association.

Lyons, J. A. (2007a). Commentary on MHAT-IV: Struggling to reduce the psychological impact of war. *Traumatology, 13*, 40–45.

Lyons, J. A. (2007b). The returning warrior: Advice for families and friends. In C. R. Figley & W. P. Nash (Eds.), *Combat stress injury: Theory, research, and management* (pp. 311–324). New York: Routledge.

Lyons, J. A., & Root, L. P. (2001). Family members of the PTSD veteran: Treatment needs and barriers. *National Center for Posttraumatic Stress Disorder Clinical Quarterly, 10*(3), 48–52.

Manguno-Mire, G., Sautter, F., Lyons, J., Myers, L., Perry, D., Sherman, M., Glynn, S., & Sullivan, G. (2007). Psychological distress and burden among female partners of combat veterans with PTSD. *Journal of Nervous and Mental Disease, 195*, 144–151.

Mason, P. H. C. (1990). *Recovering from the war: A woman's guide to helping your Vietnam vet, your family, and yourself.* New York: Viking Penguin.

Mason, P. H. C. (1998). *Recovering from the war: A guide for all veterans, family members, friends and therapists* (2nd ed.). High Springs, FL: Patience Press.

Matsakis, A. (1996). *Vietnam wives* (2nd ed.). Lutherville, MD: Sidran Press.

Matsakis, A. (1998). *Trust after trauma: A guide to relationships for survivors and those who love them*. Oakland, CA: New Harbinger Publications.

McCarroll, J. E., Ursano, R. J., Liu, X., Thayer, L. E., Newby, J. H., Norwood, A. E., & Fullerton, C. S. (2000). Deployment and the probability of spousal aggression by U. S. Army soldiers. *Military Medicine*, 165, 41–44.

Menendez, A. M., Molloy, J., & Magaldi, M. C. (2006). Health responses of New York City firefighter spouses and their families post-September 11, 2001 terrorist attacks. *Issues in Mental Health Nursing*, 27, 905–917.

Mental Health Advisory Team. (2006). *Mental Health Advisory Team (MHAT) IV: Operation Iraqi Freedom 05-07*, Final Report. Washington, D.C.: Office of the Surgeon Multinational Force–Iraq and Office of the Surgeon General U.S. Army Medical Command (http://www.scribd.com/doc/134591/mhat-iv-report; alternative source, www.armymedicine.army.mil/news/mhat/mhat_iv/MHAT_IV_Report_17NOV06.pdf).

Milliken, C. S., Auchterlonie, J. L., & Hoge, C. W. (2007). Longitudinal assessment of mental health problems among active and reserve component soldiers returning from the Iraq war. *Journal of the American Medical Association*, 298, 2141–2148.

Monson, C. M., Guthrie, K. A., & Stevens, S. P. (2003). Cognitive–behavioral couple's treatment for posttraumatic stress disorder. *Behavior Therapist*, 26, 393–402.

Monson, C. M., Schnurr, P. P., Stevens, S. P., & Guthrie, K. A. (2004). Cognitive–behavioral couples treatment for posttraumatic stress disorder: Initial findings. *Journal of Traumatic Stress*, 17, 341–344.

Nelson, B. S., Wangsgaard, S. Yorgason, J., Kessler, M. H., & Carter-Vassol, E. (2002). Single- and dual-trauma couples: Clinical observations of relational characteristics and dynamics. *American Journal of Orthopsychiatry*, 72, 58–69.

Newby, J. H., McCarroll, J. E., Ursano, R. J., Fan, Z., Shigemura, J., & Tucker-Harris, Y. (2005). Positive and negative consequences of a military deployment. *Military Medicine*, 170, 815–819.

Peebles-Kleiger, M. J., & Kleiger, J. H. (1994). Re-integration stress for Desert Storm families: Wartime deployments and family trauma. *Journal of Traumatic Stress*, 7, 173–194.

Pesonen, A. K., Räikkönen, K., Heinonen, K., Kajantie, E., Forsén, T., & Eriksson, J. G. (2007). Depressive symptoms in adults separated from their parents as children: A natural experiment during World War II. *American Journal of Epidemiology*, 166, 1126–1133.

Pfefferbaum, B., Tucker, P., North, C. S., Jeon-Slaughter, H., Kent, A. T., Schorr, J. K., Wilson, T. G., & Bunch, K. (2006). Persistent physiological reactivity in a pilot study of partners of firefighters after a terrorist attack. *Journal of Nervous and Mental Disease*, 194,128–131.

Pueschel, M. (2007). Army hiring more mental health providers. *U.S. Medicine*, 43(12), 1, 31, 35.

Rentz, E. D., Marshall, S. W., Loomis, D., Casteel, C., Martin, S. L., & Gibbs, D. A. (2007). Effect of deployment on the occurrence of child maltreatment in military and nonmilitary families. *American Journal of Epidemiology*, 165, 1199–1206.

Riggs, D. S. (2000). Marital and family therapy. In E. B. Foa, T. M. Keane, & M. J. Friedman (Eds.), *Effective treatments for PTSD: Practice guidelines from the International Society for Traumatic Stress Studies* (pp. 280–301). New York: Guilford Press.

Riggs, D. S., Byrne, C. A., Weathers, F. W., & Litz, B. T. (1998). The quality of the intimate relationships of male Vietnam veterans: Problems associated with posttraumatic stress disorder. *Journal of Traumatic Stress*, 11, 87–101.

Ruger, W., Wilson, S. E., & Waddoups, S. L. (2002). Warfare and welfare: Military service, combat, and marital dissolution. *Armed Forces and Society*, 29, 85–107.

Sareen, J., Cox, B. J., Afifi, T. O., Stein, M. B., Belik, S. L., Meadows, G., & Asmundson, G. J. (2007). Combat and peacekeeping operations in relation to prevalence of mental disorders and perceived need for mental health care: Findings from a large representative sample of military personnel. *Archives of General Psychiatry*, 64, 843–852.

Sautter, F., Lyons, J., Manguno-Mire, G., Perry, D., Han, X., Sherman, M., Myers, L., Landis, R., & Sullivan, G. (2006). Predictors of partner engagement in PTSD treatment. *Journal of Psychopathology and Behavioral Assessment*, 28(2), 123–130.

Schumm, W. R., Bell, D. B., Knott, B., & Rice, R. E. (1996). The perceived effect of stressors on marital satisfaction among civilian wives of enlisted soldiers deployed to Somalia for Operation Restore Hope. *Military Medicine*, 161, 601–606.

Seal, K. H., Bertenthal, D., Miner, C. R., Sen, S., & Marmar, C. (2007). Bringing the war home: Mental health disorders among 103,788 U.S. veterans returning from Iraq and Afghanistan seen at Department of Veterans Affairs facilities. *Archives of Internal Medicine*, 167, 476–482.

Sherman, M. D., Sautter, F., Lyons, J., Manguno-Mire, G., Han, X., Perry, D., & Sullivan, G. (2005). Mental health treatment needs of cohabiting partners of veterans with combat-related PTSD. *Psychiatric Services*, 56(9), 1150–1152.

Sherman, M. D., Sautter, F., Jackson, H., Lyons, J., & Han, X. (2006). Domestic violence in veterans with posttraumatic stress disorder who seek couples therapy. *Journal of Marital and Family Therapy*, 32(4), 479–490.

Smith, L., & Folkard, S. (1993). The perceptions and feelings of shiftworkers' partners. *Ergonomics*, 36, 299–305.

Smith, M. E. (2005). Female sexual assault: The impact on the male significant other. *Issues in Mental Health Nursing*, 26, 149–167.

Smith, T. C., Ryan, M. A. K., Wingard, D. L., Slymen, D. J., Sallis, J. F., Kritz-Silverstein, D., for the Millennium Cohort Study Team. (2008). New onset and persistent symptoms of post-traumatic stress disorder self reported after deployment and combat exposures: Prospective population based U.S. military cohort study. *BMJ Online*, January 15 (http://www.bmj.com/cgi/content/full/bmj.39430.638241.AEv1).

Spanier, G. B. (1976). Measuring dyadic adjustment: New scales for assessing the quality of marriage and similar dyads. *Journal of Marriage and the Family*, 38, 15–28.

Suris, A., Lind, L., Kashner, T. M., Borman, P. D., & Petty, F. (2004). Sexual assault in women veterans: An examination of PTSD risk, health care utilization, and cost of care. *Psychosomatic Medicine*, 66, 749–756.

Tarrier, N., Sommerfield, C., & Pilgrim, H. (1999). Relatives' expressed emotion (EE) and PTSD treatment outcome. *Psychological Medicine*, 29, 801–811.

Toomey, R., Kang., H. K., Karlinsky, J., Baker, D. G., Vasterling, J. J., Alpern, R., Reda, D. J., Henderson, W. G., Murphy, F. M., & Eisen, S. A. (2007). Mental health of U.S. Gulf War veterans 10 years after the war. *British Journal of Psychiatry*, 190, 385–393.

Walsh, F. (2007). Traumatic loss and major disasters: Strengthening family and community resilience. *Family Process*, 46, 207–227.

Weathers, F. W., Litz, B. T., Herman, D. S., Huska, J. A., & Keane, T. M. (1993). The PTSD checklist: Reliability, validity, and diagnostic utility. Paper presented at the annual meeting of the International Society for Traumatic Stress Studies, San Antonio, TX.

Weber, E. G., & Weber, D. K. (2005). Geographic relocation frequency, resilience, and military adolescent behavior. *Military Medicine*, 2005, 638–642.

Wood, D., Halfon, N., Scarlata, D., Newacheck, P., & Nessim, S. (1993). Impact of family relocation on children's growth, development, school function, and behavior. *Journal of the American Medical Association, 270*, 1334–1338.

Wright, K. M., Bliese, P. D., Thomas, J. L., Adler, A. B., Eckford, R. D., & Hoge, C. W. (2007). Contrasting approaches to psychological screening with U.S. combat soldiers. *Journal of Traumatic Stress, 20*, 965–975.

Yaeger, D., Himmelfarb, N., Cammack, A., & Mintz, J. (2006). DSM-IV diagnosed posttraumatic stress disorder in women veterans with and without military sexual trauma. *Journal of General Internal Medicine, 21*(S3), S65–S69.

Zarit, S. H., Reever, K. E., & Bach-Peterson, J. (1980). Relatives of the impaired elderly: Correlates of feelings. *Gerontologist, 20*, 649–655.

Zvonkovic, A. M., Solomon, C. R., Humble, A. M., & Manoogian, M. (2005). Family work and relationships: Lessons from families of men whose jobs require travel. *Family Relations, 54*, 411–422.

Zvonkovic, A. M., Moon, S., & Manoogian-O'Dell, M. (2007). *Fishing marriages over time*. Corvallis, OR: Oregon Sea Grant Publications (http://seagrant.oregonstate.edu/sgpubs/onlinepubs/g97007.html).

Recommended Reading

Lanham, S. L. (2005). *Veterans' and families' guide to recovering from PTSD* (3rd ed.). Annandale, VA: Purple Heart Service Foundation (while supplies last, a free copy should be available through local veteran centers; visit http://www.vetcenter.va.gov/ to find the nearest one).

Lyons, J. A. (2007). The returning warrior: advice for families and friends. In C. R. Figley & W. P. Nash (Eds.), *Combat stress injury: Theory, research, and management* (pp. 311–324). New York: Routledge.

Lyons, J. A. (2008). Using a life span model to promote recovery and growth in traumatized veterans. In S. Joseph & P. A. Linley (Eds.), *Trauma, recovery and growth: Positive psychological perspectives on posttraumatic stress* (pp. 233–258). Hoboken, NJ: John Wiley & Sons.

20
Military Children:
The Sometimes Orphans of War

JUDITH A. COHEN, ROBIN F. GOODMAN, CAROLE CAMPBELL,
BONNIE L. CARROLL, and HEATHER CAMPAGNA

Contents

Introduction

The military, like all cultures, has its own language, attire, social norms, and attitudes. Excellent overviews have been provided of the military culture and lifestyle and how these affect service members and their children (see, for example, Knox & Price, 1999, 2006). Stress, trauma, and loss are normative parts of the military culture and come in many forms via transitions, deployments, and permanent losses, as in the case of bereavement. This chapter describes how the military culture and military-related experiences affect family members, and what can be done to prevent, alleviate, and address the impact of these events on children.

Whether in training, on noncombat missions, or deployed on the battle-field, military personnel are professionals engaged in challenging and honor-able work. It may seem easier or even necessary to separate work and family, but many military men and women have another full-time job as a parent: "An estimated 1.5 million school-age children have military parents on active duty. About 49,000 U.S. military families include two parents on active duty" (Lamberg, 2004). Thousands more children are the siblings, grandchildren, and cousins of military personnel. For many, a joyous permanent homecoming eventually arrives, but for other children various emotional battle scars form.

Transitions and Deployment

Separations are a customary part of military life; they can be considered routine stressors (Hardaway, 2004), and stem from multiple causes, such as training exercises, temporary assignments of duty (with or without temporary relocation of the service member's family), permanent changes of duty station, and deployments. Relocation stress has been described in adults transferring from one environment to another, a condition that would likely affect military spouses and children (Barnhouse, Brugler, & Harkulich, 1992). Transitions and deployments can be unpredictable and occur with little warning, some-times allowing less than a week for the child (and nonmilitary caregiver) to prepare. Although dangerous circumstances do not generally surround rou-tine separations, such separations can affect a child's attachment to his or her caregiver and significantly disrupt the child's routine. In situations where the service member is the sole primary caretaker, routine separations can also disrupt the child's living situation, which is also regularly disrupted when a service member changes duty station, typically occurring on a 3-year cycle. These moves have been noted to become increasingly difficult as children enter school age and adolescence because they are accompanied by losses of friends, teachers, social clubs and activities, and homes. A 1999 Army survey found that 71% of officers and 78% of enlisted personnel "reported that their children had problems due to changing school as a result of a move" (Military Family Resource Center, 2001, p. 7), and a 2001 survey of Army spouses found that 25 to 44% of high school students had difficulty with such issues as los-ing credit for courses completed and making social adjustments (U.S. Army Community and Family Support, 2001).

The losses associated with deployments have become increasingly familiar to the American public, with images of service members leaving and return-ing from recent conflicts permeating the media. Researchers note that the distress associated with deployments is expected to be greater for service mem-bers with children due to concerns about the child and the child's separation anxiety (Medway, Davis, Cafferty, & Chappell, 1995). Mothers remaining when the fathers deploy report significant symptoms of depression (Kelley, 1994), which may reduce their capacity to provide adequate support and exacerbate any problems the child experiences. When a parent is deployed, a necessary change occurs in family patterns, routines, parent availability, and possibly even parenting style—for example, with a parent becoming more lax, strict, or protective.

Research consistently finds that children experience significant emotional and behavioral problems when a parent is deployed (Kelley et al., 2001), partly in relation to parent functioning (Black, 1993; Kelley, 1994). Children do better when parents are positive and prepared for deployment (Goodman, 2004). Although it is hoped that a loss associated with deployment is not permanent, the time of the deployment is not always certain and often changes, and deploying parents may be in a dangerous area or cannot often disclose their mission or location. This uncertainty can engender significant anxiety in children and has been theorized to be related to symptoms of grief (Huebner, Mancini, Wilcox, Grass, & Grass, 2007).

Trauma and Grief

The sudden death of a military parent is but one service-related experience that puts a child or teen at risk for a stress or trauma reaction. In general, children can experience either *acute trauma* or *chronic trauma* when they are: (1) threatened by, experience, or witness serious injury or death to themselves or others, or (2) are violated physically (NCTSN, 2007a). Whereas children of military parents may encounter potential traumatic experiences as do any children (e.g., fires or car accidents), certain unique or newly recognized circumstances can pose risks to military children. With regard to military children, examples of acute trauma would include a combat or noncombat (i.e., accidental) death or living in a remote and hostile foreign country with threats or acts of terror (Hardaway, 2004). Maltreatment of military children may be either acute, such as upon return of a parent from the Armed Forces, or chronic, as is the case when parental abuse was recurring before a tour of duty or is ongoing. The incidence of certain forms of child abuse has been found to increase during parental deployment (Gibbs, Martin, Kupper, & Johnson, 2007) and may be related to the child's age, gender, or parent rank (Jellen, McCarroll, & Thayer, 2001; McCarroll, Ursano, Fan, & Newby, 2004; Raiha & Soma, 1997). As an example of chronic trauma, some military children witness ongoing domestic violence (Jellen et al., 2001).

Increasing evidence suggests that psychological functioning, particularly posttraumatic stress disorder (PTSD), of the returning veteran is a source of stress affecting all family members (Beckham et al., 1997; Galovski & Lyons, 2004; Suozzi & Motta, 2004). Reintegration of the family unit can be the most difficult phase of the deployment cycle, when the military member faces a "culture shock due to quick foxhole to front porch transitions" (Hobfoll et al., 1991, p. 849). Three to four months after coming home, "as many as 17% of those exposed to combat in Iraq and about 11% of those who served in Afghanistan reported symptoms of PTSD, depression, or anxiety" compared to just 9% prior to deployment and none having mental health problems prior to being surveyed (Lamberg, 2004). Combat-related PTSD symptoms are often observable and concerning, if not frightening, to children. Behavior such as anger, irritability, withdrawal, and avoidance can greatly impact child–parent relationships. Cognitive PTSD symptoms can impair or interfere with a parent's ability to read cues, interpret behavior, and have appropriate expectations of their child's behavior. A review of the literature (Galovski & Lyons, 2004) concludes that a military parent's PTSD can result in family dysfunction on a number of levels and "directly impact the ability of the veteran to parent his child and interrupt the development of a positive parent–child relationship" (pp. 487–489). Upon the service member's return home, PTSD and depression can also be a factor in death by suicide, leaving a child to confront a complex mix of traumatic circumstances: a tragic cause of death, parental mental illness, and military-related stresses.

Secondary Adversities

Less dramatic but not necessarily less significant, children are also faced with secondary adversities related to military life (Meyers-Wall, unpublished manuscript). In essence, a domino effect of changes is set in motion by a military-related event that leads to concrete problems, frustration, and complex emotions. Further complicating transitions, deployments, and bereavement situations, family adjustment can be hindered by changes in

- Income (a reserve parent losing a job or a remaining parent being forced to leave a job to care for children; for grieving families, claiming military benefits is a complex process and unlike any other in our society)
- Childcare arrangements (stress on caregivers and children from finding or changing caregivers)
- Home and school (moving out of base housing or relocating)
- Identity and status (loss of a formal military identity following a death)
- Social network (dislocated and dispersed social connections and friendships due to change in duty station or death)
- Privacy (military deaths are very public)

Existing research on military-related stress and trauma in children is limited, with some being focused on children directly exposed to war (Shaw, 2003). Even fewer studies have systematically examined the effect of deployment on children, with the U.S. Army Secondary Education Transition Study (2001), which addresses the educational needs of high school students, being a rare exception. Despite this, a picture is emerging of potential risk and protective factors related to children's adjustment and reactions (Amen, Jellen, Merves, & Lee, 1988; Jensen, 1999; Jensen & Shaw, 1996). Although the degrees of risk for the following categories are not yet well identified, these domains should be considered when assessing for trauma reactions (Blount, Curry, & Lubin, 1992; Goodman, 2004; Jensen, 1999; Kelley, 1994; Otto et al., 2007; Pierce, Vinokur, & Buck, 1998; Rosen, Teitelbaum, & Westhuis, 1993; Shaw, 2003; Webb, 2004).

High-Risk Situations

- *Military-specific circumstances*—Number of previous transitions and deployments; phase in the transition and deployment cycle (longer being more difficult)
- *Previous separations*—Reaction to separations, frequency, length of time (longer being more difficult)
- *Preparation*—Sudden vs. anticipated
- *Distance and timing*—Proximity to a previous home or physical distance from the deployed parent, time in child's life
- *Dangerous time and areas*—Climate where parent is stationed as well as where family may be living (war time being more difficult than peace time)
- *Injury*—Extent of injury and the treatment and rehabilitation required, particularly with respect to traumatic brain injury and amputations resulting in permanent changes
- *Bereavement*

High-Risk Persons (Children)

- *Age*—Younger children (less than 5 years old) may be at higher risk due to limited cognitive and communication ability and coping resources.
- *Gender*—Boys are more prone to external manifestations such as behavior problems; girls tend to internalize problems and feel sad, for example.
- *Health problems*—Identify previous or current mental health, learning, physical, or medical problems.
- *Previous traumas*—New stress may trigger past reactions or increase vulnerability to problems.
- *Cognitive ability*—Understanding and coping are influenced by the child's cognitive ability.
- *Guilt*—Child may feel responsible for family well-being or may feel guilty about causing or not preventing a death.
- *Temperament, functioning, coping style, and locus of control*—Children with different styles require different support.

Past and Current Military and Nonmilitary Family and Social Environment

- *Support*—Family, social, and outside support is best. Parents who feel better supported provide a better role model for children and are better able to parent effectively. Good family cohesion and support are protective, and higher levels of community support are beneficial.
- *Past and current parent and family functioning*—Parent and child functioning are strongly related; for example, during deployment, children do better when the remaining parent is positive and prepared. Prior positive functioning helps with individual and family functioning afterward.
- *Satisfaction with military*—Positive parent attitudes about the military translate to better family satisfaction and adjustment.
- *Media*—Details about world events are immediately and constantly available in multiple formats, making it difficult for children to escape information related to their parents' safety or trauma-related triggers.
- *Politics and language*—Military families are faced with ever-present political commentary, both anti-war and pro-war. Terminology is also used indiscriminately (e.g., "He gave his life in service to his country" or "They made the ultimate sacrifice").
- *National days of remembrance*—Public military-related holidays such as Memorial Day can be triggers for personal reactions and reminders.

Trauma Reactions

No one-to-one cause-and-effect relationship exists between a specific child or military family situation and a resulting stress or trauma reaction. Furthermore, factors are interactive; for example, the issues facing the family of a young Marine deployed during peacetime for a third tour, leaving behind a toddler and wife, differ from those confronted by a single mother Army reservist going to an overseas combat location for the first time and leaving a teenage daughter alone with relatives.

Understandably children experience stress from military-related situations, such as when the family must transition to a new community, when a parent is deployed in a dangerous area, when the family is reunited after a protracted deployment, when a parent is injured, or when a parent dies. Although some reactions are expected and children generally do well (Jensen, Xenakis, Wolf, & Bain, 1991; Rosen et al., 1993; Ryan-Wenger, 2001), some children are overwhelmed by the stress, as revealed by their thoughts, feelings, behaviors, and physical reactions. Children of Vietnam, Korean, and World War II veterans suffering current PTSD had more past psychiatric treatment for such problems as attention deficit disorder and academic and behavior problems compared to a control group (Galovski & Lyons, 2004). Military children may be more prone to depression (Jensen, Martin, & Watanabe, 1996; Kelley et al., 2001) and other affective disorders related to parental PTSD. It is hypothesized that veteran trauma is transmitted intergenerationally to children via: (1) direct traumatization, such as through violence; (2) a child's identification with the parent's trauma reaction or symptoms; (3) family dysfunction resulting from the veteran parent's symptoms and behavior; (4) avoidance of the parent and feeling fearful due to the parent's symptoms and diminished comfort seeking; (5) exposure to details of the military parent's source of trauma; or (6) the parent's use of the child in some type of trauma reenactment (Galovski & Lyons, 2004).

Grief Reactions

Given the rise of military operations since 2001 and the danger inherent to military service, the risk of traumatic bereavement is very real for children of military personnel. At the start of 2008, the Department of Defense reported that 4374 military members had died as a result of participation in Operation Iraqi Freedom or Operation Enduring Freedom; an additional 30,721 military personnel have been wounded in action (DoD, 2008). Military children who lose a parent may develop a traumatic grief reaction (Cohen, Mannarino, & Deblinger, 2006). The shocking nature of a military death is exemplified by the image of a chaplain and notifying officer arriving at a family's front door. A traumatic reaction may follow a characteristically sudden and horrific death such as from an improvised explosive devise (IED) or accident. Yet, a child may have a traumatic reaction from other types of death such as an acute injury with a prolonged hospitalization even without exposure to gruesome aspects of the course of treatment and death. Research is still determining the length of time since death and the severity, duration, and interference of symptoms to distinguish the condition, but the current literature focuses on children with childhood traumatic grief (CTG) having symptoms characteristic of PTSD. Such children are stunned and overwhelmed by emotions and reactions that interfere with normative, healthy grief reactions. Characteristic PTSD symptoms include the following (NCTSN, 2007b)

- *Intrusive memories about the death*—A common example is nightmares. Also, military children may try to imagine what the death was like (e.g., based on news or movies or stories from others), possible suffering, or the notification.
- *Avoidance and numbing*—Reminders of the person who died and the death itself trigger painful and overwhelming emotions, hence the child may cope by becoming numb to all emotions or may avoid people, places, and events that could provoke reminders. Military children may withdraw from military peers or resist military ceremonies.

- *Increased arousal*—Physical and emotional reactivity can be recognized by symptoms such as difficulty sleeping, poor concentration, irritability, anger, always being on alert, being easily startled, and having new fears.

Reactions and long-term adjustments of military children vary by a wide range of factors, and by no means do all military children suffer for the sacrifices made by their parents for our country. But, the risk of acute, cumulative, and chronic trauma and stress is never far removed on the home front. The following sections describe existing programs aimed at helping military children.

Resiliency Interventions for Military Youth

Efforts to increase resiliency among military youth have been comprised predominantly of general psychoeducational programs and community-based social support interventions, which had mild success in improving coping skills for families during the Persian Gulf War (Knox & Price, 1999). Multiple branches of service have developed psychoeducational programs for families which are typically offered at the command level by a Family Readiness Group (FRG). Research into FRG interventions, however, suggests that these are utilized by a minority of families and that only approximately 25% of those who do participate are satisfied with the program (Orthner, 2002a,b). There is a great need for continued development of and outcome research on resiliency programs for military youth. One model, Project FOCUS (Lester, 2007; Lester, Saltzman, Beardslee, & Pynoos, 2007), addresses the need in this area and has been implemented with promising results with Marine Corps families.

Project FOCUS

- *Name of model*—Project FOCUS (Families Overcoming and Coping Under Stress)
- *Target for model*—Enhancing resiliency before, during, and after caregiver deployment
- *Format of model*—An 8-week program utilizing a family approach that begins with individual sessions with the children and parents, followed by conjoint family sessions
- *Provider*—Mental health professionals
- *Components of intervention*—Trauma-informed, developmentally appropriate psychoeducation; family skills building, including coping skills training, problem solving, and emotional regulation; family deployment timeline (aimed at linking skills to family experience and creating a shared family narrative); addressing adversities; and enhancing social support

- *Use for military families*—Successfully implemented with Marine Corps families; currently being adapted for families of injured service members
- *Empirical support*—Based on solid theoretical and empirical grounds, with developers citing promising preliminary results for developing more positive coping skills to address challenges (Lester et al., 2007), but randomized controlled trials (RCTs) or published open trials examining the efficacy of the program are lacking

A notable weakness in the arena of resiliency programs for military youth is the apparent lack of communication and consistency of programs and program development across service branches; for example, Project FOCUS has been successfully implemented with Marine Corps families, but a different command-level approach has been widely adopted by the Army active duty, reserves, and National Guard (Operation READY) (Texas Cooperative Extension, 2002), although this model is also lacking in empirical support. Little information is available with regard to what other branches of the service are implementing. It is possible that these programs are not yet well organized or defined, or that they are a part of the aforementioned FRG or Family Advocacy Services. Given the research suggesting that the majority of service members who use these services do not find them helpful and that military youth are at risk for experiencing problems as a result of separation and deployment, continued research into resiliency programs for these youth is greatly needed (Martin, Mancini, Bowen, Mancini, & Orthner, 2004).

Case Example Sonia and Manuel are a young Hispanic couple who have two children, ages 3 and 6. Sonia contacted the Family Advocacy Program and completed a phone intake, when she reported that her husband recently learned that he would be headed out on his second deployment with the Marine Corps. Manuel returned 18 months ago from his first deployment, which lasted 16 months and involved combat in Iraq. Sonia reported that he "hasn't been the same" since coming back from his first tour. He left for his first deployment when their older son, Manuel, Jr., was 3 and the younger son, Antonio, was an infant. Sonia reported that Manuel was an attentive and patient father who supported her in many of the caretaking responsibilities. She stated that, since returning, he angered more quickly and has been "really irritable." She indicated that she thinks he is having nightmares, but he won't talk about them. In addition, Sonia informed the intake worker that Manuel, Jr., has had behavior problems since starting school. She stated that these have worsened, as have his behaviors in home, since hearing that his father would be leaving. Finally, she expressed having significant worries of her own, stating that she has been crying more since learning of her husband's deployment. She reported that, when her husband was first deployed, they lived close to her sister and Manuel's mother, both of whom provided support for her and helped care for the children. She said that they have since changed duty stations and she feels isolated from this support system. She stated that she is very concerned that Antonio "will not know his father" due to the many absences in his early years and that she has not brought this up with Manuel because she does not "know how he will react to anything" since returning from his initial deployment.

Sonia and Manuel spent the first two sessions meeting individually with the counselor. The goals of these sessions focused on improving their ability to communicate openly with each other and to share feelings related to the previous and upcoming deployment. In addition, they received developmentally appropriate education about trauma reactions in children and parents. For Manuel, it was difficult, at first, to discuss how he had changed after returning from his first deployment. For some parents, this may be complicated by post-traumatic stress symptoms or other psychiatric problems that warrant referrals. Within the context of the sessions, the potential for trauma reminders to impact parenting and family life was discussed, and Manuel identified positive ways to cope with this (including planning on checking in with a counselor after returning from his scheduled deployment). Manuel also discussed his own worries related to Sonia and his children living farther from his mother while he will be deployed. A plan was developed to foster support by scheduling regular phone contact with family members throughout the deployment. The second two sessions were spent individually with the children. In these sessions, rapport was established, the children completed activities aimed at educating them about deployment, developing a family deployment timeline, and acquiring emotion regulation skills (at their developmental levels). Sonia and Manuel again met individually with the counselor for the fifth session, where they discussed parenting practices and family routines. They also created family deployment timelines, which were compared with those of Manuel, Jr., and Antonio. The last two sessions were family sessions that focused on creating a plan for increasing social support for the whole family at each stage of the deployment timeline. This step was particularly important given both Sonia's and Manuel's concerns related to isolation from their close-knit family support network. Faith-based coping resources as well as cultural centers that could improve social support connections were identified. In addition, the coping skills were reviewed and reinforced. Finally, adversities that might arise were discussed and plans created to help minimize the impact these would have on family functioning.

Interventions for Bereaved Military Children

As is true for bereaved children in the general population, most bereaved military children do not receive mental health interventions related to the death of their loved one. For those who do, the most common intervention is peer support groups. Peer support models are based on the view that bereavement reactions are normal, nonpathological responses and that enhancing social support, particularly from peers who are experiencing similar reactions, is both helpful and adequate for most bereaved children to successfully move through the typical tasks of bereavement. These tasks include accepting the permanence of the death and experiencing the pain associated with this loss; reminiscing about the deceased person—the good and the bad; transforming the relationship from one of interaction to one of memory; incorporating important aspects of the deceased into one's own identity; committing to relationships with living people; and reestablishing a healthy developmental trajectory after the loss (Wolfelt, 1991; Worden, 1996). Military-focused peer support groups are now available through organizations such as the Tragedy Assistance Program for Survivors (TAPS), described below.

Targeted treatments for traumatically bereaved children have also been developed and these interventions have been used with traumatically bereaved military children. Such interventions focus on resolving children's psychiatric symptoms (e.g., PTSD) as well as improving adaptive and family functioning. Being able to identify and provide treatment to military children who can benefit from this type of treatment requires overcoming the stigma that many military families have about seeking and receiving formal mental health services (Department of Defense Task Force on Mental Health, 2007).

TAPS Peer Support Model

- *Name of model*—TAPS Peer Support Model.
- *Target for model*—Bereaved children and adults who have lost a loved one who was serving in the Armed Forces.
- *Format of model*—Weekend peer support groups for children and adults who have lost a loved one in the military.
- *Provider*—Groups of same-aged children are led by persons with a mental health or education background with some type of military affiliation or paired with a military co-group leader during both regional and national camps. The TAPS program also offers an innovative military mentor peer support program in which children are paired with members of the military who are often in the same branch of service as the child's deceased loved one during the annual Good Grief Camp over Memorial Day weekend. Mentors who volunteer for the Good Grief Camp receive training from TAPS in companioning children experiencing traumatic grief reactions.
- *Components of intervention*—TAPS provides support to both children and adults who have lost a loved one while serving in the Armed Forces. TAPS offers peer support groups at both national and regional levels. The national Annual TAPS Military Survivor Seminar is held each year in Washington, D.C., over Memorial Day

and includes the Good Grief Camp, which is held concurrently with the adult seminar. At the national TAPS Good Grief Camp, military mentors serve as "big brothers" or "big sisters" for the children and assist in keeping the military connection for the grieving child (Wolfelt, 2006). In addition to the annual Good Grief Camp, regional Good Grief Camps are held throughout the United States. During the regional Good Grief Camps, support is provided within the group from various military personnel. At some locations, this includes support from military members in the Wounded Warrior Battalion who have been injured yet remain in the military. To continue the care offered in these programs, TAPS also provides a Teen Chat to support adolescents who have lost a loved one. These teens come together in a safe, facilitated forum with an adult moderator present to guide discussions.

- *Use for military families*—Developed specifically by and for military families to support all those who have lost a loved one serving in the Armed Forces, regardless of circumstance or location of the death.
- *Empirical support*—None.

Case Example Michael is a 15-year-old male who experienced the loss of his father. When Michael was 12 years old, he and his two younger sisters were playing outside when uniformed soldiers came to his home and notified his family of the death of his father in an IED attack in Afghanistan. Michael knew immediately when he saw the car pull up in his driveway that his dad was dead, although did not want to believe it. When Michael's father was on a previous deployment in Afghanistan, he was injured when ambushed by an IED and lost his hearing in both ears. Because he returned from that deployment, Michael felt that he would return from this deployment and that his father was invincible.

Michael came to his first TAPS regional seminar when he was 13. Although apprehensive and a tad shy, he began to open up about his feelings with other teenagers who had lost a loved one in the military. He discussed how difficult it was losing his dad, that he wanted him to be there to watch him play football, and how he felt sorry for his sisters who did not have as many memories as he did of his dad, whom he described as his hero. He appeared to cling to any military service member he came in contact with during the weekend and described how helpful it was for him to know he was not the only one who had lost a dad in the military. Michael also attended the annual TAPS Good Grief Camp. There, he was able to discuss the difficulties he had experienced since the death of his father and found great comfort in hearing from others similar to him who were able to say they understood … and they truly did. When Michael attended his third TAPS event, a regional Good Grief Camp, there was a remarkable difference in his comfort level in sharing precious memories of his father. He was no longer ashamed to share the tears as he had been before. At the end of the regional TAPS Good Grief Camp that day, he spoke of a euphoric feeling—he just wanted to scream because of the heavy weight that had been lifted off of his chest. Michael and others in his region, with the help of adults, began their own local teen peer support group to support others like them who have lost a precious loved one in the military.

Trauma-Focused Cognitive–Behavioral Therapy for Childhood Traumatic Grief

- *Name of model*—Trauma-Focused Cognitive–Behavioral Therapy for Childhood Traumatic Grief (TF-CBT-CTG)
- *Target for model*—Children with traumatic grief following a death that was unexpected or violent children who have developed trauma/PTSD symptoms that interfere with typical grieving
- *Format of model*—Parallel individual child and parent therapy sessions of 30 to 45 minutes over 12 to 16 weeks, including approximately 6 joint child–parent sessions
- *Provider*—Licensed mental health professionals trained in this model (free online training course for professionals is available at www.musc.edu/ctg)
- *Components of intervention*—TF-CBT-CTG was developed from TF-CBT, an evidence-based model for treating traumatized children. TF-CBT-CTG consists of sequential trauma- and grief-focused components. The trauma-focused component (TF-CBT) aims to resolve children's PTSD, depressive, anxiety, and behavioral problems. The grief-focused component then assists children with beginning a more typical grieving process. Some children may have experienced other traumas besides the death that may also require therapeutic attention. The trauma-focused component is summarized by the acronym PRACTICE: psychoeducation about the trauma and traumatic symptoms; parenting skills to address behavioral dysregulation; relaxation skills to reverse traumatic physiologic hyperarousal; affective identification and modulation skills to address affective dysregulation; cognitive coping (connecting thoughts, feelings, and behaviors and correcting inaccurate or unhelpful thoughts); trauma narrative creation and processing of dysfunctional thoughts about the death or other traumatic experiences; *in vivo* mastery of trauma reminders; conjoint child–parent sessions; and enhancing safety and future development. The grief-focused component includes psychoeducation about death and grief; grieving the loss and resolving ambivalent feelings about the deceased (e.g., "What I miss and what I don't miss"); preserving positive memories of the deceased; redefining the relationship and committing to present relationships; and treatment closure.

- *Use for military families*—Used with bereaved military children without significant adaptations
- *Empirical support*—TF-CBT-CTG has been studied in two pilot studies and one small RCT. A small randomized controlled trial (RCT) was conducted of children of uniformed service providers (police, firemen, Port Authority) who died in the terrorist attacks on the World Trade Center on September 11, 2001 (Goodman & Brown, 2008). TF-CBT has strong empirical support from six randomized controlled trials (RCTs). In the pilot studies, a minimum level of PTSD and CTG symptoms was required for inclusion. Children experienced significant improvement in CTG and PTSD symptoms after treatment; parents also experienced significant improvement in PTSD symptoms (Cohen, Mannarino, & Knudsen, 2004; Cohen, Mannarino, & Staron, 2007).

Case Example   Tara is a 14-year-old African–American girl living in a northeastern city. Her father, an Army Reservist, had recently been redeployed to Iraq when he was killed by an IED. She has no friends with military parents. Tara and her best friend, Sela, were listening to music in Sela's apartment when Sela's brother ran in yelling to Tara, "Hey, there's guys in uniforms at your door!" Tara ran to her own apartment. She found her mother sobbing with the Casualty Assistance Call Officers. Tara refused to speak with the officers or her mother. She went to her room and would not come out for the rest of the day. She had avoided Sela since then. Tara did not want to look at her father's picture or go to the cemetery. She refused to watch television, saying that "everything on TV is bad." She became increasingly angry and irritable at home and was having trouble sleeping. One of the teachers at her school said within Tara's hearing that, "People like that [meaning members of the Armed Forces] caused this war. They deserve to die." Tara walked out of the school and did not return for the rest of the day. She told her mother what she heard the teacher say and her mother reported this to the school officials, but the principal did not believe Tara's report. He told the mother that it was time for Tara to get over father's death and move on. Tara's grades fell from a B average to Cs and Ds, and she had increasing trouble concentrating. She began to skip school frequently. The mother was also concerned because Tara had not once cried over her father's death. This is what led the mother to call a child bereavement support program in the community. This program spoke with the mother on the phone and referred Tara to a child trauma program for an evaluation.

At the initial evaluation Tara displayed PTSD symptoms related to her father's death and the way she learned about his death. She reported having intrusive thoughts about learning of the death and avoided reminders of how she learned about his death (e.g., her friend Sela) as well as possible reminders about his death (e.g., school, where people might discuss it; cemetery;

pictures) and its cause (e.g., television, where the war in Iraq was covered frequently). She also had hyperarousal symptoms, including increased vigilance (e.g., checking several times to make sure the doors were locked at night), new onset of irritability, and poor sleep. Tara and her mother agreed to participate in TF-CBT-CTG treatment.

Tara told her therapist that she would "stay strong and never cry." She explained that this was part of the military culture and she wanted to be part of this now, even though she did not know much about it, to honor her father. Her therapist told her that she respected her values and would try to help her with the reasons she was coming to treatment. One of the therapist's first interventions was to ask Tara questions about the military to help the therapist understand more about it. It turned out that Tara really did not know much about the military, because (typical of many Reservist children living in urban areas) she had not participated in most aspects of this culture prior to her father's death. The therapist suggested that they learn something about the military together, which over time allowed Tara to more easily tolerate military-related reminders. Tara was able to identify a range of feelings associated with different situations (e.g., sad and embarrassed if she got a bad grade; excited if she was asked to a party). She developed a relaxation plan for falling asleep at night, which included listening to her favorite calming music on her iPod®. Tara and her mother worked together to develop cognitive coping skills for use in their daily life—for example, with the episode involving the principal. In response to what the principal said, Tara's initial thought was, "Something is really wrong with me." This made her feel angry and hurt. When she felt this way, she had trouble concentrating or gave up and skipped school altogether. Tara's therapist asked, "Is this thought accurate? Is it unreasonable for you to still feel sad or upset after what happened to your father?" Tara considered this and then responded, "No." The therapist then asked, "Then what is a more accurate thought?" Tara replied that, "The principal doesn't understand what it's like to lose your father like I did." Her therapist asked, "And when that's your thought, how do you feel?" Tara said, "I feel angry at him for saying what he did, but I don't feel like there's anything wrong with me." Her therapist asked, "And if you feel like this, how will you act?" Tara said, "I'll try my best to not let him get to me. I won't give up so easily." The mother also developed cognitive coping strategies for herself to challenge unhelpful or inaccurate thoughts.

After completing the PRAC components, Tara agreed to write a trauma narrative. She started by describing her family prior to her father leaving for Iraq, including her relationship with her father, and she then described his deployment, her communications with him while he was away, and the day she found out about his death. She also included a chapter about what had occurred since then. After completing the narrative section, Tara processed dysfunctional thoughts—for example, that she should have "known" that he would be killed and that she should have warned him "not to be a hero and do risky things to save other soldiers' lives." When she completed her narrative, Tara shared it with her mother during joint child–parent sessions. Because the therapist had been sharing Tara's narrative with the mother each week as Tara was writing it (with Tara's permission), her mother was prepared for the contents and was able to be very supportive during Tara's reading of the story. Although it was very emotional and both cried, Tara felt very relieved to be able to talk about her feelings openly with her mother, and her mother praised Tara for her ability to do so, telling her how proud her father would be of her.

After completing the trauma-focused component, Tara was willing to begin grief-specific activities. She and her therapist played the "Grief Game" (Searle & Streng, 1998) to begin talking about grief, bereavement, and mourning. Tara was tearful during some parts of this game, and the therapist asked whether it was normal to cry when talking about someone who died. Tara said, "I'm afraid if I let myself start being sad I will cry and cry and never stop." Her therapist assured Tara that this is how many grieving people feel, even people in the military, and that starting to cry is the beginning of healing. Tara began to sob for the first time since her father's death. Afterwards she felt "tired and empty, but a little better too." Her therapist suggested that this might happen many more times and that this is how grieving people experience the pain of losing someone they love so much. In the following session Tara reported that she had cried more, and that she was no longer so afraid of doing so; she had gone to her mother and this helped her feel better. Tara developed an acronym of her father's name that listed things she missed about him. When asked to write a letter to her father about "unfinished business," Tara eagerly agreed to do so. In her letter she asked whether her father had suffered before he died and whether he had had time to say goodbye to Tara and her mother before he died, and she told him that she loved him and would make him proud. Tara's imagined response letter from her father to herself said that he had not felt any pain, that his last thoughts were of Tara and that he was very proud of her, and that he wanted her to go to college, get married, and have a happy life.

Tara and her mother collected memorabilia to make a memory scrapbook of her father. This was painful but also fun for Tara as she learned many things about her father's childhood and her parents' courtship that she had not previously known. She wrote a narrative to accompany the memorabilia titled "Freedom Is Never Free: A Soldier's Life and Death" and submitted this as a civics project at school. She received an "A" for the project, which Tara and her mother viewed as redemption for Tara's previous negative school experience related to father's death. Tara next created two balloon drawings, with one balloon tethered to the ground, representing things about her father she could hold onto, and one floating into space, representing things about her father she had to let go of. She included "happy memories, pictures, my love for him, and what he taught me to respect" among the things she could still hold onto. Among things to let go of, she initially included "being with him, taking walks with him, hearing his voice, hugging him, and taking pride in him as a soldier because he is a fallen soldier." Tara was able to process the last item with her therapist to come to a clearer understanding. Her pride in her father, who gave his life fighting in service to his country, was something she could and should hold onto.

When discussing committing to new relationships, Tara said, "I don't want new friends; I just want my old one back." Tara said that she missed Sela and blamed herself for what happened; she doubted that Sela would ever want to be friends with her again. Tara used the best friend role play technique ("If this had happened to Sela and you were her best friend, what would you want her to do to convince you to be friends with you again?"). This helped her decide to call Sela and ask her if she would listen to Tara's trauma narrative in an attempt to help her previous best friend understand why the rift had occurred. Tara called Sela from the therapist's office because she was nervous about the response she would get. To her surprise, after initial standoffishness, Sela was eager to see Tara and hear her narrative. The two girls met, read Tara's story, and cried together about Tara's father's death. They quickly resumed their friendship and have grown closer than ever. Tara and her mother met together to identify difficult dates (e.g., Father's Day, her father's birthday, Tara's and her mother's birthdays, the anniversary of her father's death, Memorial Day) that might serve as future trauma or grief reminders, and they made specific plans for how to cope with these dates in optimal ways (e.g., make a plan and anticipate difficulties). At the end of treatment, all of Tara's symptoms had abated. Tara and Sela were planning a special memorial service for Tara's father for when they visited his grave. When asked what she would tell other teens whose loved one had died traumatically, she said, "I thought my father would want me to be tough. But I learned it takes courage to cry. How much you cry is how much you loved someone."

Synthesis and Conclusions

Children in military families are typically resilient. These children also have stressors and exposures to trauma and losses that are unique to military families. Research is slowly contributing to increased knowledge about potential risk and protective factors for military children developing mental health problems. Interventions are being developed and in some cases evaluated for military children, both to increase resiliency and to address the particular needs of bereaved and traumatically bereaved military children. As with civilian children, challenges remain in terms of identifying children who are at highest risk for developing problems related to trauma and loss, developing and testing optimal treatments specific for military children, optimally matching treatments to individual children, and overcoming the numerous obstacles to getting these treatments to the children who need them. Each of these is briefly addressed below.

Identifying High-Risk Children

As noted above, most military children are resilient even after experiencing significant trauma or death. Many others will not be; several risk factors for developing traumatic reactions are listed in this chapter. Because there is not a one-to-one correspondence between risk and the development of symptoms and because trauma-related disorders such as PTSD and depression carry significant long-term risk, it makes sense to screen all children whose military parents are severely wounded or killed for these symptoms. In addition, due to the influence of caregiver symptoms and functioning on children, it is advisable to assess or know the mental health of the remaining and returning caregivers before, during, and after deployments or death. Assessment of children and caregivers can be done by primary care providers (Cohen, Kelleher, & Mannarino, 2008) or by schools or can be carefully coordinated with military resources.

Developing and Testing Optimal Treatments Specific for Military Children

As described in this chapter, two alternative approaches have been taken in this regard. The first is to design novel programs specifically for military children (either specific to the needs of these children, as with Project FOCUS, or adapted from civilian programs for similar populations, such as the TAPS Good Grief Camp). The second is to start with a model that has been developed and tested for civilian children and either use it as is or adapt it for military children (such as TF-CBT-CTG). Either approach has benefits and disadvantages. Both require funding, input from military family members and military child mental health professionals, time and collaboration with the military hierarchy, and acknowledgment and support of such efforts from military branches. Optimally there would be coordination across different branches of the military to minimize duplication and in recognition of the severe shortage of military child mental health providers (DoD, 2007).

Matching Treatments to Children

Given the shortages of child treatment providers and in recognition of the stigma associated with mental health treatment among many military families (DoD, 2007), it is optimal to match children with the required level of services and neither over- nor under-treat. Some children will do best with resiliency-based services, others with group support intervention, and still others with individual therapy. Identifying which children need which level of care requires developing a matching algorithm. Efforts are underway to develop an algorithm for children traumatized by disasters; the military has the opportunity and access to develop a similar algorithm for children impacted by parental trauma and death.

Overcoming Obstacles to Access to Services

The DoD Task Force on Mental Health (2007) made clear that numerous obstacles exist for military families attempting to access mental health services. These obstacles are particularly onerous for families seeking specialized services for children and may be even greater for children whose military parent is deceased. TRICARE providers are increasingly turning away military children, and few have special training in the treatment models described in this chapter. As new treatments are developed and tested for military children, consideration should be given to requiring TRICARE and other military providers to be trained in these models or to show competency in their delivery. Service members deserve to know that their country will meet their children's important needs. When members of the military are deployed, severely wounded, or killed, some of their children will need mental health services. In order to meet these needs, an obligation exists to develop and test appropriate treatments, to screen and identify children in need of services, and to provide appropriate mental health services to military children in need.

References

Amen, D. G., Jellen, J., Merves, E., & Lee, R. E. (1988). Minimizing the impact of deployment separation on military children: Stages, current preventive efforts, and system recommendations. *Military Medicine*, 153(9), 441–446.

Barnhouse, A. H., Brugler, C. J., & Harkulich, T. J. (1992). Relocation stress syndrome. *Nursing Diagnosis*, 3(4), 166–168.

Beckham, J. C., Braxton, L. E., Kudler, H. S., Feldman, M. E., Lytle, B. L., & Palmer, S. (1997). Minnesota Multiphasic Personality Inventory profiles of Vietnam combat veterans with posttraumatic stress disorder and their children. *Journal of Clinical Psychology*, 53(8), 847–852.

Black, W. G. (1993). Military-induced family separation: A stress reduction intervention. *Social Work*, 38(3), 273–280.

Blount, W., Curry, Jr., A., & Lubin, G. I. (1992). Family separations in the military. *Military Medicine*, 157, 76–80.

Brown, E. J., Goodman, J. F., Cohen, J. A., & Mannarino, A. P. (2004). A randomized controlled treatment outcome study for childhood traumatic grief. Paper presented at the 20th Annual Meeting of the International Society for Traumatic Stress Studies: Conceptualization, Measurement, and Treatment of Childhood Traumatic Grief Symposium, New Orleans, LA.

Cohen, J. A., Mannarino, A. P., & Knudsen, K. (2004). Treating childhood traumatic grief: a pilot study. *Journal of the American Academy of Child and Adolescent Psychiatry*, 43, 1225–1233.

Cohen, J. A., Mannarino, A. P., & Deblinger, E. (2006). *Treating trauma and traumatic grief in children and adolescents.* New York: Guilford Press.

Cohen, J. A., Mannarino, A. P., & Staron, V. (2007). Modified cognitive behavioral therapy for childhood traumatic grief (CBT-CTG): A pilot study. *Journal of the American Academy of Child and Adolescent Psychiatry*, 46, 811–819.

Cohen, J. A., Kelleher, K., & Mannarino, A. P. (2008). Identifying, treating and referring traumatized children: The role of pediatric providers. *Archives of Pediatric and Adolescent Medicine*, 162, 447–452.

DoD. (2008). *Defenselink casualty report 2008*. Washington, D.C.: Department of Defense (http://www.defenselink.mil/news/casualty.pdf).

DoD Task Force on Mental Health (2007). *An achievable vision: Report of the Department of Defense task force on mental health*. Falls Church, VA: Defense Health Board.

Galovski, T., & Lyons, J. A. (2004). Psychological sequelae of combat violence: A review of the impact of PTSD on veteran's family and possible interventions. *Aggression and Violent Behavior*, 9(5), 477–501.

Gibbs, D. A., Martin, S. L., Kupper, L. L., & Johnson, R. E. (2007). Child maltreatment in enlisted soldiers' families during combat-related deployments. *Journal of the American Medical Association*, 298(5), 528–535.

Goodman, R. F. (2004). *Educating the military child in transition and deployment* (unpublished online course). Department of Defense Educational Opportunities Directorate (http://www.militarystudent.org).

Goodman, R. F., & Brown, E. J. (2008). Service and science in times of crisis: Developing, planning and implementing a clinical research program for children traumatically bereaved after 9/11. *Death Studies*, 32, 154–180.

Hardaway, T. (2004) Treatment of trauma in children of military families. In N. B. Webb (Ed.), *Mass trauma and violence* (pp. 259–282). New York: Guilford Press.

Hobfoll, S. E., Spielberger, C. D., Breznitz, S., Figley, C., Folkman, S., Leper-Green, B., Meichenbaum, D., Milgram, N. A., Sandler, I., Srason, I., & van der Kolk, B. (1991). War-related stress: Addressing the stress of war and other traumatic events. *American Psychologist*, 46(8), 848–855.

Huebner, A. J., Mancini, J. A., Wilcox, R. M., Grass, S. R., & Grass, G. A. (2007). Parental deployment and youth in military families: Exploring uncertainty and ambiguous loss. *Family Relations*, 56(2), 112–122.

Jellen, L. K., McCarroll, J. E., & Thayer, L. E. (2001). Child emotional maltreatment: A 2-year study of U.S. Army cases. *Child Abuse and Neglect*, 25(5), 623–639.

Jensen, P. S. (1999). Mental health in military children: Military risk factors, mental health, and outcomes. In P. McClure (Ed.), *Pathways to the future: A review of military family research* (pp. 155–182). Scranton, PA: Military Family Institute, Marywood University.

Jensen, P. S., & Shaw, J. A. (1996). The effects of war and parental deployment upon children and adolescents. In R. J. J. Ursano & A. E. Norwood (Eds.), *Emotional aftermath of the Persian Gulf War* (pp. 83–109). Washington, D.C.: American Psychiatric Press.

Jensen, P. S., Xenakis, S. N., Wolf, P., & Bain, M. W. (1991). The "military family syndrome" revisited: "By the numbers." *Journal of Nervous and Mental Disease*, 179(2), 102–107.

Jensen, P. S., Martin, D., & Watanabe, H. (1996). Children's response to parental separation during Operation Desert Storm. *Journal of the American Academy of Child and Adolescent Psychiatry*, 35(4), 433–441.

Kelley, M. L. (1994). The effects of military-induced separation on family factors and child behavior. *American Journal of Orthopsychiatry*, 64(1), 103–111.

Kelley, M. L., Hock, E., Smith, K. M., Jarvis, M. S., Bonney, J. F., & Gaffney, M., (2001). Internalizing and externalizing behavior of children with enlisted Navy mothers experiencing military-induced separation. *Journal of the American Academy of Child and Adolescent Psychiatry*, 40(4), 464–471.

Knox, J., & Price, D. (1999). Total force and the new American military family: Implications for social work practice. *Families in Society: The Journal of Contemporary Human Services*, 80, 128–136.

Knox, J., & Price, D. (2006). Revisiting social work and the American military family. *Families in Society: The Journal of Contemporary Social Services*, 87, 1–12.

Lamberg, L. (2004). Military psychiatrists strive to quell soldier's nightmares of war. *JAMA*, 292(3), 1539–1540.

Lester, P. (2007). Project FOCUS: Families overcoming under stress. Paper presented at the Supporting Military Kids—A Day of Awareness Conference, Washington State University, Tacoma, WA.

Lester P., Saltzman, W., Beardslee, W., & Pynoos, R. (June, 2007). Strengthening military families. Paper presented at the First Annual U.S. Marine Corps Combat/ Operational Stress Control Conference, Arlington, VA.

Martin, J. A., Mancini, D. L., Bowen, G. L., Mancini, J. A., & Orthner, D. K. (2004). *Building strong communities for military families*, NCFR policy brief. Minneapolis, MN: National Council on Family Relations.

McCarroll, J. E., Ursano, R. J., Fan, Z., & Newby, J. H. (2004). Classification of the severity of U.S. Army and civilian reports of child maltreatment. *Military Medicine*, 169(6), 461–464.

Medway, F. J., Davis, K. E., Cafferty, T. P., & Chappell, K. D. (1995). Family disruption and adult attachment correlates of spouse and child reactions to separation and reunion due to Operation Desert Storm. *Journal of Social and Clinical Psychology*, 14(2), 97–118.

Meyers-Wall, J.A. (unpublished manuscript). *Children as victims of war and terrorism.*

Military Family Resource Center & The Association of the United States Army. (2001). *Educating our military's children … Are we closing the gaps?* Washington, D.C.: Institute of Land Warfare, U.S. Army.

NCTSN. (2007a). *Defining trauma and child traumatic stress.* Los Angeles, CA: The National Child Traumatic Stress Network (http://nctsn.org/nccts/nav.do?pid=faq_def).

NCTSN. (2007b). *Childhood traumatic grief: Educational materials.* Los Angeles, CA: The National Child Traumatic Stress Network (http://nctsn.org/nccts/nav.do?pid =typ_tg).

Orthner, D. (2002a). *Deployment and separation adjustment among Army civilian spouses*, Survey Report. Washington, D.C.: U.S. Army (http://www.armymwr. com/corporate/docs/planning/SAFIVSeparation.pdf).

Orthner, D. (2002b). *Family readiness support and adjustment among Army civilian spouses*, Survey Report. Washington, D.C.: U.S. Army (http://www.armymwr. com/corporate/docs/planning/SAFIVAFTB-FRG.pdf).

Otto, M. W., Henin, A., Hirschfeld-Becker, D. R., Pollack, M. H., Biederman, J., & Rosenbaum, J. F. (2007). Posttraumatic stress disorder symptoms following media exposure to tragic events: Impact of 9/11 on children at risk for anxiety disorders. *Journal of Anxiety Disorders*, 21(7), 888–902.

Pierce, P. F., Vinokur, A. D., & Buck, C. L. (1998). Effects of war-induced maternal separation on children's adjustment during the Gulf War and two years later. *Journal of Applied Social Psychology*, 28(14), 1286–1311.

Raiha, N. K., & Soma, D. J. (1997). Victims of child abuse and neglect in the U.S. Army. *Child Abuse and Neglect*, 21(8), 759–768.

Rosen, L. N., Teitelbaum, J. M., & Westhuis, D. J. (1993). Children's reactions to the Desert Storm deployment: Initial findings from a survey of Army families. *Military Medicine*, 158(7), 465–469.

Ryan-Wenger, N. A. (2001). Impact of the threat of war on children in military families. *American Journal of Orthopsychiatry*, 71(2), 236–244.

Searle, Y., & Streng, I. (1998). *The grief game.* London: Jessica Kingsley Publishers.

Shaw, J. A. (2003) Children exposed to war/terrorism. *Clinical Child and Family Psychology Review,* 6(4), 237–246.

Suozzi, J. M., & Motta, R. W. (2004). The relationship between combat exposure and the transfer of trauma-like symptoms to offspring of veterans. *Traumatology,* 10(1), 17–37.

Texas Cooperative Extension. (2002). *The soldier/family deployment survival handbook: Family deployment readiness for the active Army, the Army National Guard, and the Army Reserve.* Washington, D.C.: U.S. Army Community Services (www.myarmylifetoo.com).

U.S. Army. (2001). *U.S. Army secondary education transition study.* Arlington, VA: Military Family Resource Center.

U.S. Army Community and Family Support. (2001). *Survey of Army families IV: Highlights.* Washington, D.C.: U.S. Army (http://www.armymwr.com/corporate/docs/planning/SAFIVExecutiveSummary.pdf).

Webb, N. B. (2004). *Mass trauma and violence: Helping families and children cope.* New York: Guilford Press.

Wolfelt, A. (1991). Children. *Bereavement Magazine,* 5, 38–39.

Wolfelt, A. (2006). *Companioning the bereaved: A soulful guide to companioning.* Bishop, CA: Companion Press.

Worden, J. W. (1996). *Children and grief: When a parent dies.* New York: Guilford Press.

21
Community Response to Returning Military

WALTER ERICH PENK and NATHAN AINSPAN

Contents

In this chapter, we examine how clinicians providing professional services may promote community resources to support changes undertaken by service members and veterans as they return from war to live at home. To understand how clinicians can use community resources to address the needs of service members and veterans who are clients, we begin with the understanding that such interactions between clients and their environments are taking place as societal institutions are transforming. In earlier papers predicting evolutions in healthcare organizations for the next decade, one of us (Miller, DeLeon, Magaletta, Morgan, & Penk, 2006) predicted that health care will require clinicians to increase interventions augmenting self-care and self-reliance among clients while simultaneously maneuvering community resources to enhance treatment and rehabilitation provided by clinicians.

We review here how clinicians can network with community resources to foster changes that their clients now are undertaking. It is noted that clinicians are called upon to intervene with clients most often at those times when clients are most distressed and making changes. Now added to conventional changes that clients try for themselves, their families, and their friends are further complications building up in an age of terrorism and massive changes in societal institutions. We also examine, for clinicians, the resources in the community now available for service members and for veterans as they return home to cope with everyday realities complicated in a time of war. The terrorist attacks on America on September 11, 2001, have changed lives and everyday living for many Americans. Challenging terrorism has brought huge increases in the American military to fight the Global War on Terror at home and abroad. Many nations are now amassing resources to fight terrorists. Living and surviving in harm's way means that all of us must now live by coping with possibilities, sometimes actualities, of societies experiencing increased trauma among their civilians. And, as we fight terrorism, we must also support those among us who respond to call for actions to protect and to serve their fellow citizens.

In this chapter, we focus upon resources in communities that clinicians can access to facilitate adjustments for warriors returning home. By "communities" we refer primarily to services from governmental agencies providing the support that warriors have earned by agreeing to protect the nation. As we summarize supports and services in communities that clinicians need to know for their clients, we quickly learn what all warriors already have mastered: that the support that was earned comes not only from governmental agencies in the federal system, such as the Department of Defense (DoD) and Department of Veterans Affairs (VA), but also from state and local governments, from many volunteer organizations created by fellow warriors sworn to leave no one behind, from businesses and corporations, and from many civilians volunteering to work with and for warriors. In so many different ways, terrorist attacks on the United States have transformed our culture and changed our attitudes and behaviors to address the possibility of more terrorist attacks. Threats of terrorism complicate lives already caught up in coping with demands for ordinary living. So, now we are challenged further to change the structures in our democratic society to protect ourselves while supporting those who carry out military expeditions in the Global War on Terrorism.

We focus our attention on community resources that can be accessed through the Internet. A major change that has taken place in the last few decades is the growth of the Internet to create linkages to major resources in the community available to both clinicians and their clients. Such Internet resources include sharing information about difficulties associated with trauma, accessing manualized practices for improving home and community adjustment by promoting ways to cope with trauma, and communicating with those who have been traumatized. Bandura (2006) recently rated resources on the Internet as a major repository that clinicians can use for clients to promote self-efficacy in their efforts to access community resources. The Internet now is a primary way to learn about community-based resources for warriors, so, to better serve clinicians delivering services to clients in transition from war to home, this chapter is organized around websites describing services delivered in the community by community-based organizations.

Alarms About Community-Based Resources on the Internet

For clinicians and for clients alike, we must sound alarms about websites advertising resources in the community. Our basic warning about websites is that ideas on websites are never responsible for the people who read them, believe in them, and take action from them. Answers from websites are like the ancient story preserved for warriors, a story about a king who seeks counsel from the Delphi Oracle before battle by asking "Who is going to win the war?" And the Delphi (a spiritual person thought to be accurate in predicting outcomes) answers, "A king will lose." The answer pleases the king, who starts the fight only to discover that the Delphi meant that the other king would win. What this story about the Delphi means for websites is this: When resources in the community are listed, clinicians are not able to control how and in what ways clinicians and clients are going to be successful in using such services. It is the role of the clinician to collaborate with clients to learn how best to profit from using the Internet to access resources in the community. This notion means that both clinician and client must master community resources, especially those listed on the Internet.

No one can guarantee safety and security in using ideas or agencies presented in websites. What helps, helps. But harm also might result. Moreover, clinicians cannot control how those who are listed now will continue to help in the future, simply because they helped in the past. The most we can do is write that, once upon a time in the past, some clinicians and clients found websites about community resources listed here to be beneficial for the questions they had about their situations. And, we now hope that such benefits once experienced can indeed later be used by others for positive rewards in their lives. But, to repeat, websites and ideas on websites about community resources can never be guaranteed to be beneficial by clinicians who recommend their use for clients.

Other alarms concern the frequency and intensity of website use. Both underuse and overuse of the Internet as resources pose risks. The problem is that clinicians are still struggling with what underuse or overuse of the Internet by clients seeking resources in the community really means. Overuse and underuse relate to frequency, but too intense or not enough use likewise poses risks. Variations in frequency and intensity always are a threat, as clinicians warn their clients about community resources. Certainly, there is some combination of frequency and intensity of Internet use that might be agreed upon as risky or dangerous, just as there are situations in which the right amount of use leads to good outcomes.

Truth to tell, clinicians are striving to create criteria by which to operationally classify for clients the risks of using the Internet for community resources, as indicated by the development of what some clinicians are now calling Internet addiction disorders (IADs). Jennifer R. Ferris (jferris@vt.edu) has summarized the debate about whether or not IADs indeed exist; she concludes what any average person in an average situation on an average day will tell another who asks whether Internet use may be addictive—namely: "Anything you can moderate is not an Internet addiction disorder." So, clinicians know at least two things about accessing community resources via the Internet: (1) what we can control (in the sense of reduce) may not be an addiction, and (2) we begin taking steps toward abuse when we spend too much time using the Internet but never changing our behaviors. Using the Internet to access community resources is itself a challenge but, at the same time, the Internet offers the potential of improving access to resources in the community that enhance self-efficacy (Bandura, 2006).

There are other forms of addiction beyond doing too much. Another form of misuse is precisely the opposite—one in which we fail to do what we know needs to be done. That is, sometimes we may fail to use the Internet even though we should use it. Many of us confess such faults on a near daily basis or once a week in church when we pray for forgiveness for what we have done and then add for what we have failed to do. Stated another way, then, not using the Internet as a resource for locating resources in the community that clinicians advocate for their clients might be considered as much a psychological disorder as is overuse of the Internet.

Approaches to Community-Based Internet Resources

Websites are like any other source of information that clinicians use in developing community-based resources for their clients—the center is the client, not the website—and that is what clinicians promote about community resources for their clients. It is not what clients interpret the website to say, but, rather, it is a matter of what clients want to do with the new information that clients have learned from community resources listed on the Internet. Both clinicians and clients can learn from websites but neither clinicians nor clients obtain all that is needed to change behaviors suggested through a website. Among the first of many things that clinicians know that clients, and clinicians as well, must face when using community-based resources is that each client must confront something within themselves to overcome before using facts from websites.

Clients and clinicians bring all their weaknesses and their strengths when they approach websites. The great Viennese neuroanatomist Sigmund Freud once wrote, 100 years ago, in his 1910 classic essay, "Wild Psychoanalysis," that printed information about psychoanalysis published in newspapers and weekly magazines was like giving a menu to a starving man in the desert. What good can come from the menu when there is no food in the desert?,

Freud asked. What was missing in answers to the starving man was the man's realization about his resistances. In other words, it is forces within ourselves that influence how we use information. To understand how clinicians and clients may use websites, we always need to start with first understanding ourselves or at least our goals. What are resistances within ourselves that will influence our abilities to make good use of information from websites about the community-based resources that are available?

As clinicians and clients collaborate to use websites for information about community-based resources, each client should understand the goals and that acquiring new information usually means that new behaviors will occur. There are many ways to answer questions about resistances and how they influence ways of using websites. We focus on two here, based upon extensive research—the first concerns stages in the process of deciding to change behaviors when accessing community resources and the second deals with classifying types of websites for community resources by needs met in clinicians and clients as people.

How Personal Resistances Influence Use of Community-Based Resources

Websites are supposed to present information by which people can make changes, but we know that many people use websites not to obtain advice on what to change but rather they are searching for justification to keep on doing what they already are doing. As some clients use clinicians to find reasons for not changing, so they are likely to use websites to justify reasons for not changing behaviors. Still other people divide websites into groups favoring and not favoring changes and then use those websites that confirm their opinions.

The psychologists Prochaska, DiClemente, and Norcross long ago developed a system for classifying the readiness of people to make changes. They noted at least four basic approaches, what they called *precontemplative, contemplative, readiness*, and *changing*. Applied to websites, we as clinicians can say that some of our clients are precontemplative in their use of community-based resources on the Internet. Precontemplative clients are those who might simply check out websites without admitting that there may be problems. These clients are the deniers, those who deny that any change is ever needed. All such clients seek is to justify what they already are doing.

A second category of clients includes the contemplators, those clients who have advanced somewhat in feelings and in thoughts—clients who agree in principle that change is needed, those who say, intellectually, that one must change. But, besides imagining changes, these are clients who never do anything to change. They just think they need to change but never do. These are clients, for example, who say, yes, I have a headache and the headaches come and go and then return, but such individuals never seek advice about what causes the headaches or what new behaviors need to be adopted to assuage the headaches. They admit the problem but do not change behaviors.

Then there are clients who belong to category three, those clients who are ready, clients who have worked through denying problems, have gone beyond intellectually assessing whether problems exist, and now they are ready for change. They admit that undertaking change is far more important than not changing, yet still such clients do not undertake the behaviors to carry out changes.

Finally, we have clients who have resolved their denying, have thought about changing, have weighed the pros and cons about changing, and have begun to search for websites that will provide information on the basis of which they can begin to experiment in behaving differently, with a view toward changing their behaviors and their conflicts and situations.

We have written this chapter on community-based resources for clinicians who work with service members and veterans as clients at any of these four levels in change-making behaviors: (1) precontemplators, who deny the need and simply want to prove their point that they deserve to continue to live as they are; (2) contemplators, who wish to learn about the pros and cons of change upon return home from war zones; (3) clients who are ready, the ones who wish to assess the negatives and the positives about dimensions involved in changing their lives as they transition from the military to civilian status; and (4) changers, who are trying new ways of behaving to resolve agreements and disagreements they are experiencing upon their return home.

We proceed on the assumption that clinicians can use the information presented in this chapter on community-based resources to help anyone wherever they are in their development as they face the extreme challenges of returning home from war. It is a mistake to assume that all clients are in the same place in their minds, in their behaviors, and in their readiness to change their minds and their behaviors. Clinicians know that clients vary in their readiness for change. No client is the same in understanding and confronting the challenges of leaving a war zone and returning home, but clinicians can teach their clients how to access these community-based resources on the Internet to facilitate their transitions from life in the military to living at home. Clinicians and clients can collaborate in accessing these community-based resources to grow in strength and resiliency.

Websites of Community-Based Resources as Seen From Maslow's Hierarchy of Needs

There is a second approach to websites about community-based resources: Just as we mentioned that people who check websites differ, so also are there differences in websites about community-based resources. This second approach classifies types of community-based resources using a framework of classifications with which each clinician is quite familiar, one that recognizes that each community-based resource is unique and each is likely to differ from another. In this section of the chapter, we classify ways in which resources are different.

The approach that we use here is one that is very old in psychology; it was written in 1943 by Abraham Maslow during the middle of World War II, when he was striving to understand human motives and what kind of resources drove people to undertake changes in behaviors. What Maslow created at that time, 67 years ago, was a *theory of motivation* that now has been well supported by subsequent research. It has been so well documented by actions that followed that most clinicians use Maslow's theories to explain human behavior without even knowing the source of such ways to classify human motivations for actions.

Maslow concluded that people take actions based on five domains of needs arranged in a hierarchy within the person. The first, in the order of their appearance, is *biological and physiological needs*, defined as basic life needs for air, food, drink, shelter, warmth, and sex. It is interesting that many of these needs are basic in the earliest stages of military training and that much time is spent making sure that all people meet these needs. Once biological and physiological needs are met, then people begin addressing the second set of needs: *safety needs*. These include protection, security, order, law, social limits, and social stability. Again, examined from the standpoint of life in the military, this is what law and order in the military attempt to achieve, not only within the military but also when the military moves into war and takes over a country as an occupying force. It is important to establish law and order wherever the military has control, and law and order are especially needed at home as our nation now prepares for a Global War on Terror within the United States.

As both biological and safety needs are met, it is then possible to address other kinds of motives for human actions, according to Maslow's hierarchy of needs. The third category has to do with *belongingness* and *love*. Motives for such actions concern family, affection, relationships, and work groups. Again, when a service member has begun to train and when biological and safety needs have been met, then attention is focused on unit command, the military group being able to work together as a group directed by military command. The fourth category of motives that people share is the need for *esteem*. Human motivation consists of needs for achievement, for status, for responsibilities, for reputation. The fifth set of needs includes *self-actualization, personal growth*, and *fulfillment*.

Websites can be classified by the extent to which they meet each one of these needs separately or some combination of these needs together. When considering websites for community-based resources, clinicians collaborating with their clients must consider that both individuals and resources differ. Clinicians should assist clients in deciding where clients are in the process of developing as clients change their goals from those for a person in the military who is moving to living as a civilian. Which skills learned in combat as a warrior can be brought to living at home as a spouse, parent, neighbor, coworker, contributor to the community? What must clients change about themselves? How much of a doer is the client, as distinguished from a thinker? How quickly can clients leave behind denying problems as they transition from warrior to civilian and begin addressing a whole set of new problems in a new world still to be known? These are but a few of the hundreds of questions that clinicians work out with clients as they access community resources to support growth and resiliency in adjusting to trauma.

Community-Based Resources on the Internet

Websites on the Internet about community-based resources consist of many different ways in which clinicians and clients can learn how to conduct transitions from life in the military to living as a civilian. So, in listing Internet-based services, clinicians and clients will find websites that offer many different kinds of outlines, recipes, and checklists about how to improve. Basically, there are two communities—one within one's body (i.e., the person) and another one outside of one's body (i.e., the environment). It takes the client working together with the clinician and the environment to use resources in the community; it always requires an interaction of one's body working together with the community to achieve the goals that the warrior has set forth to achieve.

Clinicians usually agree that the first place for clients to start, when accessing community-based resources, is within the client's mind and the client's body. Both mind and body must be secure and must be thriving to make the transition out of the military and return home and to function effectively. Most who have served in the military have been exposed to trauma, have been threatened with death, have seen the dying and the injured, and have experienced life-threatening experiences. Just as clients prepared their bodies to go to war, so likewise must clients constantly prepare their bodies to meet the challenges and demands of living as a civilian. The battles in war and peace may be different but the body must always be fit to fight these battles whether in war or at peace.

So, among the best ways to return home is to start by writing out a strategic plan on how clients are going to live as a civilian. Resuming life as a civilian in the community is not merely a matter of confronting the ordinary realities of everyday living; rather, it is a matter of balancing earlier life-threatening experiences with the new life that the client is beginning as each returns home and begins a new life in their community. Returning home means clients, as civilians, must answer such questions as: What is your mission? What are your goals and objectives? What are the problems that prevent you from achieving your goals? What are the solutions that will help you achieve your goals? What are the strategic initiatives that you need to undertake to achieve your goals and objectives? What are you targeting to achieve? What are the measures that you will use to know that you have achieved what you have targeted to do? These are the questions that clinicians must collaborate with clients to answer using community-based resources. A reality about the Global War on Terror is that the fight goes on whether or not the client remains in the military. Clinicians and clients both know that everyday living is now complicated by the possibility that all citizens are now vulnerable to terrorist attacks in our nation.

The Community of the Body

The best place to start is by making sure that clients have a body that is performing as effectively as it can function. A primary concern, then, is to discover how to make sure that the client's body is fit, that the client is functioning physically to do the best the client can do. In keeping with Maslow's hierarchy of needs, among the first community-based resources that clients need to access are those that promote effectiveness in bodily functioning with as much safety and as much health as one can achieve. Living is promoting health and avoiding risks. As Bessel van der Kolk reminded us, having faced trauma in the military, one's body and one's mind always keep score. So, among the first goals to achieve are those that make both the mind and the body as fit as possible.

What are the community-based resources for physical fitness? Again, there are many but among the best are resources provided by the Department of Defense (DoD) and the Department of Veterans Affairs (VA). Both federal organizations are mandated to provide care and rehabilitation for service members returning from battle. Among the first steps toward physical and mental health that all active-duty service members and military veterans with families must take is to establish eligibility for DoD and VA services that each warrior has earned by serving in the military. Such resources in the community include those that promote a healthy body and a healthy mind.

The VA has developed many websites for veterans to ensure care. Once eligibility is determined, veterans can begin to establish medical health information and body fitness on the My HealtheVet. More information can be obtained by establishing eligibility for VA services. Once eligibility is established, service members and veterans can contact the VA at http://www.myhealth.va.gov and begin a medical record that establishes well-being in one's mind and in one's body. It is important to remember that, once traumatized, it is essential to build up a record of the condition of one's body and one's mind to be able to access all the services that one has earned. The clock is always ticking in the VA. There comes a point when some services, once available, no longer can be accessed. Those in the military and those who are veterans must remember that eligibility sometimes is limited with regard to which DoD and VA services are available. Hence, it is vital to register for such services, to determine eligibility, and to build up a record of one's medical and mental health. One worthwhile resource that clinicians can promote for clients is the VA's My HealtheVet website at www.va.gov.

Of course, private resources are available in the community that can help develop one's body and the peace of mind that goes with keeping physically fit. Throughout the nation are YMCAs, YWCAs, and many other private organizations devoted to providing services to keep the body fit. The Internet is filled with checklists of activities to make the body fit. Among the hundreds of excellent resources is www.fitboot.com/assess.shtml. Fitboot provides instructions for carrying out a panel of routine exercise for keeping fit, as well as standards of good physical health. A basic point, the same basic point that one learned when entering military service, is that clients need to keep their bodies fit and functioning not just while in the military but also now that they are civilians in the community.

The Community of the Mind

The Department of Defense provides many guidelines, resources, and checklists all dwelling on ways to cope with struggles in the mind that arise as warriors transition into the community. Among the best such resources is Army OneSource (www.armyonesource.com). This website connects active-duty service members and military veterans to Military OneSource via e-mail or by phone (1-800-342-9647). This one contact opens the client up to all kinds of resources, from how to cope with personal problems associated with memories of trauma to solving everyday difficulties such as filing taxes or reducing health-threatening habits such as smoking. Living as a civilian is just like living in the military: One must increase behaviors that promote health and decrease behaviors that threaten the mind and body. The point of this comprehensive resource is to help service members and military veterans resolve the many challenges that are highly likely to have increased while one was serving in combat (or noncombat) zones while in the military. The idea governing the operation of such a website as Military OneSource or Army OneSource is to provide "one-stop shopping," a website or telephone site where all sorts of problems can be addressed, beginning with the community of the body and the community of mind.

The VA provides another one-stop resource for all services in the community beyond www.va.gov; this resource was established 20 years ago and its values keep growing and growing. It is the National PTSD Center, headquartered in White River Junction, Vermont, with cooperating sites in Boston, Massachusetts; West Haven, Connecticut; Palo Alto, California; and Honolulu, Hawaii. The National Center for PTSD provides many resources to keep both body and mind at the height of their functioning. There are many resources for those with posttraumatic stress disorder (PTSD) that provide training on how to reduce anxieties about past traumas in combat, including the *PTSD Guide for Military Members* and a self-help work entitled *Returning from the War Zone: A Guide for Military Personnel*, available at http://www.ncptsd.va.gov/ncmain/ncdocs/manuals/GuideforMilitary.pdf. Workbooks that teach family members how to work with service members suffering from PTSD are available at http://www.ncptsd.va.gov/ncmain/ncdocs/manuals/GuideforFamilies. pdf. The National Center for PTSD has become one of the largest centers of excellence for providing care and rehabilitation for individuals, their families, and communities who have suffered trauma, not just the trauma of combat but trauma associated with all kinds of life-threatening events, such as accidents, hurricanes, and crimes. The National Center for PTSD has grown into a national treasure to help those who have been traumatized in either civilian or combat, situations. Such aid is not about being sick; rather, it centers on regaining resiliency. It is not surprising that, when national disasters occur, the National Center for PTSD website is one location most frequently contacted for help and for support.

The Department of Veterans Affairs, like the Department of Defense, is also able to provide technical information and technical services for those who have suffered trauma. For the many veterans declared eligible there are free private and confidential online and phone mental health self-assessments to help families and service members identify their symptoms and obtain assistance. A phone number that thousands of veterans call each day is 877-222-VETS, an outlet like www.va.gov. The VA also has established more than 300 outpatient clinics in many communities across the nation, extending VA services outside medical centers into the heart of communities. See, for example, the Veteran Center Readjustment Counseling programs at www.va.gov/rcs which represent hundreds of outpatient clinics throughout the United States. The DoD likewise provides similar comprehensive coverage through the TRICARE Behavioral Health Plan. Soldiers and families are eligible for 6 months of mental health care after discharge from active duty by contacting www.tricare.osd.mil or 877-TRICARE (1-877-874-2273). Likewise, an often-called resource for improving both mind and body is www.MilitaryMentalHealth. org (phone: 1-877-877-3647). A valued resource to improve the mind as well as body through formal education is the Fund for Veterans' Education at www. veteransfund.org. In addition to DoD and VA resources available in the community, all branches of the military have created websites for specialty services. An abbreviated list follows:

- The Military Officers of America (www.moaa.org; 1-800-245-8762) provides TOPS (The Officer Placement Service) transition service, support valued by so many military officers.
- The Non-Commissioned Officers Association (www.ncoausa.org; 1-210-653-6161) provides employment services for veterans.
- Other organizations providing consultation found to be beneficial include the Reserve Enlisted Association (www.reaus.org), Reserve Officers Association (www.roa.org), Association of the U.S. Army (www.ausa.org), U.S. Army Warrant Officers Association (www.

usawoa.org), Navy League of the United States (www.navyleague.
org), Naval Enlisted Reserve Association (www.nera.org), the Fleet
Reserve Association (www.fra.org), Armed Forces Communications
and Electronics Association (www.afcea.org), Marine Executive
Association (www.marineea.org), Marine Corps Reserve Officers
Association (www.roa.org), Marine Corps League (www.mcleague.
org), Air Force Association (www.afa.org), Air Force Aid Society
(www.afas.org), and Air Force Communications Electronics
Association (www.afcea.org).

- In addition to military groups practicing the doctrine of making sure
that all veterans are supported and that no one is left behind, nonprofit groups for veterans long ago established include the American
Legion (www.legion.org), the Veterans of Foreign Wars (www.vfw.
org), and the Disabled American Veterans (www.dav.org). Such
groups provide a wide variety of support, ensuring that veterans and
their families receive the services they have earned.

It should be noted that the website resources presented here can add up
quickly into a formidable accounting with so many groups actively functioning to support military and veterans, along with their families, who are in
need. Our communities, indeed, are filled with resources. What most may be
missing, however, are the decisions within individuals to use such resources.
Communities are filled with resources; what now are needed are personal
resolve and will to access those resources. Clinicians can play extremely
important roles by teaching clients how to harness willingness and resolve
to access community-based resources on the Internet and then to use what is
learned to improve their physical and mental well-being.

The Community of Work

Veterans of World War I and World War II long ago established that returning
to work, or to some form of continuing productivity, was essential to recovering from the wounds of war. The best example of such resources can be found
in the Servicemen's Readjustment Act of 1944. The lead author of what became
known as the GI Bill was Edith Nourse Rogers of Massachusetts, who served
in the U.S. House of Representatives from 1927 until 1960, longer than any
woman has ever served in the U.S. House. She knew what trauma was. In the
Salem Witch Trials of the 1690s, her great-great-great-grandmother, Goody
Nurse, was charged as being a witch and was pressed to death. Ms. Rogers
learned from her work in World War I that, if you survived trauma, your best
means for recovery was work. So, with her colleagues in Congress, she wrote
the GI Bill which provided education and work training for military returning
from World War II war zones. The GI Bill created educational work-related
experiences for veterans, both with and without disabilities.

Based on the principles that guided development of the GI Bill, there are now new opportunities for the military in the Montgomery Bill, and Congress is writing new bills to expand training and education. Among the many websites that focus on increasing work opportunities for veterans are www.monster.com, www.careerbuilder.com, www.careers.org, www.craigslist.com, www.employmentguide.com, www.hotjobs.com, www.idealist.org, www.indeed.com, www.jobster.com, www.snagajob.com, www.thingamajob.com, and www.usajobs.opm.gov. For wounded combat veterans who need to work at home, support groups and job locators include www.nticentral.org, www.virtualvocations.com, www.tigerfish.com/employment.html, www.tjobs.com, www.workathomeinfo.org, www.wahm.com/jobs.html, www.disaboom.com, and www.tabinc.org.

Principles about how work is central to transitioning from the military to civilian living can be found in a seminal report from the American Psychological Association, a group that has been advocating support for the military and for veterans (see www.apa.org). Psychologists such as Michelle Sherman, who served on the APA's Task Force for services for military veterans, have likewise developed resources for families, specifically for children learning to live with family members who have been in combat (see the SAFE program at www.ouhsc.edu/safeprogram and Sherman's book for adolescents whose parents have experienced trauma at www.seedsofhopebooks.com).

The Department of Veterans Affairs, likewise, has one major division devoted strictly to veterans benefits, including education, job training, and housing for veterans at risk for becoming homeless. Many other organizations in the community promote education and training as primary supports in the community, such as the U.S. Department of Labor programs detailed at www.dol.gov. Likewise, new private groups are offering services for veterans, such as Iraq and Afghanistan Veterans of America (www.iava.org) and groups that train veterans for civilian occupations (www.veterantraining.com). Groups are being developed to train military veterans to form their own businesses; see, for example, whitman.syr.edu/eee/veterans. Veterans are reporting these links as being extremely helpful in making the transition from military status to civilian; see a complete list at the Official State Directors of Veteran Affairs website (www.nasdva.com) as well as www.tvc.state.tx.us/States.

Other organizations known to be helpful to veterans include Hope for the Warriors™ (www.hopeforthewarriors.org), the Injured Marine *Semper Fi* Fund (www.semperfifund.org), the Naval Special Warfare Foundation (www.nswfoundation.org), Reserve Aid (www.reserveaid.org), the Center for Deployment Psychology (www.cdp.gov), and a center to support employers who work with veterans in recovery (www.esgr.org).

Another example of new programs being developed is a U.S. Small Business Administration (SBA) initiative called the SBA's Patriot Express Pilot Loan Initiative for veterans and members of the military community wanting to establish or expand small businesses. This exciting new program is designed to foster development of new businesses by combat veterans and is open to (1) veterans; (2) service-disabled veterans; (3) active-duty service members eligible for the military's Transition Assistance Program; (4) reservists and National Guard members; (5) current spouses of any of the above; and (6) widowed spouses of service members or veterans who died during service or of a service-connected disability. The SBA and its resource partners are focusing additional efforts on counseling and training to augment this loan initiative, and more information can be found at www.sba.gov/patriotexpress.

Some resources offer employment guidance for service members transitioning into the community (see www.militaryhomefront.dod.mil). Other specialty programming includes the Military Spouse Resource Center, Military Spouse Career Center, Army Career and Alumni Program (ACAP), Army Spouse Employment Program, Marine Corps Transition Assistance Management Program (TAMP), Marine Corps Family Member Employment Assistance Program (FMEAP), Navy Transition Assistance, and Navy Spouse Employment Assistance Program (SEAP).

Communities for War's Wounded

Whereas the community resources listed above are available for military and for veterans in general, those that follow below are specialty services for veterans who have been wounded. Those wounded in war usually require more complicated services; hence, we list organizations beyond the Department of Defense and the Department of Veterans Affairs who likewise offer specialty services for those recovering from wounds.

In addition to the VA, where disability classifications are established (www.va.gov), other organizations provide consultation about services availabilities, such as the Military Severely Injured Center (1-888-774-1361). The Military Severely Injured Center is available to soldiers and their families 24 hours a day, 7 days a week. And, to repeat, representatives of Military OneSource (www.militaryonesource.com; 1-800-342-9647) are available to soldiers and their families 24 hours a day, 7 days a week. Information and referral services, as well as short-term counseling (up to 6 sessions), can be provided for those eligible. Services are also available to National Guard members 6 months after discharge.

Other resources for war's wounded can be found at Troop and Family Counseling Services for National Guard and Reserves (1-888-755-9355), where counseling services, private and confidential, are offered 24 hours a day, 7 days a week. The Department of Health and Human services provides up to six sessions of counseling, at no cost, around the clock, through their Employee Assistance Program for civilians, active-duty service members, National Guard, and reserves, as well as their families (www.foh.dhhs.gov; 1-800-222-0364).

Other services are provided through Post Deployment Health Reassessment (PDHRA), which helps veterans identify deployment-related mental health concerns and makes appropriate referrals between 3 to 6 months after returning from war zones (www.fhp.osd.mil/phrainfo/index.jsp). Likewise, specialty consultation is available from Walter Reed Medical Center Deployment Health Clinical Center for Specialized Care Programs (SCP) for service members experiencing difficulties readjusting to life after service in Operation Iraqi Freedom (OIF)/Operation Enduring Freedom (OEF) (www.pdhealth.mil; 1-202-782-6563). Chaplain Services specializes in relationship-enhancement programs for married couples in the military (1-800-634-1790, option 1). The U.S. Army's premier program designed to take care of wounded soldiers and their families is the Army Wounded Warrior Program (1-800-337-1336).

A manualized treatment and recovery program, *Our Hero Handbook*, can be obtained at www.washingtonpost.com/wp-srv/nation/documents/walter-reed/HeroHandbook.pdf. See also the Army's Wounded Warrior Program at www.aw2.army.mil for another program geared toward providing important benefits for service members wounded in war.

The Community of Transitioning Home

Transitioning home can be an extremely complex process. Again, the Department of Veterans Affairs has developed many resources for service members and veterans. Services are available for all eligible veterans through the Seamless Transition Assistance Program (www.seamlesstransition.va.gov/SEAMLESSTRANSITION/index.asp), including:

- *Compensation and pensions*—VA website hosting benefits information for veterans with disabilities
- *Education*—Information on the VA education benefits available for veterans
- *Home loan guaranty*—VA home loan guaranty eligibility website
- *Vocational rehabilitation and employment*—Rehabilitation counseling and employment advice for veterans who are disabled and in need of help readjusting
- *Insurance*—VA life insurance program for disabled veterans
- *Burial*—Information on burial benefits for certain qualified veterans
- *Women veteran benefits and the Center for Women Veterans*—Two separate websites featuring benefits issues and other programs unique to women veterans
- *Health and Medical Services*—VA website providing complete health and medical services information
- *Medical care for combat theater veterans*—VA website providing specific information for veterans of combat theater of operations

- *Special health benefits programs for veterans of Operations Enduring Freedom/Iraqi Freedom*—VA health information website for OEF/OIF veterans specific to environmental agent issues
- *HealtheVet Web portal*—VA's new health portal developed for veterans and their families to provide information and tools to enable them to achieve the best health
- *My Recovery Plan* (in development)—Excellent VA resource for career counseling, guiding growth in resiliency for work development
- *Civilian Health and Medical Program of the Department of Veterans Affairs (CHAMPVA)*—A federal health benefits program administered by the VA that is a fee-for-service (indemnity plan) program providing reimbursement for most medical expenses, including inpatient, outpatient, mental health, prescription medication, skilled nursing care, and durable medical equipment (DME); available to certain veterans' family members who are not eligible for TRICARE
- *Transitioning from war to home*—VA websites describing Vet Center Readjustment Counseling Service; provide war veterans and their family members quality readjustment services in a caring manner, assisting them toward a successful post-war adjustment in or near their communities
- *State benefits*—Benefits offered by states for veterans (see, for example, www.texvet.org); www.va.org provides links to state programs

Reuniting with colleagues long after returning home is another kind of experience that many regard as beneficial. Again, many websites have been set up to facilitate such contacts, such as My Army Life Too (www.myarmylifetoo.com), Army Families Online (www.armyfamiliesonline.org), Military OneSource (www.militaryonesource.com), Military Homefront (www.militaryhomefront.dod.mil), and National Military Family Association (www.nmfa.org).

The Community of Support Groups Voluntarily Helping Veterans

Nonprofit groups voluntarily helping veterans number in the hundreds. Among the nonprofit organizations designed to help veterans are AMVETS (www.amvets.org), Blinded Veterans Association (www.bva.org; 1-800-669-7079), Disabled Veterans of America (www.dav.org; 202-554-3501 or 1–877–426–2838), the American Legion (www.legion.org; 202-861-2700); the Military Order of the Purple Heart (www.purpleheart.org; 703-642-5360), the National Amputation Foundation (amps@aol.com,www.nationalamuptation.org; 516-887-3600), Paralyzed Veterans of America (info@pva.org, www.pva.org), Veterans of Foreign Wars (www.vfw.org; 202-453-5230), United Spinal Association (info@unitedspinal.org, www.unitedspinal.org; 1-800-807-0192), Wounded Warrior Project (info@woundedwarriorproject.

org, www.woundedwarriorproject.org; 1-540-342-0032), Computer/Electronic Accommodations Program (www.tricare.osd.mil/cap), National Military Family Association (www.nmfa.org), America Supports You (www.armerica-supportsyou.mil), Coalition to Salute America's Heroes (www.saluteheroes.org), Operation First Response (www.operationfirstresponse.org), Army Emergency Relief (www.aerhq.org), Serving Those Who Serve (www.serving thosewhoserve.org), Helping our Heroes Foundation (www.hohf.org), and the Fallen Patriot Fund (www.fallenpatriotfund.org).

Building BHAGs: How Clinicians Can Link Clients to the Environments That Build Recovery

Communities, then, are filled with public and private structures designed to serve the military and veterans. Communities of support and service, such as those listed above, all arise from the principle that veterans exposed to trauma must have access to the resources that they have earned by their actions to preserve and protect their fellow citizens (Glynn, Drebing, & Penk, 2008; Penk & Flannery, 2000). What now is needed is for clients who are eligible to take action to get such services. Clinicians can expand their practices for their clients both by developing services made available on the Internet as well as by training clients how to use other services developed by others. As noted above, structures of care and rehabilitation are available for those in the military, in the National Guard, and in the Reserves and for veterans and their families. Many general and specialized services have been created by federal and state programs as well as by many nonprofit groups volunteering to help families with support and with information to access the services that they need (Glynn et al., 2008).

Knowledge about community services is as far away as one's computer or a telephone. Ours is a new age in which a new military receives support immediately, on the battlefield, for the stresses of trauma in fighting terrorists. This support continues, fostering recovery, as service members return home and take on lives as veterans to live at peace in their homes and communities. The Global War on Terror is teaching us that we must fight terrorists in lands abroad and we must change our nation to protect our people here at home.

Both military and veterans must now know, guided by clinicians cooperating with their clients, that resources in their communities bring persons in need together with organizations that can help. Clinicians are well placed to link the military and veterans to support in the community designed to foster well-being and adjustments. The Global War on Terror is teaching us that our response, as a culture, must be that we must fight and, as we fight, we must support those who fight, both outside and within our own country. The Global War on Terror is being fought with expertise and with supports for those who respond to the nation's call to fight these battles. What is needed now is for clinicians to link military and veterans to discover that organizations in the community have been created to meet their needs as they carry on these battles.

Much has changed since the Vietnam War with regard to services being provided by the DoD and VA (Penk & Flannery, 2000). PTSD and other disorders are being treated today in war zones as well as back at home (Glynn et al., 2008). The DoD and the VA have best practices in place for the treatment of combat-related disorders among service members, veterans, and their families (Ainspan & Penk, 2008).

This is a new age with a new military, new veterans, new opportunities. Part of the newness in our opportunities is the fact that both clinicians and clients must change how we go about working together to generate new behaviors in a time of terror. Military and veterans, as well as all citizens, now must develop new ideas about how to live productively and as colleagues. Ours is an age to develop new ideas and new ways for networking with each other. We need to live our lives generating what Jim Collins called BHAGs—big hairy audacious goals. And, BHAGs are what clinicians are skilled at developing for their clients. Clinicians have a long history of working with clients to develop new ideas as ways to solve problems (Baker & Pickren, 2007).

How do clinicians work with clients to develop BHAGs? Clinicians do this through rehabilitation, by asking clients to answer such questions as "Why do I exist?" and "Why have I survived?" In answering these questions, clients become coparticipants in their treatment and rehabilitation. In turn, the clinicians show their clients how to develop new BHAGs that will define for the clients their new missions in life. They will specify new goals and objectives that clients will strive to achieve and will use the clients' community of support to generate the strategic initiatives necessary for clients to reach their chosen targets. The communities, such as those listed above, exist to support the clients' core purposes in living. The client's task, supported by clinicians, is to leverage strengths and opportunities from the community of the client to fulfill the BHAGs that clients have chosen to actualize and bring into being. Clinicians working with clients to generate new BHAGS at home and in their communities will facilitate a client's transition from war zones back to life as a civilian. It is a time of living and surviving in harm's way, and clinicians can now link clients to communities that provide support and services necessary to make sure that the BHAGs of veterans can be achieved while they live as civilians, at home, and in neighborhoods, in peace, in a time of terror.

References

Ainspan, N. D., & Penk, W. E. (Eds.) (2008). *Returning wars' wounded, injured, and ill: A handbook*. Westport, CT: Praeger Security International.

Baker, R. R., & Pickren, W. E. (2007). *Psychology and the Department of Veterans Affairs*. Washington, D.C.: American Psychological Association.

Bandura, A. (2006). Toward a psychology of human agency. *Perspectives on Psychological Science*, 1, 164–180.

Glynn, S., Drebing, C., & Penk. W. (2008). Psychosocial rehabilitation. In E. Foa, T. Keane, M. Friedman, & J. A. Cohen (Eds.), *Effective treatments for PTSD* (2nd ed., pp. 388–426). New York: Guilford Press.

Miller, T., DeLeon, P., Magaletta, P., Morgan, R., & Penk, W. E. (2006). The psychologist with 20/20 vision. *Professional Psychology: Research and Practice*, 37, 831–836.

Penk, W., & Flannery, R. (2000). Psychosocial rehabilitation. In E. Foa, T. Keane, & M. Friedman (Eds.), *Effective treatments for PTSD* (pp. 347–349). New York: Guilford Press.

Penk, W., Drebing, C., & Schutt, R. (2002). PTSD in the workplace. In J. C. Thomas & M. Hersen (Eds.), *Handbook of mental health in the workplace* (pp. 215–248). Thousand Oaks, CA: Sage.

22

Issues of Grief, Loss, Honor, and Remembrance: Spirituality and Work With Military Personnel and Their Families

KENT D. DRESCHER, MARISSA BURGOYNE,
ELIZABETH CASAS, LAUREN LOVATO,
ERIKA CURRAN, ILONA PIVAR, and DAVID W. FOY

Contents

War is a human experience that is life changing for many military service personnel. As outlined in previous chapters, much is known about the biological and psychological impact of war on human experience. The purpose of this chapter is to provide an overview of the role that spirituality plays in the recovery environment following military service during wartime. Many veterans are finding the transition from deployed to nondeployed status a serious challenge. Some, still serving on active duty, have returned to station and are preparing with their families for the possibility of another deployment; some, on active reserve status, have returned home to family and career, only to find a shortage of supportive resources to meet their needs. Although new programs such as the Warrior Transition Units have been established, they are not as yet fully staffed and capable of fully meeting current needs. Many of the personnel that have discharged from their military service have found long waiting lists for health services at the Department of Veterans Affairs (VA). The goal of this chapter is to acquaint the reader with ways in which spirituality can be a potential resource for coping with the many varied losses and stressors of deployment and homecoming.

Combat duty frequently brings service personnel into close contact with traumatic life-threatening events of many sorts. In addition to the infantry battles commonly envisioned when one thinks about the historical experience of war, in Iraq improvised explosive devices have become the most common cause of death. Current data suggest that 63% of combat deaths have been caused by these devices during the war in Iraq and Afghanistan (Defense Manpower Data Center, 2008). These traumatic experiences etch scenes of horrific carnage and death into the memories of those who survive these events—memories that for many are brought back to awareness in recurrent nightmares and through everyday reminders of service. Not everyone survives war-zone service, and those who do frequently carry a large burden of remembrance for those who died.

Loss during war is a broad, multifaceted, and likely universal experience. Usually we associate loss with death—particularly the deaths of those close to us; however, it is important to acknowledge other serious forms of loss, as well. In addition to death, many survivors experience devastating wounds, resulting in permanent disability both physical and mental. There may be loss of relationship with close friends as service personnel leave the cohesion of their units to return home, and sometimes that loss of relationship extends to intimate relationships with spouses, partners, children, parents, and others as the veteran returns "different" from the person who first deployed. Service personnel may experience loss of self as they find themselves changed within and unable to see and live in the world in the same way as they did when they left. It may include loss of innocence that comes about through exposure to the horrors and carnage of war. Along with each of these different forms of loss, comes a range of emotional experiences—sadness and grief, sometimes rage and bitterness, and often, perhaps, confusion and questioning about how best to recover.

For many years health professionals have utilized the biopsychosocial model as a way of understanding the interrelated factors associated with optimal physical and mental health functioning. In recent years, several theorists have suggested incorporating the domain of spirituality into the biopsychosocial model (Sulmasy, 2002). For many individuals, spirituality is a primary life domain that affects and is affected by everyday experience. It is also a domain frequently ignored or under-recognized by health professionals due, in part, to differences in perspective between health providers and patients, as well as to lack of training for health professionals in this area; however, it is an area that an increasing amount of research suggests is highly relevant for supporting service personnel transitioning from war-zone service.

Definitions of Religion and Spirituality

For the purposes of this chapter we define religion as a system of beliefs, values, rituals, and practices shared in common by a social community as a means of experiencing and connecting with the sacred or divine. And, we broadly define spirituality as an individual's understanding of, experience with, and connection to that which transcends the self. The object of that understanding, experience, and connection may be God, or nature, or a higher power, or something else relevant and important to a particular individual. A person's spirituality may be realized in a religious context, or it may be entirely separate and distinct from religion of any sort. In most cases, however, religion can be understood as being spiritual in nature, with spirituality serving as a more generic way of describing the experience. Kenneth Pargament (1997) has suggested a brief but powerful definition of religion as a "a search for significance in ways related to the sacred." Others have defined spirituality as "multidimensional space in which everyone can be located" (Larson, Swyers, & McCullough, 1997). A review by Miller and Thoresen (1997) cites a number of useful definitions of these two terms. The authors call particular attention to the individual vs. corporate nature of the two terms. From a mental health perspective, both these constructs can be conceptualized as practices, beliefs, and experiences.

In the first section of this chapter, we provide an overview of the research encompassing the domains of spirituality, posttraumatic stress disorder (PTSD), traumatic bereavement loss, and loss from physical disability, including traumatic brain injury (TBI) and spinal cord injury (SCI). In the second section of the chapter, we summarize the trajectories that postdeployment adjustment may take, identify some unique risks of combat service, and attempt to synthesize ideas for health providers and families for ways that spirituality can be incorporated into recovery from the many and varied losses experienced by military personnel during wartime.

Several comments should be made before proceeding with a summary of the studies examining spirituality, trauma, disability, and grief. First, there is a general lack of research within mental health and medical journals examining the role of spirituality. Studies that looked at this during the 1980s and early 1990s found that only 1 to 3% of published journal articles in psychology, psychiatry, and medicine include spiritual variables (Larson, Pattison, Blazer, Omran, & Kaplan, 1986; Weaver et al., 1998). Additionally, within this small percentage of articles, most have a problem of using "home-grown," often unvalidated, measures or measures that consist of a single item as the means of measuring spirituality. This results in a body of literature where it is difficult to replicate findings because of the lack of common measures across studies. In 1998, the National Institutes of Health (NIH) created a working group and gave it the task of finding consensus among researchers as to the best ways of measuring various aspects of spirituality for health research (Abeles et al., 1999). The resulting consensus document describes 12 different domains of spirituality potentially useful for measurement and makes recommendations for long and short versions of scales within each domain.

An earlier review of the trauma and spirituality literature (Drescher et al., 2004) indicated the presence of two trajectories for the interaction of spiritual experience in the context of traumatic experience. First, there is at least mixed support for the general hypothesis that the presence of healthy spirituality is associated with several positive outcomes in the aftermath of trauma. There is also, however, some indication that traumatic experiences may shift a person's spiritual beliefs, behavior, and outlook in unhelpful directions and that these negative faith appraisals are sometimes associated with worse health outcomes.

Traumatic Experiences, PTSD, and Spirituality

To obtain a better understanding of the relationship between traumatic experiences and spirituality, a new review of this literature was undertaken using several computer databases (PsychInfo, Proquest, Medline, Pilots, Academic Search Elite) to identify studies that met search criteria relevant to the constructs of spirituality, religion, and trauma. Only empirical studies with 30 or more participants were included in an attempt to ensure scientifically accurate results; 55 studies met these inclusion criteria and encompass a wide variety of traumatic experiences. Studies included participants who experienced physical and sexual assault in childhood or adulthood, combat-related trauma, war exposure, terrorist events, natural disasters, unexpected/violent death of a family member, and life-threatening illness.

Some of this literature examines the relationships among trauma exposure, spirituality, and quality of life. These studies generally indicate that spirituality is linked to increased quality of life, better overall mental health, and decreased psychopathology among particular populations that have endured traumatic experiences. In two studies of military veterans in VA treatment, spirituality is associated with increased quality of life and decreased psychopathology in groups of veterans exposed to sexual assault and sexual abuse. Scores on measures of overall mental health have been positively associated with religiosity among female veterans with and without a history of sexual assault (Chang,

Skinner, & Boehmer, 2001). More frequent religious attendance among male veterans with a history of sexual abuse has been reported (Chang, Skinner, Zhou, & Kazis, 2003). Furthermore, endorsing religious beliefs as a source of strength and comfort has been associated with lower depression among male veterans.

In a nonveteran study, results indicate that African–American female sexual assault survivors who experienced an increase in spirituality after sexual assault were more likely to reestablish well-being, whereas those with no increase in spirituality remained at reduced levels of well-being (Kennedy, Davis, & Taylor, 1998). Examination of specific types of spiritual coping and their relationship to distress severity suggests that among male and female adult survivors of child abuse, negative spiritual coping (punitive appraisals of God, anger at God) is associated with greater distress, whereas positive spiritual coping (attitudes and behaviors that enrich spiritual experience) is associated with decreased distress (Gall, 2006). Higher levels of spirituality have been associated with fewer reported eating disorder symptoms among female sexual abuse survivors, and higher spiritual well-being has been associated with lower psychopathology (Krejci et al., 2004).

The literature examining spirituality in the context of life-threatening illness reveals somewhat mixed findings regarding spirituality. In most studies, spirituality is associated with increased quality of life, life satisfaction, and mental health for patients with cancer. Among cancer patients, spiritual well-being and strength of spiritual beliefs have been linked to a better quality of life (Cotton, Levine, Fitzpatrick, Dold, & Targ, 1999; Wan et al., 1999). Spirituality has been associated with the ability to enjoy life while experiencing illness symptoms (Brady, Peterman, Fitchett, Mo, & Cella, 1999); religious beliefs, activity, and relationships have been associated with life happiness, life satisfaction, and lower pain levels (Yates, Chalmer, St. James, Follansbee, & McKegney, 1981); and higher hope and lower negative mood states have been indicated among those with high spiritual, religious, and existential well-being (Fehring, Miller, & Shaw, 1997). Religious and existential beliefs have also been associated with higher levels of family adjustment, social adjustment, and psychological health (Schnoll, Harlow, & Brower, 2000). Furthermore, spiritual well-being has been associated with less state-trait anxiety (Kaczorowski, 1989); however, religious well-being was not found to be related to lower state-trait anxiety, and in one study no relationship was found between religious coping and quality of life and the physical impact of cancer (Nairn & Merluzzi, 2003).

Some studies have suggested that cultural factors may influence the association between spirituality and health outcomes among breast cancer survivors. Whereas intrinsic religiosity (commitment to religion for its own sake) was more predictive of hope and spiritual well-being for Caucasian females, no single type of religiosity was more salient in predicting hope and spiritual well-being among Hispanic females (Mickley & Soeken, 1993). Among long-term cancer patients, increased quality of life was associated with positive spiritual change, as well as with hopefulness and having a purpose in life

(Ferrell, Dow, & Leigh, 1995). Furthermore, female long-term cancer survivors with a positive philosophical or spiritual outlook were more likely to have good health habits and be supportive of others, resulting in positive psychological states and life satisfaction (Kurtz, Wyatt, & Kurtz, 1995).

Although results are mixed for studies of spirituality and health outcomes for patients positive for human immunodeficiency virus (HIV) and patients with acquired immune deficiency syndrome (AIDS), most studies show spirituality to be associated with increased quality of life and life satisfaction. Among individuals who are HIV positive or have AIDS, spirituality or religiousness has been linked to a better quality of life (Brady et al., 1999; Flannelly & Inouye, 2001). Spiritual well-being has been associated with the ability to enjoy life while experiencing illness symptoms (Brady et al., 1999), religiousness has been linked with health and functioning among individuals with HIV (Flannelly & Inouye, 2001), and religious coping has been related to lower levels of depression (Woods, Antoni, Ironson, & Kling, 1999). Additionally, in one study, a relationship was not found between religious coping and quality of life among predominately African–American females who were HIV positive (Weaver et al., 2004).

The association between spirituality and well-being among terminally ill patients is not well understood at present. Whereas some research indicates that higher religiousness is more associated with female gender and illness status than to well-being among terminally ill adults (Reed, 1986), other research does indicate a relationship between spirituality and well-being among terminally ill hospitalized patients (Reed, 1987). Hospital systems may help patients with their spiritual needs by organizing clergy visits, assisting with chapel visits, assisting in the creation of spiritually friendly settings, and by being open to patients' spirituality related discussion (Reed, 1991). Spiritual meaning and stronger faith have also been shown to have an inverse relationship with depression and anxiety, in association with more education, among students following the September 11, 2001, terrorist attacks (Ai, Cascio, Santangelo, & Evans-Campbell, 2005).

Spirituality as Protective Factor and Coping Method

Spirituality serves several functions in relation to traumatic experiences. First, it can be a protective factor, helping to buffer the effects of the traumatic experience, which may result in better mental health (Chang et al., 2001; Doxey, Jensen, & Jensen, 1997) or increased resilience to complex PTSD (Jones, 2007). As a means to buffer trauma, spirituality may perhaps engender hope, resulting in a greater sense of well-being (Mickley & Soeken, 1993; Mickley, Soeken, & Belcher, 1992). Second, spirituality can be used as a method for coping with traumatic events by helping the individual ascribe meaning to the trauma, feel understood, or provide the means to enlist additional social support (Green, Lindy, & Grace, 1988; Jones, 2007; Maton, 1989; Patel, Shah, Peterson, & Kimmel, 2002; Thompson & Vardaman, 1997). In a number of different traumatized groups, positive religious coping has been found to be positively related to increased psychological health, and negative spiritual coping was positively related to poorer psychological health (Fallot & Heckman, 2005; Pargament et al., 1990; Smith, Pargament, Brant, & Oliver, 2000). In general, individuals from various religious denominations have been shown to utilize their religious and spiritual beliefs to cope positively with traumatic events (Alferi, Culver, Carver, Arena, & Antoni, 1999; Elliott, 1994; Gillard & Paton, 1999); however, unique characteristics of a particular religious perspective may still prove to be related to its helpfulness as a coping mechanism. Religion, for example, was shown to serve as a stressor rather than a buffer for the effects of stress for Catholic women with breast cancer and female incest survivors in conservative Christian families (Alferi et al., 1999; Elliott, 1994).

Taken together these findings suggest that spirituality can have positive effects on a traumatized individual by enhancing quality of life, serving as a protective factor against adverse reactions to trauma, and assisting as a coping mechanism to deal with the traumatic event. Trauma, however, may have an impact on spirituality itself and can either enhance or compromise its availability as a positive mechanism for dealing with trauma. The impact of trauma can affect an individual's spirituality in two ways. On the one hand, it can increase reliance on religion and strengthen spiritual practices, but, conversely, it can decrease religious practice and result in a negative shift in the individual's views about the goodness of God and the meaning of life.

Positive aspects of surviving trauma (including spirituality) are frequently referred to as *posttraumatic growth* or *adversarial growth*. In the face of trauma, people may demonstrate an increase in their spiritual and religious practices and beliefs. In a study conducted shortly after the 9/11 terrorist attacks, 90% of respondents reported "turning to prayer, religion, or spiritual feelings" as a way of coping (Schuster et al., 2001). Studies of medical patients have demonstrated the growth of spirituality in the face of life-threatening illnesses, with patients exhibiting increases in their frequency of prayer, church attendance, and spiritual faith (Moschella, Pressman, Pressman, & Weissmann, 1997; Reed, 1991) and demonstrating more religious and spiritual growth than those who were healthy (Cordova, Cunningham, Carlson, & Andrykowski, 2001). Spiritual growth has also been reported in studies of former prisoners who experienced torture (Salo, Quota, & Punamaki, 2005); however, additional resiliency factors such as secure adult attachment, positive affiliation to others, and personal strength were also found to be associated with posttraumatic growth in this population.

On the negative side, many people who experience traumas are so changed by their experiences that they report "losing faith" and turn away from their religious practices. A recent study of combat veterans showed that certain types of war-zone experiences were associated with loss of faith in the present, and loss of faith was in turn associated with increased utilization of VA mental health services (Fontana & Rosenheck, 2004). A survey conducted with Vietnam veterans in residential treatment for PTSD found that nearly

60% reported "abandoning their religious faith in the war zone," and nearly 80% reported "difficulty reconciling war zone experiences with their religious faith (Drescher & Foy, 1995). Green and colleagues (1988) reported that veterans exhibit difficulty generating meaning from their war experiences. Those veterans who have been the victims of military sexual trauma have also been found to have lower levels of spiritual faith and practice (Chang et al., 2001).

In addition, traumatic experiences can cause individuals to have more negative perceptions of God as wrathful and disapproving. For those who have experienced childhood sexual abuse this is particularly true, with childhood sexual abuse survivors reporting more negativistic beliefs about God and decreased spiritual participation and practice than those who had not been abused (Fallot & Heckman, 2005; Finkelhor, Hotaling, Lewis, & Smith, 1990; Hall, 1995; Kane, Cheston, Greer, Pritt, & Lawson, 1993; Pritt, 1998). These negative appraisals comprise what is termed *negative religious coping* and were associated with more severe PTSD and depression in a population of treatment-seeking combat veterans (Witvliet, Phillips, Feldman, & Beckham, 2004).

Good research support suggests that spirituality is associated with better quality of life for individuals who have experienced trauma and can also be a useful method of coping with those experiences; however, traumatic experiences themselves can substantially impact spirituality in either a positive or negative direction. If spirituality is reduced by trauma, an important coping tool may no longer be available to the individual, whose quality of life and mental health outcomes may be compromised. Current studies have several limitations, though, as the definitions of spirituality and religion are inconsistent across studies, thereby drawing into question whether the studies measure a similar construct. Major problems exist in the field as regards measure selection. These problems limit the comparability of findings across studies and diminish the statistical power to detect significant results. Very few studies focus specifically on the veteran population. Unique features of combat-trauma exposure may prove to impact spirituality differently, so more research needs to be conducted in this area.

Traumatic Bereavement, Disability, and Spirituality

Death is inherent in war, and most service members are touched by it either directly or indirectly. One early study of soldiers and Marines after the initial phase of the war in Iraq and Afghanistan indicated that 92% reported knowing someone seriously injured or killed, 75% reported seeing dead bodies, and 42% reported handling human remains during their service (Hoge et al., 2004). Attachment loss is perhaps one of the most painful human experiences. Separation distress may actually be biologically hardwired for our species. Feelings of grief and loss are normal when death occurs in the war zone, but for some these feelings will become prolonged. Much theoretical discussion about grief has taken place in the mental health field in recent years. For the last decade or so, researchers have determined that a predictable pattern of problems arises in some who fail to recover from loss, especially loss by traumatic means. Early on this was referred to as *traumatic grief*, then *complicated grief* when it was recognized that others not exposed to traumatic events had difficulty recovering as well. Most recently, the field is moving toward recognizing a disorder called *prolonged grief disorder.* Pivar and Field (2004) discuss the unique patterns of traumatic guilt, differentiated from PTSD, that are prevalent in Vietnam-era combat veterans and continue to be endured without specific grief-oriented treatment.

Another aspect of the loss burden carried by returning military service personnel is the need to cope with injury and disability. A recent report to Congress illustrates the extent of that burden. As of August 2007, 36,471 personnel had been medically evacuated from Iraq. The reasons reported for medical evacuation included battle injuries (22.4%), non-battle injuries (20.5%), and disease (57.2%). Serious injuries in the battlefield included 1005 individuals who were amputees and 3294 individuals suffering traumatic brain injuries (TBIs). Explosive blasts accounted for 48.6% of amputations and 69% of TBIs; 16% of amputees lost multiple limbs. Additionally, service personnel experience spinal cord injury (94 individuals), blindness (48 individuals; battle eye injuries account for 15.8% of all medical evacuations), and severe burns (598 individuals) (Carlock, 2007). Other estimates of the number of TBIs suggest that actual numbers may considerably exceed the numbers reported by the Department of Defense (DoD). Traumatic brain injury has been considered the signature injury among those involved in combat in Operation Iraqi Freedom (OIF)/Operation Enduring Freedom (OEF). Data from prior wars indicate that approximately 20% of the wounded military personnel had primary or concurrent head injuries, and rates of TBIs in OIF/OEF are thought to be even higher (McCrea et al., 2008).

A recent article (Ahlstrom, 2007) identified a number of categories of loss experienced by individuals disabled by chronic illness. Most of these are applicable to those disabled through war zone trauma. The loss types identified include loss of bodily functions, loss of relationships, loss of autonomous life, loss of the life imagined, loss of roles, loss of activities, loss of identity, and loss of uplifting emotions. Hewson (1997, p. 1134) summarized this perspective well: "Both bereavement and loss of ability can be conceptualized as events that are often appraised by an individual as stressful. The degree and nature of the stress depend upon factors such as personal meaning, context, and perceived coping resources. Rather than conceptualizing loss responses as a pattern of grieving, a person's responses to appraised stress can be viewed as coping strategies (emotion-focused and problem-focused) that attempt to manage the demands of the stress."

Because of the impact of permanent injury and disability on the lives of returning veterans and their families and the potential role that spirituality may play in the recovery of life quality, if not full function, we will summarize the small body of literature surrounding traumatic bereavement, disability, and spirituality. It should be recognized that very few research studies in these areas have examined military veterans. Caution should always be utilized in attempting to generalize research findings from one population to another.

The literature reviewed in this section is comprised of studies specific to bereavement due to traumatic death, TBIs, SCIs, and general disability. Many studies are conducted within rehabilitation facilities and report data from a wide variety of disabilities. Several computer databases were used to locate relevant literature for these types of loss and spirituality. Inclusion criteria for the review vary across these different forms of loss due to the number of available studies. Fifteen studies of spirituality were identified that examined individuals who experienced sudden bereavement following traumatic events, and a comprehensive review of all bereavement and spirituality studies was recently published (Wortmann & Park, 2008). In the area of disability, 34 studies and 1 literature review were found that included individuals suffering from rheumatoid arthritis, multiple sclerosis, amputation, traumatic brain injury, spinal cord injury, stroke, cancers, organ transplants, pulmonary disease, generalized weakness, and neurological syndromes.

Due to the dearth of research on TBI, both quantitative and qualitative studies with any number of participants were included. This review identified six quantitative studies and one qualitative study. No articles were found that studied TBI acquired through combat; rather, the majority of participants suffered from traumatic brain injury due to automobile accidents or stroke. Two studies sampled rehabilitation inpatients, and the rest sampled those with TBI living in the community. Participants represented both genders with a wide age span (from 16 to 65 years of age) and with average length of disability ranging from 1-1/2 to 11-1/2 years.

In 2000, the Veterans Health Administration reported 19,238 patients with spinal cord injuries, 51% of whom were veterans with service-related injuries (Spinal Cord Injury Quality Enhancement Research Initiative, 2000). Spinal cord injury is a prevalent and high-risk condition among veterans (VA Office of Research and Development, 2007). Study selection criteria included all quantitative and qualitative studies, with any sample size, that examined the relationship between spinal cord injury and spirituality. Of the 18 studies found to meet inclusion criteria, 12 studies were quantitative, 5 studies were qualitative, and 1 study utilized mixed methods. Populations primarily represented include inpatient, outpatient, and community-dwelling adult Caucasian males and females. Ages of populations sampled ranged widely across the span of adulthood, years after injury varied from 0 to more than 10 years, and the level of injury generally varied within and across studies including individuals with paraplegia and quadriplegia/tetraplegia.

Finally, a review was conducted that focused on "general disability." By use of this term we mean studies in which participants presented with a wide variety of physical limitations. Eight quantitative studies, one mixed methods study, and one literature review were found that met the above inclusion criteria. Of the nine research studies found, five were longitudinal and four were cross-sectional. These studies primarily used cohort samples recruited from settings such as hospices, ambulatory facilities, outpatient care facilities, private and public housing, and rehabilitation centers. Whereas seven of the nine sources evaluated participants presenting with a wide range of disabilities, including amputation, spinal cord injury, traumatic brain injury, stroke, cancers, organ transplants, pulmonary disease, generalized weakness, and neurological syndromes, two studies focused on specific forms of disability by limiting their study to participants who were seeking treatment for rheumatoid arthritis as well as individuals who were diagnosed with multiple sclerosis. Three studies focused solely on elderly participants, and three studies had samples that were comprised predominantly of female participants. Furthermore, several studies were comprised of primarily Caucasian participants who identified themselves as Christian, Protestant, or Roman Catholic.

Presence of Spirituality Among the Bereaved and Disabled

Among individuals with varied disabilities and conditions, including SCI, the majority considered God, faith, or religion to be significant in their lives (Kim, 2002). Among these individuals, faith has been identified as a method for coping and counteracting negative feelings; as a general form of help; as a promoter of strength, endurance, and interrelatedness; and as a provider of a sense of community. Most individuals with an SCI indicated having spiritual beliefs or engaging in some type of religious practice including existential, religious, or nonpreferential spiritual-based coping reported as being moderate or higher in intensity (Matheis, Tulsky, & Matheis, 2006; Nissim, 2003). Among individuals who had experienced loss of a loved one due to homicide, the presence of religious faith was associated with less treatment seeking, and grief and distress were lower among non-treatment seekers (Rynearson, 1995). Among individuals who had lived with an SCI for more than 20 years, most indicated having a strong belief in a God that provides personal communication and helps them cope with disability (Rossiter, 1992). In a study evaluating terminally ill individuals, it was demonstrated that the vast majority of this population states having a strong spiritual connection. Research looking more closely at this spiritual connection suggests that individuals facing terminal illness feel that they have a personal relationship with a higher power and often believe themselves to depend on this higher power (Kutner et al., 2003). Similarly, research suggests that faith is an important part of life for the majority of individuals who have been afflicted with some form of disability (Kim, 2002). There are several physical and psychological benefits for amputees who maintain some form of spiritual or religious connection. More specifically, research suggests that amputees who are religious experience greater social, functional, physical, and emotional well-being than their nonreligious counterparts (Chally & Carlson, 2004).

Adjusting to One's Disability

With regard to physical and psychological adjustment, the role of spirituality as a potential facilitator in this adjustment appears to be uncertain. Magyar-Russell (2005), for example, suggests that positive and negative religious coping acts as a significant partial mediator, thus linking spiritual appraisals to psychological adjustment following trauma. Similarly, results suggest that disabled individuals who have a limited reliance on spiritual well-being often experience lower levels of psychological adaptation to their physical condition (McNulty, Livneh, & Wilson, 2004). Results from Fitchett, Rybarczyk, DeMarco, and Nicholas (1999), however, failed to support the notion that spirituality facilitates physical or psychological adjustment after being afflicted with a debilitating condition.

Religious Appraisals Related to Disability

Research has demonstrated that many individuals commonly make positive religious attributions to aversive physical events (Magyar-Russell, 2005); for example, a significant portion of those experiencing a debilitating life event, such as amputation or a motor vehicle accident, report beliefs that a sanctified object has either been lost or violated. When looking at the time course of the spiritual appraisals made by disabled individuals, research suggests that the perception of a loss of a sanctified object is highest immediately following the onset of the disability; however, this sense of loss significantly decreases when a patient is discharged from rehabilitation. One hypothesis as to why patients' perceptions regarding the loss of a sacred object are greatest at admission to rehabilitation is that they may be both anxious and uncertain about their physical limitations as well as the likelihood of recovery. Thus, by the time they are discharged from their rehabilitation facility, those with disability may feel healthier, stronger, and more confident in their ability to function with greater independence (Magyar-Russell, 2005).

Spirituality as Social Support

When considering the social benefits of spirituality, disabled individuals who engage in religious practices experience more frequent social interaction with friends as well as more frequent holiday observance (Chally & Carlson, 2004). Among bereaved parents coping with the death of a child from sudden infant death syndrome (SIDS), church attendance was related to better outcomes through increased social support and meaning (McIntosh, Silver, & Wortman, 1993). For widows who lost their husbands in a mine accident, religious social involvement was associated with better quality of life, and religious intensity was associated with increased happiness (Bahr & Harvey, 1980). In a predominantly female and African–American sample of individuals who lost loved ones to homicide, religious support coping was related to lower distress, but not PTSD, while religious pleading and discontent were associated with worse

PTSD (Thompson & Vardaman, 1997). In fact, Potter and Zauszniewski (2000) found that the social impact of the disabling condition serves as a significant and independent predictor of an individual's level of spirituality (Chally & Carlson, 2004). Interestingly, research suggests that religious service attendance for those with a disability serves more important social rather than religious functions.

Spirituality in Relation to Meaning and Purpose

Spirituality within the context of TBI is often defined as a sense of purpose and an investment in personal relationships. Those who return to the workplace after TBI, particularly those who experience difficulties, report an increased sense of personal meaning, suggesting that triumph through adversity may create an increased sense of purpose. Those with TBI also report an increased sense of purpose due to the near-death nature of their traumatic events, suggesting an increased understanding of the self as mortal and finite (Dawson, Schwartz, Winocur, & Stuss, 2007). Among bereaved parents who lost a child to accident or violence, religious coping was related to meaning and through meaning to better health and relationship outcomes (Murphy, Johnson, Lohan, & Tapper, 2003). Similarly, a positive, significant relationship between religious conviction and having life purpose and meaning was found among individuals with a SCI (Schultz, 1985). Furthermore, an increase in the sense of purpose in life and life events was also indicated after disability or impairment (Kim, 2002; McColl et al., 2000; Povolny, Kaplan, Marme, & Roldan, 1993). Individuals who considered their disability to have a purpose were more likely to attribute the disability to God (Kim, 2002).

Changes in Spirituality Following Disability

Among individuals with a SCI or other disability or condition, most indicate a change in spirituality after the disability or impairment (Kim, 2002; Kim, Heinemann, Bode, Sliwa, & King, 2000; Magyar-Russell, 2005; Povolny et al., 1993). Nearly half of individuals who experienced a SCI or other disability reported strengthening of their faith, others experienced their disability as an impediment to their faith, and a minority indicated no change in faith following their injury (Kim et al., 2000). Studies of TBI and spirituality have demonstrated these divergent spiritual paths. Hallett, Zasler, Maurer, & Cash (1994) found that over half of their participants cited an unchanged role of religious participation following TBI, but 21.4% cited increased religious participation. A qualitative study by McColl et al. (2000) found that the TBI group contained participants with increased and decreased spirituality following their traumatic event. Both increases and decreases in faith practices, such as prayer and church attendance (Kim et al., 2000; McColl et al., 2000), and spiritual appraisals, including sacred loss and desecration (Magyar-Russell, 2005), were indicated after disability among individuals with a SCI and other disabilities or conditions. Sacred loss pertains to the perception that an object considered sacred has been lost, whereas desecration refers to the perceived violation of such an object. Among some individuals with SCI or traumatic brain injury, spiritual changes included increased self-awareness and appreciation, increased awareness of one's own vulnerability, and mortality (McColl et al., 2000).

Life Satisfaction and Quality of Life

In recent research, spirituality has also been associated with greater life satisfaction and quality of life in rehabilitation patient populations (Chally & Carlson, 2004). Furthermore, those disabled individuals who engage in negative religious coping methods appear to have greater difficulty returning to activities of daily living (Fitchett et al., 1999). In several samples of individuals with TBI, spirituality or religiousness has been shown to be inversely related to depression, fear, and anxiety (Hallett et al., 1994; Magyar-Russell, 2005). Conversely, religiousness or spirituality has been found to be positively related to happiness, belonging, and increased quality of life (Hallett et al., 1994; Kalpakjian, Lam, Toussaint, & Merbitz, 2004). Spirituality has been associated with greater quality of life and life satisfaction among individuals with a SCI and other disabilities or conditions (Brillhart, 2005; Ford, 1994; Kim et al., 2000; Nissim, 2003; Tate & Forchheimer, 2002). Among individuals with a SCI or other disability or condition, nonspiritual well-being has been associated with lower life satisfaction, quality of life, and general health than for those with religious or existential beliefs (Riley et al., 1998). In other studies, existential spirituality has generally been associated with greater quality of life, life satisfaction, and general health than religious spirituality (Matheis et al., 2006; Nissim, 2003).

Spirituality and Physical or Mental Health Outcomes

Studies show that many of those suffering from disability believe that their spiritual health positively contributes to their physical health (Kutner et al., 2003). In fact, engaging in faith practices has been demonstrated to produce long-term effects on health and well-being, thus resulting in better functioning within disabled populations (Chally & Carlson, 2004). Studies have also evaluated the association between spirituality and psychological variables. In the psychological arena, research suggests that disabled individuals who participate in religious activities experience increased levels of optimism and greater positive affect (Chally & Carlson, 2004); however, the relationship between spirituality and religious practices and depressive symptoms is not well understood. Some research has indicated that low levels of spirituality are predictive of depression severity (Gitlin, Hauck, Dennis, & Schulz, 2007), but other research has found no association of these variables for disabled individuals (Jackson, 2006). Finally, several samples of individuals with TBI have shown inverse relationships between spirituality and depression, fear, and anxiety (Hallett et al., 1994; Magyar-Russell, 2005).

Posttraumatic or Adversarial Growth

The literature suggests that posttraumatic growth after traumatic brain injury increases over time. A study comparing those who were recently injured with those who had been living with their injury for some time found a greater degree of posttraumatic growth in the latter group (Powell, Ekin-Wood, & Collin, 2007). Of even greater weight is a longitudinal study by Magyar-Russell (2005) that found an increase in posttraumatic growth over time. In addition to spirituality, other factors have been identified as being highly influential in the healthy healing of those with traumatic brain injury. Social support seems to be the most influential element of mental health after traumatic brain injury, as well as positive affect, feeling of self-efficacy, hope, community integration, and productivity, either as a return to the workplace or as a resumption of meaningful roles (Dawson et al., 2007; Kim et al., 2000; McColl et al., 2000; Powell et al., 2007).

Spirituality as a Form of Coping

Among some individuals with a SCI or other disabilities or conditions, spirituality was considered a coping mechanism that helped survivors maintain hope and a positive perspective, to be thankful for what they had, and to let go of control, as well as counteracting feelings of doubt (Kim et al., 2000). According to the perceptions of social workers and case managers who work with the SCI population, two primary coping strategies utilized among individuals with a SCI include family affiliation and religion or spirituality. Nearly half indicated using religion or spirituality to cope with their experiences at least sometimes; however, some individuals with a SCI have indicated that spirituality did not significantly contribute to their perceived ability to cope (DeGraff, 2007), and spirituality has also been found to be associated with PTSD reactions among veterans and civilians with a SCI (Martz, 2004). For individuals adapting to the loss of a child due to accident or violence, religious coping at 4 months was related to more acceptance and, through acceptance, to lower overall distress and less PTSD (Murphy et al., 2002). Similarly, for those who lost a child to accident or homicide, positive religious coping was associated with lower grief among those using high levels of task-oriented coping (Anderson, Marwit, Vandenberg, & Chibnall, 2005).

Suggestions for Helping Returning Veterans Find a Growth and Resilience Trajectory

The process of return from deployment and reintegration with family and friends can be one of great joy and relief for many, but for others it can be quite difficult and stressful. The positive side of this is the fact that most returnees will successfully adapt and reintegrate and continue on with the portions of their lives interrupted during deployment; however, early estimates (Hoge,

Auchterlonie, & Milliken, 2006) suggest that as many as one third of returnees will encounter significant physical or psychological problems that interfere with their ability to function successfully. We mentioned earlier that the continuum of losses includes death at the far extreme, physical and mental disability due to injury and the cumulative effect of war-zone exposure, and the sometimes less visible changes to the individual's view of self, humanity, and the world.

Whereas early theories of adaptation to grief and loss described a more or less linear progression through several identifiable stages of adaptation, more recent theories have better acknowledged the complexity and individuality of human experience, and some theorists have proposed that a stress and coping model of adaptation to bereavement and disability may better fit with research observations. Coping models suggest that situational appraisals are directly related to the type and intensity of emotional reactions during times of stress as well as to the style of behavioral coping responses. Stressors can be placed on a continuum ranging from the hassles of everyday life to major life changes to life-threatening traumatic events at the far extreme. In light of the variety of types of losses experienced by military service personnel during and following deployment and the individual nature of the timing for those losses, a model of understanding response to loss that allows for this variability is very useful.

The literature review of the relationship between spirituality and PTSD, traumatic bereavement, and disability showed evidence for two trajectories for the role spirituality plays relative to stressful and traumatic experiences in the war zone and following deployment. These two trajectories can be conceptualized as a growth and resilience trajectory vs. one of tension, struggle, and increased risk for problems. Clear evidence suggests that negative spiritual outcomes such as negative appraisals of God and faith, loss of faith, and difficulties with forgiveness can and do arise after traumatic events and that these experiences are often associated with negative mental health outcomes such as more severe PTSD or depression. There is also support for the notion that spirituality can thrive in the posttrauma environment in ways that are associated with better life quality and sense of well-being, as well as less need for mental health services.

Moral Injury Defined: Its Overlap with and Differences From PTSD

To inflict suffering and win the battle of human will, warriors must become callous to the pain and horror they wield. In a guerrilla war, where insurgents hide within the civilian population, that callousness can begin to extend to everyone. Through military training and then through exposure to horrific events for many months across multiple deployments, combatants may be hardened in a way that can be very difficult to reverse when they return home. Even though conscience, and even perhaps morality, can be ignored for a time in the midst of battle to ensure personal survival and protect one's friends and allies, when the war finally ends and reflection begins soldiers may find that they have changed, perhaps been "morally injured" by their experiences and actions.

For many years, clinicians, researchers, and chaplains have recognized that some individuals participating in combat develop problems and symptoms that extend even beyond the diagnostic criteria for PTSD (Ford, 1999). Combat is uniquely an activity where behaviors that are proscribed in other contexts (e.g., killing) are sanctioned and even celebrated when performed in accord with rules of engagement. Interestingly, the PTSD criteria that define the nature of traumatic events does not include the sanctioned infliction of trauma in combat within its definition. Some theorists have suggested that killing in combat may have inherent emotional and psychological consequences that extend beyond the diagnostic criteria for PTSD and that might better fit what one could call "moral injury" (Grosman, 1995; MacNair, 2005).

If an injury is defined as damage or harm done to or suffered by a person, then a moral injury could be construed as damage or harm to one's moral center as a result of things experienced, seen, and done in the war zone. Some traumatic war-zone experiences have the power to damage individuals' views of themselves as worthwhile human beings and leave them shackled with distorted views of themselves and their enemies that are harmful to their life function after they leave the war zone. Brett Litz, PhD, along with the chapter's first and last author and another colleague (Drescher, Litz, Rosen, & Foy, 2007), have developed a working conceptual definition of moral injury as follows: "disruption in an individual's confidence and expectations about their own or others' motivation or capacity to behave in a just and ethical manner brought about by bearing witness to perceived immoral acts, failure to stop such actions, or perpetration of immoral acts, in particular actions that are inhumane, cruel, depraved, or violent, bringing about pain and suffering of others or their death." Changes to a person's sense of self occur on a broad spectrum and may be seen as a diminished self-worth at the mild end of the spectrum to seeing oneself as a pariah, unworthy of even living in the midst of civilized society, at the most extreme end. Oftentimes, social isolation ensues in an attempt not to inflict oneself upon others.

The risk for moral injury may be high in a war of insurgency where enemy combatants pose as civilians and are not distinguishable from them. Moral injury is associated with inner turmoil, withdrawal, shame, and concealment. This may entrench the impact of moral conflict because service members cannot get feedback from others that might correct distorted self-appraisals. It is believed by the authors that these disruptions in moral directedness and expectancies may lead to some of the following: (1) negative changes in ethical attitudes and behavior; (2) change in or loss of spirituality, including negative attributions about God; (3) guilt, shame, and alienation; (4) anhedonia and dysphoria; (5) distrust in social and cultural contracts; (6) aggressive behaviors; and (7) poor self-care or self-harm. These are symptoms not part of the PTSD diagnostic criteria that are frequently reported by combat veterans under clinical care.

Combat situations, particularly in war theaters where one is fighting insurgent forces not easily distinguished from civilians, compel service personnel to make quick decisions and take actions in ambiguous situations. These actions may result in deaths, both intentionally and unintentionally, of enemy fighters, civilians (including women and children), and even friendly forces. Sometimes accompanying these decisions and actions are powerful emotions of grief, loss, rage, and hatred that stem from previous experiences in the war zone. Even in situations where the apparently correct action or decision was made, personnel can later come to question or doubt the appropriateness of their actions or decisions. Such second-guessing may lead them down a path of harsh judgment about their own character and hopelessness about the very nature of humankind.

A recent survey of service personnel in the war zone illustrates that moral choices and ethical decision making may be influenced by the presence of strong emotions such as anger or loss, and by the presence of combat-stress injury already experienced. The results of this survey indicated that a substantial percentage (approximately 10%) of soldiers and Marines reported "mistreating noncombatants (i.e., damaged/destroyed Iraqi property when not necessary or hit/kicked a noncombatant when not necessary)." Data also indicate that personnel with strong anger or high levels of combat trauma exposure or who had a mental health problem were twice as likely to report having mistreated noncombatants. The number of deployments and length of deployment were significantly associated with higher rates (Multinational Force Iraq–Surgeon General's Office, 2006). Some of these findings may be early indicators of moral injuries that may endure and contribute to a negative trajectory after return from the war zone.

In terms of spirituality, initial research suggests that several risk indicators may be associated with worse mental health outcomes. First is the potential that exposure to the carnage of war and the ambiguity of combat in a war of insurgency may lead to serious spiritual questioning, sometimes leading to a loss of faith. Spiritual tensions that arise for many combat veterans attempting to come to terms with their war-zone experiences may reduce their use of spiritual resources as part of reentry and may in turn lead to worsening psychiatric symptoms and greater medical service utilization. Additionally, signs of negative religious coping or negative attributions about God (e.g., "God has abandoned me," "God is persecuting/punishing me for my sins") may appear, and these can be associated with more severe PTSD and depression in some veterans. Finally, difficulties with forgiveness and higher levels of both hostility and guilt may be associated with more severe problems later on. It is notable that much of our current knowledge about relationships between trauma and spirituality comes from studies conducted years after those traumatic experiences occurred. When these risk indicators are present, efforts should be made to provide a forum where these issues related to spiritual experience can be discussed or explored.

Fostering a Growth Trajectory: Guiding Principles for Helping in the Transition Home

The American Psychological Association has identified five factors frequently associated with psychological resilience across a variety of studies (APA, 2004). First and foremost is strong social support both within the family and in the community at large. Social support is a clear area where spirituality can play an important role. Spirituality for many people is experienced with others in healthy communities. Not only do people experience emotional support through friendships with others sharing similar beliefs and values, but in many cases there is opportunity for instrumental support, as well (e.g., help with jobs, parenting issues, even finances). Social support is listed first because it can be extremely important in strengthening the other four factors. A second resilience factor is the capacity to manage strong feelings and impulses. Finding avenues for emotional expression—verbal expression to peers with common experiences, to family members, or to therapists; written expression through journaling or blogging—and finding measured and nonexplosive means of expressing what are sometimes strong emotions can be important. The third resilience factor is effective communication and problem solving. The transition back home from war-zone service can require very different sets of communication and problem-solving skills from those required in the military. Nonmilitary society is a bit less hierarchically organized and as such requires a sometimes more subtle style of communicating (fewer orders) and a different array of problem-solving activities. Attempting to use familiar military styles of behavior during the homecoming transition can result in serious distress and problems for service members and their families.

What Can We Do to Help? Answering Family Questions

A helpful set of tools, originally designed to teach basic helping skills to disaster and relief workers in the immediate aftermath of natural disasters, is collectively referred to as *psychological first aid* (Brymer, 2006). A training manual for the use of psychological first aid designed specifically for clergy engaged in post-disaster help has also been recently developed. Though originally designed for use in disasters, the U.S. Navy/Marine Corps has recently elected to use psychological first aid training to enhance and integrate the helping skills of chaplains and other health providers in the war zone. The five basic helping principles of psychological first aid can provide guidance for families in knowing how to help their loved ones and also help direct the clinical activities of therapists.

The first of these five principles is *safety*. For individuals who have been exposed to death and life-threatening situations, safety is a primary concern. It can be difficult to turn off the reactions learned in the war zone, and many returnees find it difficult to feel safe in a variety of everyday situations; for example, driving a car is something that most people here at home do not think too much about, but driving in Iraq is one of the most hazardous activities. Many service personnel feel uncomfortable driving after deployment due to the very different driving conditions and hazards they have encountered. Even activities such as walking on a crowded street or shopping in a mall may be perceived as extremely difficult and threatening for service personnel following deployment. Normalizing those issues and concerns and helping family members work through them can be a great aid in the recovery process.

Arousal reduction is a second basic principle. Perceived life threat brings with it a great deal of physiological arousal (i.e., fear and anxiety). Not only is a high level of arousal uncomfortable, but it can also greatly interfere with day-to-day functioning. Many self-help resources are aimed at relaxation and stress management, and they often provide instruction in meditation techniques or relaxation exercises, sometimes via audiotapes or CDs. From a spiritual perspective, evidence suggests that both prayer and meditation can bring about similar levels of arousal reduction for many people. Helping service members find resources that can teach them necessary relaxation skills is something anyone can assist with.

Social support has been much mentioned already in this chapter. Helping family members reconnect, engage, and interact effectively with other people is a very important way families can assist returning personnel with their transition. *Self-efficacy* is another basic principle. Self-efficacy is confidence in one's own abilities and capability in overcoming the tasks at hand. Military personnel frequently have great self-efficacy around their core military duties. The transition home to an environment that now feels unfamiliar can sometimes shake one's confidence in one's own capacity. Helping service members recognize the skills that they still have and supporting them as they recover confidence in their own abilities will be very useful.

The final core principle is one that is highly consistent with spirituality: *hope*. Hope for the future is key to making progress in life. Although despair keeps a person stuck, hope motivates an individual to move toward a better future. Relationships and activities that stimulate and build hope can be foundational in helping ensure an effective transition following deployment. Serving others and volunteering to help rebuild the lives of others through one's own efforts can play a strong role in rebuilding hope for oneself. Service and volunteering are things that family members can do together as an activity that promotes connection within the family and awareness and empathy for the needs of others.

Maintaining a Healthy Family: Suggestions for Identifying Potential Problems

Military personnel and veterans returning from the war zones in Iraq and Afghanistan have family roles as sons and daughters; husbands, wives, or partners; mothers and fathers; brothers and sisters. They will have been away from their families for varying lengths of time, sometimes repeatedly. Some may have received mental health treatment even before returning home, but for others the decision as to whether additional help is needed will be addressed later. For many, the prospect of redeployment may still loom as a major family concern. Homecoming is a time of significant family transition for all family members.

The primary source of support for the returning soldier is likely to be his or her family. At the same time, professional clinical care can be a valuable source of support for both the service member and his or her family. We know from veterans of previous wars that there can be a risk of disengagement from family at the time of return from a war zone. We also know that emerging problems with PTSD, TBI, depression, or substance abuse can wreak havoc with the support and comfort the returning soldier seeks within the family. Whereas the returning soldier who seeks mental health care clearly needs the clinician's attention and concern, that attention and concern can and should be extended to include his or her family as well. Support for veterans and their families can increase the potential for a veteran's successful reintegration back into daily life. Clinicians can also help the family work on solving any emerging problems related to the veteran's difficulties with reintegration. Early support and help for the family can reduce the potential for longer term, more damaging problems in the future.

The questions listed below can be used as guides in helping families or clinicians decide what issues might suggest the need for additional support or care:

1. Has the family been able to resume routines or develop new, equally satisfying ones?
2. Is the family connected with a healthy supportive community beyond the extended family that is available to provide emotional and instrumental support?
3. Did the returning partner/child/parent miss any significant family events during deployment (e.g., births, deaths, marriages, graduations)?
4. Were there any major family/life changes in the veteran's absence?
5. Did the family put anything on hold during the family member's absence that now requires urgent attention?
6. Have the spouses/partners eased back into making decisions together?
7. Have there been times since returning home that the returning family member has given the impression that he/she wants to be left alone?

8. Does it feel to parents/spouse/partner like the returning family member is still a bit of an outsider to the family?
9. If the service member is in an intimate relationship, do the partners as a couple feel closer or more distant than before?
10. Has the family experienced more disagreements than usual, and are there more unresolved areas of disagreement or increased tension since the veteran returned from the war?
11. Do family members or either spouse have concerns about risks associated with the level of tension or conflict in their relationship?
12. Has anyone outside the immediate family who helped out during the veteran's absence still feel more a part of the family than the couple would ordinarily want them to be?
13. Is the returning parent feeling more protective toward the children or needing to resist the urge to be more protective?
14. Has the family (including children) had a chance to talk about what their worries and fears were while the family member was away?
15. Has the returning family member been able to talk with his or her partner about his or her war experiences?
16. Is the returning parent able to answer their children's questions about their war experiences?
17. Has the returning family member been able to talk with his/her extended family about his/her war experiences?
18. How will you all know that your returning family member is really home?

Answers to these and other questions may highlight the need for additional support for the family to help the transition home proceed more smoothly. This information can also help clarify immediate areas of concern and begin to illuminate healthy directions for growth to support the family's reorganization and return to stability in coordination with the veteran's work on his or her own personal transition goals. Clergy support may help a family recognize that a returning veteran may be in spiritual or moral conflict that is causing her or him to distance and isolate from the family as well as religious community.

The murders of four military wives by their returning husbands at Fort Bragg in the summer of 2002 sadly reminded us all of the importance of assessing for potential tension, conflict, and unresolved issues that might have either preceded deployment or emerged during that veteran's absence from home. It is important to be aware that there is a chance that the returning soldier and his family will need assistance in identifying compatible goals or that one or both might have differing goals compared to what they shared before the soldier's departure for the war zone. Couples and families can be offered a therapeutic forum for communicating and negotiating their goals. If members are identifying high tension or strong levels of disagreement or their goals are markedly incompatible, then issues related to safety must be assessed and plans might have to be made that support safety for all family members.

Conclusion

This chapter has reviewed what is known about the role spirituality can play in the aftermath of dramatic events, disability, and loss. Key themes have been identified in terms of ensuring a healthy recovery trajectory. It is important to recognize the tension that exists in the hearts of many returning veterans between isolation and connection. Spirituality as we have defined it is connection—connection to that which transcends oneself. It is our hope that spiritual resources can play a role in facilitating health for families and service personnel as they resume their lives together again following deployment. For many veterans, their experiences have created within them a tension between living in the past (reexperiencing traumatic events) and living for the future (finding meaning and purpose). Helping service personnel to live in the present with an eye toward building a strong and healthy future should be the goal of therapists, families, and friends. Current research is identifying positive aspects of many spiritual traditions such as acceptance, compassion, hope, and gratitude that are frequently associated with the best of human experience.

References

Abeles, R., Ellison, C. G., George, L. K., Idler, E. L., Krause, N. et al. (1999). *Multidimensional measurement of religiousness/spirituality for use in health research.* Kalamazoo, MI: Fetzer Institute/National Institute on Aging Working Group.

Ahlstrom, G. (2007). Experience of loss and chronic sorrow in persons with severe chronic illness. *Journal of Clinical Nursing,* 16(3), 76–83.

Ai, A. L., Cascio, T., Santangelo, L. K., & Evans-Campbell, T. (2005). Hope, meaning, and growth following the September 11, 2001, terrorist attacks. *Journal of Interpersonal Violence,* 20(5), 523–548.

Alferi, S. F., Culver, J. L., Carver, C. S., Arena, P. L., & Antoni, M. H. (1999). Religiosity, religious coping, and distress: A prospective study of catholic and evangelical Hispanic treatment for early-stage breast cancer. *Journal of Health Psychology,* 4(3), 343–356.

Anderson, M. J., Marwit, S. J., Vandenberg, B., & Chibnall, J. T. (2005). Psychological and religious coping strategies of mothers bereaved by the sudden death of a child. *Death Studies,* 29, 811–826.

APA. (2004). *Resilience factors and strategies.* Washington, D.C.: American Psychological Association (http://apahelpcenter.org/featuredtopics/feature.php?id=6andch=3).

Bahr, H. M. and Harvey, C. D. (1980). Correlates of morale among the newly widowed. *Journal of Social Psychology,* 110, 219–233.

Brady, M. J., Peterman, A. H., Fitchett, G., Mo, M., & Cella, D. (1999). A case for including spirituality in quality of life measurement in oncology. *Psychooncology,* 8(5), 417–428.

Brillhart, B. (2005). A study of spirituality and life satisfaction among persons with spinal cord injury. *Rehabilitation Nursing,* 30(1), 31–34.

Brymer, M., Jacobs, A., Pynoos, R., Ruzek, J., Steinberg, A., Vernberg, E., & Watson, P. (2006). *Psychological first aid: Field operations guide* (2nd ed.). Los Angeles, CA: National Child Traumatic Stress Network and National Center for PTSD.

Carlock, D. (2007). Guide to resources for severely wounded Operation Iraqi Freedom (OIF) and Operation Enduring Freedom (OEF) veterans. *Issues in Science and Technology Librarianship*, No. 51.

Chally, P. S., & Carlson, J. M. (2004). Spirituality, rehabilitation, and aging: A literature review. *Archives of Physical Medicine and Rehabilitation*, 85(7, Suppl. 3), S60–67.

Chang, B. H., Skinner, K. M., & Boehmer, U. (2001). Religion and mental health among women veterans with sexual assault experience. *International Journal of Psychiatry in Medicine*, 31(1), 77–95.

Chang, B. H., Skinner, K. M., Zhou, C., & Kazis, L. E. (2003). The relationship between sexual assault, religiosity, and mental health among male veterans. *International Journal of Psychiatry in Medicine*, 33(3), 223–239.

Cordova, M. J., Cunningham, L. L., Carlson, C. R., & Andrykowski, M. A. (2001). Posttraumatic growth following breast cancer: A controlled comparison study. *Health Psychology*, 20(3), 176–185.

Cotton, S. P., Levine, E. G., Fitzpatrick, C. M., Dold, K. H., & Targ, E. (1999). Exploring the relationships among spiritual well-being, quality of life, and psychological adjustment in women with breast cancer. *Psychooncology*, 8(5), 429–438.

Dawson, D. R., Schwartz, M. L., Winocur, G., & Stuss, D. T. (2007). Return to productivity following traumatic brain injury: Cognitive, psychological, physical, spiritual, and environmental correlates. *Disability and Rehabilitation*, 29(4), 301–313.

Defense Manpower Data Center. (2008). *Global war on terrorism casualty statistics by reason*. Washington, D.C.: Defense Manpower Data Center/Data, Analysis, and Programs Division (http://siadapp.dmdc.osd.mil/personnel/CASUALTY/gwot_reason.pdf).

DeGraff, A. H. (2006). Religion and spirituality related to ability to cope for people living with spinal cord injury. Doctoral dissertation, University of Northern Colorado (ProQuest Database, AAT 3231300).

Doxey, C., Jensen, L., & Jensen, J. (1997). The influence of religion on victims childhood sexual abuse. *International Journal for the Psychology of Religion*, 7(3), 179–186.

Drescher, K. D., & Foy, D. W. (1995). Spirituality and trauma treatment: Suggestions for including spirituality as a coping resource. *National Center for Post-Traumatic Stress Disorder Clinical Quarterly*, 5(1), 4–5.

Drescher, K. D., Ramirez, G., Leoni, J. J., Romesser, J. M., Sornborger, J., & Foy, D. W. (2004). Spirituality and trauma: Development of a group therapy module. *Group Journal*, 28(4), 71–87.

Drescher, K. D., Litz, B., Rosen, C., & Foy, D. W. (2007). *An examination of moral injury among veterans of combat*. Unpublished manuscript.

Elliott, D. M. (1994). The impact of Christian faith on the prevalence and sequelae of sexual abuse. *Journal of Interpersonal Violence*, 9(1), 95–108.

Fallot, R. D., & Heckman, J. P. (2005). Religious/spiritual coping among women trauma survivors with mental health and substance use disorders. *Journal of Behavior and Health Services & Research*, 32(2), 215–226.

Fehring, R. J., Miller, J. F., & Shaw, C. (1997). Spiritual well-being, religiosity, hope, depression, and other mood states in elderly people coping with cancer. *Oncology Nursing Forum*, 24(4), 663–671.

Ferrell, B. R., Dow, K. H., & Leigh, S. (1995). Quality of life in long-term cancer survivors. *Oncology Nursing Forum*, 22(6), 915–922.

Finkelhor, D., Hotaling, G., Lewis, I. A., & Smith, C. (1990). Sexual abuse in a national survey of adult men and women: Prevalence, characteristics, and risk factors. *Child Abuse and Neglect*, 14(1), 19–28.

Fitchett, G., Rybarczyk, B. D., DeMarco, G. A., & Nicholas, J. J. (1999). The role of religion in medical rehabilitation outcomes: A longitudinal study. *Rehabilitation Psychology*, 44(4), 333–353.

Flannelly, L. T., & Inouye, J. (2001). Relationships of religion, health status, and socioeconomic status to the quality of life of individuals who are HIV positive. *Issues in Mental Health Nursing*, 22(3), 253–272.

Fontana, A., & Rosenheck, R. (2004). Trauma, change in strength of religious faith, and mental health service use among veterans treated for PTSD. *Journal of Nervous and Mental Disease*, 192(9), 579–584.

Ford, J. D. (1999). Disorders of extreme stress following war-zone military trauma: Associated features of posttraumatic stress disorder or comorbid but distinct syndromes? *Journal of Consulting and Clinical Psychology*, 67(1), 3–12.

Ford, M. R. (1994). Relationship between spiritual well-being and perceived quality of life in the older spinal cord injured patient. Master's thesis, California State University (ProQuest Database, AAT 1360377).

Gall, T. L. (2006). Spirituality and coping with life stress among adult survivors of childhood sexual abuse. *Child Abuse and Neglect*, 30(7), 829–844.

Gillard, M., & Paton, D. (1999). Disaster stress following a hurricane: The role of religious differences in the Fijian Islands. *The Australasian Journal of Disaster and Trauma Studies*, 3(2).

Gitlin, L. N., Hauck, W. W., Dennis, M. P., & Schulz, R. (2007). Depressive symptoms in older African–American and white adults with functional difficulties: The role of control strategies. *Journal of the American Geriatrics Society*, 55(7), 1023–1030.

Green, B. L., Lindy, J. D., & Grace, M. C. (1988). Long-term coping with combat stress. *Journal of Traumatic Stress*, 1(4), 399–412.

Grosman, D. (1995). *On killing: The psychological cost of learning to kill in war and society*. New York: Little, Brown and Company.

Hall, T. (1995). Spiritual effects of childhood sexual abuse in adult Christian women. *Journal of Psychology and Theology*, 23(2), 129–134.

Hallett, J. D., Zasler, N. D., Maurer, P., & Cash, S. (1994). Role change after traumatic brain injury in adults. *American Journal of Occupational Therapy*, 48(3), 241–246.

Hewson, D. (1997). Coping with the loss of ability: "Good grief" or episodic stress responses? *Social Science and Medicine*, 44(8), 1129–1139.

Hoge, C. W., Castro, C. A., Messer, S. C., McGurk, D., Cotting, D. I., & Koffman, R. L. (2004). Combat duty in Iraq and Afghanistan, mental health problems, and barriers to care. *New England Journal of Medicine*, 351(1), 13–22.

Hoge, C. W., Auchterlonie, J. L., & Milliken, C. S. (2006). Mental health problems, use of mental health services, and attrition from military service after returning from deployment to Iraq or Afghanistan. *Journal of the American Medical Association*, 295(9), 1023–1032.

Jackson, K. I. (2006). *The impact of church attendance and prayer on the physical disability–depressive symptomatology relationship for older African–Americans*. University Park: The Pennsylvania State University.

Jones, J. M. (2007). Exposure to chronic community violence. *Journal of Black Psychology*, 33(2), 125–149.

Kaczorowski, J. M. (1989). Spiritual well-being and anxiety in adults diagnosed with cancer. *The Hospice Journal*, 5(3–4), 105–116.

Kalpakjian, C. Z., Lam, C. S., Toussaint, L. L., & Merbitz, N. K. (2004). Describing quality of life and psychosocial outcomes after traumatic brain injury. *American Journal of Physical and Medical Rehabilitation*, 83(4), 255–265.

Kane, D., Cheston, S. E., Greer, J., Pritt, A., & Lawson, R. (1993). Perceptions of God by survivors of childhood sexual abuse: An exploratory study in an underresearched area. *Journal of Psychology and Theology*, 21, 228–237.

Kennedy, J., Davis, R., & Taylor, B. (1998). Changes in spirituality and well-being among victims of sexual assault. *Journal for the Scientific Study of Religion*, 37(2), 322–329.

Kim, J. J. (2002). Spirituality and the disability experience: Faith, subjective well-being, and meaning and purpose in the lives of persons with disabilities. Ph.D. dissertation. Chicago, IL: Northwestern University.

Kim, J. J., Heinemann, A. W., Bode, R. K., Sliwa, J., & King, R. B. (2000). Spirituality, quality of life, and functional recovery after medical rehabilitation. *Rehabilitation Psychology*, 45(4), 365–385.

Krejci, M. J., Thompson, K. M., Simonich, H., Crosby, R. D., Donaldson, M. A., Wonderlich, S. A. et al. (2004). Sexual trauma, spirituality, and psychopathology. *Journal of Child Sexual Abuse*, 13(2), 85–103.

Kurtz, M. E., Wyatt, G., & Kurtz, J. C. (1995). Psychological and sexual well-being, philosophical/spiritual views, and health habits of long-term cancer survivors. *Health Care for Women International*, 16(3), 253–262.

Kutner, J. S., Nowels, D. E., Kassner, C. T., Houser, J., Bryant, L. L., & Main, D. S. (2003). Confirmation of the "disability paradox" among hospice patients: Preservation of quality of life despite physical ailments and psychosocial concerns. *Palliative & Supportive Care*, 1(3), 231–237.

Larson, D. B., Pattison, E. M., Blazer, D. G., Omran, A. R., & Kaplan, B. H. (1986). Systematic analysis of research on religious variables in four major psychiatric journals, 1978–1982. *American Journal of Psychiatry*, 143, 329–334.

Larson, D. B., Swyers, J. P., & McCullough, M. E. (1997). *Scientific research on spirituality and health: A consensus report*. Rockville, MD: National Institute for Healthcare Research.

MacNair, R. (2005). *Perpetration-induced traumatic stress: The psychological consequences of killing*. New York: Authors Choice Press.

Magyar-Russell, G. M. (2005). A longitudinal study of sacred loss and desecration among adults in rehabilitation hospitals. Ph.D. dissertation. Bowling Green, KY: Bowling Green State University.

Martz, E. (2004). Death anxiety as a predictor of posttraumatic stress levels among individuals with spinal cord injuries. *Death Studies*, 28(1), 1–17.

Matheis, E. N., Tulsky, D. S., & Matheis, R. J. (2006). The relation between spirituality and quality of life among individuals with spinal cord injury. *Rehabilitation Psychology*, 51, 265–271.

Maton, K. I. (1989). The stress-buffering role of spiritual support: Cross-sectional and prospective investigations. *Journal for the Scientific Study of Religion*, 28(3), 310–323.

McColl, M. A., Bickenbach, J., Johnston, J., Nishihama, S., Schumaker, M., Smith, K. et al. (2000). Spiritual issues associated with traumatic-onset disability. *Disability Rehabilitation Journal*, 22(12), 555–564.

McCrea, M., Barth, J., Cox, D., Fink, J., French, L., Hammeke, T. et al. (2008). Official position of the Military TBI Task Force on the role of neuropsychology and rehabilitation psychology in the evaluation, management, and research of military veterans with traumatic brain injury. *The Clinical Neuropsychologist*, 22(1), 10–26.

McIntosh, D. N., Silver, R. C., & Wortman, C. B. (1993). Religion's role in adjustment to a negative life event: Coping with the loss of a child. *Journal of Personality and Social Psychology*, 65(4), 812–821.

McNulty, K., Livneh, H., & Wilson, L. M. (2004). Perceived uncertainty, spiritual well-being, and psychosocial adaptation in individuals with multiple sclerosis. *Rehabilitation Psychology*, 49(2), 91–99.

Mickley, J., & Soeken, K. (1993). Religiousness and hope in Hispanic- and Anglo-American women with breast cancer. *Oncology Nursing Forum*, 20(8), 1171–1177.

Mickley, J. R., Soeken, K., & Belcher, A. (1992). Spiritual well-being, religiousness and hope among women with breast cancer. *Image: Journal of Nursing Scholarship*, 24(4), 267–272.

Miller, W. R., & Thoresen, C. E. (1997). Spirituality and health. In W. R. Miller (Ed.), *Integrating spirituality into treatment: Resources for practitioners* (pp. 179–198). Washington, D.C.: American Psychological Association.

Moschella, V. D., Pressman, K. R., Pressman, P., & Weissmann, D. E. (1997). The problem of theodicy and religious response to cancer. *Journal of Religion and Health*, 36(1), 17–20.

Multinational Force Iraq–Surgeon General's Office. (2006). *Mental Health Advisory Team (MHAT) IV Operation Iraqi Freedom 05-07*, Final Report. Washington, D.C.: Office of the Surgeon General U.S. Army Medical Command.

Murphy, S. A., Johnson, L. C., Lohan, J., & Tapper, V. J. (2002). Bereaved parents' use of individual, family, and community resources 4 to 60 months after a child's violent death. *Family and Community Health*, 25, 71–82.

Murphy, S. A., Johnson, L. C., & Lohan, J. (2003). Finding meaning in a child's violent death: A five-year prospective analysis of parents' personal narratives and empirical data. *Death Studies*, 27, 381–404.

Nairn, R. C., & Merluzzi, T. V. (2003). The role of religious coping in adjustment to cancer. *Psychooncology*, 12(5), 428–441.

Nissim, N. E. (2003). The impact of spirituality on the quality of life of spinal cord injury patients. Doctoral dissertation, Fairleigh Dickinson University (ProQuest Database, AAT 3081401).

Pargament, K. I. (1997). *The psychology of religion and coping: Theory, research, practice.* New York: Guilford Press.

Pargament, K. I., Ensing, D. S., Falgout, K., Olsen, H., Barbara Reilly, K., & Van Haitsma Warren, R. (1990). God help me. I. Religious coping efforts as predictors of the outcomes to significant negative life events. *American Journal of Community Psychology*, 18(6), 793–824.

Patel, S. S., Shah, V. S., Peterson, R. A., & Kimmel, P. L. (2002). Psychosocial variables, quality of life, and religious beliefs in ESRD patients treated with hemodialysis. *American Journal of Kidney Diseases*, 40(5), 1013–1022.

Pivar, I. L., & Field, N. P. (2004). Unresolved grief in combat veterans with PTSD. *Journal of Anxiety Disorders*, 18(6), 745–755.

Potter, M. L., & Zauszniewski, J. A. (2000). Spirituality, resourcefulness, and arthritis: Impact on health perception of elders with rheumatoid arthritis. *Journal of Holistic Nursing*, 18(4), 311–331.

Povolny, M. A., Kaplan, S. P., Marme, M., & Roldan, G. (1993). Perceptions of adjustment issues following a spinal cord injury: A case study. *Journal of Applied Rehabilitation Counseling*, 24(3), 31–34.

Powell, T., Ekin-Wood, A., & Collin, C. (2007). Post-traumatic growth after head injury: A long-term follow-up. *Brain Injury*, 21(1), 31–38.

Pritt, A. (1998). Spiritual correlates of reported sexual abuse among Mormon women. *Journal for the Scientific Study of Religion,* 37(2), 273–285.

Reed, P. G. (1986). Religiousness among terminally ill and healthy adults. *Research in Nursing and Health,* 9(1), 35–41.

Reed, P. G. (1987). Spirituality and well-being in terminally ill hospitalized adults. *Research in Nursing and Health,* 10(5), 335–344.

Reed, P. G. (1991). Preferences for spiritually related nursing interventions among terminally ill and nonterminally ill hospitalized adults and well adults. *Applied Nursing Research,* 4(3), 122–128.

Riley, B. B., Perna, R., Tate, D. G., Forchheimer, M., Anderson, C., & Luera, G. (1998). Types of spiritual well-being among persons with chronic illness: Their relation to various forms of quality of life. *Archives of Physical Medicine and Rehabilitation,* 79, 258–264.

Rossiter, C. L. (1992). Personal adjustments and concerns of successful, long-term (more than twenty years) traumatic spinal cord injured persons. Doctoral dissertation, University of Cincinnati (ProQuest Database, AAT 9232360).

Rynearson, E. K. (1995). Bereavement after homicide: A comparison of treatment-seekers and refusers. *British Journal of Psychiatry,* 166, 507–510.

Salo, J. A., Quota, S., & Punamaki, R. (2005). Adult attachment, posttraumatic growth and negative emotions among former political prisoners. *Anxiety, Stress and Coping,* 18(4), 361–378.

Schnoll, R. A., Harlow, L. L., & Brower, L. (2000). Spirituality, demographic and disease factors, and adjustment to cancer. *Cancer Practice,* 8(6), 298–304.

Schultz, R. C. (1985). Purpose in life among spinal cord injured males. *Journal of Applied Rehabilitation Counseling,* 16(2), 45–51.

Schuster, M. A., Stein, B. D., Jaycox, L., Collins, R. L., Marshall, G. N., Elliott, M. N. et al. (2001). A national survey of stress reactions after the September 11, 2001, terrorist attacks. *New England Journal of Medicine,* 345(20), 1507–1512.

Smith, B. W., Pargament, K. I., Brant, C., & Oliver, J. M. (2000). Noah revisited: Religious coping by church members and the impact of the 1993 midwest flood. *Journal of Community Psychology,* 28(2), 169–186.

Spinal Cord Injury Quality Enhancement Research Initiative. (2000). *VA health care atlas FY 2000.* Washington, D.C.: Department of Defense/Veterans Affairs (http://www1.va.gov/rorc/atlas/chapter_13_spinal_cord_injury.pdf).

Sulmasy, D. P. (2002). A biopsychosocial–spiritual model for the care of patients at the end of life. *The Gerontologist,* 42, 24–33.

Tate, D. G., & Forchheimer, M. (2002). Quality of life, life satisfaction, and spirituality: Comparing outcomes between rehabilitation and cancer patients. *American Journal of Physical Medicine and Rehabilitation,* 81, 400–410.

Thompson, M. P., & Vardaman, P. J. (1997). The role of religion in coping with the loss of a family member to homicide. *Journal for the Scientific Study of Religion,* 36(1), 44–51.

VA Office of Research and Development. (2007). *Quality enhancement research initiative: Spinal cord injury.* Washington, D.C.: Health Services Research and Development Service (http://www.hsrd.research.va.gov/queri/).

Wan, G. J., Counte, M. A., Cella, D. F., Hernandez, L., McGuire, D. B., Deasay, S. et al. (1999). The impact of socio-cultural and clinical factors on health-related quality of life reports among Hispanic and African–American cancer patients. *Journal of Outcome Measurement,* 3(3), 200–215.

Weaver, A. J., Kline, A. E., Samford, J., Lucas, L. A., Larson, D. B., & Gorsuch, R. L. (1998). Is religion taboo in psychology? A systematic analysis of research on religious variables in seven major American Psychological Association journals: 1991–1994. *Journal of Psychology and Christianity*, 17(3), 220–232.

Weaver, K. E., Antoni, M. H., Lechner, S. C., Duran, R. E., Penedo, F., Fernandez, M. I. et al. (2004). Perceived stress mediates the effects of coping on the quality of life of HIV-positive women on highly active antiretroviral therapy. *AIDS and Behavior*, 8(2), 175–183.

Witvliet, C. V. O., Phillips, K. A., Feldman, M. E., & Beckham, J. C. (2004). Posttraumatic mental and physical health correlates of forgiveness and religious coping in military veterans. *Journal of Traumatic Stress*, 17(3), 269–273.

Woods, T., Antoni, M., Ironson, G., & Kling, D. (1999). Religiosity is associated with affective and immune stats in symptomatic HIV-infected gay men. *Journal of Psychosomatic Research*, 46(2), 165–176.

Wortmann, J. H., & Park, C. L. (2008). Religion and spirituality in adjustment following bereavement: An integrative review. *Death Studies*, 32(8), 703–736.

Yates, J. W., Chalmer, B. J., St. James, P., Follansbee, M., & McKegney, F. P. (1981). Religion in patients with advanced cancer. *Medicine in Pediatric Oncology*, 9(2), 121–128.

23

Future Directions: Trauma, Resilience, and Recovery Research

ALAN L. PETERSON, JEFFREY A. CIGRANG, and WILLIAM C. ISLER

Contents

Introduction

Throughout history, military troops deployed into battle have been found to be one of the most high-risk populations for the development of trauma-related stress disorders (Harvey, Bryant, & Tarrier, 2003; Kessler, Sonnega, Bromet, Hughes, & Nelson, 1995; Prigerson, Maciejewski, & Rosenheck, 2002). Posttraumatic stress disorder (PTSD) is a frequent and significant mental health consequence of exposure to violence and trauma (Kessler et al., 1995). Today, exposure to extreme combat violence and trauma by military personnel who have served in Iraq and Afghanistan is a primary contributor to PTSD among Americans (Hoge et al., 2004, 2006, 2007; Milliken, Auchterlonie, &

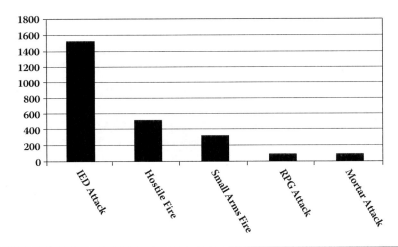

Figure 23.1 Weaponry cause of death in Iraq from July 2003 to December 2007. (From iCasualties. org., http://www.icasualties.org/oif/default.aspx.)

Hoge 2007). Since the terrorist attacks on the United States on September 11, 2001, almost 2 million U.S. military personnel have deployed to Afghanistan and surrounding areas in support of Operation Enduring Freedom (OEF) and to Iraq as part of Operation Iraqi Freedom (OIF). These deployments of U.S. service members are the largest since the Vietnam War and have resulted in exposure of many troops to a variety of deployment-related stressors.

The most frequent causes of death and injury in Iraq (see Figure 23.1) are roadside bombs from improvised explosive device (IED) attacks on vehicle convoys and foot patrols (iCasualties.org, 2007). The frequency of IED attacks and the rate of injury and death from them have increased as the war in Iraq has progressed (see Figure 23.2). IEDs are often hidden under trash or other debris or buried under the ground and then remotely detonated by a trigger person who waits and watches for the victims. IEDs are also sometimes detonated by suicide bombers who carry the IED either on themselves or in a car loaded with hundreds of pounds of explosives. An often overlooked fact is that for each individual bombing that occurs there may be dozens of physically uninjured bystanders and first responders who are exposed to a high level of perceived life threat and intense negative emotional arousal during and immediately after the trauma. These uninjured bystanders often witness horrific and grotesque injuries in those who have been hit by the bomb shrapnel; however, these uninjured bystanders are usually overlooked because they did not suffer any physical injuries that required medical care. These peritraumatic factors have consistently been linked to higher rates of PTSD in exposed populations (Brewin, Andrews, & Valentine, 2000; Ozer, Best, Lipsey, & Weiss, 2003). Thus, the eventual magnitude of mental health problems among service

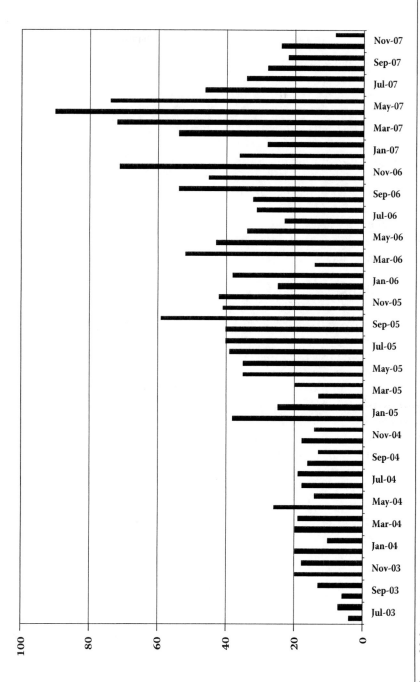

Figure 23.2 Improvised explosive device fatalities in Iraq by month from July 2003 to December 2007. (From iCasualties.org., http://www.icasualties.org/oif/default.aspx.)

members exposed to combat duty in Iraq will very likely far exceed the physical health costs of the war. Additionally, exposure to IEDs and the devastation and terror they inflict will very likely be a primary cause of PTSD in U.S. military personnel returning from Iraq.

Hoge and colleagues at the Walter Reed Army Institute of Research have conducted seminal epidemiological research on PTSD in U.S. military personnel after return from deployments in support of OIF/OEF. In the first study (Hoge et al., 2004), four U.S. combat infantry units (three Army units and one Marine unit) were given an anonymous survey, the Posttraumatic Stress Disorder Checklist (PCL), either before their deployment to Iraq ($n = 2530$) or in a different cohort of troops 3 to 4 months after their return from combat duty in Iraq or Afghanistan ($n = 3671$). The results indicated that combat exposure was significantly greater among those who were deployed to Iraq compared to Afghanistan. Army personnel returning from Iraq reported significantly more PTSD symptoms (12.9% with PCL $\geq$ 50) than the cohort that was surveyed prior to an Iraq deployment (5%). In addition, participants who endorsed more mental health symptoms were twice as likely to report concern about the possible stigma and other potential barriers associated with seeking mental health care.

The second study (Hoge, Auchterlonie, & Milliken, 2006) evaluated the relationship between combat deployment and mental health symptoms during the first year after return from deployment to Iraq ($n = 222,620$), Afghanistan ($n = 16,318$), and other locations ($n = 64,967$). The survey evaluated was the Post-Deployment Health Assessment (PDHA), which is completed either immediately before returning from the deployed location or within 1 to 2 weeks after returning. It should be noted that the PDHA is not anonymous and becomes part of the individual's military medical record. The PDHA includes several mental health measures, including a four-item PTSD screener based on the Primary Care PTSD screener (PC-PTSD) (Prins et al., 2004). The results indicated that the prevalence of reporting a mental health problem was 19.1% among service members returning from Iraq compared to 11.3% from Afghanistan and 8.5% from other locations. Using the ≥ 2 threshold for PTSD symptoms on the PC-PTSD measure, 9.8% of troops returning from Iraq endorsed PTSD symptoms as compared to 4.7% of those returning from Afghanistan and 2.1% from other locations.

The third study (Milliken et al., 2007) compared the results of the PDHA with a Post-Deployment Health Reassessment (PDHRA) completed 3 to 6 months after returning from a deployment by a cohort of 88,235 U.S. soldiers. This study included active-duty Army soldiers as well as those serving in the Army National Guard and Reserves. The results indicated that more soldiers reported mental health concerns on the PDHRA than on the PDHA. Using the same ≥ 2 threshold for PTSD symptoms on the PC-PTSD as in the previous study (Hoge et al., 2006), the results for active-duty soldiers indicated that the rates of PTSD increased from 11.8% on the PDHA immediately after returning from deployment to 16.7% on the PDHRA completed 3 to 6 months later. For National Guard and Reserve soldiers, the rates were even higher: 12.7% on the PDHA and 24.5% on the PDHRA. It is interesting to note that about half of those with threshold PTSD symptoms on the PDHA no longer reported these symptoms at the point of the PDHRA. Similarly, a large proportion initially scoring negative for PTSD on the PDHA were positive 3 to 6 months later, indicating a possible delayed onset of symptoms when soldiers settled in after returning from their deployment.

Posttraumatic stress disorder is one of the few mental health disorders where the etiology is clear—it is caused by exposure to one or more traumatic events; however, the relationship between the specific details of the traumatic event itself and individual victim variables is complex and poorly understood. Why is it that some individuals develop PTSD after only cursory exposure to minor traumatic events whereas other individuals remain resilient despite exposure to extreme traumatic stressors? Why is it that most individuals who are exposed to traumatic events recover naturally, without any formal psychological treatment? Why is it that certain types of traumatic events result in PTSD in the majority of individuals who are exposed to them?

In this chapter, we review the research on factors related to risk, resilience, and recovery in individuals exposed to traumatic stress with a specific emphasis on combat-related stress in U.S. military personnel. The majority of trauma research has been done with civilian populations; however, military research is reviewed when available. The authors are OIF/OEF veterans who have completed a total of six U.S. military deployments since September 11, 2001, in support of the Global War on Terrorism, and in some cases we present our expert opinions if appropriate military research data are not available. We highlight some of the seminal epidemiological research that has been conducted with active-duty OIF/OEF veterans as well as the dearth of clinical research on the treatment of acute stress disorder (ASD) and PTSD. Finally, we outline important future clinical research that is necessary to better understand risk, resilience, and recovery from combat-related stress and to develop evidence-based treatments adapted to the unique needs of military personnel.

Risk, Resiliency, and Recovery

A variety of factors contributes to risk for the development of combat-related PTSD, resiliency to maintain physical and psychological health despite exposure to extreme trauma, and recovery back to normal functioning after exposure to trauma. The relationship between trauma exposure and the development of PTSD is complex. Although most people are exposed to traumatic events at some point in their lives, the majority do not develop PTSD and continue to be resilient and have positive emotional experiences with only minor and transient disruptions in their ability to function (Bonanno, 2004). As time passes, some individuals even report that their lives are somehow improved because of their exposure to a traumatic event, a concept referred to as *posttraumatic growth* (Tedeschi & Calhoun, 1996). For certain types of trauma, however, the risk for developing PTSD is very high. Research by Kessler and colleagues (1995) found that 61% of men in the U.S. adult population and 51% of women reported a lifetime history of exposure to at least one traumatic event. The estimated lifetime prevalence of PTSD in this population was 7.8% (10.4% of women, 5.0% of men). This study found that the three most common types of trauma experienced included witnessing someone being badly injured or killed (36% of men, 14% of women); being involved in a fire, flood, or natural disaster (19% of men, 15% of women); and being involved in a life-threatening accident (25% of men, 14% of women). More men reported exposure to each of these three types of trauma as well as physical attacks and combat exposure. More women reported rape and sexual assault (Kessler et al., 1995).

Risk Factors for PTSD

Three primary types of risk factors for the development of PTSD have been identified. The first is the risk related to the type of trauma (or traumas) and its severity. The second is related to the individuals themselves and includes factors such as gender, age, socioeconomic status, education, intelligence, race, psychiatric history, and previous trauma exposure. The third factor is the individual's environment, which includes the individual's network of social support and life stress after trauma exposure.

The type of trauma appears to be related to the risk for development of PTSD. In a national sample of women, 32% developed PTSD after being raped and 38% had PTSD after being physically assaulted (Resnick, Kilpatrick, Dansky, Saunders, & Best, 1993). The combination of multiple trauma factors can have an additive effect in increasing risk. Kilpatrick et al. (1989) evaluated the percentage of crime victim groups with and without rape, life threat, and physical injury who developed crime-related PTSD (see Figure 23.3). The results showed a significant additive impact of these different aspects of trauma exposure. Of those individuals who were raped, were physically injured, and feared for their lives during the assault, 79% developed PTSD. These results highlight the importance of the type of trauma in the risk for development of PTSD.

Although similar research has not been conducted in military combat environments, it is likely that some types of combat trauma exposure during deployments have a high likelihood of resulting in PTSD. In a recent review of risk factors for PTSD, Vogt, King, and King (2007) observed that a dysfunctional response to traumatic exposure is more likely when the traumatic event involved physical injury, more malicious and grotesque events, active vs. passive involvement of the individual, subjective distress, dissociation, and the presence of other life stresses.

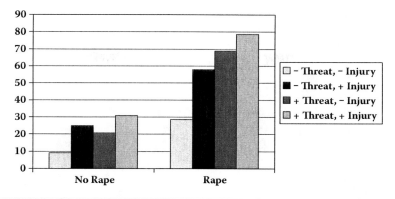

Figure 23.3 Percentage of crime victim groups with and without rape, life threat, and physical injury that developed crime-related PTSD. (From Kilpatrick, D.G. et al., *Behavior Therapy*, 20, 199–214, 1989. With permission.)

The relationship between physical injury and the development of PTSD has been examined in two studies of OIF/OEF veterans. One study evaluated military personnel who sustained combat-related physical injuries during their deployment (Koren, Norman, Cohen, Berman, & Klein, 2005) and found that 16.7% met diagnostic criteria for PTSD as compared to 2.5% in the uninjured comparison group. Another study found that a large percentage of combat-injured personnel had a delayed onset of combat-stress symptoms (Grieger et al., 2006). This finding is important considering the large proportion of injured OIF/OEF veterans who are surviving after sustaining severe injuries. The survival rate for combat wounded personnel in Iraq (see Figure 23.4) is approximately 90% (Gawande, 2004). This survival rate is the highest in recorded history and is a significant increase from the 74 to 75% survival rate of wounded personnel in the wars in Korea, Vietnam, and the Persian Gulf.

From a population perspective, the majority of individuals exposed to traumatic events do not go on to develop PTSD; therefore, there is increasing acceptance that exposure to the trauma alone may not sufficiently explain the development of PTSD and that individual vulnerability and environmental factors may also play a role (Brewin et al., 2000; Yehuda, 1999; Yehuda & McFarlane, 1995).

Brewin et al. (2000) conducted a metaanalysis of risk factors for the development of PTSD in trauma-exposed adults. For military populations, lack of social support, subsequent life stress, adverse childhood events, trauma severity, and childhood abuse were the most significant risk factors. Childhood abuse is a risk factor for both early attrition from military service and the development of PTSD (Cabrera, Hoge, Bliese, Castro, & Messer, 2007; Garb & Cigrang, 2008), suggesting that traumatic experiences in childhood can instill a generalized vulnerability in military members.

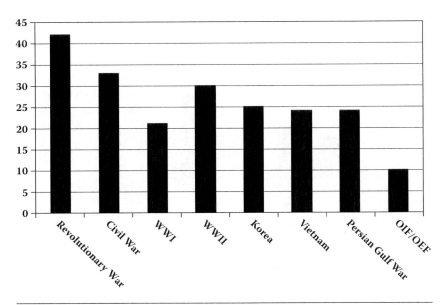

Figure 23.4 Lethality of war wounds. (From Gawande, A., *N. Engl. J. Med.*, 351, 2471–2475, 2004. With permission.)

Lack of social support after trauma has consistently been associated with increased risk for development of combat-related PTSD. Much of the research on PTSD in military samples has used Vietnam-era veterans who returned from war at a time when the concept of combat-related stress disorders was poorly understood by the public at large. American society as a whole was much divided on the war, and many Vietnam veterans experienced a lack of acceptance and support. Although public opinions on the wars in Iraq and Afghanistan are certainly divided, there is generally strong public support for American men and women currently serving in the military. It is feasible that this greater public support at a national level could help buffer to some degree the psychological consequences of trauma exposure, but currently no published studies have examined the relationship between social support and PTSD among OIF/OEF veterans.

Resiliency

Resiliency is defined as: "(1) an occurrence of rebounding or springing back; and (2) the physical property of a material that can return to its original shape or position after deformation that does not exceed its elastic limit" (Webster's, 2008). The idea of psychological resiliency has gained research popularity in the last two decades. An Ovid database search revealed that the terms "resilient," "resilience," and "resiliency" were used in article titles only 11 times before 1987. This increased to 69 times during the next 10 years and then grew to 464 uses by 2008. The bulk of resiliency research has focused on children and families, and little is known about the factors that build resiliency in military service members.

One might hypothesize that the opposite of the factors that contribute to the onset of PTSD symptoms would protect or create a resilient buffer for individuals experiencing traumatic events. Bonanno (2004) proposes many pathways to resilience, including hardiness, self-enhancement, repressive coping, and positive emotion and laughter. Another dimension of resilience is repressive coping, in which unpleasant thoughts, emotions, and memories are avoided (Weinberger, 1990). Bonanno pointed out that, although this type of emotional dissociation is typically thought of as maladaptive, it actually appears to foster adaptation to extreme adversity.

Others have speculated that social support plays a more prominent role in resiliency. King and colleagues (1999) noted large associations for both men and women between hardiness (sense of control, commitment to self, viewing change as challenges) and functional social support (perceived emotional sustenance and instrumental assistance), which may serve to offset the consequences of stressors on PTSD. The consequences of any traumatic exposure may lead to a decrease in available resources which in turn leaves the service member more vulnerable to a stress reaction when faced with another significant traumatic event or loss.

Recovery

Recovery refers to the process whereby individuals who have been exposed to traumatic events return back to a level of functioning that is similar to the way they were functioning prior to the traumatic event. For some traumas, especially those that are relatively minor in severity, individuals may indeed be very similar to the way they were prior to the trauma exposure. Other individuals who have been exposed to severe trauma, especially multiple, prolonged traumatic events, may never return to their original level of functioning. Perhaps the most important factor is whether or not the individual continues to have symptoms that interfere with functioning.

A recent Institute of Medicine report (IOM, 2007) highlighted that there is no accepted definition of recovery from PTSD. In clinical and research settings where patients are treated for PTSD, recovery is often defined as the point where they have been treated to the level of symptom remission. Another definition is when symptoms are reduced to the point where they no longer meet the DSM-IV diagnostic criteria for PTSD. Recovery can also be defined as when an individual's score on a standardized PTSD measure drops below the diagnostic threshold.

For the majority of individuals who experience trauma, their recovery occurs naturally without any type of formal mental health intervention. The specific process and factors related to natural recovery are not well understood.

As previously mentioned, the specific factors related to the traumatic event are clearly important, especially for certain types of trauma such as military combat exposure, rape, and physical assault (Resnick et al., 1993). Most individuals, however, recover naturally without any formal intervention, and the lifetime prevalence of PTSD is less than 10%, even though more than half of the U.S adult population has been exposed to trauma (Kessler et al., 1995, 2005).

One factor that may explain part of this natural recovery phenomenon is that most trauma research lumps all trauma together without differentiation to severity of the trauma. Clearly, all trauma exposure is not alike, and many factors related to the intensity, duration, and frequency of the trauma can make a significant difference in the impact of traumatic events. The majority of types of traumatic events experienced by most individuals are recovered from without any formal treatment; however, the more severe the traumatic event, the more likely an individual is to develop PTSD.

One hypothesis to explain why most individuals recover naturally is that the traumatic events are not so severe as to interfere with behaviors that promote natural recovery. The avoidance of thoughts, feelings, and behaviors associated with the traumatic event is thought to increase the probability of developing PTSD. Thus, the two behaviors that are believed to be most helpful in promoting natural recovery are talking about the trauma and natural, *in vivo* exposure to triggers that have become linked to the traumatic event. By talking about the event and not avoiding environmental symptom triggers, individuals naturally habituate to the trauma memory. Most traumatic events are ones that individuals can talk about with spouses, friends, family, coworkers, neighbors, etc. The more individuals talk about traumatic events, the more likely they will naturally process the cognitive and emotional memories of these events, thereby promoting natural recovery. Similarly, the more individuals continue to function and naturally expose themselves to nondangerous environmental stimuli, the more likely these symptoms are to dissipate over time.

It is interesting to note that lack of social support is one of the strongest predictors for the development of PTSD after exposure to a traumatic event. From this perspective, lack of social support might actually mean lack of a natural support network that allows an individual to talk about an event and to continue to function in life despite their exposure to a traumatic event. Very little is known about the natural recovery process after exposure to traumatic events. We hypothesize that there is a threshold in the trauma severity continuum beyond which individuals are very unlikely to talk with others about the traumatic event. We think this is because the specific details of the traumatic event are so horrific that describing the details to someone else would be extremely difficult and distressing. The more extreme, disgusting, terrifying, or horrific the event, the more likely the individual will clam up and avoid talking about, thinking about, or reexperiencing anything related to the traumatic event. This extreme avoidance is thought to be a key contributor to the development of PTSD.

This theory of natural recovery may help explain why most individuals exposed to combat trauma do not develop PTSD; for example, one can reasonably argue that virtually 100% of U.S. military personnel are exposed to some form of combat-related trauma during a deployment to Iraq. At almost every deployed location, mortar and rocket attacks are a regular occurrence. Compared to normal life in America, being attacked with mortars and rockets is certainly outside the range of usual human experience. Although exaggerated startle responses are common after exposure to repeated attacks, most deployed military members do not develop PTSD after exposure to mortars and rockets; however, if mortars and rockets land near them and they are exposed to individuals who have been killed or severely wounded, the likelihood of developing PTSD is significantly elevated. Hence, all traumatic events are not alike, even if the basic type of trauma exposure is similar.

Treatment of PTSD in Nonmilitary Populations

The IOM Report (2007) provides a timely, up-to-date review of the scientific evidence on the treatment of PTSD. The IOM Committee completed a comprehensive review of the scientific literature on the use of both pharmacotherapy and psychotherapy for the treatment of PTSD. The Committee reviewed 37 pharmacotherapy studies and divided them by class of medications. For all drug classes and specific drugs reviewed in each of the classes, the Committee concluded that the scientific evidence to date is inadequate to determine the efficacy of pharmacotherapy in the treatment of PTSD. This does not mean that medications do not work for the treatment of PTSD, but current evidence is insufficient to support this sole treatment approach. It should also be noted that the initiation of psychotropic medications can limit eligibility for deployment (Department of Defense & Department of Veterans Affairs, 2006), and the potential impact on functioning in deployed settings is not known. Military combat deployments require rapid decisions to identify true danger signals accurately and those decisions may have life or death consequences.

The IOM Committee also reviewed the use of psychotherapy treatment approaches for PTSD in civilians and military veterans treated after discharge or separation from active duty. The Committee reviewed 17 randomized clinical trials (RCTs) of eye movement desensitization and reprocessing (EMDR), cognitive restructuring, and coping skills training. The Committee concluded that the evidence is inadequate to determine the efficacy of these treatments of PTSD. Exposure-based treatments, including prolonged exposure (PE) and cognitive processing therapy (CPT), were the only treatments with strong scientific evidence for their efficacy. The IOM reviewed 24 RCTs of exposure-based treatments, some of which included in the same treatment condition (or arm) exposure plus cognitive restructuring or exposure plus coping skills training. The Committee found that the scientific evidence is sufficient to conclude the efficacy of exposure therapies for the treatment of PTSD.

Treatment of Acute Stress Disorder in Nonmilitary Populations

Harvey and Bryant (1998) assessed motor vehicle accident survivors for acute stress disorder (ASD) within 1 month of the trauma and then reassessed them for PTSD 6 months later. At follow-up, 78% of ASD participants and 60% of subclinical ASD participants met criteria for PTSD. These same patients were reassessed 2 years later, and 63% who had met the initial criteria for ASD and 70% who met the criteria for subsyndromal ASD were diagnosed with PTSD (Harvey & Bryant, 1998). These data suggest that individuals who become clinically symptomatic after exposure to significant trauma are at high risk for the development of PTSD. Fortunately, three randomized clinical trials from Bryant and colleagues have demonstrated that early interventions with prolonged exposure can reduce the rate of PTSD to about 8 to 15% as compared to 63 to 83% of those provided supportive counseling (Bryant et al., 1998, 1999, 2006).

The results obtained by Bryant's group are very consistent with the only published paper to date to evaluate the use of prolonged exposure as an early intervention for the prevention of combat-related PTSD in deployed settings (Cigrang, Peterson, & Schobitz, 2005). In this study, prolonged exposure was used to treat three active-duty military members with significant combat operational stress symptoms consistent with acute stress disorder while deployed to Iraq. All three individuals had significant reductions in symptoms to levels below diagnostic threshold, were able to return to duty in Iraq, and successfully completed their deployments.

Treatment of ASD and PTSD in Military Personnel

Most previous research on PTSD in military personnel has been conducted with Veterans Affairs (VA) patients after they have been discharged from active duty. This evaluation and treatment of military personnel usually occurs many years after the initial exposure to trauma and well after the onset and establishment of chronic PTSD. Little research has been conducted on early interventions with active-duty military personnel for the prevention and treatment of PTSD in theater (i.e., in the deployed setting).

Most previous military research has targeted PTSD surveillance among U.S. military personnel, with the assumption that once PTSD sufferers are identified they can be referred for appropriate treatment; however, stigma associated with the PTSD label has limited early identification of problems (Department of Defense Task Force on Mental Health, 2007; Hoge et al., 2004), and no clinical trials of the treatment of combat-related PTSD in active-duty military personnel have been published. A recent report by the IOM (2007) assessed the evidence to support the treatment of PTSD in civilians and discharged military veterans. The results from several studies of non-combat-related PTSD indicated that a large proportion of patients (>50%) can be treated to the point of recovery, remission, or the loss of diagnosis with early exposure therapy interventions (Bryant et al., 1998, 1999, 2006; Foa, Zoellner, & Feeny, 2006); however, the IOM Report (2007) also noted that strikingly few studies have been conducted in populations of veterans. Studies of discharged military veterans with combat-related PTSD have reported much more modest symptom reductions (Boudewyns, Stwertka, Hyer, Albrecht, & Sperr, 1993; Bradley, Greene, Russ, Dutra, & Westen, 2005; Glynn, 1999; Keane, Fairbank, Caddell, & Zimmering, 1989; Monson et al., 2006; Schnurr et al., 2003, 2007). The IOM report (2007) also highlighted the scant evidence exploring the unique aspects of treating combat-related PTSD in veterans, and it concluded that well-designed research is needed to answer the key questions regarding the efficacy of treatment modalities in combat veterans. Even more importantly, virtually nothing is known about the treatment of combat-related PTSD in active-duty military personnel.

It is unknown whether we can simply transport existing treatment protocols into military settings without significant modifications given the many unique aspects of the military environment; therefore, treatment-outcome evaluations must include specific measures of long-term military functioning and not just symptom reduction. In addition, there are dramatic differences in the treatment options available and the types of patients who present for treatment in deployed vs. nondeployed settings. Although research findings with civilians and discharged combat veterans can inform our approach to treating active-duty military, many factors make the military population, particularly combat-exposed military personnel, unique. The stigma of seeking mental health treatment, concerns about the potential impact on one's military career, higher requirements for mental and physical fitness than in most civilian populations, limited military confidentiality, and the risk of reexposure to combat trauma during and after treatment all present potentially complex obstacles. Perhaps most importantly, the ultimate long-term follow-up evaluation of the efficacy of treatments for combat-related PTSD in active-duty military personnel is the assessment of functioning and resilience during future deployments to combat environments with a high likelihood of additional trauma exposure.

Several approaches have been used to treat psychological trauma reactions in deployed military settings with limited empirical support. Starting in the early 1990s, Critical Incident Stress Debriefing (CISD) has gained almost universal acceptance as a brief treatment for trauma-exposed military personnel; however, recent studies evaluating the efficacy of CISD as a treatment for PTSD have indicated either no reduction of PTSD symptoms or even an increased risk for those undergoing debriefing (Litz, Gray, Bryant, & Adler, 2002). Other commonly used interventions for military personnel with combat stress disorders in the deployed setting include treatments referred to as BICEPS, PIES, and PIE. In theater, BICEPS is an acronym for the way military mental health professionals are trained to treat combat stress disorders—they should be treated with brevity, immediacy, at a central location, with expediency, and in proximity to where their military unit is located. Similarly, PIE and PIES refer to similar approaches where only three or four of these principles are used. None of these approaches recommends specific evidence-based interventions or techniques to be used but they do provide a general description of a therapeutic recovery milieu to be established in a combat theater.

One study to date has evaluated the immediate (Solomon & Benbenishty, 1986) and long-term (Solomon, Shklar, & Mikulincer, 2005) effectiveness of the PIE principles. A quasi-experimental design was used with deployed Israeli military during the Lebanon War in 1982. The first study reported relatively high rates of return to duty and low rates of PTSD at the 1-year follow-up point for those service members who received treatment in the deployed setting as compared to a group evacuated to Israel for treatment (Solomon & Benbenishty, 1986). At a 20-year follow-up (Solomon et al., 2005), the results continued to support the finding that PIE treatment delivered near the front line resulted in lower rates of PTSD than among those who did not receive frontline treatment. The primary limitations of both of these evaluations are that separate cohorts were evaluated based on where they were treated and it is possible that pretreatment cohort differences were present. Similar studies are needed but have not been conducted to evaluate the effectiveness of mental health treatments delivered in deployed settings for U.S. service members.

Future Research on Risk, Resilience, and Recovery

In terms of clinical trials and studies of evidence-based interventions, the U.S. military has entered the 21st century with basically the same scientific knowledge base on treating combat stress disorders in active-duty military personnel as existed during the Vietnam War. Many factors make it difficult to conduct research, especially clinical trials research, in military settings. Nevertheless, clinical trials research can and must be done. Such research will likely require significant research funding and the collaborative efforts of military, civilian, and VA researchers and clinicians.

Early Intervention Approaches for the Prevention of Chronic PTSD

One of the most promising future research areas is to evaluate the effectiveness and efficacy of early intervention treatments for ASD and PTSD. Early intervention refers to secondary prevention intervention approaches to reduce the prevalence of PTSD by shortening the duration of the disorder and reducing chronicity (IOM, 2007). It also includes tertiary prevention to target the reduction of symptom burden and disability associated with PTSD.

By intervening early and before symptoms have become chronic, there is the potential for treatments to result in: (1) large effect sizes, (2) a large proportion of patients being treated into remission or to the point of loss of diagnosis, (3) many patients remaining fit for military duty, and (4) reduced medical discharges and disability.

The most practical first approach to conducting clinical trials with active-duty military personnel is to evaluate treatments that have been demonstrated to be effective in civilian and already discharged military populations (e.g., Vietnam veterans). Prolonged exposure (PE) and cognitive processing therapy (CPT) are the two leading candidates for adaptation for use in active-duty military service members. Additionally, research studies are needed to evaluate novel adaptations of PE and CPT that can be delivered in deployed and nondeployed settings; in time-limited intensive outpatient settings; to individuals, couples, groups, or families; in primary care settings; and in individuals with the most common comorbid disorders, including chronic pain, alcohol abuse or dependence, traumatic brain injury, burns, and amputations.

Early Intervention in Deployed Settings

The earliest opportunity for intervention to prevent PTSD after combat-trauma exposure is in the deployed setting; however, a number of obstacles must be overcome in the completion of a program of evaluation, treatments and outcomes, and mental health clinical trials in deployed settings. The location of treatment, the frequency of treatment sessions, the duration of individual treatment sessions, and the availability of both patients and providers are all treatment components that may have to be modified for use in deployed settings.

It may seem logical that evidence-based treatments that have been found effective in civilian populations might be easily adapted for use in deployed military settings; however, because of the many modifications that may be necessary to adapt an already established treatment approach for use in the deployed setting, it is not clear how well these treatments work in combat settings. In addition, the military patient population and the treatment options that are available in deployed settings are considerably different than patients seen in garrison. In deployed settings, most individuals exposed to some form of trauma are resilient, continue to function adequately, and do not require any type of mental health intervention. A small but noteworthy percentage of military members, however, do become symptomatic enough after exposure to a significant combat-related trauma that they seek out mental health care. Although data are not currently available from the deployed setting, civilian data suggest that individuals who become highly symptomatic after exposure to a traumatic event (e.g., acute stress disorder) are at significant risk for developing PTSD (Bryant et al., 1998, 1999, 2006; Harvey & Bryant, 1998).

Intensive Outpatient Treatment Approaches

Exposure therapy is a well-studied and highly efficacious treatment approach for PTSD (IOM, 2007), and prolonged exposure (PE) is the single most researched and best supported exposure therapy protocol (Foa et al., 1991, 1999, 2005; Resick, Nishith, Weaver, Astin, & Feurer, 2002; Rothbaum et al., 2005, 2006; Schnurr et al., 2007). The standard PE treatment protocol requires 10 to 12 outpatient treatment sessions, each lasting 60 to 90 minutes, conducted over a 5- to 12-week period. Research on other types of anxiety disorders such as agoraphobia (Foa, Jameson, Turners, & Paynee, 1980), panic disorder (Deacon & Abramowitz, 2006, 2007), and specific phobias (Hellstrom, Fellenius, & Ost, 1996; Ost, 1989; Ost, Brandberg, & Alm, 1997; Ost, Alm, Brandberg, & Breitholtz, 2001) has demonstrated that massed practice approaches often yield similar or even better outcomes than traditional outpatient treatment approaches. These massed practice approaches compress treatment into several hours in one day or several consecutive days of treatment. However, no study to date has evaluated the potential to treat PTSD in a massed practice format.

Individual vs. Group Treatments

Group therapy for the treatment of PTSD is being used widely in both military and VA settings. Research is needed to evaluate the efficacy of exposure therapy delivered in individual and group formats. Some forms of exposure therapy, such as prolonged imaginal exposure, are probably not appropriate for group interventions because of the potential risk of secondary exposure to other group members; however, cognitive processing therapy (CPT) is a prime candidate for a study comparing individual and group intervention approaches. CPT is an evidence-based treatment for PTSD first developed as a group treatment (Resick & Schnicke, 1992, 1993). A PTSD study that has compared group vs. individual therapy has never been performed. If both treatments are equivalent, then group treatment would be a much more efficient therapy modality in most cases. On the other hand, if large differences exist between the two modalities, the DoD may need to invest greater resources in ensuring that there are adequate therapists available to provide individual therapy.

Individuals vs. Couples Treatment

Not only can PTSD have a significant impact on individuals, but it can also have a major impact on relationships. Studies reveal both veterans and their partners report significant marital and family problems associated with veterans' PTSD symptoms, including lower relationship satisfaction, less cohesive relationships, less emotional expression and intimacy in their relationships, and more conflict in the relationships (Carroll, Rueger, Foy, & Donahoe, 1985; Gold et al., 2007; Jordan & Marmar, 1992; MacDonald, Chamberlain, Long, & Flett, 1999; Riggs, Byrne, Weathers, & Litz, 1998). Studies also have found elevated levels of anger, hostility, conflict, and violence in the families of veterans with PTSD (Byrne & Riggs, 1996). Delivering a PTSD intervention to couples offers a number of potential advantages compared to the standard individual treatment approaches. Allowing a spouse or partner to have a better understanding of factors related to PTSD and intervention approaches that may be helpful in reducing symptoms has the potential to significantly enhance treatment efficacy. In addition, behavioral couple therapy approaches may help improve relationship functioning and eliminate sources of stress that may mitigate against the success of exposure therapies.

Primary-Care Treatment

Most mental health patients, including those with PTSD, are seen in primary-care settings rather than specialty mental health settings. To date, most primary-care interventions for PTSD have targeted medication treatments. With the recent IOM Report (2007) that found only limited support for pharmacological treatments for PTSD, the use of brief exposure-based therapies delivered in primary-care settings is an approach that should be investigated (Forbes et al., 2007). In recent years, behavioral health consultants have been integrated into primary-care clinics in military and civilian settings (Runyan, Fonseca, Meyer, Oordt, & Talcott, 2003), which allows for early identification and intervention of mental health problems before symptoms or conditions become chronic. Early interventions with less symptomatic patients may allow for evidence-based treatments for PTSD to be adapted to fit within the time constraints of primary-care clinics and still obtain clinically significant effects; however, no published reports have evaluated exposure therapy protocols for use in primary care.

Treatment of PTSD with Comorbid Conditions

Another area that warrants future research is evaluating PTSD in individuals with comorbid conditions that may make treatment of PTSD difficult unless additional treatment components or modules are developed to address these conditions. It is hypothesized that failure to adequately address many of these comorbid conditions may be one factor that has contributed to the large number of Vietnam veterans with chronic and difficult-to treat PTSD.

Within the current military population, there are four types of injuries in OIF/OEF veterans returning from deployments that are considered signature injuries: PTSD, mild traumatic brain injury (mTBI), burn, and amputation. For individuals with comorbid PTSD and mTBI, the diagnostic picture is often very complicated because of the overlapping symptoms of both PTSD and mTBI (Hoge et al., 2008). Standard clinical recommendations are that PTSD should not be treated for 3 to 6 months after the initial injury when the brain has had a chance to recover from the mTBI. There is little research to support this supposition, and it is reasonable to believe that delaying treatment for

PTSD may make it more complicated to treat or even make it difficult to fully recover from the TBI. Only one study on treating comorbid PTSD and mTBI has been conducted to date (Bryant, Moulds, Guthrie, & Nixon, 2003), and the results indicate that TBI does not appear to interfere with the effect of exposure therapy treatments for PTSD. In addition, recent research indicates that many of the symptoms of mTBI might actually be related to PTSD (Hoge et al., 2008). Larger and better controlled studies are needed, especially ones that target the large percentage of OIF/OEF veterans returning from deployments with comorbid PTSD and mTBI.

Similarly, a large number of OIF/OEF veterans have comorbid PTSD and traumatic medical injuries such as burns and amputations. Despite the high comorbidity of PTSD in individuals who have sustained these traumatic injuries, no clinical trials have been published to evaluate the treatment of these comorbid disorders.

Perhaps one of the most common comorbid disorders with PTSD is alcohol abuse and alcohol dependence. It is thought that the use of alcohol is one of the most common self-medication approaches for individuals with PTSD. Comorbid PTSD and alcohol abuse and dependence are high in the Vietnam veteran population and considered to be difficult to treat (Riggs, Rukstalis, Volpicelli, Kalmanson, & Foa, 2003). Although data are currently lacking on the prevalence of comorbid PTSD and alcohol abuse or dependence in OIF/OEF veterans, it is anticipated that this will be a common comorbid disorder, as well. As described previously for the treatment of PTSD, early intervention approaches might also result in greater effect sizes if treatment is initiated prior to alcohol abuse and dependence becoming chronic.

One final disorder that is also very common in PTSD is chronic pain. In a study of civilian orthopedic trauma cases, about 50% of the population was found to have significant symptoms of PTSD (Starr et al., 2004). Chronic pain conditions are the most common cause of medical discharge from active-duty military service during peace time as well as during times of military conflict. Because both pain and PTSD after military trauma can evolve into chronic conditions, the financial costs associated with treatment exceed hundreds of millions of dollars annually (Berkowitz, Feuerstein, Lopez, & Peck, 1999; Greer, Miklos-Essenburg, & Harrison-Weaver, 2006; Huang, Berkowitz, Feuerstein, & Peck, 1998; Zouris, Walker, Dye, & Galarneau, 2006). Unfortunately, recent research suggests that individuals suffering from comorbid chronic pain and traumatic stress may respond poorly to treatment targeting only one diagnosis, contributing to the chronicity and severity of chronic PTSD and chronic pain diagnoses (Bosse et al., 2002). Research is needed to identify the comorbidity of orthopedic trauma and traumatic stress in an active-duty military population and to evaluate whether behavioral health treatment approaches can prevent the development of PTSD and/or chronic pain syndromes.

Combined Medication and Psychotherapy Treatments

To date, no published military or civilian study has directly compared the efficacy of medication and psychotherapy, either alone or in combination, for the treatment of PTSD. One study (Rothbaum et al., 2006) evaluated whether the addition of PE plus sertraline for PTSD would result in greater improvement than continued treatment with sertraline alone. After 10 weeks of open-label sertraline treatment, patients were randomly assigned to 5 additional weeks of sertraline alone ($n = 31$) or sertraline plus 10 sessions of twice-weekly PE ($n = 34$). The results indicated that sertraline led to a significant reduction in PTSD severity after 10 weeks but did not result in further symptom reductions after 5 more weeks. Those participants who received PE showed further reduction in PTSD severity, but this augmentation effect was observed only for those participants who had initially shown a partial response to the medication. Future research is needed to directly compare medications alone or in combination with exposure therapy. The clinical use of psychotropic medication for the treatment of PTSD in military personnel is very common, although there are complicating factors because of the potential deployment-limiting implications of psychotropic medications in active-duty military.

Next-Generation Treatment Alternatives

After first establishing the efficacy of PE and CPT for military populations delivered in deployed and nondeployed settings, additional research should evaluate innovative next-generation treatment alternatives. These research studies might include the use of virtual reality exposure, Internet-delivered interventions, and the addition of D-cycloserine to exposure therapies. Although technology-delivered and medication-enhanced interventions have potential as alternatives to therapist-based exposure therapies, we believe a more critical first step is to adapt currently established treatments to develop the most effective treatments possible before looking at next-generation approaches.

Promoting Resiliency and Preventing PTSD

One area of research that has received very little attention is in the development of approaches to improve trauma resiliency. Research programs to enhance resiliency seem feasible for military service members who are preparing to deploy to face a known traumatic stressor, such as working in graves registry or the DoD mortuary. Promising approaches include stress inoculation training and programs that incorporate graduated exposure to potential traumas. Most branches of the US military have pre-exposure preparation programs in an attempt to build resiliency and prepare individuals for facing particular stressors and deployed settings. However, data to support these approaches are lacking.

The BATTLEMIND program (Military Operational Medical Research Program, 2007) is a relatively new approach that is being used by the Army to help prepare soldiers for deployment, redeployment, and postdeployment. The focus of BATTLEMIND training is how skills that help soldiers survive in combat may cause problems if not changed or adapted when they get home. The BATTLEMIND program has good face validity and is well received by most soldiers, but published data are not yet available to support the efficacy of this approach.

Another approach to the secondary prevention of PTSD is the use of medications, such as propranolol, that are administered immediately after trauma exposure. The goal of this approach is to block memory consolidation and to foster a therapeutic forgetting; however, this research has raised numerous ethical concerns about erasing memories, because it is difficult to determine how important memories might be in determining who we are and how we relate with others.

The use of prophylactic medications prior to trauma exposure may be another way to minimize the impact of traumatic events. There is some evidence from animal research that the prophylactic use of selective serotonin reuptake inhibitors (SSRIs) may reduce the traumatic stress response after exposure to laboratory-induced stressors (Bondi, Rodriguez, Gould, Frazer, & Morilak, 2008).

Questions remain regarding how best to train military personnel who are most likely to encounter potentially traumatic events. Studies should be designed to demonstrate if leaders can be trained to build resiliency in individuals within an organization. Research is also needed to evaluate the use of desensitization to build resiliency. Studies could be designed to examine current military training programs (e.g., basic training, training for advanced surgical teams, or POW resistance training) to incorporate low-level sensitization to potentially traumatic events and then follow the trainees to examine their responses to deployment and exposure using the current Post-Deployment Health Reassessment process. Overall, more research is necessary to identify, train, and evaluate resiliency building factors in military service members.

Discussion

The American public and the military community have made enormous commitments and sacrifices in OIF/OEF. Recent studies have found that about one of every six returning active-duty and one in four Guard and Reserve veterans have symptoms of PTSD. U.S. service members and others see things that no humans should have to see, and as a result even some of the healthiest individuals can have traumatic memories that last a lifetime. Research with civilians with non-combat-related PTSD (e.g., rape or motor vehicle accidents) has shown that most can be treated successfully with exposure therapies, and their symptom reductions are large and fairly permanent. In contrast, counseling with Vietnam veterans indicates that treating combat-related PTSD can be difficult when it is treated after discharge from active duty and after a significant amount of time has passed between when the trauma occurred and counseling is initiated. Research is needed to determine what intervention is best for the prevention and early intervention treatment for PTSD and other combat-related stress disorders. Unless things are done differently, we face the possibility of a new generation of combat veterans who are at significant risk of chronic mental health disorders with the accompanying psychosocial, occupational, healthcare, and financial burdens.

References

Berkowitz, S. M., Feuerstein, M., Lopez, M. S., & Peck, C. A. (1999). Occupational back disability in U.S. Army personnel. *Military Medicine, 164,* 421–428.

Bonanno, G. A. (2004). Loss, trauma, and human resilience: Have we underestimated the human capacity to thrive after extremely aversive events? *American Psychologist, 59,* 20–28.

Bondi, C. O., Rodriguez, G., Gould, G. G., Frazer, A., & Morilak, D. A. (2008). Chronic unpredictable stress induces a cognitive deficit and anxiety-like behavior in rats that is prevented by chronic antidepressant drug treatment. *Neuropsychopharmacology, 33,* 320–331.

Bosse, M. J., MacKenzie, E. J., Kellam, J. F., Burgess, A. R., Webb, L. X., Swiontkowski, M. F., Sanders, R. W., Jones, A. L., McAndrew, M. P., Patterson, B. M., McCarthy, M. L., Travison, T. G., & Castillo, R. C. (2002). An analysis of two-year outcomes of reconstruction or amputations of leg-threatening injuries in level 1 trauma centers. *The New England Journal of Medicine, 347,* 1924–1931.

Boudewyns, P. A., Stwertka, S., Hyer, L., Albrecht, J., & Sperr, E. (1993). Eye movement desensitization for PTSD of combat: A treatment outcome pilot study. *The Behavior Therapist, 16,* 29–33.

Bradley, R., Greene, J., Russ, E., Dutra, L., & Westen, D. (2005). A multidimensional meta-analysis of psychotherapy for PTSD. *The American Journal of Psychiatry, 162,* 214–227.

Brewin, C. R., Andrews, B., & Valentine, J. D. (2000). Meta-analysis of risk factors for posttraumatic stress disorder in trauma-exposed adults. *Journal of Consulting and Clinical Psychology, 68,* 748–766.

Bryant, R. A., Harvey, A. G., Dang, S. T., Sackville, T., & Basten, C. (1998). Treatment of acute stress disorder: A comparison of cognitive–behavioral therapy and supportive counseling. *Journal of Consulting and Clinical Psychology, 66,* 862–866.

Bryant, R. A., Sackville, T., Dang, S. T., Moulds, M., & Guthrie, R. (1999). Treating acute stress disorder: An evaluation of cognitive behavior therapy and supportive counseling techniques. *The American Journal of Psychiatry, 156,* 1780–1786.

Bryant, R. A., Moulds, M., Guthrie, R., & Nixon, R. D. (2003). Treating acute stress disorder following mild traumatic brain injury. *American Journal of Psychiatry, 160,* 585–587.

Bryant, R. A., Moulds, M. L., Nixon, R. D., Mastrodomenico, J., Felmingham, K., & Hopwood, S. (2006). Hypnotherapy and cognitive behaviour therapy of acute stress disorder: A 3-year follow-up. *Behaviour Research and Therapy, 44,* 1331–1335.

Byrne, C. A., & Riggs, D. S. (1996). The cycle of trauma; relationship aggression in male Vietnam veterans with symptoms of posttraumatic stress disorder. *Violence and Victims, 11,* 213–225.

Cabrera, O. A., Hoge, C. W., Bliese, P. D., Castro, C. A., & Messer, S. C. (2007). Childhood adversity and combat as predictors of depression and post-traumatic stress in deployed troops. *American Journal of Preventative Medicine*, 33, 77–82.

Carroll, E. M., Rueger, D. B., Foy, D. W., & Donahoe, Jr., C. P. (1985). Vietnam combat veterans with posttraumatic stress disorder: Analysis of marital and cohabitating adjustment. *Journal of Abnormal Psychology*, 94, 329–337.

Cigrang, J. A., Peterson, A. L., & Schobitz, R. P. (2005). Three American troops in Iraq: Evaluation of a brief exposure therapy treatment for the secondary prevention of combat-related PTSD. *Pragmatic Case Studies in Psychotherapy*, 1, 1–25.

Deacon, B., & Abramowitz, J. (2006). A pilot study of two-day cognitive–behavioral therapy for panic disorder. *Behavior Research and Therapy*, 44, 807–817.

Deacon, B., & Abramowitz, J. (2007). Two-day, intensive cognitive behavioral therapy for panic disorder. *Behavior Modification*, 31, 595–615.

Department of Defense Task Force on Mental Health. (2007). *An achievable vision: Report of the Department of Defense Task Force on Mental Health.* Falls Church, VA: Defense Health Board.

Department of Defense & Department of Veterans Affairs. (2006). *The continuum of care for post traumatic stress disorder (PTSD)*, Serial No. 109-19. Washington, D.C.: Committee on Veterans' Affairs, House of Representatives.

Foa, E. B., Jameson, J. S., Turners, R. M., & Paynee, L. L. (1980). Massed vs. spaced exposure sessions in the treatment of agoraphobia. *Behavior Research and Therapy*, 18, 333–338.

Foa, E. B., Rothbaum, B. O., Riggs, D., & Murdock, T. (1991). Treatment of post-traumatic stress disorder in rape victims: A comparison between cognitive–behavioral procedures and counseling. *Journal of Consulting and Clinical Psychology*, 59, 715–723.

Foa, E. B., Dancu, C. V., Hembree, E. A., Jaycox, L. H., Meadows, E. A., & Street, G. P. (1999). A comparison of exposure therapy, stress inoculation training, and their combination for reducing posttraumatic stress disorder in female assault victims. *Journal of Consulting and Clinical Psychology*, 67, 194–200.

Foa, E. B., Hembree, E. A., Cahill, S. P., Rauch, S. A., Riggs, D. S., Feeny, N. C., & Yadin, E. (2005). Randomized trial of prolonged exposure for posttraumatic stress disorder with and without cognitive restructuring: Outcome at academic and community clinics. *Journal of Consulting and Clinical Psychology*, 73, 953–964.

Foa, E. B., Zoellner, L. A., & Feeny, N. C. (2006). An evaluation of three brief programs for facilitating recovery after assault. *Journal of Traumatic Stress*, 19, 29–43.

Forbes, D., Creamer, M. C., Phelps, A. J., Couineau, A. L., Cooper, J. A., Bryant, R. A., McFarlane, A. C., Devilly, G. J., Matthews, L. R., & Raphael, B. (2007). Treating adults with acute stress disorder and post-traumatic stress disorder in general practice: A clinical update. *Medical Journal of Australia*, 187, 120–123.

Garb, H. N., & Cigrang, J. A. (2008). Psychological screening: Predicting resilience to stress. In B. Lukey & V. Tepe (Eds.), *Biobehavioral resilience to stress* (pp. 3–24). New York: Taylor & Francis.

Gawande, A. (2004). Casualties of war: Military care for the wounded from Iraq and Afghanistan. *New England Journal of Medicine*, 351, 2471–2475.

Glynn, S. M., Eth, S., Randolph, E. T., Foy, D. W., Urbaitis, M., Boxer, L., Paz, G. G., Leong, G. B., Firman, G., Salk, J. D., Katzman, J. W., & Crother, J. (1999). A test of behavioral family therapy to augment exposure for combat-related posttraumatic stress disorder. *Journal of Consulting and Clinical Psychology*, 67, 243–251.

Gold, J. I, Taft, C. T., Keehn, M. G., King, D. W., King, L. A., & Samper, R. E. (2007). PTSD symptom severity and family adjustment among female Vietnam veterans. *Military Psychology*, 19, 71–81.

Greer, M. A., Miklos-Essenburg, E., & Harrison-Weaver, S. (2006). A review of 41 upper extremity war injuries and the protective gear worn during Operation Enduring Freedom and Operation Iraqi Freedom. *Military Medicine*, 171, 595–597.

Grieger, T. A., Cozza, S. J., Ursano, R. J., Hoge, C., Martinez, P. E., Engel, C. C., & Wain, H. J. (2006). Posttraumatic stress disorder and depression in battle-injured soldiers. *American Journal of Psychiatry*, 163, 1777–1783.

Harvey, A. G., & Bryant, R. A. (1998). The relationship between acute stress disorder and posttraumatic stress disorder: A prospective evaluation of motor vehicle accident survivors. *Journal of Consulting and Clinical Psychology*, 66, 507–512.

Harvey, A. G., Bryant, R. A., & Tarrier, N. (2003). Cognitive behaviour therapy for posttraumatic stress disorder. *Clinical Psychology Review*, 23, 501–522.

Hellstrom, K., Fellenius, J., & Ost, L. (1996). One versus five sessions of applied tension in the treatment of blood phobia. *Behavior Research and Therapy*, 34, 101–112.

Hoge, C. W., Lesikar, S. E., Guevara, R., Lange, J., Brundage, J. F., Engel, Jr., C. C., Messer, S. C., & Orman, D. T. (2002). Mental disorders among U.S. military personnel in the 1990s: Association with high levels of health care utilization and early military attrition. *American Journal of Psychiatry*, 159, 1576–1583.

Hoge, C. W., Castro, C. A., Messer, S. C., McGurk, D., Cotting, D. I., & Koffman, R. L. (2004). Combat duty in Iraq and Afghanistan, mental health problems, and barriers to care. *The New England Journal of Medicine*, 351, 13–22.

Hoge, C. W., Auchterlonie, J. L., & Milliken, C. S. (2006). Mental health problems, use of mental health services, and attrition from military service after returning from deployment to Iraq or Afghanistan. *Journal of the American Medical Association*, 295, 1023–1032.

Hoge, C. W., Terhakopian, A., Castro, C. A. et al. (2007). Association of posttraumatic stress disorder with somatic symptoms, health care visits, and absenteeism among Iraq war veterans. *American Journal of Psychiatry*, 164, 150–153.

Hoge, C. W., McGurk, D., Thomas, J. L., Cox, A. L., Engel, C. C., & Castro, C. A. (2008). Mild traumatic brain injury in U.S. soldiers returning from Iraq. *New England Journal of Medicine*, 358, 453–463.

Huang, G. D., Berkowitz, S. M., Feuerstein, M., & Peck, C. A. (1998). Occupational upper-extremity-related disability: Demographic, physical, and psychosocial factors. *Military Medicine*, 163, 552–558.

iCasualties.org. (2007). *Iraq coalition casualty count*, http://www.icasualties.org/oif/default.aspx.

Institute of Medicine (IOM). 2007. *Treatment of posttraumatic stress disorder: An assessment of the evidence.* Washington, D.C.: The National Academies Press.

Jordan, B. K., & Marmar, C. R. (1992). Problems in families of male Vietnam veterans with posttraumatic stress disorder. *Journal of Consulting and Clinical Psychology*, 60, 916–926.

Keane, T. M., Fairbank, J. A., Caddell, J. M., & Zimmering, R. T. (1989). Implosive (flooding) therapy reduces symptoms of PTSD in Vietnam combat veterans. *Behavior Therapy*, 20, 245–260.

Kessler, R. C., Sonnega, A., Bromet, E., Hughes, M., & Nelson, C. B. (1995). Posttraumatic stress disorder in the National Comorbidity Survey. *Archives of General Psychiatry*, 52, 1048–1060.

Kessler, R. C., Demler, O., Frank, R. G., Olfson, M., Pincus, H. A., Walters, E. E., Wang, P., Wells, K. B., & Zaslavsky, A. M. (2005). Prevalence and treatment of mental disorders, 1990 to 2003. *New England Journal of Medicine*, 352, 2515–2523.

Kilpatrick, D. G., Saunders, B. E., Amick-McMullan, A., Best, C. L., Veronen, L. J., & Resnick, H. S. (1989). Victim and crime factors associated with the development of crime-related post-traumatic stress disorder. *Behavior Therapy*, 20, 199–214.

King, D. W., King, L. A., Foy, D. W., Keane, T. M., & Fairbank, J. A. (1999). Posttraumatic stress disorder in a national sample of female and male Vietnam veterans: Risk factors, war-zone stressors, and resilience-recovery variable. *Journal of Abnormal Psychology*, 108, 164–170.

Koren, D., Norman, D., Cohen, A., Berman, J., & Klein, E. M. (2005). Increased PTSD risk with combat-related injury: A matched comparison study of injured and uninjured soldiers experiencing the same combat events. *American Journal of Psychiatry*, 162, 276–282.

Litz, B. T., Gray, M. J., Bryant, R. A., & Adler, A. B. (2002). Early intervention for trauma: Current status and future directions. *Clinical Psychology: Science and Practice*, 9, 112–134.

MacDonald, C., Chamberlain, K., Long, N., & Flett, R. (1999). Posttraumatic stress disorder and interpersonal functioning in Vietnam war veterans: A mediational model. *Journal of Traumatic Stress*, 12, 701–707.

Military Operational Medical Research Program. (2007). BATTLEMIND, http://www.battlemind.org/.

Milliken, C. S., Auchterlonie, J. L., & Hoge, C. W. (2007). Longitudinal assessment of mental health problems among active and reserve component soldiers returning from the Iraq war. *Journal of the American Medical Association*, 298, 2141–2148.

Monson, C. M., Schnurr, P. P., Resick, P. A., Friedman, M. J., Young-Xu, Y., & Stevens, S. P. (2006). Cognitive processing therapy for veterans with military-related posttraumatic stress disorder. *Journal of Consulting and Clinical Psychology*, 74, 898–907.

Ost, L. (1989). One session treatment for specific phobias. *Behavior Research and Therapy*, 1, 1–7.

Ost, L., Brandberg, M., & Alm, T. (1997). One versus five sessions of exposure in the treatment of flying phobia. *Behavior Research and Therapy*, 35, 987–996.

Ost, L., Alm, T., Brandberg, M., & Breitholtz, E. (2001). One vs. five sessions of exposure and five sessions of cognitive therapy in the treatment of claustrophobia. *Behavior Research and Therapy*, 39, 167–183.

Ozer, E. J., Best, S. R., Lipsey, T. L., & Weiss, D. S. (2003). Predictors of posttraumatic stress disorder and symptoms in adults: A meta-analysis. *Psychological Bulletin*, 129, 52–73.

Prigerson, H. G., Maciejewski, P. K., & Rosenheck, R. A. (2002). Population attributable fractions of psychiatric disorders and behavioral outcomes associated with combat exposure among U.S. men. *American Journal of Public Health*, 92, 59–63.

Prins, A., Ouimette, P., Kimerling, R., Cameron, R. P., Hugelshofer, D. S., Shaw-Hegwer, J., Thrailkill, A., Gusman, F. D., & Sheikh, J. I. (2004). The primary care PTSD screen (PC-PTSD): Development and operating characteristics. *Primary Care Psychiatry*, 9, 9–14.

Resick, P. A., & Schnicke, M. K. (1992). Cognitive processing therapy for sexual assault victims. *Journal of Consulting and Clinical Psychology*, 60, 748–756.

Resick, P. A., & Schnicke, M. K. (1993). *Cognitive processing therapy for rape victims: A treatment manual*. Newbury Park, CA: Sage.

Resick, P. A., Nishith, P., Weaver, T. L., Astin, M. C., & Feurer, C. A. (2002). A comparison of cognitive-processing therapy with prolonged exposure and a waiting condition for the treatment of chronic posttraumatic stress disorder in female rape victims. *Journal of Consulting and Clinical Psychology*, 70, 867–879.

Resnick, H. S., Kilpatrick, D. G., Dansky, B. S., Saunders, B. E., & Best, C. L. (1993). Prevalence of civilian trauma and posttraumatic stress disorder in a representative national sample of women. *Journal of Consulting and Clinical Psychology*, 61, 984–991.

Riggs, D. S., Byrne, C. A., Weathers, F. W., & Litz, B. T. (1998). The quality of the intimate relationships of male Vietnam veterans: Problems associated with posttraumatic stress disorder. *Journal of Traumatic Stress*, 11, 87–101.

Riggs, D. S., Rukstalis, M., Volpicelli, J. R., Kalmanson, D., & Foa, E. B. (2003). Demographic and social adjustment characteristics of patients with comorbid posttraumatic stress disorder and alcohol dependence: Potential pitfalls to PTSD treatment. *Addictive Behaviors*, 28, 1717–1730.

Rothbaum, B. O., Atsin, M., & Marsteller, F. (2005). Prolonged exposure versus eye movement desensitization and reprocessing (EMDR) for PTSD rape victims. *Journal of Traumatic Stress*, 18, 607–616.

Rothbaum, B. O., Cahill, S. P., Foa, E. B., Davidson, J. R., Compton, J., Connor, K. M., Astin, M. C., & Hahn, C. G. (2006). Augmentation of sertraline with prolonged exposure in the treatment of posttraumatic stress disorder. *Journal of Trauma Stress*, 19, 625–638.

Rundell, J. R. (2006). Demographics of and diagnoses in Operation Enduring Freedom and Operation Iraqi Freedom personnel who were psychiatrically evacuated from the theater of operations. *General Hospital Psychiatry*, 28, 352–356.

Runyan, C. N., Fonseca, V. P., Meyer, J. G., Oordt, M. S., & Talcott, G. W. (2003). A novel approach for mental health disease management: The Air Force Medical Service's interdisciplinary model. *Disease Management*, 6, 170–188.

Schnurr, P. P., Lunney, C. A., Sengupta, A., & Waelde, L. C. (2003). A descriptive analysis of PTSD chronicity in Vietnam veterans. *Journal of Traumatic Stress*, 16, 545–553.

Schnurr, P. P., Friedman, M. J., Engel, C. C., Foa, E. B., Shea, M. T., Chow, B. K., Resick, P. A., Thurston, V., Orsilla, S. M., Haug, R., Turner, C., & Bernardy, N. (2007). Cognitive–behavioral therapy for posttraumatic stress disorder in women: A randomized clinical trial. *Journal of the American Medical Association*, 297, 820–830.

Solomon, Z., & Benbenishty, R. (1986). The role of proximity, immediacy, and expectancy in frontline treatment of combat stress reaction among Israelis in the Lebanon War. *American Journal of Psychiatry*, 143, 613–617.

Solomon, Z., Shklar, R., & Mikulincer, M. (2005). Frontline treatment of combat stress reaction: A 20-year longitudinal evaluation study. *American Journal of Psychiatry*, 162, 2309–2314.

Starr, A. J., Smith, W. R., Frawley, W. H., Borer, D. S., Morgan, S. J., Reinert, C. M., & Mendoza-Welch, M. (2004). Symptoms of posttraumatic stress disorder after orthopaedic trauma. *Journal of Bone and Joint Surgery*, 86, 1115–1121.

Tedeschi, R. G., & Calhoun, L. G. (1996). The posttraumatic growth inventory: Measuring the positive legacy of trauma. *Journal of Traumatic Stress*, 9, 455–471.

Vogt, D. S., King, D. W., & King, L. A. (2007). Risk pathways for PTSD: Making sense of the literature. In M. J. Friedman, T. M. Keane, & P. A. Resick (Eds.), *Handbook of PTSD: Science and practice* (pp. 55–76). New York: Guilford Press.

Webster's. (2008). *Webster's online dictionary*, http://www.websters-online-dictionary.org.

Yehuda, R. (Ed.) (1999). *Risk factors for posttraumatic stress disorder*. Washington, D.C.: American Psychiatric Press.

Yehuda, R., & McFarlane, A. C., (1995). Conflict between current knowledge about posttraumatic stress disorder and its original conceptual basis. *American Journal of Psychiatry*, 152, 1705–1713.

Zouris, J. M., Walker, G. J., Dye, J., & Galarneau, M. (2006). Wounding patterns for U.S. Marines and sailors during Operation Iraqi Freedom, major combat phase. *Military Medicine*, 171, 246–252.

Index